XII Congreso Internacional Terroir 2018

E3S Web of Conferences Volume 50 (2018)

Zaragoza, Spain
18 - 22 June 2018

Editor:

Andres Martinez Burgos

ISBN: 978-1-5108-7006-2

Printed from e-media with permission by:

Curran Associates, Inc.
57 Morehouse Lane
Red Hook, NY 12571

Some format issues inherent in the e-media version may also appear in this print version.

Printed by Curran Associates, Inc. (2018)

For additional information, please contact EDP Sciences – Web of Conferences at the address below.

EDP Sciences – Web of Conferences
17, Avenue du Hoggar
Parc d'Activité de Courtabœuf
BP 112
F-91944 Les Ulis Cedex A
France

Phone: +33 (0) 1 69 18 75 75
Fax: +33 (0) 1 69 28 84 91

contact-edps@webofconferences.org

Additional copies of this publication are available from:

Curran Associates, Inc.
57 Morehouse Lane
Red Hook, NY 12571 USA
Phone: 845-758-0400
Fax: 845-758-2633
Email: curran@proceedings.com
Web: www.proceedings.com

TABLE OF CONTENTS

Comité Organizador

Presidente: Joaquín Olona

Vicepresidente: Enrique Novales

Director: Ernesto Franco

Vocales: Luis Miguel Albisu, Alfonso Bordonaba, Vicente Gómez, Miguel Lorente, Miguel Ángel Mainar, Vicente Sotés y Salvador Congost.

Comité Científico

Vicente Sotés, Presidente

Vicente Gómez-Miguel, Secretario

Borbála Balo, Hungría

Luigi Bavaresco, Italia

Benjamin Bois, Francia

Jan Booysen, Sudáfrica

Ernesto Franco, España

Enrique García-Escudero, España

María Luisa González –Sanjosé, España

Peter Hayes, Australia

Gregory Jones, Estados Unidos

Stefanos Koundouras, Grecia

Cornelis Van Leeuwen, Francia

Carlos Lopes, Portugal

Vittorino Novello, Italia

Hernán Ojeda, Francia

Giuliano Pereira, Brasil

Jorge Queiroz, Portugal

Hans Schultz, Alemania

Attilio Scienza, Italia

Diego Tomasi, Italia

Jorge Tonietto, Brasil

Jesús Yuste, España

Vivian Zufferey, Suiza

edp sciences

E3S Web of Conferences

Statement of Peer review

In submitting conference proceedings to E3S Web of Conferences, I certify to the Publisher that I adhere to the **Policy on Publishing Integrity** of the journal in order to safeguard good scientific practice in publishing.

1. All articles have been subjected to peer review administered by the proceedings editors.
2. Reviews have been conducted by expert referees, who have been requested to provide unbiased and constructive comments aimed, whenever possible, at improving the work.
3. Proceedings editors have taken all reasonable steps to ensure the quality of the materials they publish and their decision to accept or reject a paper for publication has been based only on the merits of the work and the relevance to the journal.

Title, date and place of the conference

XII Congreso Internacional Terroir, 18-22 de junio de 2018. Zaragoza (España)

Proceedings editor(s):

Andrés Martínez Burgos. (Responsable Técnico Sarga)

Date and editor's signature

08 Agosto 2018

 https://doi.org/10.1051/e3sconf/20185001001

Quantifying hydropedological properties of terroir at different scales. Implications in vineyard characteristics of three viticultural regions of the Iberian Peninsula.

Joaquín Cámara[1], Alberto Lázaro[2] and Vicente Gómez-Miguel[1]

[1] Universidad Politécnica de Madrid, Dpto. Producción Agraria, ETSIAAB, 28040 Madrid, Spain
[2] Universidad de Burgos, Dpto. Biotecnología y Ciencia de los Alimentos, Facultad de Ciencias, 09001 Burgos, Spain

Summary. Hydropedology is a branch of science that studies the spatial and temporal variability of the soil water content, knowledge needed for a proper application of precision agriculture techniques in viticulture. This work aims to quantify at different scales hydropedological properties of viticultural regions with different geological, climatic and cultural characteristics, and it also aims to explore the relationships between these hydropedological properties and the vineyard characteristics within the studied regions. The structure of drainage networks at landscape scale and the heterogeneity of the soil particle size distribution are the hydropedological properties considered for the purpose of this study. The three winegrowing regions are the Região Demarcada do Douro, in Portugal, and the D.O. Arribes and D.O. Campo de Borja, in Spain, involving a total area of 4,144 km^2. The datasets for this work are the soil maps of the three regions scale 1/25,000, the digital terrain model, and the spatial database with the characteristics of the vineyards within the study area. Qualitative relationships have been established between landscape attributes, vineyard characteristics and physical soil properties. Results show that quantifying hydropedological properties using complexity measures provides innovative information of terroirs that could have multiple applications in terroir zoning studies.

1 Introduction

Hydropedology is an emerging interdisciplinary science that studies interactive pedologic and hydrologic processes and properties in the thin layer of the Earth's surface and near-surface terrestrial environment from the top of the vegetation canopy to the bottom of the weathering zone [1]. Environmental processes within this layer, including mass and energy exchange, soil formation, streamflow generation, and landscape evolution are key processes to sustaining biodiversity and humanity [2], that also influence the vineyards development.

Hydropedology aims to understand the spatial and temporal variability of soil water content, which knowledge is required for implementing precision agriculture techniques in viticulture.

The objectives of this work are to quantify at different scales hydropedological properties of viticultural regions with different geological, climatic and cultural characteristics, and the exploration of the relationships between these hydropedological properties and the vineyard characteristics in the three studied regions.

The hydropedological properties quantified in this study are the structure of drainage networks at landscape scale, which represent the surficial water movements across the landscape, and the soil texture heterogeneity at pedon scale, that seems to be closely related to the storage and release of water in soils [3].

2 Material and methods

2.1 Study area

The study area involves three wine growing regions of the Iberian Peninsula: the Regiao Demarcada do Douro, in Portugal, and the D.O. Arribes and the D.O. Campo de Borja, in Spain (Figure 1). These regions cover a total area of 4,144 km^2, 461 km^2 of which are under vineyard. RD Douro and DO Arribes are in the Northwestern part of the Iberian Peninsula, within the Duero/Douro river basin, while DO Campo de Borja is in the Northeastern part, within the Ebro river basin. The geology of RD Douro and DO Arribes consists of metamorphic rocks (schists, slates, and gneisses) and

https://doi.org/10.1051/e3sconf/20185001001

igneous rocks (granites and granitoids) of the Iberian Hercynian Massif. In contrast, the geology of DO Campo de Borja is formed by Ordovician and Triassic materials, along with Quaternary materials of the Ebro river basin in the lower lands of the DO.

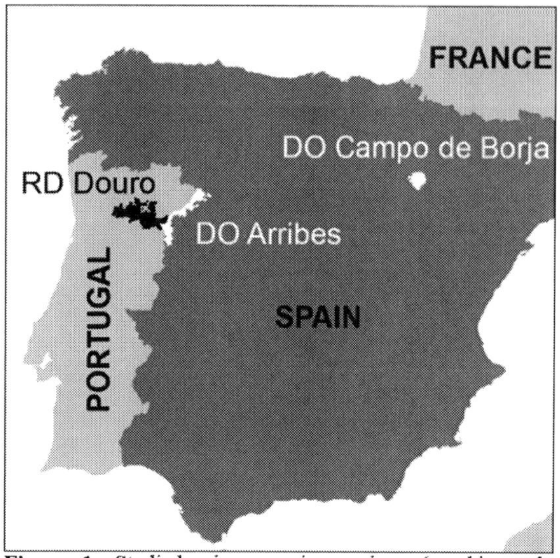

Figure 1. Studied wine growing regions (graphic scale 1/12,000,000).

2.2 Data set

Three sources of information of each region have been used in this study; the soil map (scale1/25,000), the digital terrain model (DTM), and the vineyard information contained in the official register of each winegrowing region.

Soil maps of these areas [4-6] include analytical and morphological data of 2,192 soil profiles and have been used to estimate soil texture heterogeneities.

The DTM resolution of DO Arribes and DO Campo de Borja is 5 meters, while DTM of RD Douro is 10 meters-resolution. DTMs have been used to generate the surficial drainage networks of these regions (Figure 2) by means of the Hydrology Toolbox of ArcGIS 10.1, which uses the D8 algorithm for networks extraction developed by O'Callaghan and Mark [7]. The threshold area to initiate channelization has been established at 10,000 m^2, which has been successfully used in previous works [8-9].

Vineyard information corresponds to data from the official viticultural registries of the three winegrowing regions which include individualized data of the vineyard plots such as plant density, vine variety, altitude, exposition and planting date. This information has been joined to georeferenced information of the agricultural plots for the cartographic operations carried out.

2.3 Quantifying hydropedological properties

The hydropedological properties studied in this work have been the structure of drainage networks at landscape scale and the heterogeneity of the soil particle size distribution at pedon scale.

The structure of drainage networks has been quantified by means of fractal techniques.

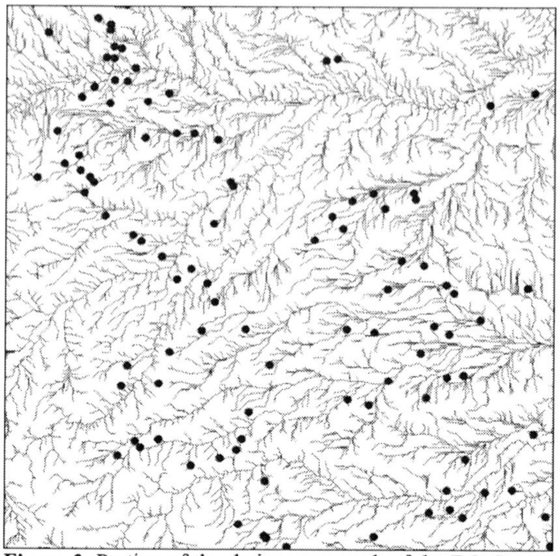
Figure 2. Portion of the drainage network of the DO Campo de Borja with the locations of same soil profiles (graphic scale 1/150,000).

The fractal dimension of the networks that overlap the winegrowing regions has been estimated using the box counting method [10-11], where the objects to be analyzed are covered by a regular grid of side r. The number of grid boxes containing the drainage network is counted $N(r)$, and then, the value r is repeteadly halved obtaining the series of N_i values. Taking a reference box size (r_1), the size factor (s_i) is calculated using the following equation.

$$s_i = r_i / r_1 \quad (1)$$

When r_i decreases the following expression converges to a finite value defined as box counting dimension (D).

$$D \approx \frac{\log N_i}{\log\left(\frac{1}{s_i}\right)} \quad (2)$$

The reference box size (r_1) has been established at 320 m, and the other box sizes used in the estimations have been 160 m, 80 m, 40 m, 20 m and 10 m.

The heterogeneity of the soil particle size distribution has been measured using the Balanced Entropy Index (BEI), based on Shannon's entropy and recently introduced in terroir studies [12]. This index was proposed by Martín et al. [13] and it is given by the following equation

$$BEI = \frac{\sum_i P_i \log P_i}{\sum_i P_i \log l_i}, \quad (3)$$

where the numbers l_i are the scale lengths of the three basic size intervals (l_1 = 0.001, l_2 = 0.024 and l_3 = 0.975) and P_i are the soil's clay-silt-sand fractions. The *BEI* above provides a continuous parameterization, from 0 to 1, of the soil texture (Figure 3) that can be easily computed from textural data.

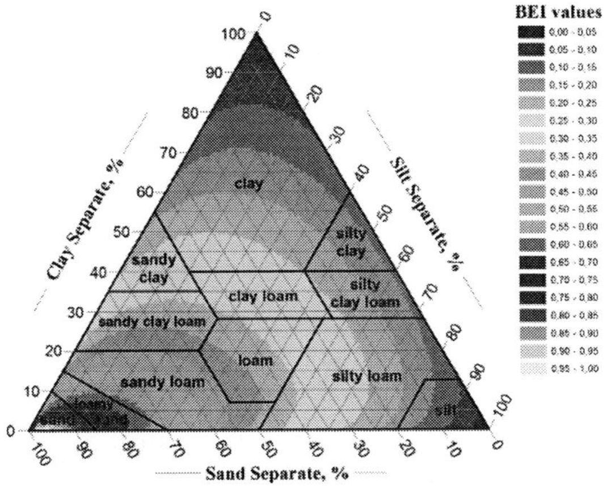

Figure 3. Distribution of BEI values on the USDA textural triangle [14].

The BEI has been calculated in the 3,018 soil samples included in the databases of the soil maps 1/25,000 of the three regions. An average value of BEI was obtained for each winegrowing region under study.

3 Results

3.1 Fractal dimension of drainage networks

Results of the fractal dimension (*D*) of the three drainage networks are shown in Table 1.

Table 1. Box counting results

Studied region	Fractal dimension (*D*)	Coefficient of determination for D estimation (R^2)
RD Douro	1.1610	0.989
DO Arribes	1.2208	0.993
DO Campo de Borja	1.2842	0.991

The fractal nature of drainage networks, and the suitability of the followed method to characterize their structure, are supported by the high values of R^2 obtained for the fractal dimension estimations, which are higher than 0.980 in the three cases.

D values of drainage networks range from 1.1610, in RD Douro, to 1.2842 in DO Campo de Borja.

Results show that fractal dimensions of drainage networks in regions formed by metamorphic and igneous rocks (RD Douro and DO Arribes) is lower than in regions where quaternary materials have a significant presence (DO Campo de Borja). Moreover, *D* value in RD Douro, where metamorphic areas are predominant, is lower than in DO Arribes, where granitic areas cover a wider part of the region. These results are consistent with those obtained in previous works [9], where fractal dimensions in granitic areas is higher than in metamorphic areas.

3.2 Soil texture heterogeneity

Results of *BEI* calculations are showed in Table 2.

Table 2. Balanced Entropy Index results

Studied region	Soil samples with textural data	*BEI* (mean ± CI)
RD Douro	1,510	0.457 ± 0.003 b
DO Arribes	711	0.530 ± 0.005 c
DO Campo de Borja	797	0.303 ± 0.004 a

Different letters in a column indicate significant differences ($P<0.05$). LSD test

Significant statistical differences have been established between the values of *BEI* obtained in the three studied regions. The regional average *BEI* values range from 0.303±0.004, in DO Campo de Borja, to 0.530±0.005, in DO Arribes. These results quantitatively show that metamorphic and granitic rocks evolve in soils with coarser soil textures than soils developed in quaternary materials.

Results show that *BEI* values are higher in areas mostly formed by granitic materials (DO Arribes) than in regions where metamorphic rocks are predominant (RD Douro), which is consistent with the results and conclusions of previous works [14-15].

Results suggest that the winegrowing regions could be characterized by a *BEI* value that reflects similarities and differences between regions. This finding complements those results obtained in the previous works cited above, which showed that *BEI* values are useful to discriminate between areas with different lithological materials within the same winegrowing region.

3.3 Hydropedological properties and vineyard characteristics

In this preliminary study, planting density (*n*) has been chosen, amongst the vineyard characteristics included in the managed databases, to explore the relationships between the hydropedological and vineyard properties of the three studied winegrowing regions. In order to compare plant density with the other variables, a regional average value of plant density has been calculated for each winegrowing region (Table 3).

Table 3. Vineyard planting density

Studied region	Plots with density data	Planting density (*n*) (mean ± CI)
RD Douro	138,786	6,670 ± 8 c

https://doi.org/10.1051/e3sconf/20185001001

DO Arribes	715	2,138 ± 57 a
DO Campo de Borja	1,824	2,406 ± 65 b

Different letters in a column indicate significant differences (P<0.05). LSD test

Unfortunately, not every vineyard plot included in the official registries has planting density information. However, the number of plots with proper information allowed the obtention of significant statistical results.

As happened with *BEI* results, significant statistical differences could be established between the average planting densities (n) of the three studied regions. *n* values range from 2,138±57, in DO Arribes, to 6,670±8, in RD Douro. These results quantitatively show that cultural practices in RD Douro are significantly different than those observed in DO Arribes and DO Campo de Borja.

Results shown in Table 1, Table 2 and Table 3 have been plot together in Figure 4.

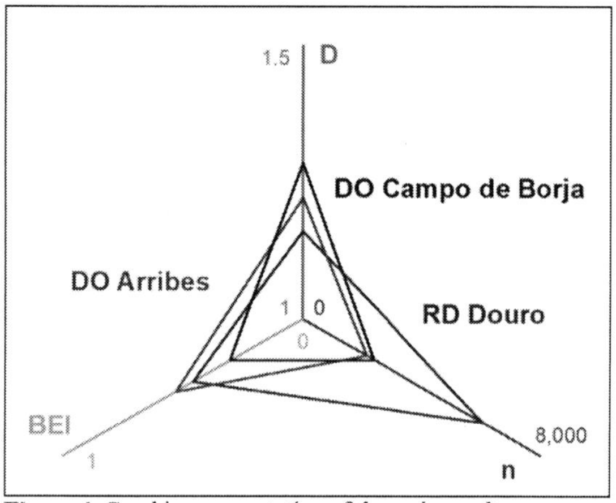

Figure 4. Graphic representation of the main results.

Results suggest that, in areas where vineyards irrigation is restricted for absence of infrastructures (DO Arribes), or water availability (DO Campo de Borja), the fractal dimension of drainage networks is linearly related to plant density in vineyards. This is not the case of RD Douro where the steep inclination of the slopes, and the availability of water from the Duero river, has resulted in specific cultural practices modifying this relationship.

Average values of *BEI* and planting density in DO Campo de Borja and DO Arribes show the same trend observed at plot scale in previous works in RD Douro [12], where discriminating among different lithologies higher values of *BEI* were related to lower values of planting densities in vineyards.

According to the obtained results, *BEI* and *D* seem to be linearly related in areas with comparable geologic materials such as RD Douro and DO Arribes, where lower values of *D* are related to lower values of *BEI*.

4 Conclusions

The relationship between the structure of drainage networks at landscape scale, the heterogeneity of the soil particle sizes distribution at pedon scale and the planting density of vineyards was explored in three winegrowing regions of the Iberian Peninsula.

The quantification of hydropedological properties in winegrowing regions provides innovative information of terroirs that, further to be useful to characterize them, could be employed to establish environmental relationships or differences between regions.

Although additional studies, including other hydropedological and vineyard properties, are still to be performed, the potential uses of complexity measures in terroir studies, such as the fractal dimension of the drainage networks or the Balanced Entropy Index, are promising.

References

1. National Council Research, *Basic research opportunities in the earth sciences* (Natl. Acad. Press, Washington, DC, 2001)
2. H. Lin, J.W. Hopmans, D. Richter, Vadose Zone J. **10**, 781-782 (2011)
3. M.A. Martín, Y.A. Pachepsky, J.M. Rey, J. Taguas, W.J. Rawls, Soil Science **170**, 759-766 (2005)
4. V. Gómez-Miguel, V. Sotés, J. Cámara, G. Hernando, *Delimitación de las zonas vitícolas de la DO Arribes* (Univ. Politécnica de Madrid, Madrid, 2011) Technical report not published.
5. V. Gómez-Miguel, V. Sotés, J. Cámara, G. Hernando, *Zonagem e Ordenamento da cultura da vinha do Vale do Douro(Portugal).* (Univ. Politécnica de Madrid, Madrid, 2011) Technical report not published.
6. V. Gómez-Miguel, V. Sotés, J. Cámara, A. Martínez, *Zonificación de los terroir vitícolas de la DO Campo de Borja* (Univ. Politécnica de Madrid, Madrid, 2015) Technical report not published.
7. J. O'Callaghan, D.M. Mark, Comput. Vision Graph. **3**, 323-344 (1984)
8. A.B. Ariza-Villaverde, F.J. Jiménez-Hornero, E. Gutiérrez, Geomorphology, **197**, 85-95 (2013)
9. J. Cámara, V. Gómez-Miguel, M.A. Martín, Pure Appl. Geophys. **173**, 945-961 (2016)
10. D. Russel, J. Hanson, E. Otto, Physical Rev. Letters **45**, 1175-1178 (1980)
11. I. Rodríguez-Iturbe, A. Rinaldo, *Fractal river basins: Chance and self-organization* (Cambridge Univ. Press, Cambridge, 1997)
12. J. Cámara, A. Lázaro, V. Gómez-Miguel, *Proceedings of the XI International Terroir Congress*, 365-370 (Southern Oregon University, Ashland, USA, 2016)
13. M.A. Martín, J.M. Rey, J. Taguas, Ecol. Model. **182**, 221-228 (2005)
14. J. Cámara, V. Gómez-Miguel, M.A. Martín, Geoderma **287**, 157-163 (2017)
15. J. Cámara, M.A. Martín, V. Gómez-Miguel, Fractals **23**, 1540007 (2015)

Variación del clima de un Terroir y su consecuencia sobre la respuesta de la vid

Variation of the climate of a Terroir and its consequence on the vine's response

Milka Ferrer[1,a], Gerardo Echeverría[1], Gustavo Pereyra[1], Julia Salvarrey[1], Leandro Arrillaga[1] y Mercedes Fourment[1]

[1]Departamento de Producción Vegetal, Facultad de Agronomía, Av. Garzón 780 Montevideo,Uruguay
[a] mferrer@fagro.edu.uy

Resumen

El clima vitícola, un componente mayor del Terroir, fue identificado en la región sur del Uruguay: sequía moderada, templado cálido, noches templadas, sometido a la influencia de la brisa marina. En el período 1994-2016, se analizó el régimen térmico e hídrico y su influencia sobre el rendimiento, sanidad y composición de la uva, de viñedos comerciales de Tannat. ACP mostró 3 grupos de años: **Grupo 1** (húmedo), precipitaciones del ciclo del cultivo superiores a la media, sanidad limitante, acidez y rendimiento mayores a la media y azúcares inferiores, cosechas tardías. **Grupo 2** (cálido y seco), condiciones térmicas superiores al promedio y la componente hídrica por debajo, sanidad superior a la media, azúcares y acidez inferiores, cosechas adelantadas. **Grupo 3** (fresco, húmedo en primavera y seco en maduración), condiciones térmicas inferiores a la media, precipitaciones superiores a la media en brotación –cuajado e inferiores en el mes previo a cosecha, superiores a la media azúcares y tamaño de baya, cosecha atrasada. Se establecieron correlaciones entre las variables del clima, rendimiento y la calidad de la uva. En la serie de años estudiada, la variabilidad climática de la región fue muy marcada y la respuesta de Tannat acompañó dicha variabilidad.

Palabras clave: Variabilidad climática, rendimiento, sanidad, composición de la uva, Tannat

Summary

Viticultural climate, a major component of Terroir, in Uruguay southern region was classified as moderately dry, temperate-warm, with temperate nights and influenced by sea breeze. During the 1994-2016 period, thermal and water regimes and their influence on grapevine yield, sanitary status and berry composition was analyzed for Tannat cultivar grown in commercial vineyards. Principal component analysis (PCA) separated years in three groups: Group 1 (wet), rainfall over the growing season higher than the average, sanitary status is limiting, acidity and yield higher than the average, whereas sugar content is lower, late harvest. Group 2 (warm and dry), thermal conditions greater and water component lower than the average, better sanitary status, sugar contents and acidity lower than the average, early harvest. Group 3 (fresh, wet in spring and dry during ripening), thermal conditions lower than the average, rainfall higher during budbreak-fruitset and lower than the average in the month previous to harvest, berry size and sugar contents greater than the average, late ripening. Correlations among climate, yield and berry quality variables were established. In the studied years, climate variability within the region was high and the response of the Tannat cultivar was in accordance to this variability.

Keywords: climate variability, yield, sanitary status, berry composition, Tannat

1 Introducción

El clima es uno de los componentes mayores del Terroir. El clima vitícola de la región sur del Uruguay fue identificado como de sequía moderada, templado cálido, noches templadas (CCM) [1], sometido a la influencia de la brisa marina delestuario del Rio de la Plata [2, 3]. En el contexto de cambio climático, para el Uruguay se prevé una mayor variabilidad interanual de las precipitaciones, en particular durante el período de maduración de la uva, así como un aumento de la temperatura media [4]. Esta situación condiciona el rendimiento y la composición de la uva y favorece la aparición de enfermedades fúngicas que determinan pudriciones a nivel de racimos ocasionando distorsión en la producción y en la calidad de la uva [5, 6].

Las variaciones que se registran en el rendimiento dependen principalmente del número de racimos que se indujeron en la temporada anterior. Este proceso fisiológico dentro de la yema, ocurre durante el período de floración de la temporada en curso, por lo que es necesario un buen aporte de las reservas y cuya movilización, depende fundamentalmente de un aporte hídrico no limitante [7, 8]. Uno de los componentes del rendimiento mayormente afectado por la temperatura y por disponibilidad hídrica es la baya. Condiciones térmicas durante la maduración o déficit hídricos en la primera etapa de su desarrollo determinan fuertemente su tamaño [9-12].

Las condiciones meteorológicas explican más del 80% de la variabilidad en la composición de la uva [13]. La temperatura, en particular durante el período de maduración, tiene una incidencia directa sobre la degradación de los ácidos orgánicos. Rangos de temperaturas superiores a los 25 - 30 °C pueden ser limitantes para la fotosíntesis y por lo tanto para la acumulación de azúcares en la baya [11,14]. Las precipitaciones según la etapa del ciclo del cultivo tienen un efecto negativo o positivo en el contenido de azúcares y de los ácidos [13,14].

En nuestro ambiente, *Botrytis cinerea* es de las enfermedades que determinan mayores pérdidas en rendimiento y calidad de la uva [12,15]. Los factores climáticos predisponentes para la infección de este hongo, son las precipitaciones abundantes durante la maduración y las condiciones térmicas templadas [12,16].

En nuestros diferentes terroirs, la variedad Tannat, considerada insignia del país, aporta vinos con gran tipicidad y originalidad y expresa a su gran potencial enológico [17].

El objetivo de este trabajo fue analizar en un contexto de un clima cambiante y en función de las condiciones agro-meteorológicas de un período de 22 años (1994-2016), la variación del régimen térmico e hídrico y su influencia sobre el rendimiento, sanidad y composición de la uva variedad Tannat.

2 Materiales y Métodos

2.1 Ubicación del ensayo

La información de campo se obtuvo de tres parcelas de viñedos comerciales de la variedad Tannat injertada sobre SO4, conducidas en espaldera alta y con poda Guyot, con filas orientadas N-S, ubicadas en el Departamento de Canelones (Lat.:34° 37′ S; 56° 17′ W). El clima vitícola de esta región fue calificado como de *sequía moderada, templado cálido, con noches templadas*, según la "Clasificación Climática Multicriterios" [1,18]. La información del cultivo provino de proyectos de investigación del grupo de Viticultura de la Facultadde Agronomía, correspondientes al periodo 1994 a 2016, siguiendo protocolos específicos para cada variable.

La información de las variables climáticas se tomó de la base de datos de la estación meteorológica del Instituto Nacional de Investigación Agropecuaria (INIA) Las Brujas (Lat 34°40', Long 56°20') conforme a las normas de la OMM. La estación está situada en laregión de estudio y abarca un radio de 14 km. Las variables climáticas diarias consideradas se presentan en la Tabla 1. Se calculó los Grados Días base 10 y el número de días desde el 1° de Enero hasta la fecha de cosecha de cada año. La escala temporal fue de 23 años (1994 a 2016 inclusive), correspondientes a la misma serie de registros del cultivo.

2.2 Evaluaciones sobre la planta de cada parcela

Componentes del rendimiento (kg/ha): en cosecha, en 30 plantas se pesó individualmente: uva sana y enferma y se calculó el porcentaje del total. El peso de baya (g) (balanza Ohaus Scout, Ohaus Corp., USA) se obtuvo sobre una muestra de 250 bayas por duplicado tomadas según protocolo propuesto por Carbonneau [19]. Para la determinación del estado sanitario (enfermo) se tuvo en cuenta los racimos con más de 5% de incidencia de podredumbres de racimo.

Tabla 1: Variables Climáticas

Variable	Sigla
Temperatura Max Enero (°C)	TME
Temperatura Max Febrero (°C)	TMF
Temperatura min Febrero (°C)	TmF
Grado Días (Enero cosecha, °C)	GDD
Grado Días Diario (Enero cosecha, °C)	GDDD
Número de días T > a 30 °C (1Noviembre – cosecha)	N°D>30°C N-co
Precipitaciones Setiembre – Octubre (mm)	PP St-Oc
Precipitaciones Noviembre – Diciembre (mm)	PP Nv-Dc
Precipitaciones Enero – Cosecha (mm)	PP En-Cs
Precipitaciones Ciclo del cultivo (mm)	PPCiclo
Precipitaciones del mes previo a cosecha (mm)	PP 1MC
Número de días a cosecha (1Enero – cosecha)	N°dC (1En-co)

Composición de la uva: en la cosecha, sobre las muestras de 250 bayas en las que se midió el peso y a partir del método de la O.I.V [20], se determinó: sólidos solubles por refractometría (Atago, Master-T, Japón) y acidez total (AT), por titulación (Bureta Mohr).Para el contenido de antocianos, se tomaron tres muestras de 250 bayas para determinar el potencial polifenólico (potencial total en antocianos ApH 1,0, (se presentan datos de 3 años), siguiendo la metodología propuesta por Glories y Augustin [21] modificado por González-Neves [22].

Análisis estadístico de los resultados: éstos fueron analizados utilizando las técnicas estadísticas multivariadas de análisis de componentes principales (ACP) para separar las variables asociadas yel coeficiente

https://doi.org/10.1051/e3sconf/20185001002

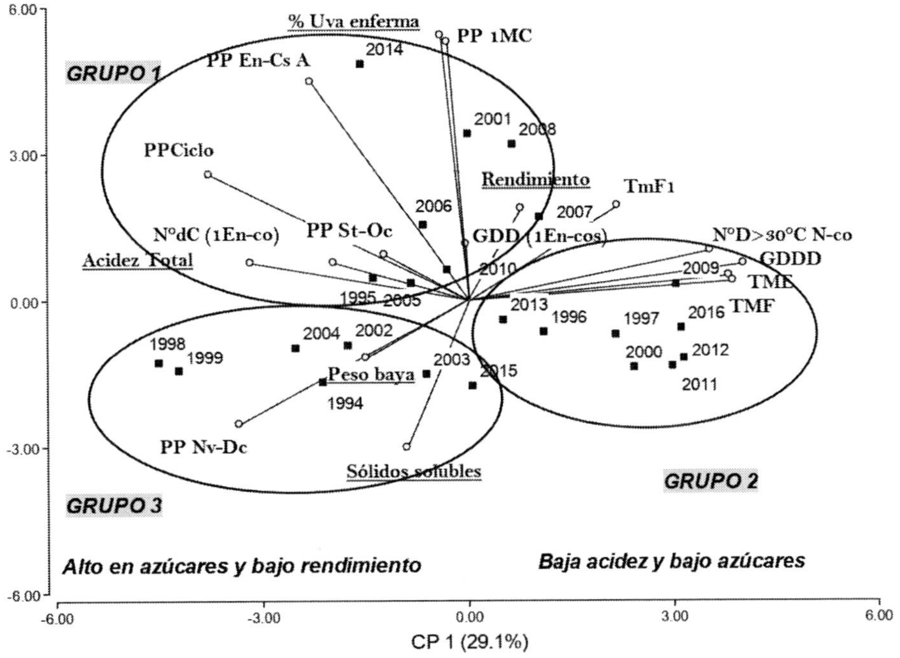

Figura 1. Análisis de Componentes Principales: Años, factores del clima y respuesta de la planta. Abreviaciones listadas en la Tabla 1

de correlación de Pearson para evaluar las posibles relaciones entre las diferentes variables consideradas. Los análisis se llevaron a cabo empleando el software InfoStat [23].

3 Resultados y discusión

3.1 Grupo de años

En el análisis de ACP, los dos primeros componentes principales explican el 48,2% de la varianza total (Figura 1). El componente principal 1 (CP1) aporta el 29,1% mientras que el componente principal 2 (CP2) aporta el 19,1%. Los vectores de carga que integran el ACP, se encuentran asociados entre sí en distinta magnitud formando 3 grupos de años que se caracterizaron en función de las medias de las variables del período analizado y contribuyendo de manera diferente a cada CP.

Los grupos comprenden prácticamente el mismo número de años (8, 8 y 7), pero con condiciones meteorológicas contrastantes, que condicionan principalmente la composición de la uva y su sanidad, en acuerdo con van Leeuwen et al. [13].

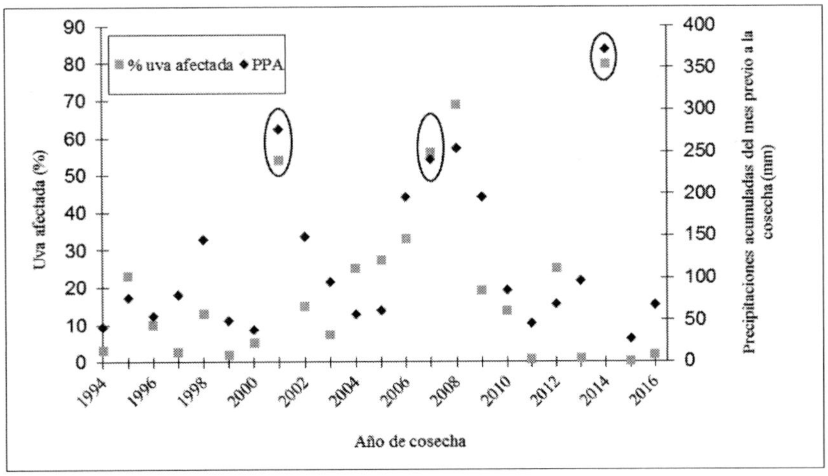

Figura 2. Relación entre las precipitaciones del mes previo a la cosecha y el Porcentaje de uva enferma

https://doi.org/10.1051/e3sconf/20185001002

El **Grupo 1 (húmedo)** comprende 8 años, asociado a precipitaciones superiores a la media durante todo el ciclo del cultivo, así como las temperaturas nocturnas del mes previo a la cosecha. Estas condiciones generaron que el estado sanitario fuera limitante, el rendimiento y la acidez del mosto fueran superiores a la media, en tanto que por debajo del promedio estuvieron los azúcares y se constató una fecha de cosecha tardía. Estos resultados, en función de las condiciones meteorológicas, son citados por numerosos autores [7, 8, 12-14, 16]. La mayor incidencia de podredumbres de racimo se correlacionó positiva y significativamente con las precipitaciones del mes anterior a la cosecha (PP1MC, $r=0,88$, $p<0,0001$), con las del período de maduración (PP En-Cs, $r=0,68$, $p=0,0004$) y con las del ciclo del cultivo (PP Ciclo, $r=0,42$. $p=0,04$). En la figura 2 se muestra la correlación entre las precipitaciones del mes previo a la cosecha y el porcentaje de uva enferma por pudriciones de racimos. Otra consecuencia de las precipitaciones es el aumento del vigor de las plantas que está asociado a mayores rendimientos y efectos sobre la composición de la uva [24,25]. En este sentido, las precipitaciones que se correlacionaron positivamente con la acidez, son las del ciclo del cultivo (PPciclo, $r=0,46$, $p=0,03$), las de maduración (PP En-Cs, $r=0,34$, $p=0,1$) y las del mes previo a la cosecha, negativamente con la cantidad de azúcares en la baya ($r=-0,42$, $p=0,04$). Este último resultado, probablemente se asocie al efecto de dilución en esta etapa [25]. Las temperaturas también juegan un rol sobre la acidez [26]. Las temperaturas durante la maduración (TME y TMF) estuvieron en valores inferiores a la media del período analizado (29 y 28 °C respectivamente), a la vez que éstas se correlacionaron negativa y significativamente con la acidez (TME, $r=-0,53$, $p=0,0092$, TMF, $r=-0,37$, $p=0,08$). Esta degradación más lenta, en particular del ácido málico, prolongó el período de maduración ($r=0,61$, $p=0,002$), en la medida que la acidez total fue uno de los índices de cosecha considerado (Figura 3).

En el **Grupo 2 (cálido y seco)**, que comprende 8 años, las condiciones térmicas estuvieron por encima del

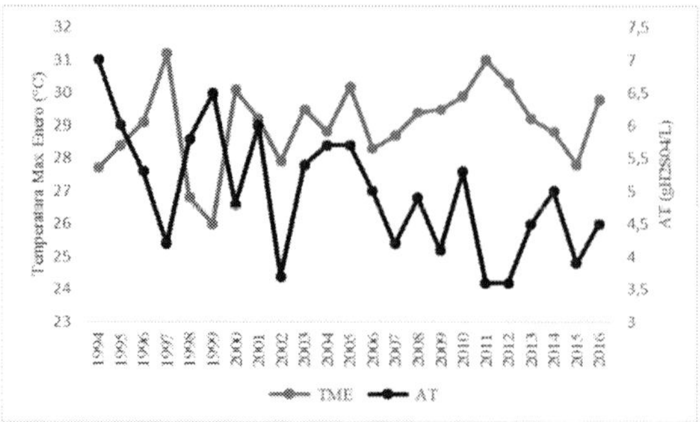

Figura 3. Relación entre la acidez total y la temperatura máxima del mes más cálido

promedio y la componente hídrica por debajo de lo esperado durante todo el ciclo del cultivo. Estas condiciones tuvieron como consecuencia un buen estado sanitario, en tanto que azúcares y acidez fueron inferiores a la media y como era de esperar con cosechas adelantadas. Los bajos contenidos de acidez se corresponden, como ya fue citado, a su degradación como consecuencia de las altas temperaturas durante la maduración. Por otra parte, también influyeron negativamente en el proceso de fotosíntesis. Como es comunicado por varios autores, en rangos de temperaturas superiores a los 25-30 °C puede ser limitante para la fotosíntesis y por lo tanto para la acumulación de azúcares en la baya [11, 14]. En nuestro estudio, la mayoría de los años de este grupo la temperatura máxima superó los 30 °C. Condiciones de estrés hídrico son reportadas en la bibliografía como causa del bloqueo de la fotosíntesis [28]. En el caso de nuestro trabajo, las precipitaciones registradas durante el ciclo del cultivo, cuando se las compara con la media de los años, en 7 de los 8 años estuvo por debajo de esta media (643mm), con extremos de -131mm a – 333 mm. El balance hídrico del 2016, según los valores propuestos [1], se catalogó de Índice de sequía fuerte (-177 mm) cuando el clima de la región lo cataloga de Sequía moderada con valores históricos del Índice de 50mm.

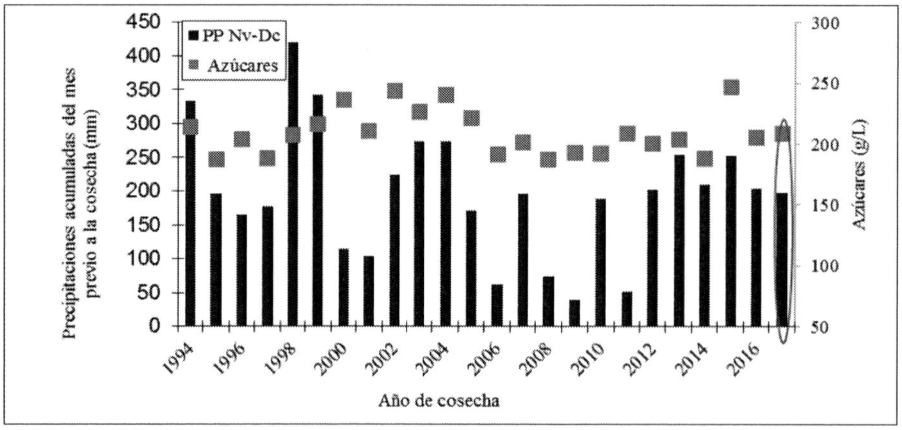

Figura 4. Relación entre las Precipitaciones del mes previo a la cosecha y el contenido de azúcares en la baya

Tabla 2. Año representativo de cada grupo

	Grupo 1 (cálido en maduración y húmedo) **Año 2014**	**Grupo 2** (cálido y seco) **Año 2016**	**Grupo 3** (fresco, húmedo en primavera y seco en maduración) **Año 2002**
Acidez Total (gH2SO4/L)	5,0	3,9	4,1
Azúcares (g/L)	188	205	243
ApH1	784	1756	3632
Rendimiento (kg/ha)	23430	22140	16900
Peso de la baya (g)	1,53	1,31	1,62
Uva enferma (%)	79,8	0,75	1,5

El **Grupo 3 (fresco, húmedo en primavera y seco en maduración),** comprende 7 años, asociado a condiciones térmicas por debajo de la media, temperaturas nocturnas bajas en el mes previo a la cosecha, y las precipitaciones fueron superiores a la media en el período brotación –cuajado e inferiores en el mes previo a cosecha. Estas condiciones favorecieron el mayor contenido de azúcares así como el tamaño de baya, los que estuvieron por encima de la media. El rendimiento por hectárea fue inferior a la media, la cosecha en estas condiciones fue atrasada y no se registraron problemas sanitarios. El componente hídrico de período floración – cuajado, que se correlacionó positivamente con el contenido de azúcares (PP Nv.Dc, r=0,48, p=0,04), ya que permitió una buena instalación de la superficie foliar, que favoreció la producción de fotosintatos, e influyó positivamente en el tamaño de la baya en acuerdo con lo comunicado por varios autores [9, 29], (Figura 4). El tamaño de la baya fue afectado negativamente por las temperaturas elevadas durante la maduración (TMF, r=-0,42, p=0,04, N°D>30°C N-co r=-0,46, p=0,02). En este grupo de años, las temperaturas TMF y N°D>30°C estuvieron por debajo de la media, por lo que no causaron su deshidratación en la etapa final de la maduración de la baya [10]. Del punto de vista enológico este grupo de años fue el que se destacó por la calidad de sus vinos.

En la Tabla 2 se muestra el rendimiento, sanidad y la composición de la uva de un año representativo de cada grupo.

4 Conclusiones

Para la serie de años estudiada, la variabilidad climática de la región fue muy marcada y, en consecuencia, se observó una respuesta de Tannat acorde a dicha variabilidad. Esto muestra la importancia que tiene el clima para nuestras condiciones de cultivo, por lo que para evaluar el potencial del terroir vitícola del sur de Uruguay, se deben continuar los estudios del rol de los diferentes componentes del clima. Las condiciones meteorológicas del año tienen marcada influencia en la calidad enológica de la uva, siendo las del grupo 2 las más favorables.

5 Referencias

1. J. Tonietto y A. Carbonneau Agric. Forest Meteorol. **124**, 81–97 (2004)
2. M. Fourment, M. Ferrer, G. González-Neves, G. Barbeau, V. Bonnardot, H. Quénol, Int. J. Biomet. **61**, 1617-1628 (2017)
3. G. Echeverría, Tesis doctoral. Montevideo Uruguay, Facultad de Agronomía, Universidad de la República. 233p (2017)
4. G. Tiscornia; A. Cal; A. Gimenez Rev Invest. Agrop - RIA, 42 1, p.: 66 - 71 (2016)
5. R. Mira de Orduña, Food Res. Inter., **43,** 1844 – 855 (2010)
6. M. Ferrer, G. Gonzalez-Neves; G. Echeverría, A. Carbonneau, Progrès Agric. Vitic. **128**, 367-371 (2011)
7. N. Guilpart, A. Metay, CH. Gary. Eur. J. Agron. **54**, 9-20 (2014)
8. M. Mendez-Costabel, K. Wilkinson, S. Bastian, C. Jordans, M. McCarthy, C. Ford, N. Dokoozlian, Aust. J. Grape Wine Res. **20**, 100-110 (2014)
9. H. Ojeda, C. Andary, E. Kraeva, A. Carbonneau, A. Deloire, Am. J. Enol. Vitic. 53 **4**, 261 – 267 (2002)
10. S. Rogiers, J. Hatflied, V. Guntajaudzems, R. White, M. Keller. Am. J. Enol. Vitic. **55**, 121-127 (2004)
11. M. Ferrer, G. Echeverria, A. Carbonneau, S. Afr. J. Enol. Vitic. **35**, 103-113 (2014)
12. M. Ferrer, G. Echeverría, J.M. Miras-Avalos, Inter. J. Env. Agric. Res. **3**, 16-27 (2017)
13. C. van Leeuwen, P. Friant, X. Choné, O. Tregoat, S. Koundouras, D. Dubourdieu, Am. J. Enol. Vitic. **55**, 207-217 (2004)
14. J. Hunter, V. Bonnardot, S. Afr. J. Enol. Vitic. **32**, 137-154 (2011)
15. V. Gepp, S. Vero, M. Cassanello, G. Romero, E. Silvera, P. González, J. Rebellato, Y. Ferreira, O. Bentancur, Agrocien. **16**, 97-107 (2012)
16. E. González-Domínguez, T. Caffi, N. Ciliberti, V. Rossi, PlosOne 10 (2015)
17. G. González-Neves, G. Gil, L. Barreiro, M. Ferrer, J. Franco, Agrocien. **10**, 1-14 (2006)
18. M. Ferrer, Tesis doctoral. Montpellier, Francia. Université Montpellier II, 388p (2007)

19. A. Carbonneau, A. Moueix, N. Leclair, J. Renoux, Bull. OIV **727**, 679-690 (1991)
20. O.I.V. Office Internationale de la Vigne et du vin. Paris, France, 388p (2009)
21. Y. Glories; M. Agustin, *Proc. Journée Technique CIVB*, Bordeaux, France pp 56-61 (1993
22. G. González-Neves, Tesis Doctoral, Montpellier, Francia. Université de Montpellier II. 279p (2005)
23. J.A. di Rienzo, F. Casanoves, M.G. Balzarini, L. González, M. Tablada, C.W. Reboledo, *Infostat versión 2012*, Universidad Nacional de Córdoba (2013)
24. I. Filippetti, G. Allegro, G. Valentini, C. Pastore, E. Colucci, C. Intrieri, J. Int. Sci. Vigne Vin, **47**, 21-33 (2013)
25. M.C. Ramos, J.A. Martínez-Casanovas, Plant Soil 333, 375-389 (2010)
26. R.R. Walker, D.H. Blackmore, P.R.Clingeleffer, G.H. Kerridge, E.H. Ruth, P.R. Nicholas, Aust. J. Grape Wine Res, **11**, 2–8 (2005).
27. M. Keller, Aust. J. Grape Wine Res. **16**, 56–69 (2010)
28. J. Bota, O. Stayk, J. Flexas, H. Medrano, Func. Plant Biol. **31**, 697-708 (2004)
29. P. Costenza, B Tisseyre, JJ. Hunter, A. Deloire, S. Afr. J. Enol. **25**, 43-47 (2004)

Climate zoning of the Burgundy winegrowing region

Catinca Gavrilescu[1], Yves Richard[1], Daniel Joly[2], Benjamin Bois[1,3]

[1] *CRC – UMR Biogeosciences, Université Bourgogne Franche-Comté / CNRS, 6 Bd Gabriel, 21000 Dijon, France*
gavrilescu_catinca@yahoo.com, yves.richard@u-bourgogne.fr

[2]*Université de Bourgogne Franche-Comté, THEMA-CNRS, France*
daniel.joly@univ-fcomte.fr

[3] *Institut Universitaire de la Vigne et du Vin, Université Bourgogne Franche-Comté, rue Claude Ladrey, 21000 Dijon, France*
benjamin.bois@u-bourgogne.fr

Abstract. As one of the most emblematic wine regions of cool climate terroir viticulture, Burgundy is endowed with a set of very specific natural features suitable to the production of high quality wines, where climate is arguably one of the main factors to profoundly influence vine physiology/phenology and grape composition. These environmental nuances have led to a wide variety of styles in Pinot noir and Chardonnay wines that have been largely acknowledged and appreciated by the international market and vitivinicultural industry. However, individual grape varieties optimum quality is known to be closely related to well-defined climate and geographical ranges. Climate change and global warming latest trends make them more susceptible to undergo modifications in terms of berry ripening processes and advancements in harvest dates due to short-term and long-term spatiotemporal fluctuations in climate variability. The impact of air temperature on grapevine development and harvest outcomes has been widely documented by the scientific community, its influence translating as quality and quantity fluctuations in space (*"terroir"* effect) and time (*"vintage"* effect). Through this study we aim to assess the extent of these threats by means of modelling and spatializing the regional climate variations based on 5 agricultural climatic indices: the number of days with temperatures equal or greater than 35°C (heat stress), the number of days indicating a frost risk (equal or greater than -1°C), the mid-flowering, the mid-véraison and the theoretical grape maturity (200g/l of sugar) occurrence dates. Mid-flowering, mid-véraison as well as the theoretical grape maturity were estimated through the summation of temperatures over 10°C (starting 1st of March) based on the GFV (Grapevine Flowering Véraison) linear phenological model and were calculated for the 2 prime varieties cultivated in Burgundy (Chardonnay and Pinot noir). Daily minimum and maximum temperatures issued from a network of 64 weather stations scattered throughout the main 9 wine production subregions of Burgundy were spatially interpolated on a grid with a 75m resolution over a 41751ha area (74556 pixels). Spatial interpolations were performed at a daily time step integrating various topographical features through a regression-kriging model for the 2011-2015 period. Daily grid minimum and maximum temperatures were further used to calculate the 5 agroclimatic indices for each of the years of the study period. The entirety of the 74556 pixels were later classified at regular intervals in 6 groups which were assigned to each of the three phenological stages: *"very early"*, *"early"*, *"intermediate"*, *"late"*, *"very late"* and *"variable"*. The number of heat stress days as well as those presenting a frost risk were equally classified based on their occurrence as *"very rare"*, *"rare"*, *"intermediate"*, *"frequent"* and *"very frequent"*. The annual spatial structure of the individual classes was very similar due to temperature distribution being strongly governed by terrain features. We were able to identify observable differences between the north and the south subregions of Burgundy with a potential variation ranging from 7 to 15 days in terms of phenological and theoretical maturity occurrence dates. Côte de Nuits and Côte de Beaune vineyards indicate similar climate characteristics with early phenological timing (97% and 82% respectively of the area classified as *"early"*) and little frost and heat risks. The number of days with a frost risk is a lot more elevated in the Côte Châtillonnaise and the Chablis subregions, while the number of heat stress days was larger in the subregions located in the south of Burgundy.

https://doi.org/10.1051/e3sconf/20185001003

1. Introduction

Viticulture is highly depended on weather and climate conditions, particularly during the grapevine's growth cycle [1, 2]. As a terroir component, climate has been observed to have the largest influence on both grape and wine quality, followed by cultivar and soil [3]. Climate variability of wine regions is strongly related to their location, each geography producing different types of climate and meteorological events that vary both in frequency and intensity [4]. Given its importance for the wine industry, a fine scale spatial structure analysis of climate is indispensable when trying to understand and describe the potential of wine regions [5]. To this end, multiple viticultural climate zoning studies have been carried out [6, 7, 8] in view of comparing the world winegrowing regions' climate profiles [9, 10], as well as to describe more locally and accurately the climate typicity of each individual wine region [11, 12, 13]. Climate discrepancies between the world viticultural areas are mainly attributed to differences in thermal regimes [2, 14, 15]. Temperature has been reported to have a determinant role in the grapevine's different developmental stages [16, 17, 18] and the berries' maturity [19, 20, 3].

Thermal fields fine scale spatial analysis and characterization requires the use of an integrated approach by including the topographic component [21], which has been shown to have a very strong influence on the spatial distribution of climate parameters [22, 23].

The present study aims at describing and comparing the climate features of several Burgundy wine subregions while analyzing their thermal subtleties at a fine scale resolution (75m). This was achieved through a regression-kriging interpolation of minimum and maximum daily temperatures relative to the different topographic parameters derived from a digital elevation model (DEM). Finally, a typology of the viticultural subregions was developed as a result of the classification made on the basis of agroclimatic indices calculated from daily interpolated temperatures.

2. Data and methods

2.1. Area of study and land use data

This study has been carried out across the Burgundy winegrowing region, located in the east-central part of France. Burgundy's viticultural climate has *oceanic* and *meridional* influences [24, 25], which results in hot and dry periods during summer and rainy periods in winter [26].

The climate analysis was performed for the areas actually planted in grapevines and not for the entire extent of the appellations of origin (AOC), the later including the whole vineyard potential areas of Burgundy. The 9 following wine producing subregions were studied: Côte de Nuits and Hautes Côtes de Nuits, Côte de Beaune and Hautes Côtes de Beaune, Côte Châtillonnaise, Chablis, Côtes du Couchois, Côte Chalonnaise, Mâconnais (Figure 1a). The vineyard area delimitation used for this study was that of CORINE Land Cover 2012 (CLC) European land use database [27]. The polygons depicting the vineyard areas (represented in violet in Figure 1a) were identified and retrieved based on a unique code (221). The CLC database vineyard polygons cover an area of 31468 ha, whereas the institutional organisms (e.g. BIVB - Bureau Interprofessionnel des Vins de Bourgogne) declared in 2014 an area of only 29000 ha (8.5% overestimation). This spatial inaccuracy has a very limited impact on the proposed zoning and is probably due to the difficulty of entirely dissociating the non-vineyard (residential zones, vegetation, etc.) of the actual vineyard areas.

2.2. Climate data and cross validation

The observed climate data used for this study were retrieved from two sources: the CLIMEO (owned and managed by the Chambre Régionale d'Agriculture de Bourgogne-Franche Comté and the BIVB) and the Météo-France weather stations networks. Based on this data, two databases were compiled, one with the daily minimum (tn) and another with the daily maximum (tx) temperatures. This research was carried out for the 2011-2015 period and it analysed a total of 3652 daily tn and tx (1826 each) recovered from 64 distinct weather stations spread across our study area (Figure 1b). However, these databases are incomplete due to missing values which account for 17% of the total data (with 17.5% of tn and 16.6% of tx missing). Missing data was unavoidable as it corresponds to stations that were installed either after the beginning date of this study or they were shut down during it. Nevertheless, missing values were later estimated by interpolation of the daily tn and tx. The temperatures were gridded at a 75m resolution (pixel size of the DEM) across

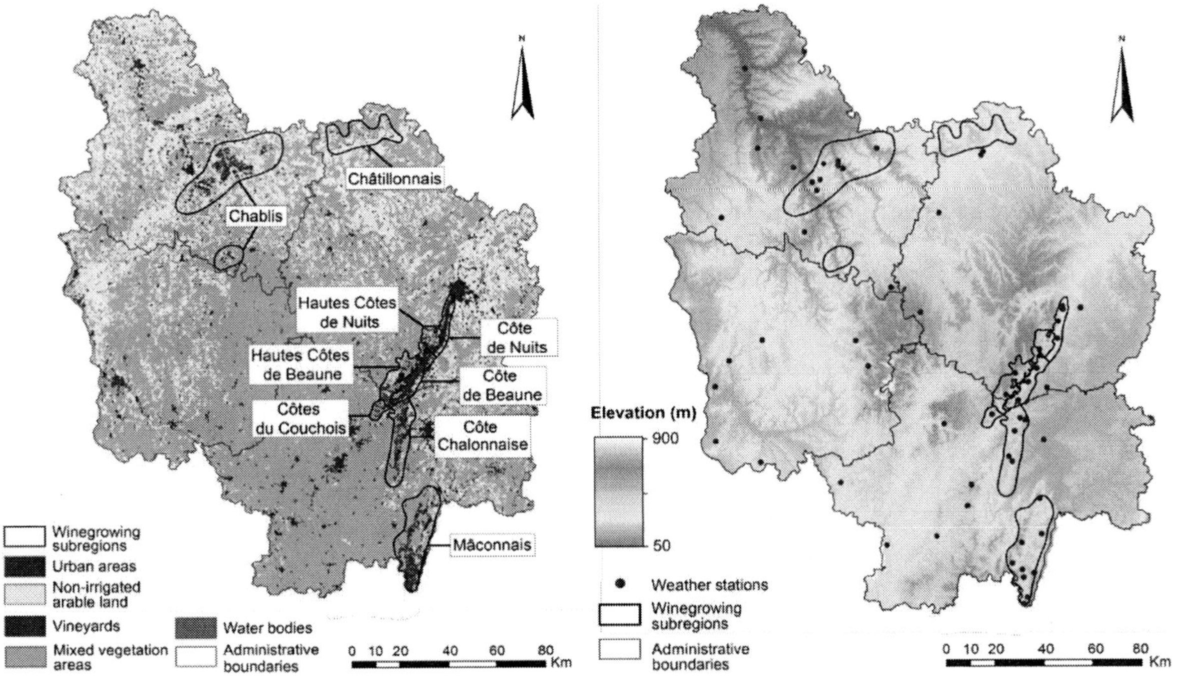

Figure 1: a) Geographical representation of Burgundy and its main land uses (CLC 2012). The vineyard area is represented in violet and it was separated in 9 subregions (thick black lines); b) Geographical location of the 64 weather stations.

the whole study area and were later extracted for the vineyard areas only by using the CLC vineyard polygons. The number of estimated pixels corresponding to the vineyards was of 74556 which account for an area of 41751 ha. The interpolation method applied was that of the regression-kriging [28, 29], which is divided in three steps: 1) establishing the correlation between a predictor (topographic parameters) and a predictant (tn and tx); 2) interpolating the regression residuals and 3) adding the interpolated residuals (2) to the prediction model (1). By following these steps, 3652 gridded daily temperatures were obtained (1826 for each) and cross validated using the LOOCV (*Leave-one-out cross validation*) method. Model performance was assessed for all stations as well as for those installed exclusively in the vineyard areas and we based it on three statistical indices: the R^2 (determination coefficient), the RMSE (root mean square errors) and the EF (model efficiency).

$$EF = 1 - \frac{\sum_{i=1}^{n}(Pi-Oi)^2}{\sum_{i=1}^{n}(Pi-\bar{O})^2} \qquad (1)$$

The efficiency is given by the above formula (1), where P_i are the simulated values and O_i are the observed values [30]. Interpolated daily temperature data were systematically tested for each of the 3652 created models. In terms of topographical parameters, the estimation of continuous tn and tx surfaces was done using the elevation, slope, orientation, humps, depressions and landscape roughness. These parameters were found to be the ones that best predicted tn or tx distribution for any given day of our study period.

2.3. Agroclimatic indices and the GFV phenological model

For each year of the selected timeframe 5 agroclimatic indices were calculated by cumulating the daily interpolated tn and tx. These indices were later spatialized and amongst them we mention the number of heat stress and that of frost days, the mid-flowering, the mid-véraison and the theoretical grape maturity (i.e. day when berries' sugar concentration attains 200g/L). The heat stress days are defined as the days where the weather stations' tx are equal or greater than 35°C (Apr to Sep), whereas the days that count as frost days are those with tn equal or inferior to -1°C (Mar to Jun).

The three indices related to the grapevine phenology were simulated for the Chardonnay and the Pinot noir varieties with the GFV (Grape Flowering Veraison) model developed by [31]. They were estimated through the summation of temperatures over 10°C (counting from the 60th day of the year which is either the 29 of Feb. or the 1st of Mar.). Since these two are very similar in terms of heat requirements their thresholds were averaged (Table 1), giving us a mean heat cumulation value of 1218 to attain mid-floraison, 2529 for the mid-véraison and 2990 for the theoretical maturity. The model was validated with observed phenological data and it was found to simulate the phenology occurrence dates very close to those measured on the field, indicating a mean error of 5 days for the mid-flowering and 7 days for mid-véraison and theoretical grape maturity respectively.

3. Results and discussion

3.1. Temperature spatial accuracy evaluation

The statistics relative to the tn and tx quality of spatialization are shown in table 1. The performance indices were calculated for all stations as well as for those found in the vineyard areas only. The modelling precision (after interpolation) of daily temperatures across Burgundy was in average of 0.89°C for tx and of 1.23°C for tn (RMSE). This means that any given pixel's value has an error attached to it of ± 0.89 (± 1.23) °C for the tx (tn). The medians of efficiencies and R^2 are 0.42 (0.54) and 0.67 (0.76) respectively, for tn (tx).

		Vineyard area stations			All stations		
		R^2	RMSE	EF	R^2	RMSE	EF
Tn	mean	0.67	0.94	0.36	0.64	1.23	0.39
	med	0.73	0.86	0.45	0.67	1.17	0.42
Tx	mean	0.74	0.70	0.35	0.74	0.89	0.51
	med	0.78	0.67	0.45	0.76	0.86	0.54

Table 1: The three statistical indicators (R^2 – determination coefficient, RMSE – Root Mean Square Error, EF – modelling efficiency) used for the estimation of the interpolation quality performance, applied to all stations as well as to those stations covering only the vineyard area.

The validation (LOOCV) of temperature interpolations within the vineyard area exhibit similar performances, but with a higher R^2. This suggests that a higher stations' density in the vineyard area allows a more accurate interpolation of temperatures and also a more homogeneous landscape/climate context (Figure 1b).

3.2. Phenology occurrence dates classification

Spatializing the agroclimatic indices relative to the grapevine's phenology allowed us to establish the viticultural zoning of Burgundy's vineyards thermal potentialities. These three phenological indices were spatially estimated across the 9 subregions (74556 pixels) mentioned in the previous sections and temporally estimated across the 5 years' timeframe (2011-2015). Therefore, temperature variability was assessed for both its spatial effect (*terroir*) and temporal effect (vintage). The 74556 pixels were later classified at regular intervals in 6 groups which were assigned to each of the three phenological indices: *"very early"* (red), *"early"* (orange), *"intermediate"* (yellow), *"late"* (green), *"very late"* (blue) and *"variable"* (Figure 2).

The pixels with a small DOY (day of year counting from Mar 1st) value were classed as *"very early"*, while those having a high DOY value were classed as *"very late"*. For the *"variable"* class, the interannual standard deviation (sd) was calculated for every pixel. If sd > 1 then it means that the pixel value is very variable from one year to the next. In order to map these classes, the DOY values of each year were averaged for the 2011-2015 period (Figure 2). The same principle was applied to the heat stress and the frost (not shown in this paper) days with their occurrence being classed as *"very rare"*, *"rare"*, *"intermediate"*, *"frequent"* and *"very frequent"*.

3.3. The theoretical grape maturity zoning

Thermal variations in Burgundy have allowed for a classification of the theoretical grape maturity (200 g/L of sugar concentration) occurrence dates, with the 6 derived classes being detailed in table 2. The values illustrated are those of the number of mean days per class and per year. Classes vary from *"very early"* (red) to *"very late"* (blue) and they represent smaller or larger groups of pixels, depending on how they were classified based on their maturity value. Interclass differences can be observed, with

	Years	Very early	Early	Interm.	Late	Very late	Var.
Number of mean days by class	2011	238	244	251	257	265	254
	2012	241	248	255	262	269	263
	2013	254	261	269	275	284	283
	2014	243	249	257	262	270	263
	2015	232	238	246	253	261	243

Table 2: Theoreti[illegible] for the 2011-201[illegible] number of mean D[illegible] and range from "ve[illegible]

https://doi.org/10.1051/e3sconf/20185001003

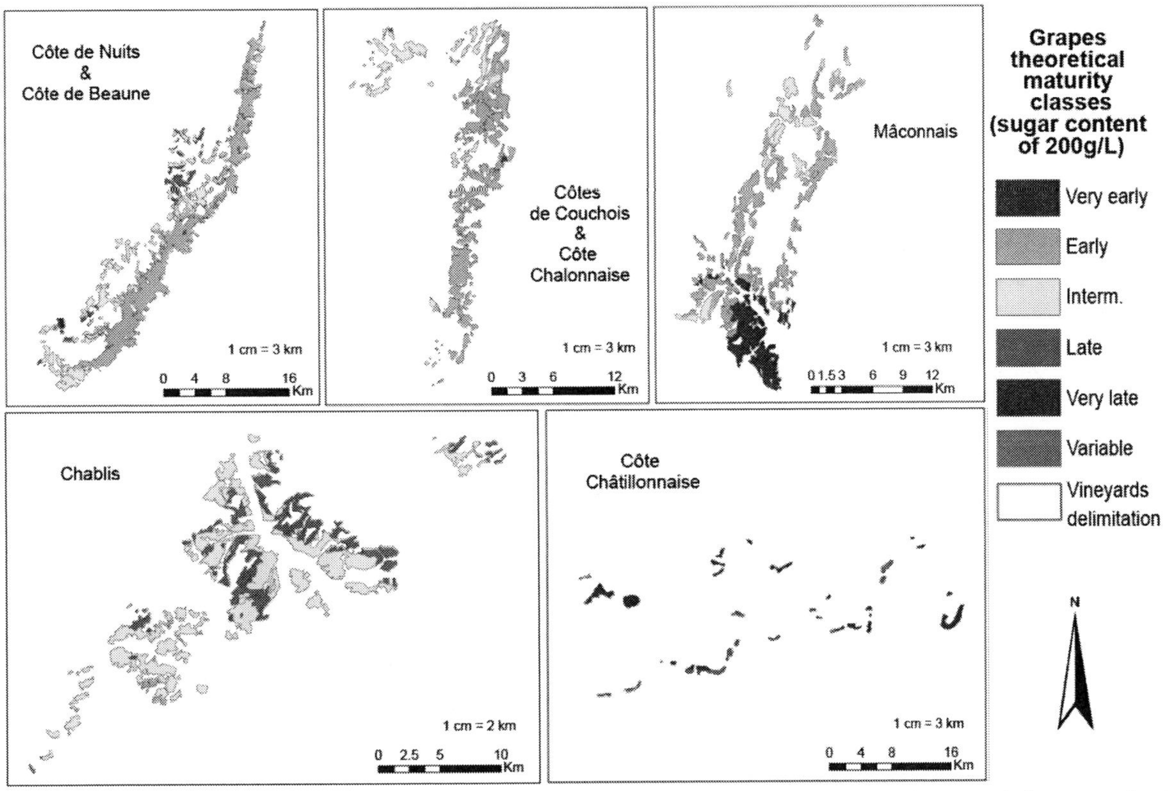

Figure 2: Spatialisation of "theoretical maturity" precocity classes for the Pinot noir and the Chardonnay varieties across the 9 subregions of Burgundy, based on the yearly cumulative temperatures of 2011-2015.

a gap ranging from 34 to 39 days (depending on year) between the earliest class and the latest. We notice that predicted theoretical maturity occurrence dates mirrors closely the field reality, the 200 g/L sugar concentration being attained around the end of August (depending on year) which is very close to the harvest date in Burgundy (beginning of September, again depending on year). As mentioned above, table 2 provides only the average values of each class, but an intraclass variation has been equally observed. Each class is defined by an interval with a minimum and a maximum value, meaning that among the pixels of the same class there will still be a difference, although not as large as that between pixels belonging to different classes. Table 2 has been spatialized and the maturity classes can be observed in figure 2 for each of the subregions. 28% of the Mâconnais subregion has been classified as *"very early"*, while Côte de Nuits, Côte de Beaune and Côte Chalonnaise have their largest area classified as *"early"* (97%, 82% et 83% respectively). Hautes Côtes de Nuits (56%), Hautes Côtes de Beaune (76%), Côtes du Couchois (97%) and Chablis (66%) are mainly classed as "intermediate", while Côte Châtillonnaise is the only one considered as *"very late"*. Furthermore, Côte de Nuits, Côte de Beaune and Côte Chalonnaise, where high added value wines are produced, exhibit limited variations in theoretical maturity timing despite a substantial South-North gradient. Finally, we would like to highlight that no zone of our study area has identified as *"variable"*, meaning that the grape theoretical maturity remains consistent throughout the years indicating that the classification used in the zoning of thermic potentialities is particularly robust.

3.4. Other phenological stages and thermal risks

The mid-flowering and mid-véraison simulated phenological stages exhibit a similar pattern as that of the theoretical ripening (results not shown in this paper).

The climate risks associated to minimum and maximum temperatures that were estimated for this study were the number of heat stress days and that of frost days. The later exhibits a greater risk frequency in Chablis and the

Châtillonnais, two subregions known for their higher predisposition to spring frost in comparison to others in Burgundy [32]. The spatial zoning of the number of heat stress days allowed us to identify winegrowing areas where the vine's thermal stress is more likely to occur during the summer period The days with temperatures equal or higher than 35°C are particularly frequent in the subregions of Chablis, the Châtillonnais and a small part of the Mâconnais which were classified as having an *"intermediate"* and *"elevated"* heat stress risk. All the other subregions are either classified with a *"very rare"* (up to 5 heat stress days/year) or a *"rare"* number of heat stress days. Among these we mention the Côte de Nuits and Hautes Côtes de Nuits, Côte de Beaune and Hautes Côtes de Beaune, as well as the Mâconnais.

4. Conclusion

Burgundy's agroclimatic potentialities have been analyzed, classified and spatialized. By using minimum and maximum temperature cumulations as well as a phenology simulation model, we were able to individually characterize Burgundy's main producing winegrowing subregions based on their thermic stress risk and their theoretical grape maturity (200 g/L sugar concentration). Grape maturity was used as a proxy in order to estimate the influence of thermal regimes on the vines' different developmental stages and it provided us with clues as to the distribution of the different estimated precocity classes as well as their associated surfaces. Furthermore, local topography was integrated to this approach to further highlight the "mosaic" nature of Burgundy's terroirs. Although climate and topography are proved to be factors that limit the geography of grapevines, they are also the ones that give wine its typicality and authenticity [6], which is what the concept of "climats" is all about. This kind of climate spatial zoning provided us with a clear scientific basis of Burgundy's diversity in climate conditions and helped delineate the climate viticultural potential of each subregion. To our knowledge, this unique approach was used to serve a viticultural purpose only once before, when it was applied to the Bordeaux wine region [12, 13]. Considering Burgundy's advancement in phenological dates due to recent climate changes [33], this kind of tool will give us a better understanding of the current terroir resilience and help us increase its adaptive capacity to climate change by offering a way to identify the appropriate strategies that will respond both to the environmental circumstances and the winemaker's needs.

Acknowledgements:

The authors would like to thank Météo-France, BIVB and Chambre Régional d'Agriculture de la Bourgogne for sharing their climate data and the Syndicat des Bourgogne for their financial support.

References

[1] J., Gladstones, 1992: Viticulture and Environment, Winetitles, Adelaide, 310 p.

[2] G.V., Jones, M.A., White, O.R., Cooper, K., Storchmann, 2005: Climate Change and Global Wine Quality. Climatic Change, 73, 319-343.

[3] C., van Leeuwen, P., Friant, X., Chone, O., Tregoat, S., Koundouras, et D., Dubourdieu, 2004: Influence of climate, soil, and cultivar on terroir. American Journal of Enology and Viticulture, 55(3): 207-217.

[4] D., Blanco-Ward, J.M., Garcia-Queijeiro et G.V., Jones, 2007: Spatial climate variability and viticulture in the Mino River Valley of Spain, Vitis 46, 63–70.

[5] A., Carbonneau, 2003: Ecophysiologie de la vigne et terroir, Terroir, Zonazione Viticol-tura, Piacenza, Italy (Phytoline: Piacenza).

[6] M.A., Amerine and A.J., Winkler, 1944: Composition and quality of musts and wines of California grapes, Hilgardia, 15, n°6.

[7] B., Bois, 2007: Cartographie agroclimatique à méso-échelle : méthodologie et application à la variabilité spatiale du climat en Gironde viticole. Conséquences pour le développement de la vigne et la maturation du raisin. Thèse de Doctorat, Université de Bordeaux II.

[8] G.V., Jones, M, Moriondo, B, Bois, A., Hall, A., Duff, 2009: Analysis of the spatial climate structure in viticulture regions worldwide. Le Bulletin de l'OIV 82(944,945,946): 507-518.

[9] J., Tonietto, A., Carbonneau, 2004: A multicriteria climatic classification system for grape-growing regions worldwide. Agric. Forest Meteorol. 124, 81-97.

[10] H., Quénol, 2014: Changement climatique et terroirs viticoles. Lavoisier, Paris, 444 pages.

[11] A., Jacquet and R., Morlat, 1998: Caractérisation de la variabilité climatique des terroirs viticoles en val

de Loire. Influence du paysage et des facteurs physiques du milieu. Agronomie, 17(9/10): 465-480.

[12] B., Bois, C., van Leeuwen, P., Pieri, J.P., Gaudillère, E., Saur, D., Joly, L., Wald, D., Grimal, 2008: Viticultural agroclimatic cartography and zoning at mesoscale level using terrain information, remotely sensed data and weather station measurements. Case study of Bordeaux winegrowing area, in: Proceedings of the VIIth International terroir Congress, pp. 455–462.

[13] B., Bois, C., van Leeuwen, 2008: Variabilité climatique dans la zone de production des vins de Bordeaux. Terroirs viticoles, ENITA Bordeaux, 45-50.

[14] D.D., Cichi, D.C., Costea, I., Olteanu, 2007: Effets des champs climatiques sur la vigne cultivée au sud de la Roumanie, Université de Craiova, Faculté d'Horticulture, Craiova.

[15] C., Bonnefoy, 2013: Observation et modélisation spatiale de la température dans les terroirs viticoles du Val de Loire dans le contexte du changement climatique, Thèse de doctorat, Université de Rennes 2.

[16] I., Chuine, P., Yiou, N., Viovy, B., Seguin, V., Daux and E., Le Roy Ladurie, 2004: Grape ripening as a past climate indicator, Nature, 432, 289-290.

[17] F., Spanik, B., Siska, and M., Galik, 2004: Changes in the phenology of grapevines (Vitis vinifera) as influenced by climate change impacts in Slovakia. Meteorol. Casopis 4:179-182.

[18] L.B., Webb, P.H., Whetton, et E.W.R., Barlow, 2008: Modelled impact of future climate change on the phenology of winegrapes in Australia. Aust. J. Grape Wine Res. 13:165-175.

[19] G., Seguin, 1970: Les sols de vignobles du Haut-Médoc. Influence sur l'alimentation en eau de la vigne et sur la maturation du raisin. Thèse de Doctorat d'Etat, 141 p., Faculté des Sciences, Université de Bordeaux.

[20] E., Lebon, A., Pellegrino, F., Tardieu, et J., Lecoeur, 2004: Shoot development in grapevine (Vitis vinifera) is affected by the modular branching pattern of the stem and intra- and inter-shoot trophic competition. Annals of Botany, 93(3): 263-274

[21] D., Joly, 2009: LISDQS: logiciel d'interpolation statistique de données quantitatives et spatiales. http://thema.univ-fcomte.fr/LISDQS-un-nouveau-logiciel-d.

[22] R., Geiger, 1969: Topoclimates, in *World Survey of Climatology*.

[23] G., Guyot, 1997: Climatologie de l'environnement, in *Masson*.

[24] D., Joly, T., Brossard, H., Cardot, J., Cavailhes, M., Hilal, P., Wavresky, 2010: Les types de climats en France, une construction spatiale. Cybergeo : European Journal of Geography: 21p.

[25] M., Madelin, B., Bois, J.P., Chabin, 2010: Modification des conditions de maturation du raisin en Bourgogne viticole liée au réchauffement climatique. EchoGéo 14 : 11p. URL: http://echogeo.revues.org/12176

[26] J.P., Chabin, M., Madelin, C., Bonnefoy, 2007: Les vignobles beaunois face au réchauffement climatique. Actes du colloque "Réchauffement climatique, quels impacts probables sur les vignobles?", Dijon, 28-30 Mars : 13p.

[27] European Environment Agency, 2016. Corine Land Cover (CLC) 2012, Version 18.5.1. URL: http://land.copernicus.eu/pan-european/corine-land-cover/clc-2012/view

[28] I., Odeha, A., McBratney, D., Chittleborough, 1994: Spatial prediction of soil properties from landform attributes derived from a digital elevation model, Geoderma.

[29] T., Hengl, G.B.M., Heuvelink, D.G., Rossiter, 2007: About regression-kriging: from equations to case studies. Comput. Geosci. 33, 1301e1315.

[30] D.G., Mayer, and D.G., Butler, 1993: Statistical validation. Ecol. Model., 68: 21-32.

[31] A.K., Parker, I., Garcia de Cortazar-Atauri, C., van Leeuwen, et I., Chuine, 2011: A general phenological model to characterise the timing of flowering and veraison of Vitis vinifera L.Aust. J. Grape Wine Res., 17, n°2, 206-216.

[32] J.F., Bazin, 2015. Le vin de Bourgogne. Dunod. 264p.

[33] C., Cuccia, 2013: Impact du changement climatique sur la phénologie du Pinot noir en Bourgogne, Thèse de doctorat, Université de Bourgogne.

Can Mediterranean terroirs withstand climate change? Case studies at the Alentejo Portuguese winegrowing region

Carlos M. Lopes[1a]; Joaquim M. Costa[1], Ricardo Egipto[1], Olfa Zarrouk[2] and Maria M. Chaves[2]

[1]LEAF, Linking Landscape, Environment, Agriculture and Food, Instituto Superior de Agronomia, Universidade de Lisboa, Tapada da Ajuda 1349-017 Lisboa, Portugal, email: carlosmlopes@isa.ulisboa.pt.
[2]LEM, ITQB-NOVA, Univ. Nova de Lisboa, Apartado 127, 2780-901 Oeiras, Portugal.

Abstract. Climate change introduced new challenges to vinegrowers in the Mediterranean areas such as the hot and dry winegrowing region of Alentejo, south Portugal. Warmer and drier conditions are harmful for grape yield and berry quality attributes and for vine's longevity, mainly when optimal thresholds are exceeded. Therefore winegrowers are forced to move from rainfed to irrigated production systems making Alentejo' wine production strongly based on available water resources for irrigation. This work aims to review and discuss ecophysiological and agronomical data obtained in irrigation trials set up at different terroirs of Alentejo. In the last four decades, classical bioclimatic temperature-based indices showed a significantly positive trend, while the dryness index present a negative trend over time. Furthermore, ecophysiological data collected in deficit irrigation experiments are reviewed and discussed, focusing on the effects of drought and heat stress on vigour, yield and berry composition. Emphasis is given on the indirect effects of leaf senescence on cluster exposure and consequences on berry temperature and composition. In order to promote the sustainability and quality of wine production in these hot and dry terroirs short- to long-term adaptation measures are suggested. The limitations and risks of using deficit irrigation during heat waves are also underlined.

1. Introduction

The incidence of drought have increased in recent years and, according to model predictions for future climate in South Mediterranean Europe, it is expected an increase in winter rains and an extended summer drought, coupled to a substantial increase in temperature [1, 2]. Consequently, water demand will increase and soil water balance during the grapevine vegetative period will become more negative [3]. Furthermore, climatic scenarios predict that heat waves tend to strike the Mediterranean region more frequently and for longer periods [4, 5]. All these changes pose new challenges to vinegrowers in the Mediterranean areas such as the hot and dry winegrowing region of Alentejo, south Portugal.

The predicted higher air temperatures and less precipitation during the grapevine growing season, and mainly during the ripening period, were already observed in Portugal in recent decades [2, 6, 7]. In Alentejo's winegrowing region the vineyard surface area has experienced a fast expansion during the last decades of the XX century attaining in 2016 about 23,400 ha [8], most of them drip irrigated.

Water stress and supra-optimal temperatures are known to decrease plant carbon assimilation, to reduce vegetative growth and yield and to affect berry composition [9] and, in extreme cases, could also affect vine's longevity. In order to overcome the negative impact of climatic changes on grape yield and composition and vine's longevity, Alentejo´s grape growers have been forced to move from rainfed to irrigated production systems and nowadays all the new vineyards are planted with a drip irrigation system, making Alentejo' wine production strongly based on available water resources for irrigation. Despite the recent building of the "Alqueva" big dam, water in Alentejo is still a scarce resource forcing the grape growers to use water more efficiently.

In those regions, deficit irrigation strategies such as sustainable and regulated deficit irrigation have been successfully adopted as management tools to control vine vegetative growth and improve fruit composition, while enhancing plant water use efficiency [10 - 12]. However, deficit irrigation induces progressive increase in water stress along the growing season [13, 14], which can promote early leaf senescence and leaf fall at the fruiting zone [8, 15]. Furthermore, in case of heat waves, due to the restriction in stomatal conductance and consequent decrease in the ability to energy dissipation, supra-optimal canopy temperatures are attained and vine vulnerability to leaf burning increased [16]. The resultant defoliation increases cluster sun exposure, leading to berry sunburn and/or to abnormal ripening, producing reduced yield, juice with low titratable acidity and a loss of aromatic compounds [17, 18]. In red varieties, the high temperature that the fully exposed berries can achieve may also affect anthocyanin concentration either through degradation, inhibition of synthesis, or, more likely, both [19 - 20].

In this context, we aimed to review and discuss ecophysiological and agronomical data obtained in two vineyard irrigation experiments conducted at different terroirs of the hot and dry region of Alentejo, south of Portugal. Furthermore, short- to long-term adaptation measures are suggested.

2. Material and Methods

Two independent field experiments were carried out at two PDO sub-regions of Alentejo, South Portugal: Évora and Reguengos. The Évora field trial was carried out in 1999 at a commercial vineyard located nearby the city of Évora. The fourteen-year-old grapevines of the red variety 'Aragonez' syn. 'Tempranillo', grafted on 1103P rootstock, were spaced 2.5 m between and 1.1 m along North-South oriented rows and trained on a vertical trellis. The soil is an Antrossoil with an ApCR profile derived from granite with 75% of sand and high percentage of gross elements. The experiment was set up in a randomized block design with four replicates per treatment. Two deficit irrigation strategies (DIv- 50% ETc withheld after veraison and DIh - 50% ETc supplied until harvest), were tested against a non-irrigated control (NI). Irrigation started at the end of May (10 days after full bloom) and was cut-off at the end of July (veraison) for the DIv treatment and one week before harvest (end of August) for DIh one. A total of 144 and 216 mm was applied once a week in DIv and DIh respectively. More detailed information could be obtained in [15].

The second irrigation trial took place in 2013 at a commercial vineyard located near Reguengos de Monsaraz. The eleven years-old grapevines of the variety 'Aragonez', grafted to 1103 Paulsen rootstock were spaced 1.5 m within and 3.0 m between north-south oriented rows. Vines were trained on a vertical shoot positioning system with one pair of movable wires and spur-pruned on a bilateral Royat Cordon system with 16 nodes per vine. Soil texture is a sandy-loam to silty-clay-loam, with a pH of 7.0 to 7.6 and low organic matter content. The experimental layout is a randomized complete block design with two irrigation treatments (Sustainable Deficit Irrigation (SDI -control; ~30% ETc) and Regulated Deficit Irrigation (RDI; ~15% ETc), and 4 replications per treatment. For more details check [21, 26].

Soil and plant assessments were similar in the two experiments. Soil water content was monitored with a neutron probe (Évora) or a capacitance probe (Reguengos) in 4 access tubes per treatment installed in the row between two contiguous vines with readings taken periodically at increments of 0.2 m (Évora) or 0.1 m (Reguengos) from soil surface to a depth of *ca* 1.0 m. The fraction of transpirable soil water was calculated according to [22]. Predawn and midday leaf water potential were measured in fully mature and well-exposed leaves per treatment, on the day before irrigation, using a pressure chamber. Diurnal curves of leaf gas exchange were done along the season with a portable infrared gas analyser (LI-6400, Li-Cor Inc., USA) in 4-6 fully mature , sunlit leaves per treatment. Light at the cluster zone was measured on sunny days at midday using a Sunflek Ceptometer inserted horizontally at the cluster zone along the row. Incident active radiation (PAR) was expressed in percentage of a reference PAR, measured over the canopy top. Berry temperature was monitored continuously with a dual channel temperature data logger with a two-junction fine wires copper-constantan thermocouples attached. These probes were positioned on berry surface of a sample of clusters located in different canopy positions (exposed east and west) of 2 vines per treatment. Cluster number and weight per vine were assessed at harvest and one berry sample per replication was collected to evaluate berry composition. Pruning weight and shoot number per vine were also recorded.

3. Results and Discussion

3.1. Pattern of bioclimatic indices in the last decades

As shown in Fig. 1A, the temperature-based Huglin index calculated for the last four decades presented a significantly positive trend over years, changing from the warm temperate class to warm and very warm classes [23]. In turn, the dryness index, which is an index based on the potential soil water balance calculated in the budburst-harvest period, shows a negative trend over years with most part of the seasons classified within the very dry class [24] (Fig. 1B).

3.2. Rainfed vs deficit irrigation at Évora

Figure 2A presents the seasonal trend of the fraction of transpirable soil water (FTSW) measured in the 0-1.0 m soil profile at the end of each irrigation cycle. The soil was almost fully refilled in winter/early spring but, after March the FTSW started a decreasing pattern till the end of the growing cycle. At the start of irrigation (end of May) all treatments presented a similar FTSW (*ca* 60%). However, in June and July (flowering to veraison) the NI treatment showed a faster decrease in FTSW, presenting significantly lower values than the two deficit irrigated treatments. During the ripening phase, the two deficit irrigation continued the decreasing pattern, attaining harvest with very low FTSW, similar to that of the NI treatment.

Regarding plant water status, the predawn leaf water potential (Ψ_{pd}) presented a parallel decreasing pattern as that described for FTSW (Fig. 2B). In NI vines Ψ_{pd} attained very low values at veraison (*ca* -0.9 MPa) and values lower than -1.0 MPa during ripening, indicative of severe water stress. In the deficit irrigation treatments Ψ_{pd} presented a similar decreasing trend until veraison, but the values attained were higher than those of NI. After arresting irrigation at veraison DIv showed a steeper Ψ_{pd} decrease than DIh vines, attaining at harvest values close to those of NI plants and lower than those of the DIh. The midday leaf water potential (Ψ_m) showed also a decreasing pattern along the season. However the relative differences between treatments were less pronounced than in Ψ_{pd} (Fig. 2B). The seasonal variation of midday stomatal conductance (gs) mirrors that reported for Ψ_{pd} but with a stronger reduction before veraison. After irrigation suspension on the DIv, the gs showed a strong decline, achieving at harvest significantly lower values than those of the DIh vines but similar to those of the NI ones (Fig. 2C).

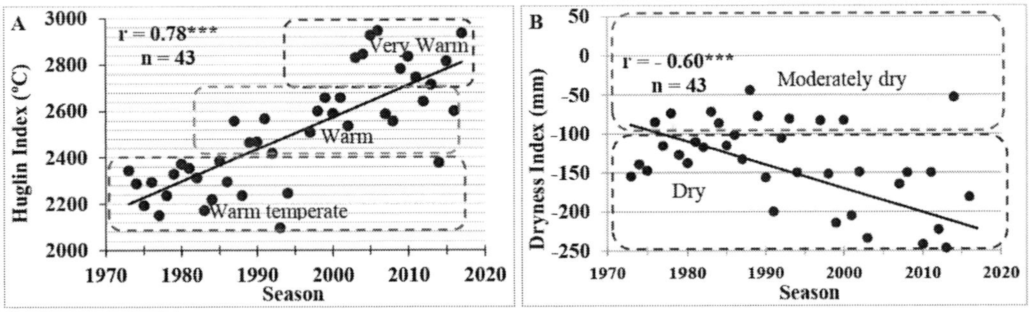

Figure 1. Evolution trend for the Huglin Index (A) and Dryness Index (B) over a 45 year period (1973-2017), calculated using data from a weather station located at Évora. The dashed rectangles indicate de boundaries of the classes.

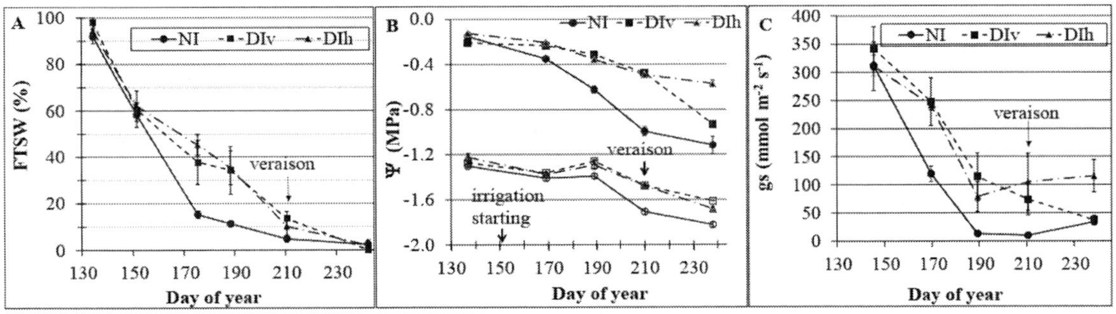

Figure 2. Seasonal pattern of the fraction of transpirable soil water (FTSW) (0-1.0 m) (A), predawn (close symbols) and midday (lower curves with open symbols) leaf water potential (B) and midday stomatal conductance (C) of cv Aragonez (syn.Tempranillo) vines subjected to different irrigation water supply and measured at the end of each drying cycle following irrigation: NI- non-irrigated; DIv - 50% ETc withheld after veraison, DIh - 50% ETc supplied until harvest. Average ± standard error of 4 access tubes (A) or 6 adult exposed leaves (B & C).

In NI vines leaf senescence started prematurely and, at veraison, about 60% of main leaf area had already been lost. At harvest (end of August) while the most irrigated plants showed only 26% of defoliation, NI vines presented 72%. In DIv treatment the irrigation cut-off at veraison induced also a fast increase in leaf senescence, which attained at harvest a similar value to that of NI vines (Fig. 3A). The high rate of defoliation observed in NI vines induced a higher interception of PAR in the cluster zone than the two irrigated treatments. The irrigated treatments showed similar values of intercepted PAR until mid-ripening but at harvest DIv presented a higher value than DIh (Fig. 3B).

Due to a significantly lower berry and cluster weight, NI vines presented the significantly lowest yield, while DIh showed the highest one and DIv an intermediate value (Fig. 4). Regarding berry composition, no significant differences were observed except for the anthocyanins content, which was higher in NI vines as compared with the deficit irrigation treatments (Fig. 4).

 https://doi.org/10.1051/e3sconf/20185001004

3.3. Deficit irrigation at Reguengos de Monsaraz

The seasonal trend of the FTSW measured in the experiment carried out at Reguengos is presented in Fig. 5A. Irrigation started middle of June enabling to almost refill the soil profile (60-80%). In the second half of June and July withholding irrigation in RDI induced a punctual strong decrease in FTSW as compared to SDI. After veraison (end of July) both irrigation treatments showed a similar trend but RDI presented lower FTSW after each irrigation event.

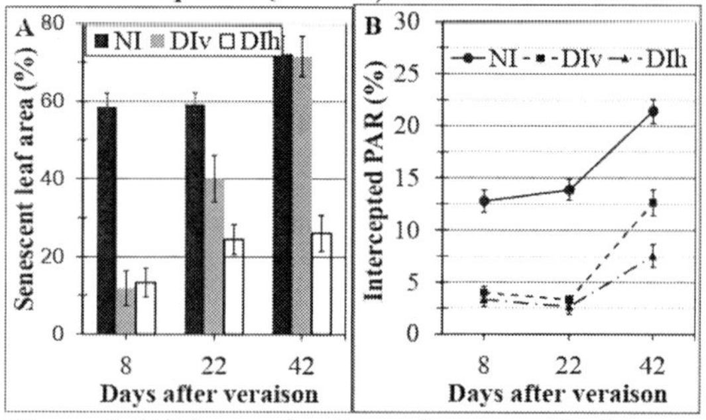

Figure 3. (A) Seasonal pattern of senescent leaf area (average ± SE, n=16 shoots). (B) Photosynthetic active radiation (PAR, as percentage of the reference intercepted at the top of the canopy) intercepted at cluster zone (average of 40 ± SE measurements).

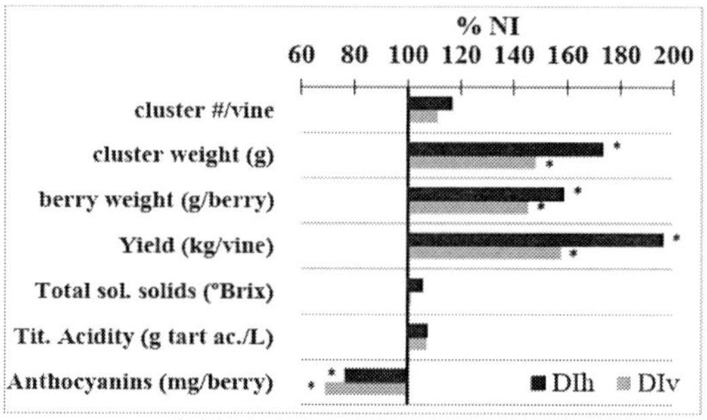

Figure 4. Yield and berry composition of 'Aragonez' (syn. 'Tempranillo') grapevines subjected to irrigation water supply. Values expressed in % of the control NI vines. Asterisks indicate statistically significant differences from NI at P<0.05 by LSD test. NI: non-irrigated; DIv: 50% ETc withheld after veraison; DIh: 50% ETc supplied until harvest.

ψ_{pd} presented a decreasing trend along the season in both irrigation treatments, with oscillations related to the several irrigation events (Fig. 5B). After the first period of withholding irrigation in RDI, SDI started to show higher ψ_{pd} values as compared to RDI and the largest differences were observed at ripening. While the lowest ψ_{pd} values observed in SDI were around -0.5 MPa (indicative of a mild to strong water stress), RDI attained significantly lower ψ_{pd} values (*ca* -0.7 MPa), indicative of a strong to severe water stress [25]. During the ripening period ψ_m attained very low values (from -1.56 to -2.18 MPa) but no significant differences were detected between treatments (data not shown). Regarding the effects of irrigation on gas exchange, in general RDI induced a significant reduction in gs and net assimilation rate [26]. Figure 5C presents an example of the gs diurnal trend in a very hot day of the ripening period, showing a decreasing pattern from early morning in both treatments, with SDI vines presenting always larger gs than RDI ones.

Deficit irrigation strategies promoted different water supply and influenced the ability of the canopy for evaporative cooling. As compared to SDI, RDI strategy induced a significant increase in canopy temperature as assessed by thermography [26]. Thermal data from individual leaf temperature assessed by thermocouples indicates also that the exposed leaves can achieve very high temperature during the day as shown in the example of Fig. 6A. The same figure also shows that at the hottest hours of the day, the west side of the canopy presents the highest leaf temperatures, which can achieve values higher than 10°C above air temperature [8, 26].

Berry temperature was also assessed with thermocouples installed in different cluster positions within the canopy and then related to berry composition (for more detailed data see [21]). Using berry temperature from berry pea-size to full maturation, the work of [21] showed that irrigation treatment has a slight effect on the temperature of berries from clusters at similar exposure. However, when comparing the maximum berry temperature per irrigation treatment, berries from west exposed clusters presented maximum temperatures 5 to 8 °C and 2.5 to 9 °C higher than those from east exposed clusters in SDI and RDI treatments, respectively. Consequently the berries from west exposed clusters accumulated a higher number of hours of temperature above 35°C than berries from east exposed clusters [21]. Figure 6B presents an example of the daily evolution of exposed berries temperature during three hot days of the ripening period for RDI treatment, showing that from mid-morning to mid-afternoon the temperature of exposed berries (East and West) was above 35°C and that the highest temperatures were attained at the west exposed berries during the afternoon. As the west-exposed berries displayed lower anthocyanin concentration than east-exposed berries, irrespective of irrigation (Fig. 6C), the same authors concluded that the higher number of heating hours could have a key role in anthocyanins biosynthesis repression and degradation.

We found no significant effect of irrigation strategy on vine vigor (data not shown). Irrigation has also no statistically significant effect on yield despite of a minor reduction (16.5%) observed in RDI vines as compared to SDI ones.

4. Discussion and conclusions

In this work we have reviewed some ecophysiological and agronomical data obtained in two irrigation trials set up at the very hot and dry winegrowing region of Alentejo, located at inland south Portugal. In the last decades the temperature-based Index of Huglin showed a positive trend, while the Dryness Index presents a negative trend over time, indicating that the incidence of heat and drought stress at the Alentejo winegrowing region has increased in recent years. Results from the Évora experiment show that the rainfed vineyards cannot withstand severe drought stress as soil water reserves cannot support adequate transpiration inducing very early and severe water deficits.

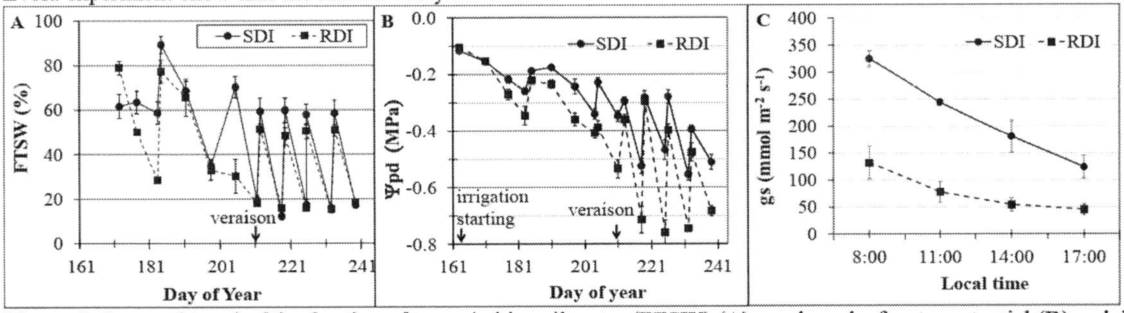

Figure 5. Seasonal trend of the fraction of transpirable soil water (FTSW) (A), predawn leaf water potential (B) and daily course of stomatal conductance to water vapor (gs) measured in a warm day of the ripening phase (19th August 2013).

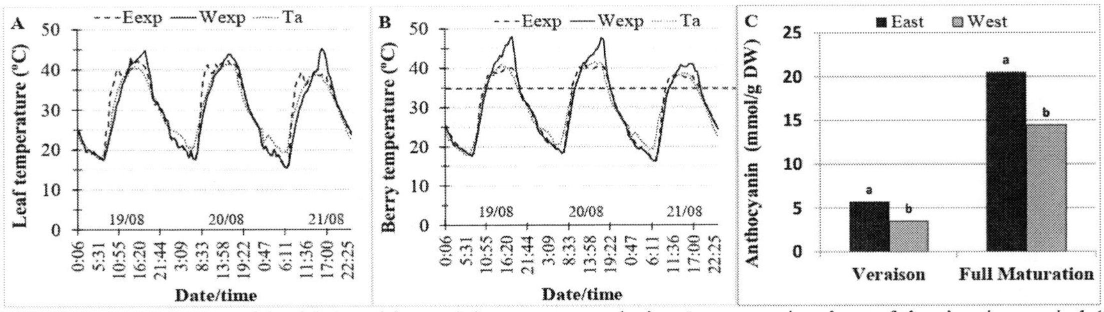

Figure 6. Hourly exposed leaf (A) and berry (B) temperature during 3 consecutive days of the ripening period (19th to 21th August). Eexp – East exposed; Wexp – West exposed; Ta – average air temperature. In plot B) the dashed horizontal line shows the berry temperature threshold above which anthocyanin biosynthesis could be compromised. The plot C) shows the averaged effect of canopy side (East *vs* West) on anthocyanin accumulation in the grape berry skin of exposed clusters The bars with different letter suffixes indicate significant differences ($p<0.05$). Adapted from [21]).

As a response to soil drying decreased stomatal closure resulted in higher leaf temperature and promoted leaf senescent and leaf burning, especially in NI vines. Increased light penetration at the cluster zone due to leaf abscission, increased the fraction of exposed clusters which can be, in most temperate wine regions, beneficial to phenols and anthocyanins accumulation since this process is strongly light dependent [19]. However, the degree of cluster exposure influences berry temperature, which can block the synthesis of anthocyanins or induce their degradation [20, 27]. Defoliated vines, like those of NI, are much more exposed to berry sunburn and/or experience the negative effects of the high temperature on berry ripening. However, surprisingly, the anthocyanin content was higher in NI berries than in the deficit irrigated ones results that can be explained by the positive effect of the smaller berry size (smaller berries would result in a higher skin/flesh ratio [28], which could have overridden the expected negative effect of high berry temperatures on anthocyanin biosynthesis.

In practical terms, Évora trials showed that deficit irrigation should be applied until harvest, as the irrigation cutoff at veraison has also induced a severe water stress and a strong leaf senescence during the ripening period.

The results from Reguengos experiment showed that the use of RDI allowed to save water, as compared to the conventional SDI used by the grower, without significant effects on yield. However, in this terroir it is very risky to reduce irrigation amounts behind a certain threshold as it can induce very stressful conditions during the ripening period. Indeed, the use of deficit irrigation strategies induced a progressive decrease in the FTSW along the growing season, which in the case of the RDI strategy, induced predawn values that are indicative of a strong to severe water stress [25]. As reported above, in these conditions the restriction in stomatal conductance and consequent decrease in the ability for energy dissipation may increase vine vulnerability to leaf burning [8, 16] mainly during heat waves, which are occurring with higher frequency and for longer periods [4]. Furthermore, the results obtained at the Reguengos trial show the negative impacts of excessive cluster exposure on berry composition and reinforce the need to promote basal leaf health during the whole ripening period in order to provide sun protection to the grapes and, consequently, avoid either the risk of sunburn and/or negative effects of excessive high berry temperature on anthocyanin concentration [19 - 21].

In order to face these challenges and mitigate the negative impact of more extreme weather conditions in South Portugal, short to long term adaptation measures must be adopted [8]. More efficient irrigation practices and more precise irrigation scheduling will help to save water, while minimizing the negative impacts of climate change on viticulture in South Mediterranean countries [8, 29, 30]. Optimized plant and soil water monitoring system, and improved decision support systems based on weather soil and plant data integrated in real time will enable the growers to better manage the irrigation and to be alerted for risky situations (e.g. lower soil water availability, heat events forecast, etc) when certain thresholds are attained [31].

Furthermore, the irrigation infrastructure should be able to deliver higher amounts of water than the ones estimated for deficit irrigation in order to allow to refill the soil profile before a forecasted heat wave. The vineyard training system and canopy management should also be chosen/adjusted in order to provide natural bunch shading particularly at the canopy side sun-exposed during the afternoon. Many other adaptations are possible, e.g. the use of drought resistant rootstocks and varieties, the choice of vineyard soils with higher water holding capacity, the use of more adequate soil management practices, the use of sun protection agents, etc. [32 - 34].

In conclusion, our results show that the winegrowers of Alentejo need to do a strong effort to adopt measures that can cope with climate change and improve the sustainability and quality of wine production in these hot and dry terroir.

Acknowledgements

This research received funding from the Portuguese project PAMAF I&ED 2007 and by European Community's Seventh Framework Programme (FP7/2007-2013) under the grant agreement n° FP7-311775, project INNOVINE.

References

1. IPCC Climate Change 2013: *The Physical Science Basis. I Contribution of Working Group I to the Fifth Assessment Report of the Intergovernmental Panel on Climate Change.* Cambridge Univ. Press, UK (2013).
2. H. Fraga, A.C. Malheiro, J. Moutinho-Pereira, J.A. Santos, Int. J. Biometeorol., **57**(6),909-925 (2013)
3. H. Shultz, Journal Wine Economics, **11**(1), 181-200 (2016)
4. R.O. Teskey, T.M. Wertin, I. Bauweraerts, M Ameye, M.A. McGuire, K. Steppe, Plant Cell Environ., **38**(9),1699–1712 (2014).
5. H. Fraga, I. García de Cortázar Atauri, JA Santos, Agric.Water Manag., **196**,66-74, (2018)
6. J.A. Santos, A.C. Malheiro, J.G. Pinto, G.V. Jones, Clim. Res., **51**(1),89-103 (2012).
7. J.M. Costa, M. Vaz, J. Escalona, R. Egipto, C. Lopes, H. Medrano, M.M. Chaves, Agr. Water Manage, **164**,5-18. (2016).
8. IVV, *Vinhos e Aguardentes de Portugal 2016.* Instituto da Vinha e do Vinho, Lisboa. (2017).
9. M.M. Chaves, T.P. Santos, C.R. Souza, M.F. Ortuño, M.L. Rodrigues, C.M. Lopes, J.P. Maroco, J.S. Pereira, Ann Appl Biol., **150**, 237-252 (2007).
10. P.R. Dry, B.R. Loveys, M.G. McCarthy, M. Stoll, J. Int. Sci. Vigne Vin, **35**,129-139 (2001).
11. B.R. Loveys, M. Stoll, W.J. Davies, *Water Use Efficiency in Plant Biology.* Bacon M.A. (Eds.), University of Lancaster, Lancaster, UK, pp. 113-141. (2004).
12. T.P. Santos, C.M. Lopes, M.L. Rodrigues, C.R. Souza J.P. Maroco, J.S. Pereira, J.R. Silva, M.M. Chaves, Funct. Plant Biol., **30**,663-671 (2003).
13. C.M. Lopes, T. Santos, A. Monteiro, M.L Rodrigues, J.M. Costa, M.M. Chaves, Scientia Horticulturae, **129**,603-612 (2011).
14. K.C. Shellie, Am. J. Enol. Vitic., **65**,197–205 (2014)
15. C.M. Lopes, M.L. Rodrigues, T. Santos, J. Vicente-Paulo, MM Chaves, *Proc. Congres on Climate and Viticulture, Centro Transferencia Agroalimentaria, Saragoza*, 10-14 April 2007: 87-92 (2007)
16. M.M. Chaves, J.M. Costa, O. Zarrouk, C. Pinheiro, C.M. Lopes, J.S. Pereira, Plant Sci., **251**,54–64 (2016)
17. S.F. Price, P.J. Breen, M. Valladao, B.T. Watson. Am. J. Enol. Vitic., **46**:187–194 (1995).
18. A. Belancic, E. Agosin, A. Ibacache, E. Bordeu, R. Baumes, A. Razungles, et al. Am. J. Enol. Vitic., **48**,181-186 (1997).
19. L. Haselgrove, D. Bottingvan, H.P.B. van Heeswijck, P.R. Dry, C. Ford, P.G. Iland, Aust. J. Grape Wine Res., **6**,141-149 (2000).
20. S.E. Spayd, J.M. Tarara, D.L. Mee, J.C. Ferguson, Am. J. Enol. Vitic., **53**,171-182 (2002).
21. O. Zarrouk, C. Brunetti, R. Egipt, C. Pinheiro, T. Genebra, A. Gori, C.M. Lopes, M. Tattini, M.M. Chaves, Frontiers in Plant Science, **7**,1-18 (2016).
22. A. Pellegrino, E. Lebon, M. Voltz, J. Wery, Plant Soil **266**,129-142 (2004).
23. G.V. Jones, A.A. Duff, A. Hall, J.W. Myers, Am. J. Enol.Vitic., **61**,313-326 (2010).
24. J. Tonietto, A. Carbonneau, Agr. Forest Meteorol., **124**,81-97 (2004).
25. H. Ojeda, C. Andary, E. Kraeva, A. Carbonneau, A. Deloire, Am. J. Enol. Vitic., **53**, 261-267 (2002)
26. I.F. García-Tejero, J.M. Costa, R. Egipto, V.H. Durán-Zuazo, R.S.N. Lima, C.M. Lopes, M.M. Chaves. Agric. Water Manag., **176**, 80–90 (2016).
27. J. Bergqvist, N. Dokoozlian, N. Ebisuda, Am. J. Enol. Vitic., **52**(1), 1-7 (2001).
28. R.L. Wample, R. Smithyman, *Deficit Irrigation Practices.* FAO (Eds.), Rome, pp. 89-100 (2002)
29. M.M. Chaves, O. Zarrouk, R. Francisco, J.M. Costa, T. Santos, A.P. Regalado, M.L. Rodrigues, C.M. Lopes, Ann Bot. **105**, 661-676 (2010).
30. H. Medrano, M. Tomás, S. Martorell, J.M. Escalona, A. Pou, S. Fuentes, J. Bota, Agron. Sust. Dev. 1-19 (2014).
31. R. Egipto, A. Barriguinha, J. ·Silvestre, N. Carvalho, M. Neto, J.M. Costa, C.M. Lopes, *Actas 10.° Simpósio de Vitivinicultura do Alentejo*, pp. 57–61 (2016).
32. A. Palliotti, S. Tombesi, O. Silvestroni, V. Lanari, M. Gatti, S. Poni, Sci. Hortic., **178**, 43–54 (2014).

33. M.R. Mozell, L. Thach, Wine Economics and Policy **3,** 81–89 (2014).
34. J.M. Costa, R. Egipto, J. Silvestre, C.M. Lopes, M.M. Chaves *Water scarcity and sustainable agriculture in semiarid environment: tools, strategies, and challenges for woody crops.* Elsevier Inc. pp. 219-245 (2018).

Implementation of climate change science in viticulture sustainable development planning in Serbia

Ana Vukovic[1], Mirjam Vujadinovic[1], Mirjana Ruml[1], Zorica Rankovic-Vasic[1], Zoran Przic[1], Zoran Beslic[1], Sasa Matijasevic[1], Dragan Vujovic[1], Slavica Todic[1], Nebojsa Markovic[1], Branislava Sivcev[1], Dragoljub Zunic[1], Ljubomir Zivotic[1] and Darko Jaksic[2]

[1]Faculty of Agriculture, University of Belgrade, Nemanjina 6, 11080 Belgrade, Serbia
[2]Ministry of Agriculture, Forestry and Water Management, Nemanjina 22-26, 11000 Belgrade, Serbia

Abstract. Growing interest in Serbia in adaptation of viticulture to climate change emerged from a recorded positive impact of summer increased draught on domestic wine quality. Another motivation is that viticulture has been recognized as one of the fastest developing agricultural sectors in Serbia and, to contain its growing potential, it is crucial to provide basis for its future sustainable development. Prioritization and implementation of adaptation measures in practice require reliable assessment of climate projections. For climate change impact assessment is used high resolution multi-model ensemble of nine regional climate models simulations, bias corrected, with two most probable future scenarios of GHG emissions RCP4.5 and RCP8.5, for the period 2016-2100. Analysis has been done for the near future, mid-century and end of the century periods. The periods are defined according to the IPCC 5th Assessment Report, which enables comparison of climate change impacts with a wider region, and preferably motivate future international collaboration and knowledge exchange.

1 Introduction

After viticulture has been recognized as one of the fastest developing agricultural sectors in Serbia and probably the only one with positive response to climate change trends over this region, local experts in collaboration with relevant governmental institutions continue to implement latest climate change knowledge in decision-making and promote climate change information among the producers. Increased summer drought [1] had good wine quality response, especially during the years 2012 and 2017, which were critical for other agricultural sectors and caused significant reduction in crop production. Flowering, veraison and harvest advanced significantly [2] because of the change in thermal conditions. Future climate change analysis showed that this trend will continue and accelerate [3], and varieties that are growing in lower altitudes will be under the stress from extreme temperature and late spring frost, which signals the need for advance planning of vineyard displacements and change of selection of varieties in present grape growing areas.

Study presented in this paper is continuation of several studies related to climate change impact on viticulture in Serbia, all using regional climate models results (RCMs) and showing similar signals of climate change relevant to viticulture. The models resolution and complexity of the analysis increased [4-8, 3]. The latest study involved RCP8.5 scenario [9] and analysis of results obtained with one RCM, Non-Hydrostatic Multiscale Model on the B grid (NMMB) on a very high horizontal resolution of about 8 km, with applied bias correction of the results, which is of crucial importance in this application of data [3].

Viticultural zoning performed in Serbia, officially entered into force in 2014 [10], showed diversity in terroir characteristics among wine-growing regions and districts. This, combined with the fact that majority of producers have relatively small scale wine-growing farms, shows necessity for high resolution approach and reliable analysis, since their incomes are highly vulnerable to weather conditions. The latest research on climate change and its impacts on viticulture development and planning, presented in this paper, is implemented according to IPCC 5th Assessment Report (AR5 [9]), using two scenarios, RCP4.5 as lower end and RCP8.5 as higher end scenario, splitting the future period, according to the planning strategy, into three periods: near future, mid-century and end of century. In order to have most reliable results as possible, high-resolution multi-model ensemble approach has been used. Changes of selected climate and bioclimatic indices, which describe climate characteristics relevant for wine-growing regions and

https://doi.org/10.1051/e3sconf/20185001005

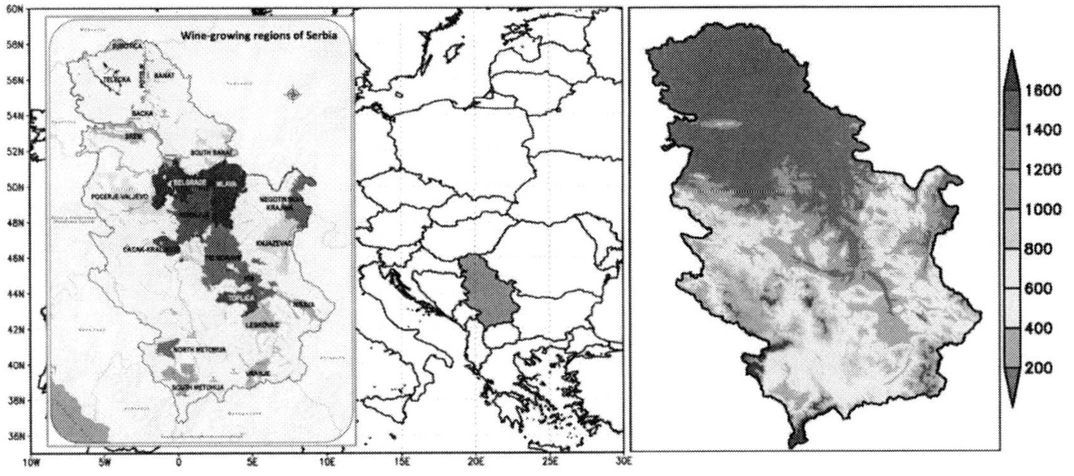

Figure 1. Wine-growing regions in Serbia (left, Viticultural zoning of Serbia, 2014; Vujadinovic et al, 2016) and terrain altitude in meters above sea level (right).

potentials risks, are tested for a degree of confidence to recognize anthropogenic global warming impact on climate change in Serbia and focus the vulnerability and risk assessment in viticulture sector on these extremes. Selection of scenarios and periods, and definition of present climate period 1986-2005, as in IPCC AR5, enables the comparison of the results for Serbia with wider region and possibility for cross-border collaboration and exchange of expertise.

2 Datasets and methodology

For future climate analysis are used results of nine RCMs from EURO-CORDEX, bias corrected, with resolution of 0.11°. Multi-model ensemble median values are chosen as representative results. Statistical significance evaluation of climate variables change is done using "t-test". If more than half of ensemble members have significant change, than it is considered with certainty that future change is far beyond natural variability. As mentioned before, analysis of the future climate change is divided in periods: 2016-2035 (near future), 2046-2065 (mid-century) and 2081-2100 (end of century), and chosen scenarios are RCP4.5 (stabilization scenario as lower end scenario) and RCP8.5 (constant increase as higher end scenario). For reference period is chosen 1986-2005, as present climate period, and E-OBS gridded daily datasets of temperature and precipitation are used, originally available on 0.25° but re-gridded on models resolution 0.11°.

Climatological indices relevant for viticulture [11, 12, 3], chosen for analysis in this paper are: Winkler index (WI), Huglin Index (HI), Cool night index (CI), Dryness index (DI), number of days with maximum temperature above 35°C (TX35), and number of dry days during the vegetation period April-October (DD).

Wine-growing regions in Serbia (Figure 1) are distributed mainly within central part of Serbia, where is hilly and mountain region with river basins, which is in contrast to northern part that is flat with altitudes lower than 100m and vineyards sparsely distributed. Other parts of Serbia are on higher altitudes and still with unsuitable climate for wine-growing.

3 Results and discussion

Future change of temperature and precipitation over Serbia (not shown here) comparing to present climate, according to RCP4.5 and RCP8.5 scenarios, show that: during the near future period mean temperature change will be within the interval 0.5-1.5°C, during the mid-century 1-3°C, and during the end of the century 1.5-5°C, with more pronounced increase over central and southern parts (lower thresholds are obtained with RCP4.5 and higher values are obtained according to RCP8.5); precipitation changes show increase of annual values by the mid-century, and during the second half of the century decrease over central and southern parts up to 20% according to RCP8.5; most critical change is obtained for summer season, when maximum temperature change is over 5°C in central and southern parts of Serbia, and significant precipitation reduction over the whole country, with values of decrease from 10% to 30% (according to RCP8.5), intensifying from north to south.

In Figure 2 are presented values for the present climate 1986-2005 of all selected indices. In Figure 3 and Figure 4 are presented values for three future periods, according to RCP4.5 and RCP8.5, respectively. In present climate within wine-growing regions HI values show that climate is temperate and temperate warm, with very cool nights according to CI, and sub-humid according to DI. During the near future period climate is changing in a way that within wine-growing regions prevail temperate warm climate over a wider area, with cool nights, and stays in sub-humid category. During the mid-century period temperate warm climate is spatially spreading over Serbia and in northern parts changing to warm category according to RCP4.5, and according to RCP8.5 warm climate also enters central parts of the country. Temperate nights category will appear over

Belgrade region, and will spread over northern parts in case of RCP8.5. DI shows decrease toward moderately dry category. During the second part of the century GHG emissions stabilize according to RCP4.5, climate change will continue, but with decreased intensity comparing to RCP8.5, which predicts further increase of GHG. In case of long term planning, analyzing end of the century period, it should be considered that present wine-growing regions will have warm climate or, in higher end case scenario, very warm climate, but current heat conditions will spread outside presently defined wine-growing regions. It is highly possible for present wine-growing regions to enter temperate nights category, and major part of the country to experience change of water availability toward moderately dry category.

Analysis of WI, T35 and DD shows that in present climate wine-growing regions are in II and III zone according to WI, T35 has values up to 5, and DD mainly 150-165. During the near future period all present wine-growing regions enter III zone, T35 increases up to 10, in central parts over 10, and DD is slightly increasing. During the mid-century period northern parts will enter IV zone, and it is possible that other wine-growing regions experience the same change in case of RCP8.5. T35 is possible to increase over 15 in some wine-growing regions, especially in central parts. DD values show increased risk of drought in vegetation period. By the end of the century RCP4.5 show the same direction of change but significantly reduced intensity of change, while according to RCP8.8 wine-growing regions enter V zone, values of T35 exceed 30, and DD shows significant increase of days without rain over the whole Serbia, even with 20-30 days more than in present climate.

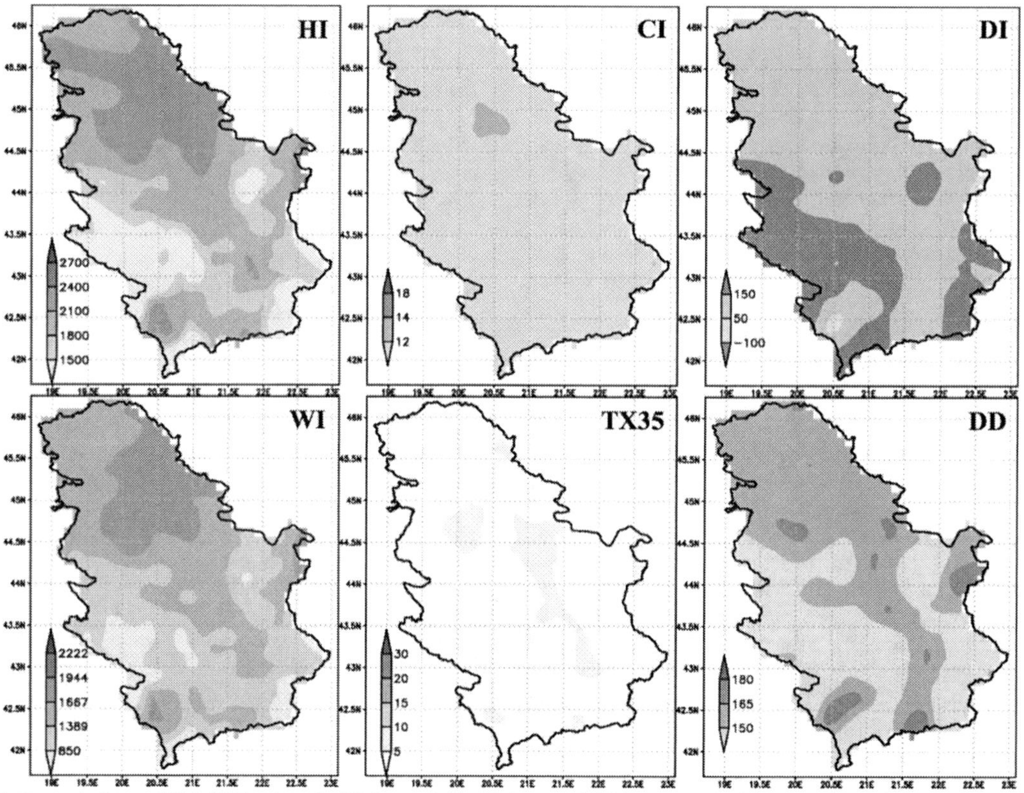

Figure 2. Present climate 1986-2005 values for HI (upper left), CI (upper middle), DI (upper right), WI (lower left), T35 (lower middle) and DD (lower right).

https://doi.org/10.1051/e3sconf/20185001005

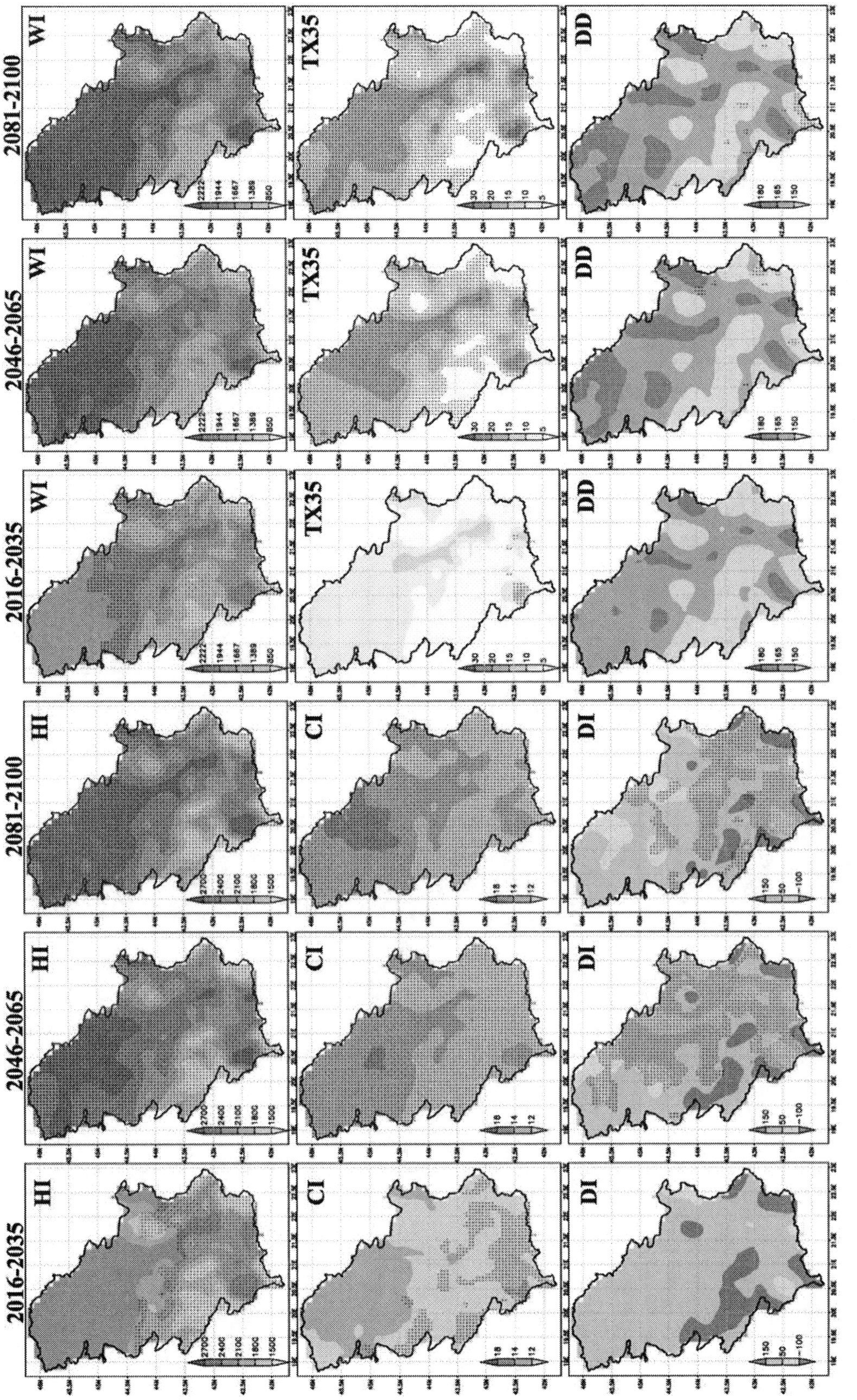

Figure 3. Values of selected indices for three future periods, according to RCP4.5.

 https://doi.org/10.1051/e3sconf/20185001005

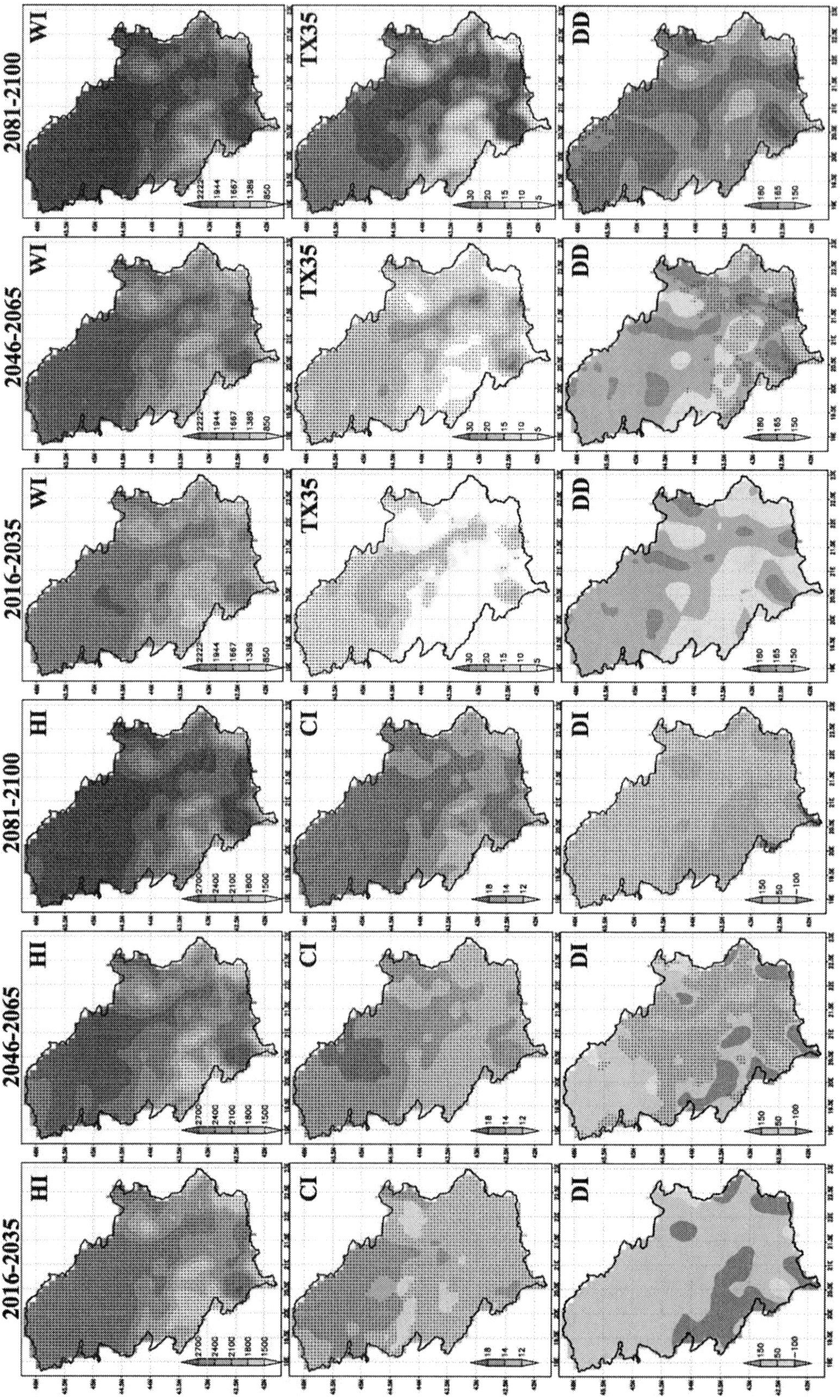

Figure 4. Values of selected indices for three future periods, according to RCP8.5.

4 Conclusions

Results discussed in this paper represent reliable base for planning of adaptation measures and mitigation of climate negative impacts on different time scales, concerning applied methodology. Wine-growing regions will gain heat conditions favorable for high production of good quality wines during the next decades, with appropriate implementation of agro-technical measures, but further in the future risk of high temperatures and significant lack of water in some regions will force displacement of wine-growing regions. Presented results prove the need for frequent change of wine-growing policy carried out through regular update of viticultural zoning, which will ensure sustainable development for this sector of economy. More detail and comprehensive analysis will be implemented through the following, third, National Communication of Republic of Serbia.

References

1. MEP, *The Second National Communication on Climate Change under the United Nations Framework Convention on Climate Change* (Ministry of Environmental Protection, Belgrade, Serbia, 2017)
2. M. Ruml, N. Korac, M. Vujadinovic, A. Vukovic, D. Ivanišević, J Agric. Sci. **154(2)**, 186-206 (2016)
3. M. Vujadinovic, A. Vukovic, D. Jaksic, V. Djurdjevic, M. Ruml, Z. Rankovic-Vasic, Z. Przic, B. Sivcev, N. Markovic, B. Cvetkovic, P. La Notte P., XI Terroir Congress, Willamette Valley, Oregon, USA (2016)
4. Vukovic, M. Vujadinović, V. Djurdjević, Z. Ranković-Vasić, N. Markovic, Z. Atanackovic, B. Sivčev, N. Petrović, VIII International Terroir Congress, Soave, Italy (2010)
5. M. Ruml, A. Vukovic, M. Vujadinovic, V. Djurdjevic, Z. Rankovic-Vasic, Z. Atancakovic, B. Sivcev, N. Markovic, S. Matijasevic, N. Petrovic, Agric. For. Meteorol. **158**, 53-62 (2012)
6. M. Vujadinovic, A. Vukovic, V. Djurdjevic, Z. Rankovic-Vasic, Z. Atanackovic, B. Sivcev, N. Markovic, N. Petrovic, Acta Horticulturae **931**, 87-94 (2012)
7. Vukovic, M. Vujadinovic, V. Djurdjevic, B. Cvetkovic, Z. Rankovic-Vasic, Z. Przic, M. Ruml M, A. Krzic, Proceedings of the 7th International Conference on Information and Communication Technologies in Agriculture, Food and Environment, Kavala, Greece, 892-901 (2015)
8. Vukovic, M. Vujadinovic, M. Ruml, Z. Przic, Z. Rankovic-Vasic, B. Cvetkovic, V. Djurdjevic, A. Krzic, Third Balkan Symposium on Fruit Growing, Belgrade, Serbia (2015)
9. IPCC, *Climate Change 2013: The Physical Science Basis. Contribution of Working Group I to the Fifth Assessment Report of the Intergovernmental Panel on Climate Change* (Cambridge University Press, New York, USA, 2013)
10. D. Ivanisevic, D. Jaksic, N. Korac, *Atlas of Viticulture (in Serbian)* (Statistical Office of the Republic of Serbia, Belgrade, Serbia, 2015)
11. J. Tonietto, A. Carbonneau, Agric. For. Meteorol. **124**, 81-97 (2004)
12. A.J. Winkler, J.A. Cook, W.M. Kliwe, L.A. Lider, *General viticulture* (University of California Press, California, USA, 1974)

 https://doi.org/10.1051/e3sconf/20185001006

Climate change and vine protection : the case of mildews management in Burgundy

S. Zito[1], A. Caffarra[2], Y. Richard[1], T. Castel[1], B. Bois[1,3]

[1]*CRC – UMR Biogeosciences, Université Bourgogne Franche Comté / CNRS, 6 Bd Gabriel, 21000 Dijon, France*
[2]*ITK, Avenue de l'Europe, 34830 Clapiers*
[3]*Institut Universitaire de la Vigne et du Vin, Université Bourgogne Franche Comté, rue Claude Ladrey, 21000 Dijon, France*

Abstract

Viticulture worldwide is currently facing two major challenges: adapting to climate change and reducing its environmental footprint. Plant protection is a central aspect of these challenges, firstly because pests and diseases development is strongly controlled by climate conditions, and secondly, because viticulture requires in many regions large quantities of pesticides. Phytosanitary protection is even more crucial for terroir-based viticulture areas, because the negative image given by excessive pesticide use impacts the whole region which reputation is partly built on environmental friendly practices. Moreover, most of terroir wines sensory properties and fame rely on specific cultivars, which makes it difficult to replace them using diseases resistant varieties. This study addresses the potential impact of climate change on pesticide use to control powdery and downy mildew in Burgundy. To assess the past evolution of diseases risk, a database composed by yearly number of applications of phytosanitary treatments for powdery and downy mildew diseases was built. This information was collected from 400 grapevine growers originating from 5 sub-regions of Burgundy. The data refer to yearly average number of treatments during the 1995-2014 period. Pesticide applications was related to climate by means of multiple linear regression models between the average number of treatments for powdery and downy mildews control and monthly temperature and monthly rainfall indices from April to July. Models providing the lowest error (estimated trough leave-one-out cross-validation) were selected for each of the 5 Burgundy wine sub-regions. According to each region, mildews yearly treatments number were significantly related to monthly climate data. In most models, May and June average temperature were selected with negative regression coefficients while April and May monthly precipitation were selected with positive regression coefficients. These models were fed with 1980-2100 temperature and rainfall projected data using CMIP5 RCP8.5 scenario from 18 GCM (General Circulation Models) statistically downscaled and debiased to match a daily 12 km target resolution. Increasing temperature and a slight ensemble rainfall increase (depending on the GCM) simulated during spring and early summer had inverse effects on models trend. However, regression models project a decreasing trend of the number of treatments for mildews control along the 21st century, based upon GCM data. Spraying numbers are expected to slightly vary in the future (-4,8 to -34.1% reduction), which strengthens the need for alternative plant protection strategies to match both consumer and policies promoting lower pesticides use for viticulture.

Keywords: downy mildew, powdery mildew, grapevine, climate change, Burgundy, pesticides

1. Introduction

Ever since the beginnings of viticulture, insects and diseases have caused significant yield and quality losses, their management often leading to pollution [1], human health deterioration [2] and economic consequences [3]. Climate change impacts on grapevine and parasites biology and their populations are obvious, therefore, plant protection is expected to evolve.

Surprisingly few researches have focused phytosanitary risks observed and projected (21st century) evolution in viticulture. Some studies based their research on empirical models in order to simulate downy mildew expected earliness and severity based on spraying observations [4,5] while others used mechanistic modelling to simulate host vs pest/pathogen interaction for European grapevine moth (*Lobesia botrana*), powdery mildew [6] and downy mildew [7]. All these study were based on projected data derived from one or two atmosphere-ocean general circulation models (AOGCM). It is highly to water related climate variables. It is highly recommended to use an ensemble approach to assess climate impacts. Yet, the epidemiology of grapevine major diseases (i.e. mildews according to [8] survey) strongly depends on water (rainfall, leaf wetness duration and air relative humidity), and when not related to temperature only, climate projections strongly vary according to the GCM parametrization [9,10]. There is consequently a need to evaluate the sensitivity of phytosanitary risks assessment simulations in response to a large number of climate projections, to better assess the possible evolution of grapevine protection in the future. Our study aims at documenting the evolution of powdery and downy mildew risks in relation with climate change, from 1980 to 2100, using 18 downscaled/debiased GCM data. This research (i) assesses mildews "pressure" over the past 20 years for the Burgundy wine region, as shown in figure 1; (ii) builts regression-based empirical models to relate the number of treatments per year necessary to control the diseases and monthly climate variables (iii) projects the number of treatments throughout the 21st century using statistically downscaled CMIP5 (Coupled Model Intercomparison Project [11]) climate projection (scenario RCP8.5).

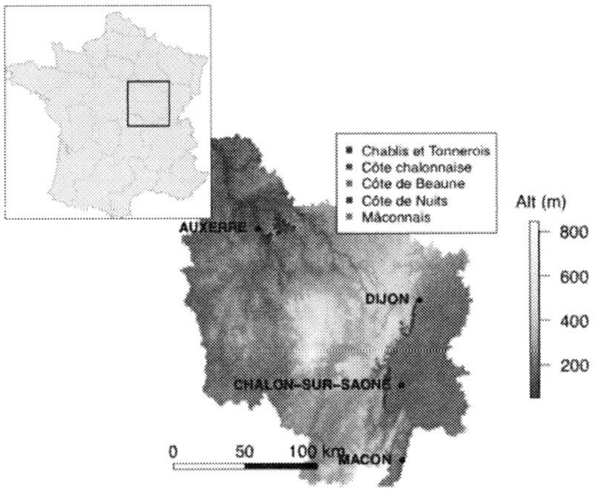

Figure 1. The studied area including the 5 wine sub regions of Burgundy France.

2. Material and methods

2.1 Spraying mildew database

Disease pressure is a fuzzy concept used by various grape growing stakeholders which assessment metrics are yet to be developed. Disease pressure can be roughly estimated by using different indicators such as the number of applied treatments, experts (e.g. consultants) or grapegrowers opinions, and disease intensity levels observed in the vineyard.

In this study, we used the number of applications done by grapegrowers each year as a bulk estimate of the disease intensity, as previously performed by [4]. We collected this information from 400 grapevine growers originating from all 5 sub regions of Burgundy. In this paper, we focus on the data of the "Côte de Beaune" collected by the Agricultural Regional Chamber of Burgundy which refer to yearly average number of powdery and downy mildew treatments during the 1995-2014 period. The number of winegrowers varies between 38 and 135 depending on year. We refer hereafter to « mildews pressure » when addressing the average number of treatment made each year for both powdery and downy mildew diseases.

2.2 Climate database

2.2.1 Observation

Monthly climate data was obtained for each pixel of a 12 km resolution grid elaborated though interpolation of daily climate data from Meteo France weather stations. Interpolation was performed by regression kriging as described by [12]. Variables included are total precipitation, minimum and maximum air temperature at daily time step. These data were used to elaborate as an empirical model in order to estimate mildew's pressure in Burgundy, for the 1995 to 2014 period.

For each Burgundy wine sub region, climate data was extracted from several 12 km grid cells where grapevine is currently grown. These correspond to five, two, ten, four and ten grid cells for "Côte de Beaune", "Côte Chalonnaise", "Chablisien", "Côte de Nuits" and "Mâconnais" respectively.

2.2.1 Simulation

In the case of future climate simulations, CMIP5 climate projections were collected at from ESGF (https://esgf-node.llnl.gov/search/cmip5/). Data from the following eigtheen GCM (General Circulation Models) are used: ACCESS, BCC, CanESM2, CCSM4, CNRM, CSIRO, GFDLM2G, GFDLM2M, INMCM, IPSLCM5A, IPSLCM5B, MIROC, MIROC5, MPILR, MPIMR, MRI, Nor. Here only the RCP8.5 (Representative Concentration Pathways) simulation, corresponding to very high greenhouse gas emissions [13] are used. The simulated climate data time period cover 120 years from 1980 to 2100. Data are statistically downscaled and debiased at a 12km grid resolution using the robust quantile mapping method [14]. Note that the quantile mapping is applied at monthly time step.

2.3 Disease model

In order to simulate the potential evolution of powdery and downy mildew pressure, we considered the mean number of fungicide sprays per year necessary to control both diseases. Analyzing each disease separately was initially considered. However, the number of powdery mildew and downy mildew treatments performed each year are strongly correlated (R = 0.98). Indeed, in a vast majority of cases, powdery and downy mildew treatments are carried out simultaneously to limit the number of spraying operation in the vineyard. The number of treatments may be the result of high pressure from one disease or another. We therefore considered the sum of treatments number (hereafter referred to as ST, for Sum of Treatments) applied for both diseases each year as a global indicator of grapevine "mildews" pressure. This indicator was compared to monthly climate data on the 1995-2014 period for each Burgundy wine regions. The climate variables used as the model predictors were monthly sum of precipitation (mm) and maximum, minimum and mean temperature, averages from April to July. We tested all the possible models within linear regression, from 65536 combinations. To avoid any multicollinearity within the models, combinations in which at least two predictors were significantly correlated (p-value ≤ 5% on a Pearson correlation t-test) were discarded. Each regression model performance was assessed by Leave-One-Out Cross Validation (LOOCV) procedure. The model providing the highest efficiency (EF, see Mayer and Butler 1994 for more information) was selected. EF is calculated as follows:

$$EF = 1 - \frac{\sum_{i=1}^{n}(P_i - O_i)^2}{\sum_{i=1}^{n}(O_i - \bar{O})^2}$$

where P_i is the predicted value of ST in the wine region for year i ; O_i is the observed ST for year i and $\bar{O}$ is the average observed ST for all the n available years.

 https://doi.org/10.1051/e3sconf/20185001006

3. Results

3.1 Simulated disease pressure

For the sake on conciseness, only the case of « Côte de Beaune » (CB) wine region is detailed below. Testing all possible combination for simulating ST using April to July monthly precipitation and average temperature resulted in the selection of the following model:

(1) $St_i = -0{,}213TX_i05 - 0{,}382TX_i06 + 0{,}026RR_i04 + 20{,}713$

where, for year i, ST_i is the sum of mildews treatments (sprayings) simulated in Côte de Beaune, TX_i05, TX_i06 the May and June maximum temperature respectively and RR_i04 the April precipitation. With this model, ST is simulated with an efficiency equals to 0.61, a determination coefficient (R^2) equals to 0.61 and a root mean squared error (RMSE) equals to 1.25 treatment (LOOCV results).

The figure 2 illustrates the relationship between the disease pressure indicator for the selected wine region and the simulated pressure indicator with observed climate data.

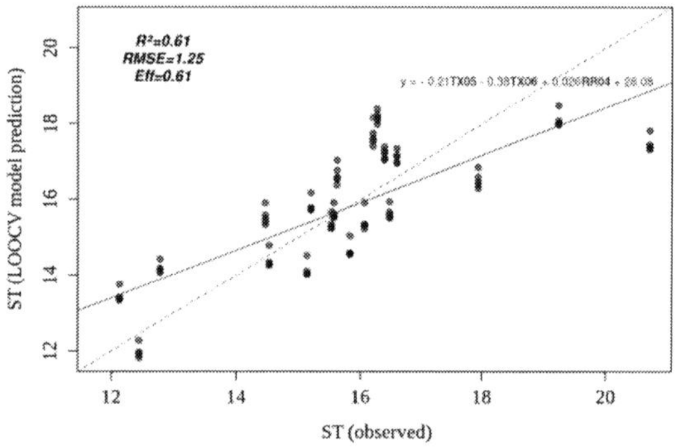

Figure.2 "Côte de Beaune" wine region: mildews spraying number ST observed and simulated through LOOCV.

Selected models and associated LOOCV performances are shown (Table 1). Efficiency ranged from 0.39 to 0.66. Maximum temperature in May and June as well as April precipitation were selected in the most efficient model of each region.

Table.1 Regressors and LOOCV efficiency (Eff) of multiple linear regression models predicting the sum of treatments for mildews control applied in each major wine region of Burgundy.

Wine region	Regressors (sign)	Eff
Chablisien	-T04 -TN07 + RR05	0.39
Côte de Nuits	-TX05 -TX06 +RR04	0.40
Côte de Beaune	-TX05 -TX06 +RR04	0.61
Côte Chalonnaise	-TN04 -TX05 -TX06 +RR04 +RR06	0.66
Mâconnais	-T05 -TX06 +RR04	0.47

3.2 Disease pressure projections

Once each empirical model based on observed climate data (1995-2014) was elaborated for the five sub-wine region of Burgundy, the 18 climate downscaled projections were used to simulate mildews pressure (ST) from 1980 to 2100.

The figure 3 illustrate the evolution of the simulated ST from 1980 to 2100 with the 18 downscaled climate models for the Côte de Beaune wine region (5 grid cells). Each curve represent the trend yearly ST simulated using climate data of one model. The trend is smoothed using the loess method, a non-parametric approach that fits multiple regression in local neighborhood. Simulated ST using projected climate data exhibit a fall for all 18 GCM during the 21st century, with more or less intensity. GCM simulations tend to overestimate ST in the present period (16 to 17.5 treatments per year to control mildews) by comparison to current observations (15.8 treatments in Côte de Beaune, table 2). The spread of simulated ST by all climate models increases throughout the 21st century.

The mean of simulated ST for twenty year was calculated and compared for three periods (1995-2014, 2046-2065, 2081-2100) for each region (Table 1). The delta represents the difference between historical and the futures simulations over the two periods of simulated mildews pressure using all climate models

For each of the 5 wine region , the decrease of mildews pressure was significant (Wilcoxon signed-rank test at alpha = 5%) for both the middle and the end of the 21st century periods.

From a geographical point of view, the more the region is located in the South and the stronger the decrease is simulated (Figure 1, Tableau 2). It can be inferred that this trend is mostly due to a projected stronger rise in maximum temperature in the southern regions than it is in the northern regions of Burgundy. For example, in the southern wine region of Mâconnais, the mean maximum temperature projected rise during spring (T04, T05, T06) is +1,7°C and +1,6°C higher than in northern Chablisien region. Futur simulated ST decreased accompanied by increased variability (increase of the standard deviation).

4. Discussion and conclusion

This paper assess change in phytosanitary treatments against mildews from 1995 and their evolution in response to climate change throughout the 21st century. To estimate future projection of treatments, we have used an empirical approach using linear/quadratic multiple regression modelling with monthly temperature and precipitation data. Following a robust predictors selection scheme (LOOCV, multicollinearity removal), between 3 and 6 variables are identified as producing the most efficient regression models according wine region. In the case of Côte de Beaune, variables selected are May and June maximum temperature and April cumulated precipitation. These variables are selected in most regions.

 https://doi.org/10.1051/e3sconf/20185001006

Table.2 Mean sum of treatments on the historical time period (1995-2014) with standard deviation (SD) and mean simulated differences between the historical simulated of ST (Delta) and 2046-2065 and 2081-2100 periods simulated ST for each Burgundy wine region with the 18 GSM.

Wine regions	Observations (1995-2014)		Simulations (2046-2065)		Simulations (2081-2100)	
	Mean	SD	Delta	SD	Delta	SD
Chablisien	15,1	1,2	**-1,9**	1,5	**-3,8**	1,7
Côte de Nuits	15,7	1,6	**-2,0**	2,0	**-4,0**	2,5
Côte de Beaune	15,8	1,6	**-1,8**	1,9	**-3,6**	2,4
Côte Chalonnaise	14,7	2,1	**-2,3**	2,7	**-4,7**	3,2
Mâconnais	15,0	1,9	**-2,4**	2,6	**-4,9**	3,3

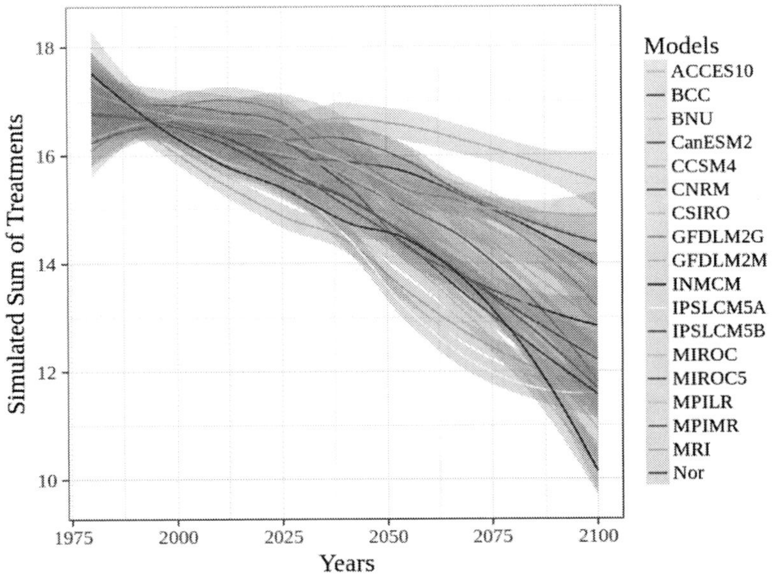

Figure 3. Simulated Sum of treatments with eighteen statistically downscaled GSM data from 1980 to 2100 for the Côte de Beaune wine region.

As primary contamination of both mildews agents (*Plasmopara viticola* and *Erysiphe necator*) requires rainfall [15,16], monthly precipitation in April (selected as ST predictors within our model) was logically selected, with a positive regression coefficient. Moreover, rainfall tend to wash out plant protection products on grapevine leaves, which might force the grapegrowers to renew the products application with shorter delays. Temperature regression coefficient are negative, suggesting that increasing air temperature during spring and early summer are detrimental to mildews epidemics outbreaks. These results are contradictory to those of [4] who found that higher temperature during spring enhanced downy mildew pressure. This is rather confusing because one could expect that in cool climate regions such as Burgundy, a raise in temperature would favour secondary cycle length and therefore increase the outbreak speed and severity. An hypothesis is that low temperature may favour higher air humidity and longer leaf wetness duration, and therefore optimum conditions for mildews development. Moreover, for Spring and Summer in France, high monthly temperature are usually correlated with sunny, hence limited number of rainy and humid days.

Because of increasing temperature throughout the 21st century, the regression model simulates globally a fall in pesticide application to control mildews. The spread within models projections increases, indicating that ST is highly sensitive to climate variations provided by both simulated interannual and AOGCM internal variability. Further research is currently being perform to assess the relative contribution to this increase of variability.

According to [17], studying climate change impacts on plant diseases must first take into consideration the potential impacts of global climate on plant population structures and dynamics, micro-evolutionary processes and plant community structures. Physiology, phenology and host/pathogen interactions are equally important factors, as they can influence, positively or negatively, the impact of climate change on diseases [18].

The empirical approach developed here does not allow to address precisely the possible changes in plant/pathogens biology and interactions. Modelling mechanistically both grapevine and mildews agents biology requires high resolution climate data in time (and space) to relevantly simulate plant and pathogens development. Such approach is prevented using monthly climate data, as done in our research. There is a trade of for climate change impact studies between climate data robustness and impact models (i.e. plant or pathogen or interaction) relevance. Using poor time resolution (i.e. monthly data) limits the variability and possibly unrealistic simulated climate values through downscaling. But yet the impact modelling is bulk and does not allow to understand the ongoing processes of estimated impacts. Such trade of is little addressed in the scientific literature, while substantial literature discussing the downscaling methods impacts on modelling is available (e.g. [19,20]), using in most case daily climate data (see for example [21]). This dilemma could be, indeed, to test the robustness and sensitivity of plant-pathogens models to climate variability.

References

[1] M. Komárek, E. Čadková, V. Chrastný, F. Bordas, J.-C. Bollinger, Contamination of vineyard soils with fungicides: A review of environmental and toxicological aspects, Environment International. **36** 138–151 (2010)

[2] Ntzani Evangelia E, Ntritsos G Chondrogiorgi M, Evangelou E, Tzoulaki I, Literature review on epidemiological studies linking exposure to pesticides and health effects, EFSA Supporting Publications. **10** 497E (2013)

[3] D. Pimentel, 'Environmental and Economic Costs of the Application of Pesticides Primarily in the United States,' Environment, Development and Sustainability. **7** 229–252 (2005)

[4] F. Salinari, G. Simona, T. Francesco Nicola, R. Andrea, R. Vittorio, S. Federico, R. Cynthia, G. Maria Lodovica, Downy mildew (Plasmopara viticola) epidemics on grapevine under climate change, Global Change Biology. **12** 1299–1307 (2006)

[5] F. Salinari, S. Giosuè, V. Rossi, F.N. Tubiello, C. Rosenzweig, M.L. Gullino, Downy mildew outbreaks on grapevine under climate change: elaboration and application of an empirical-statistical model*, EPPO Bulletin. **37** 317–326 (2007)

[6] A. Caffarra, M. Rinaldi, E. Eccel, V. Rossi, I. Pertot, Modelling the impact of climate change on the interaction between grapevine and its pests and pathogens: European grapevine moth and powdery mildew, Agriculture, Ecosystems & Environment. **148** 89–101 (2012)

[7] M. Launay, J. Caubel, G. Bourgeois, F. Huard, I. Garcia de Cortazar-Atauri, M.-O. Bancal, N. Brisson, Climatic indicators for crop infection risk: Application to climate change impacts on five major foliar fungal diseases in Northern France, Agriculture, Ecosystems & Environment. **197** 147–158 (2014)

[8] B. Bois, S. Zito, A. Calonnec, Climate vs grapevine pests and diseases worldwide: the first results of a global survey, OENO One. **51** 133 (2017)

[9] J.M. Murphy, D.M.H. Sexton, D.N. Barnett, G.S. Jones, M.J. Webb, M. Collins, D.A. Stainforth, Quantification of modelling uncertainties in a large ensemble of climate change simulations, Nature. **430** 768–772 (2004)

[10] J. Rogelj, M. Meinshausen, R. Knutti, Global warming under old and new scenarios using IPCC climate sensitivity range estimates, Nature Climate Change. **2** 248 (2012)

[11] K.E. Taylor, R.J. Stouffer, G.A. Meehl, An Overview of CMIP5 and the Experiment Design, Bull. Amer. Meteor. Soc. **93** 485–498 (2011)

[12] V. Rossi, T. Caffi, S.E. Legler, Dynamics of Ascospore Maturation and Discharge in Erysiphe necator, the Causal Agent of Grape Powdery Mildew, Phytopathology. **100** 1321–1329 (2010)

[13] T. Stocker, ed., Climate change 2013: the physical science basis: Working Group I contribution to the Fifth assessment report of the Intergovernmental Panel on Climate Change, Cambridge University Press, New York, 2014

[14] L. Gudmundsson, J.B. Bremnes, J.E. Haugen, T. Engen-Skaugen, Technical Note: Downscaling RCM precipitation to the station scale using statistical transformations – a comparison of methods, Hydrology and Earth System Sciences. **16** 3383–3390 (2012)

[15] C. Gessler, I. Pertot, M. Perazzolli, Plasmopara viticola : a review of knowledge on downy mildew of grapevine and effective disease management, Phytopathologia Mediterranea. **50** 3–44 (2011)

[16] Gadoury David M., Cadle-Davidson Lance, Wilcox Wayne F., Dry Ian B., Seem Robert C., Milgroom Michael G., Grapevine powdery mildew (Erysiphe necator): a fascinating system for the study of the biology, ecology and epidemiology of an obligate biotroph, Molecular Plant Pathology. **13** 1–16 (2012)

[17] S. Chakraborty, A.V. Tiedemann, P.S. Teng, Climate change: potential impact on plant diseases, Environmental Pollution. **108** 317–326 (2000)

[18] S.M. Coakley, H. Scherm, S. Chakraborty, Climate Change and Plant Disease Management, Annual Review of Phytopathology. **37** 399–426 (1999)

[19] A.W. Wood, L.R. Leung, V. Sridhar, D.P. Lettenmaier, Hydrologic Implications of Dynamical and Statistical Approaches to Downscaling Climate Model Outputs, Climatic Change. **62** 189–216 (2004)

[20] C. Teutschbein, J. Seibert, Is bias correction of regional climate model (RCM) simulations possible for non-stationary conditions?, Hydrology and Earth System Sciences. **17** 5061–5077 (2013)

[21] J.W. White, G. Hoogenboom, B.A. Kimball, G.W. Wall, Methodologies for simulating impacts of climate change on crop production, Field Crops Research. **124** 357–368 (2011)

Consecuencias de la aplicación de distintas dosis de riego en la producción, el desarrollo vegetativo y la calidad de uva del cv. Cabernet Sauvignon en el valle del río Duero

Consequences of the application of different irrigation doses on grape yield, vegetative development and quality in cv. Cabernet Sauvignon in the Duero river valley

Jesús Yuste[1,a], Alejandro Vicente[2] y Daniel Martínez-Porro[1]

[1]Instituto Tecnológico Agrario de Castilla y León, 47071 Valladolid, España
[2]Actualmente: actividad vitivinícola externa, España

Resumen. La influencia del riego en zonas de clara escasez pluviométrica es conocida en general, pero la cantidad de agua aplicada en una zona de cultivo determinada puede afectar de modo particular a la producción y a la calidad de la uva. A lo largo del periodo 2012-2014 se estudió la respuesta agronómica a la aplicación de 3 tratamientos de régimen hídrico: secano (P0), riego del 25% ETo (P1) y riego del 50% ETo (P2), aplicado desde tamaño guisante hasta vendimia. El ensayo se llevó a cabo con cv. Cabernet Sauvignon sobre portainjerto 110R, plantado en 2004 y conducido en espaldera en cordón Royat bilateral, con marco de plantación de 2,2 m x 1,2 m, en Valladolid. El desarrollo vegetativo aumentó progresivamente con la dosis de riego, incrementándose el peso de madera de poda un 31% al pasar del tratamiento P0 al P1, y un 27% al pasar del P1 al P2, debido al peso del sarmiento. El rendimiento en uva también aumentó con las dosis de agua, un 25% al pasar del tratamiento P0 al P1 y un 13% al pasar del P1 al P2, debido mayormente al peso del racimo. La concentración de azúcares se vió ligeramente favorecida por el riego. El pH del mosto mostró cierta tendencia a la reducción con el riego, sobre todo con el de mayor dosis, a la par que la acidez titulable, el ácido tartárico y el ácido málico aumentaron con el riego, sobre todo con el de mayor dosis. El K y el IPT no mostraron diferencias notables ni constantes entre tratamientos, aunque en general el K aumentó algo con el riego y el IPT se redujo ligeramente con el riego de mayor dosis.

Abstract. *The influence of irrigation in areas of clear pluviometric scarcity is generally known, but the amount of water applied in a given area of cultivation can have a particular effect on the production and quality of the grape. Throughout the period 2012-2014, the agronomic vine response to 3 treatments of water regime was studied: rainfed (P0), irrigation of 25% ETo (P1) and irrigation of 50% ETo (P2), from pea size state until harvest. The trial was developed with vines of cv. Cabernet Sauvignon onto 110R rootstock, planted in 2004 and vertically trellised as a bilateral Royat cordon. The vine distances were 2.2 m x 1.2 m. It was located in Valladolid (Spain). The vegetative development increased progressively with the irrigation dose, increasing the weight of pruning wood by 31% when going from treatment P0 to P1, and 27% when passing from P1 to P2, due to the shoot weight. Grape yield also increased with water doses, 25% from P0 to P1 and 13% from P1 to P2, due mainly to cluster weight. The sugar concentration was slightly favored by irrigation. The pH of the must showed a certain tendency to reduction with irrigation, especially with the higher dose, while the titratable acidity, tartaric acid and malic acid increased with irrigation, especially with the higher dose. K and IPT showed no significant or constant differences between treatments, although in general the K increased somewhat with irrigation and the IPT was slightly reduced with higher dose of irrigation.*

1 Introducción

En las zonas de cultivo en que el balance hídrico es claramente deficitario, la producción y la calidad de la uva dependen de los aportes de agua de riego [1], pero la cantidad de agua que debe ser aportada estará condicionada por los diversos factores que pueden influir en la respuesta de la vid, tales como la reserva hídrica del suelo, el estado del viñedo, la variedad y las características edafoclimáticas del lugar de cultivo [4, 15]. La escasez hídrica ha llevado a desarrollar sistemas de riego cada vez más eficientes en el uso del agua, aunque en cierto modo se haya provocado un cambio en la distribución de la humedad en el suelo y en el desarrollo radicular del viñedo [13]. En general, el riego debe ser aplicado con moderación tras la parada de crecimiento del pámpano principal, con el objetivo de que la disponibilidad hídrica ayude a mantener una actividad fisiológica adecuada que facilite la síntesis de azúcar y la translocación de sustancias elaboradas a las bayas sin fomentar el desarrollo vegetativo [7, 14].

Diversos autores han observado cómo la aplicación de riego provoca el aumento de rendimiento, en mayor o menor cuantía, debido a la modificación de los diferentes componentes del mismo, primordialmente el tamaño de baya [6, 2], así como el estrés hídrico provoca un efecto de reducción del rendimiento causado por la acción inhibidora en la diferenciación del racimo [12]. Sin embargo, Reynolds et al. [10] observaron que un manejo adecuado del déficit hídrico incrementó la concentración de compuestos de la uva, como azúcares, antocianos y fenoles, debido a la reducción del crecimiento vegetativo y del peso de la baya.

Desde el punto de vista cualitativo, el efecto del estrés hídrico en el desarrollo de la baya ha mostrado diversos efectos [5, 7], tales como una menor síntesis de los ácidos málico y tartárico, así como un mayor consumo de ácido málico [3]. Por otro lado, también ha sido observada la incidencia de la disponibilidad hídrica del viñedo en los diversos compuestos fenólicos [8], en particular en la biosíntesis y en la concentración de antocianos y taninos en la baya. La modificación de dicha concentración está relacionada con el tamaño de la baya [11], el cual se ve influido por el régimen hídrico, lo que determina la variación de la relación hollejo-pulpa de la baya y consecuentemente afecta a la concentración de los compuestos fenólicos, responsables en gran medida de la calidad del vino tinto [9, 11].

Ante la escasez de trabajos relativos al riego desarrollados en el ámbito del valle del río Duero con la variedad tinta Cabernet Sauvignon, el objetivo del trabajo es estudiar las consecuencias de la aplicación de dos dosis deficitarias de riego por goteo frente al secano, en Valladolid (España), a través de parámetros de producción de uva, desarrollo vegetativo y composición de la uva, para mejorar la eficiencia en el manejo del riego del viñedo de dicha variedad en Castilla y León.

2 Material y Métodos

El trabajo se ha llevado a cabo en los años 2012, 2013 y 2014 en un viñedo localizado en Valladolid, Castilla y León (España). El material empleado es *Vitis vinifera* L, cv Cabernet Sauvignon, sobre portainjerto 110 Richter, plantado en 2004, con un marco de 2,2 m x 1,2 m (3.788 plantas.ha^{-1}). El sistema de conducción es espaldera vertical, mediante cordón Royat bilateral y poda con 3 pulgares de 2 yemas en cada brazo (12 yemas por cepa).

El diseño experimental es en bloques al azar, con 4 repeticiones por tratamiento y una parcela elemental de 27 cepas, 7 de control, con líneas contiguas a cada lado destinadas al efecto borde. El ensayo experimental consiste en el establecimiento de 3 regímenes hídricos basados en la dosis de riego, que son: secano, 0% ETo (P0), riego de dosis baja, 25% ETo (P1) y riego de dosis media, 50% ETo (P2). El riego fue aplicado con frecuencia semanal, desde de la parada de crecimiento de pámpanos principales hasta vendimia, en la línea de cepas, con goteros autocompensantes de caudal 4,0 L.h^{-1}, colocados cada 60 cm, a 30 cm del tronco de la cepa.

En la tabla 1 aparecen reflejados los datos de riego y precipitación del periodo de estudio.

Tabla 1. Agua de riego aplicada y Fechas de inicio y fin en los tratamientos P0, P1 y P2; Precipitación anual de 1-octubre a 30-septiembre y Precipitación de 1-abril a 30-septiembre en 2012-2014.

Trat.	Riego (mm)				Año	Fechas de Riego		Precipitación (mm)	
	2012	2013	2014	Media		Inicio	Fin	Anual	1 abr - 30 sep
P0	-	-	-	-	2012	18 jul	1 oct	263	134
P1	89	97	93	93	2013	16 jul	23 sep	469	129
P2	177	193	186	185	2014	17 jul	26 sep	259	148
					Media	17 jul	27-sep	330	137

El suelo del ensayo presenta alta pedregosidad (más de 65% de elementos gruesos), lo que le confiere un buen drenaje, con una capacidad potencial de retención de agua estimada en 70 mm / m de profundidad. Es de tipo arcillo-arenoso en el 1er horizonte (20 cm) y franco-arcillo-arenoso en los dos horizontes siguientes (20-100 cm), llano, sin limitaciones físicas ni químicas para el cultivo de viñedo. En la tabla 2 aparecen reflejados los datos termopluviométricos del periodo 2012-2014.

Tabla 2. Datos medios mensuales y anuales (1-octubre a 30-septiembre) de Temperatura (ºC) máxima (Tmax), mínima (Tmin) y media (Tm), y Precipitación (mm) del periodo 2012-2014.

	T_{max}			T_{min}			T_m			P		
	2012	2013	2014	2012	2013	2014	2012	2013	2014	2012	2013	2014
Oct	22,1	18,5	19,2	5,5	6,2	8,3	13,1	12,0	13,2	17,2	72,6	23,2
Nov	12,8	12,3	11,8	5,0	3,0	2,2	8,5	7,4	6,8	64,4	60,4	3,2
Dic	9,2	9,4	8,4	0,0	1,3	-1,6	4,0	5,1	2,7	11,0	21,4	1,8
Ene	7,0	9,0	10,1	-1,4	0,7	2,3	2,2	4,6	5,9	28,4	36,8	22,8
Feb	10,1	9,8	10,0	-3,9	-0,9	1,3	2,6	4,0	5,4	1,0	31,2	48,1
Mar	17,1	12,2	15,8	0,6	2,9	2,2	8,6	7,1	8,7	7,6	117,9	11,0
Abr	13,7	15,4	20,0	3,9	3,8	6,4	8,6	9,5	13,0	66,8	28,6	22,4
May	23,4	17,5	21,5	8,3	4,8	6,6	16,0	11,0	14,1	20,2	27,9	18,8
Jun	26,7	23,9	26,4	11,3	8,7	9,9	19,2	16,3	18,5	12,6	33,6	9,6
Jul	28,8	31,9	28,0	11,7	14,0	12,8	20,4	23,0	20,3	12,2	9,4	36,2
Ago	30,2	29,9	28,6	12,7	12,4	12,4	21,5	20,9	20,6	0,0	29,0	0,2
Sep	25,2	26,2	26,0	10,1	10,3	12,3	17,4	17,6	18,5	21,8	0,0	61,2
AÑO	18,9	18,0	18,8	5,3	5,6	6,3	11,8	11,5	12,3	263,2	468,9	258,5

Se han medido los parámetros siguientes: peso de madera de poda y número de sarmientos; peso de uva, número de racimos y peso de baya; sólidos solubles totales, pH, acidez total, ácido tartárico, ácido málico, potasio e índice de polifenoles, en mosto. A partir de ellos, se han calculado otros parámetros vegetativos, productivos y cualitativos. El análisis de resultados se ha realizado mediante análisis de varianza con el programa SPSS 16.

3 Resultados y Discusión

3.1 Desarrollo vegetativo

El desarrollo vegetativo, expresado a través del peso de madera de poda, aumentó de forma creciente con la dosis de riego aplicada, aunque las diferencias sólo llegaron a ser estadísticamente significativas entre los tres tratamientos en 2013, puesto que las diferencias no fueron significativas en 2012 entre P0 y P1 ni en 2014 entre P1 y los otros dos tratamientos (tabla 3). El riego de menor dosis provocó un aumento medio interanual del peso de madera de poda del 31% respecto al secano, mientras que el aumento provocado por el riego de mayor dosis respecto al de menor dosis fue del 27%.

La tendencia observada en el peso de madera de poda se ha correspondido fielmente con la del peso del sarmiento, dado que el número de sarmientos por cepa fue similar en todos los tratamientos, debido al ajuste de carga mediante operación en verde realizada en la segunda quincena de mayo, excepto en 2013. En este año, el secano, P0, mostró un número significativamente menor de sarmientos que los otros tratamientos, debido sobre todo al mayor número de chupones de los otros tratamientos. No hubo diferencias significativas entre tratamientos en el número de sarmientos francos ni de chupones ningún año. El peso del sarmiento mostró un aumento medio interanual en el tratamiento P1 con respecto al secano del 24%, el mismo que mostró el tratamiento P2 con respecto al P1, aunque las diferencias entre los tres tratamientos sólo resultasen estadísticamente significativas en 2013.

Tabla 3. Madera de poda (kg), N° total de sarmientos, N° de sarmientos francos, N° de chupones, por cepa; Peso de sarmiento (g); Rendimiento en uva (kg) y N° de racimos, por cepa; Peso de racimo (g), Peso de baya (g) y N° de bayas por racimo, de los tratamientos (Trat.): P0, P1 y P2. Significación estadística (Sig.): -, no significativo; *, p<0,05.

	Trat.	Madera Poda	N° Sarm.	N° francos	N° chup.	Peso Sarm.	Rdto.	N° Racimos	Peso Racimo	Peso Baya	N° Bayas
2012	P0	0,518 b	12,3	10,9	1,4	43,2	1,46	18,8	77 b	0,74	106
	P1	0,580 b	12,8	11,2	1,6	45,7	2,07	19,6	106 a	0,93	115
	P2	0,738 a	12,8	11,0	1,8	57,7	2,35	19,7	118 a	0,97	121
	Sig.	*	-	-	-	-	-	-	*	-	-
2013	P0	0,741 c	14,2 b	11,6	2,7	52,1 c	2,94	24,0	123	1,09 b	112
	P1	1,101 b	15,7 a	12,2	3,5	70,7 b	3,40	24,4	137	1,29 a	106
	P2	1,501 a	15,8 a	11,6	4,2	97,1 a	3,82	24,8	153	1,30 a	118
	Sig.	*	*	-	-	*	-	-	-	*	-
2014	P0	0,855 b	12,5	11,4	1,1	70,0	2,00	18,1	111	1,36	81
	P1	1,090 ab	12,3	11,1	1,2	88,8	2,53	20,1	126	1,32	96
	P2	1,275 a	12,8	11,4	1,4	100,3	2,88	22,6	127	1,41	90
	Sig.	*	-	-	-	-	-	-	-	-	-
Media	P0	0,705	13,0	11,3	1,7	55,1	2,13	20,3	104	1,06	100
	P1	0,923	13,6	11,5	2,1	68,4	2,67	21,4	123	1,18	105
	P2	1,171	13,8	11,3	2,5	85,1	3,01	22,4	133	1,23	110

3.2 Producción de uva

El rendimiento en uva aumentó progresivamente con la dosis de agua aportada, a pesar de que las diferencias observadas entre tratamientos no llegasen a resultar estadísticamente significativas con un nivel del 95% de probabilidad (tabla 3). El incremento medio de rendimiento generado por el riego de menor dosis, P1, con respecto al secano, P0, fue del 25%, mientras que el generado por el riego de mayor dosis, P2, con respecto al riego de menor dosis, P1, fue sólo del 13%. Las diferencias en producción fueron fundamentalmente debidas al peso de racimo, que también aumentó progresivamente con la cantidad de agua aportada, un 18% al pasar de P0 a P1 y un 8% al pasar de P1 a P2, aunque las diferencias mostradas por dicho parámetro sólo llegasen a resultar estadísticamente significativas entre el secano, P0, y los tratamientos regados en 2012.

El número de racimos por cepa también contribuyó al aumento del rendimiento, pero en menor medida que el peso del racimo, pues dicho número se incrementó un 5% tanto al pasar de P0 a P1 como al pasar de P1 a P2. Las diferencias entre tratamientos en el número de racimos no fueron estadísticamente significativas.

El incremento del peso del racimo debido a la cantidad de agua aportada estuvo más relacionado con el peso de baya que con el número de bayas, pues el cambio de P0 a P2 mostró un aumento del 16% en el peso de baya frente a un aumento del 10% en el número de bayas del racimo. Las diferencias observadas en ambos parámetros solamente resultaron estadísticamente significativas entre el secano, P0, y los tratamientos regados en el peso de baya en 2013.

El Índice de Ravaz disminuyó ligeramente de forma gradual con el aumento de la cantidad de agua aportada, excepto en 2012, en que la sequía penalizó más, en el tratamiento P0, a la producción de uva que al desarrollo vegetativo (tabla 4). En todo caso, la tendencia interanual decreciente con el aumento de riego de este índice no se vio reflejada en diferencias anuales estadísticamente significativas.

3.3 Composición de la uva

La concentración de sólidos solubles se vio generalmente favorecida por el riego con respecto al secano, con un aumento interanual de más de medio grado brix al pasar de secano a regadío, aunque la tendencia no fue constante entre los tratamientos aplicados, observándose incluso un valor ligeramente más alto del secano, P0, en 2014, probablemente favorecido por su menor nivel de rendimiento. Las diferencias sólo resultaron estadísticamente significativas en 2012 entre los tratamientos P1 y P0 (tabla 4).

El pH del mosto mostró cierta tendencia a la reducción con el riego respecto al secano, P0, pero sobre todo al pasar al riego de mayor dosis, P2, como reflejaron las diferencias estadísticamente significativas observadas entre P0 y P2 tanto en 2013 como en 2014. La acidez titulable aumentó con la dosis de riego, sobre todo con la dosis mayor, del tratamiento P2, que supuso un incremento interanual de casi 1,5 g/L con respecto al secano, P0. Las diferencias entre ambos tratamientos, P2 y P0, fueron estadísticamente significativas todos los años, mientras que el tratamiento P1 sólo fue discriminado estadísticamente del P0 en 2012 y del P2 en 2014. El ácido tartárico mostró una tendencia similar, aunque con menores diferencias entre tratamientos, a la acidez titulable, ligeramente creciente con la dosis de riego, aunque las diferencias sólo resultaron estadísticamente significativas entre el tratamiento más regado, P2, y los otros dos tratamientos en 2014. El ácido málico aumentó con la dosis de riego, sobre todo con la dosis mayor, del tratamiento P2, que supuso un incremento interanual de casi 1 g/L con respecto al secano, P0. Las diferencias favorables al tratamiento P2 con respecto a los otros dos fueron estadísticamente significativas en 2012 y 2013, mientras que el tratamiento P1 sólo fue discriminado estadísticamente del P0 en 2013.

El potasio no mostró diferencias notables ni constantes entre tratamientos, aunque la media interanual mostró una concentración ligeramente mayor en los tratamientos regados que en el secano.

El índice de polifenoles totales tampoco reflejó diferencias notables ni definidas entre tratamientos, a pesar de que las diferencias observadas en 2014 resultaron estadísticamente significativas entre el P1 y el P2. En general, la concentración de polifenoles se vio ligeramente reducida por la aplicación del riego con mayor dosis, con respecto a la restricción hídrica que supusieron el riego de menor dosis y el secano.

Tabla 4. Concentración de sólidos solubles totales S.S.T. (ºbrix), pH, Acidez titulable (g ac. tartárico/L), Ácido tartárico (g/L), Ácido málico (g/L), Concentración de Potasio (ppm), Índice de polifenoles totales (I.P.T.) e Indice de Ravaz, de los tratamientos (Trat.): P0, P1 y P2. Significación estadística (Sig.): -, no significativo; *, $p<0,05$.

	Trat.	S.S.T.	pH	Acidez titulable	Ácido tartárico	Ácido málico	Potasio	I.P.T.	Indice Ravaz
	P0	22,63 b	3,94	3,63 b	6,08	0,74 b	1503	10,75	2,88
2012	**P1**	23,70 a	4,05	4,50 a	6,61	0,92 b	1653	10,25	3,58
	P2	23,30 ab	3,88	4,97 a	6,92	1,39 a	1495	10,00	3,23
	Sig.	*	-	*	-	*	-	-	-
	P0	21,63	3,49 a	5,60 b	5,81	1,55 c	1120	8,75	4,20
2013	**P1**	22,81	3,38 ab	6,36 ab	5,58	2.14 b	1135	9,00	3,09
	P2	22,95	3,29 b	7,05 a	5,16	3,17 a	1175	7,50	2,55
	Sig.	-	*	*	-	*	-	-	-
	P0	23,15	3,38 a	5,52 b	7,68 b	1,91	1563	8,00 ab	2,57
2014	**P1**	23,03	3,36 a	5,60 b	7,54 b	1,82	1538	8,50 a	2,34
	P2	22,78	3,25 b	7,08 a	8,82 a	2,16	1688	7,50 b	2,26
	Sig.	-	*	*	*	-	-	*	-
	P0	22,47	3,60	4,91	6,52	1,40	1395	9,17	3,22
Media	**P1**	23,18	3,59	5,49	6,57	1,63	1442	9,25	3,00
	P2	23,01	3,47	6,36	6,97	2,24	1453	8,33	2,68

4 Conclusiones

El desarrollo vegetativo aumentó progresivamente con la dosis de riego aplicada, tal que el riego de menor dosis provocó un aumento medio interanual del peso de madera de poda del 31% respecto al secano, mientras que el aumento provocado por el riego de mayor dosis respecto al de menor dosis fue del 27%, aunque las diferencias anuales entre tratamientos no siempre llegaron a ser estadísticamente significativas. Dicho aumento fue debido básicamente al incremento del peso del sarmiento, dado que el número de sarmientos por cepa fue similar en todos los tratamientos.

El rendimiento en uva mostró una tendencia al aumento con las dosis de agua, con un incremento medio del riego de menor dosis respecto del secano del 25% y del riego de mayor dosis respecto del riego de menor dosis del 13%, aunque las diferencias anuales entre tratamientos no resultasen estadísticamente significativas. Las diferencias en producción de uva fueron fundamentalmente debidas al peso de racimo, que también aumentó con la cantidad de agua aportada, debido mayormente al peso de baya así como al número de bayas del racimo, aunque el número de racimos por cepa también contribuyó a dicho aumento en la producción de uva.

La concentración de azúcares se vió generalmente favorecida por el riego, aunque dicha tendencia no fue intensa ni constante entre tratamientos, mientras que el pH del mosto mostró cierta tendencia a la reducción con el riego respecto al secano, pero sobre todo al aplicar el riego de mayor dosis. La acidez titulable, así como el ácido tartárico y el ácido málico, aumentó con la dosis de riego, sobre todo con la dosis mayor, observándose menores diferencias entre tratamientos en el ácido tartárico y mayores en el ácido málico. El potasio y el IPT no mostraron diferencias notables ni constantes entre tratamientos, aunque la media interanual mostró una concentración de K ligeramente mayor en los tratamientos regados que en el secano y un valor de IPT ligeramente más reducido con el riego de mayor dosis. Las respuestas observadas en la planta deben servir para enfocar el riego del viñedo hacia la optimización productiva y cualitativa del uso del agua.

Referencias

1. M.V. Alburquerque, J. Yuste. Efectos de la dosis de riego en el estado hídrico, el desarrollo y la calidad de la uva de la variedad Tempranillo en el valle del río Duero. Enoviticultura **10**: 24-31 (2011)
2. M.V. Alburquerque, R. Yuste, J. Yuste. Influencia de la dosis de riego en Tempranillo en el Valle del Duero: fisiología, superficie foliar, productividad y calidad de la uva. La Semana Vitivinícola **3476**: 1334-1340 (2016)
3. J. Blouin, G. Guimberteau. Maturation et maturité des raisins. 151 p. Éd. Féret. Bordeaux (2000)
4. J.R. Castel, M.E. Valdés, M.H. Prieto, D. Uriarte, L. Mancha, A. Montoro, F. Mañas, R. López-Urrea, P. López-Fuster, J. Yuste, M.V. Alburquerque, J.R. Yuste, E. Barajas, A. Yeves, D. Pérez, D.S. Intrigliolo. Efectos de clima y suelo sobre la respuesta al riego de Tempranillo. I. comportamiento agronómico y relaciones hídricas en cuatro localidades de España. La Semana Vitivinícola **3393**: 198-206 (2013)
5. J. Girona, J. Marsal, M. Mata, J. del Campo, B. Basile. Phenological sensitivity of berry growth and composition of Tempranillo grapevines (*Vitis vinifera* L.) to water stress. Aust. J. Grape Wine Res. **15**: 268–277 (2009)
6. D.S. Intrigliolo, J.R. Castel. Effects of irrigation on the performance of grapevine cv. Tempranillo in Requena, Spain. Am. J. Enol. Vitic. **59** (1): 30-38 (2008)
7. D.S. Intrigliolo, J.R. Castel. Response of grapevine cv. Tempranillo to timing and amount of irrigation: water relations, vine growth, yield and berry and wine composition. Irri. Sci. **28**: 113-125 (2010)
8. J.A. Kennedy, M.A. Matthews, A. Waterhouse. Effect of maturity and vine water status on grape skin and wine flavonoides. Am. J. Enol. Vitic. **53** (4): 268-274 (2002)
9. H. Ojeda, C. Andary, E. Kraeva, A. Carbonneau, A. Deloire. Influence of pre- and post-veraison water deficit on synthesis and concentration of skin phenolic compounds during berry growth of *Vitis vinifera* cv. Shiraz. Am. J. Enol. Vitic. **53** (4): 261-267 (2002)
10. A.G. Reynolds, W.D. Lowrey, L. Tomek, J. Hakimi, Ch. Savigny. Influence of irrigation on vine performance, fruit composition, and wine quality of Chardonnay in a cool, humid climate. Am. J. Enol. Vitic. **58** (2): 217-228 (2007)
11. G. Roby, J.F. Harbertson, D.A. Adams, M.A. Matthews. Berry size and vine water deficits as factors in winegrape composition: anthocyanins and tannins. Aust. J. Grape Wine Res. **10**: 100-107 (2004)
12. L.G. Santesteban, C. Miranda, J.B. Royo. Regulated deficit irrigation effects on growth, yield, grape quality and individual anthocyanin composition in *Vitis vinifera* L. cv. Tempranillo. Agric. Water Manag. **98** (7): 1171-1179 (2011)
13. C.J. Soar, B.R. Loveys. The effect of changing patterns in soil-moisture availability on grapevine root distribution and viticultural implications for converting full-cover irrigation into a point-source irrigation system. Aust. J. Grape Wine Res. **13** (1): 2-13 (2007)
14. J. Yuste, M.V. Alburquerque, R. Yuste, E. Gamero, M.E. Valdés. Respuesta cualitativa y productiva del viñedo en vaso a la variación de la dosis de riego en Tempranillo cultivado en las condiciones semiáridas del valle del Duero. Vida Rural **428**: 42-48 (2017)
15. J. Yuste, A. Vicente, E. Barajas. Efectos del tipo de riego por goteo (tradicional y deshidratación parcial de raíces) en la calidad y la producción de uva del cv. Cabernet Sauvignon (*Vitis vinifera* L.) en el valle del río Duero. Enoviticultura **40**: 28-35 (2016)

Agradecimientos

La elaboración de este trabajo ha sido posible a través de la base de proyectos INIA previos y del RTA2014-00049-C05-01, fondos FEDER y de la Junta de Castilla y León, así como la colaboración del Grupo de Viticultura y la Estación Enológica del ITACYL.

Efectos del uso de cubiertas vegetales en viñedo ecológico de la variedad Pedro Ximénez

Pilar Ramírez Pérez[1], José Manuel Cabezas Luque[1], Rosario Sánchez Marín[1] y Juan Manuel León Gutiérrez[1]

[1]IFAPA Centro de Cabra, Ctra. Antigua Cabra-Doña Mencía km 2.5 Cabra (Córdoba), España

Resumen. El manejo de los suelos vitícolas andaluces ha consistido tradicionalmente en un laboreo muy intensivo con varios pases anuales. Esta aplicación continuada de labores provoca importantes problemas de erosión que se agravan en zonas con pendiente y con suelos de textura arcillosa. El impacto negativo que se genera tanto a nivel ambiental como económico, ha impulsado la búsqueda de alternativas al laboreo tradicional, como es la implantación de cubiertas vegetales.
En este trabajo se presentan los resultados obtenidos en los años 2015 a 2017. Se ha realizado en la finca experimental del IFAPA Centro de Cabra (Córdoba, España), en una parcela de la variedad Pedro Ximénez cultivada en ecológico. Se han evaluado tres tratamientos de manejo de suelo: Laboreo tradicional (L), cubierta vegetal espontánea (CVE) y cubierta vegetal sembrada (CVS). Se ha comparado los efectos de los distintos tratamientos en parámetros relacionados con la fisiología, rendimiento y calidad de la uva.
El laboreo ha mantenido producciones significativamente superiores que los tratamientos de cubiertas vegetales, en dos de las tres campañas estudiadas El peso de madera de poda también fue significativamente mayor en las tres campañas, mientras que el contenido en sólidos solubles fue menor.

Abstract. The management of Andalusian vineyard soils has traditionally involved an intense tillage several times per year. This constant operation causes serious problems of erosion which get worse in sloping surfaces and clay soils. Alternatives for the traditional tillage have been promoted due to both the environmental and economic negative impacts, these alternatives include plant covers.
Results obtained between 2015-2017 are presented in this work. The research has been done on the experimental parcel of IFAPA Centro de Cabra (Córdoba, Spain), in Pedro Ximenez ecological vineyard. Three soil management treatments have been assessed: Conventional tillering (L), spontaneous cover crop (CVE) and seeded cover crop (CVS). Different effects of the treatments on parameters related to physiology, yield and grape quality have been compared.
Tillering treatment has had significantly higher productions during two seasons than cover crops treatments. Pruning wood weight was significantly higher too during three seasons while the content of soluble solids was lower.

1. Introducción

Andalucía es una región que, por su situación geográfica, orografía y climatología, posibilita el cultivo de la vid en muy diversos y adecuados edafoclimas. Además, factores como la alta insolación y la baja incidencia de plagas y enfermedades hacen factible la obtención de uvas y vino ecológicos de calidad. El viñedo ecológico en Andalucía es aún incipiente, pero ha ido incrementándose cada año hasta llegar a las actuales 863 ha, lo que supone un 2.92% de la superficie total de viñedo andaluz.

El laboreo es la técnica de manejo de suelo con más arraigo en las comarcas vitícolas tradicionales andaluzas como son el Marco de Jerez y de Montilla-Moriles. En estas zonas se ha labrado de forma intensiva con varios pases anuales combinándose en algunos casos con tratamientos herbicidas. Esta aplicación frecuente de labores en el terreno provoca en muchas ocasiones un serio problema de erosión y apelmazamiento que se acrecienta en las zonas con suelos de textura arcillosa y en parcelas con pendientes, donde además es común ver labores realizadas a favor de la línea de máxima pendiente. El impacto negativo que ocasiona el laboreo en estas condiciones a nivel medioambiental y económico además del propio coste energético y de recursos de esta técnica de cultivo, ha impulsado la búsqueda de alternativas al laboreo tradicional, entre las que se encuentran la implantación de cubiertas vegetales [1]. Existe, además, una demanda creciente del sector vitivinícola acerca del conocimiento sobre las técnicas de cultivo ecológicas para manejar el suelo sin utilizar productos químicos de síntesis y que además protejan y mejoren los suelos de los cultivos.

La implantación de cubiertas vegetales consiste en sembrar en las calles alguna especie cultivable o en dejar crecer la vegetación natural. Cuando existen limitaciones de agua, lo más adecuado puede ser dejar sólo cubierta vegetal durante el otoño y el invierno en las calles y eliminar la que aparezca en la línea de las cepas. En este caso, la cubierta se mantiene desde el otoño, sembrando después de la vendimia y se elimina tras el desborre o principios de primavera, antes de que empiece a competir por agua y nutrientes con la viña. En cultivo ecológico, lo ideal es eliminar la cubierta mediante siega mecánica, dejando los restos en superficie para que actúen como un "mulching" de protección frente a la erosión y como aporte de materia orgánica; aunque también se pueden eliminar mediante laboreo superficial e incorporarlas al

 https://doi.org/10.1051/e3sconf/20185001008

suelo en forma de abonos verdes. La siega mecánica no elimina la cubierta vegetal, tan solo limita temporalmente sus necesidades de agua y nutrientes, por lo que sería necesario realizar varias siegas a lo largo de la primavera incluso inicios de verano.

El objetivo de este trabajo es evaluar la influencia de una cubierta vegetal en *Vitis vinifera* cv. Pedro Ximénez cultivada en ecológico en un clima mediterráneo, sobre la producción, el vigor y la calidad de la uva.

2. Materiales y Métodos

Este trabajo se ha desarrollado los años 2015 a 2017 en el viñedo experimental del IFAPA de Cabra, situado en el centro geográfico de Andalucía, al sur de la provincia de Córdoba, entre los paralelos 37°10′ y 37°40′ y dentro de la Denominación de Origen Montilla-Moriles.

El material vegetal utilizado es de la variedad Pedro Ximénez injertada sobre el portainjerto 41-B Millardet-Grasset. El marco de plantación es de 2.5 metros entre calles y 1.2 metros entre plantas. El sistema de poda utilizado es en cordón Royat bilateral, con 4 pulgares por brazo y una carga de 16 yemas por cepa, el sistema de empalizamiento es en espaldera y el cultivo en secano. El manejo del cultivo se ha realizado siguiendo el Reglamento (CE) N° 834/2007 sobre producción y etiquetado de productos ecológicos y el Reglamento (CE) N° 889/2008 por el que se establecen las disposiciones de aplicación del Reglamento anterior.

El diseño del ensayo ha sido en bloques al azar con tres repeticiones. Cada parcela elemental consta de tres filas de ciento dieciocho cepas cada una, siendo la fila intermedia la utilizada para todos los controles. Se han aplicado tres tratamientos: Laboreo tradicional (L), cubierta vegetal espontánea (CVE) y cubierta vegetal sembrada (CVS). El laboreo tradicional se ha realizado con pases de cultivador. La cubierta vegetal espontánea se ha mantenido desde octubre hasta marzo, controlándose mediante siega mecánica. La cubierta vegetal sembrada, compuesta por una combinación de avena a una dosis de 150 kg/ha y veza a una dosis de 65 kg/ha, se sembró en octubre y se mantuvo hasta marzo, cuando fue controlada mediante siega mecánica. En los tres tratamientos las líneas de cepas se mantuvieron libres de vegetación mediante siegas mecánicas.

La zona se caracteriza por un clima mediterráneo con cierto grado de continentalidad, con veranos cálidos, largos y secos e inviernos cortos y relativamente suaves, con una temperatura media anual en los últimos 20 años de 16.6 °C, una máxima media del mes más cálido (julio) de 34.4 °C y una mínima media del mes más frío (enero) de 5.0 °C. La precipitación media anual es de 661 mm, repartidos de otoño a primavera, siendo los meses más secos junio, julio y agosto y una evapotranspiración de referencia (ET_o) anual de 1157 mm. Las precipitaciones de estas tres campañas se sitúan muy por debajo de la media, con 399 mm en 2014/15, 572 mm en 2015/16 y 396 mm en 2016/2017. La evapotranspiración de referencia (ET_o) fue de 1162, 1153 y 1207 mm respectivamente. (Figura 1). Las temperaturas medias de estas tres campañas han sido superiores a la media, situándose en 17.0, 17.0 y 17.8°C. El suelo es de tierra albariza, pobre en materia orgánica y con un 17% de caliza activa. Su textura franca le confiere un alto poder retentivo de la humedad, aunque insuficiente en las condiciones meteorológicas de estas campañas.

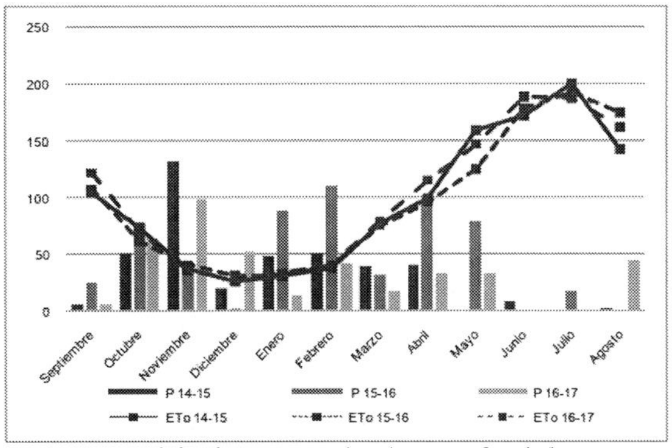

Figura 1. Precipitaciones y ET_o (mm) mensuales de las campañas 2014/15, 2015/16 y 2016/17 en la estación meteorológica del IFAPA de Cabra.

A partir del cuajado se midió semanalmente el potencial hídrico xilemático (Ψ_s) al mediodía, utilizando una cámara de presión (Arimad 3000; MRC Inc. Israel). Tras la parada de crecimiento vegetativo, se determinó la superficie foliar externa (SA) y la superficie foliar total (LAI). La determinación del área foliar se realizó con un medidor de área (CI-203; CID Inc. USA).

Tras el envero, se realizó un seguimiento semanal de la maduración y se fijó el momento de la vendimia cuando la uva alcanzó un contenido suficiente de azúcar en el conjunto de los tratamientos para la elaboración de vinos generosos. Se realizaron los siguientes controles: producción (kg/cepa), número de racimos por cepa y peso de cien bayas (g). También se realizaron las siguientes determinaciones analíticas: sólidos solubles (°Brix), acidez total (g de ác. tartárico/L), pH, ácido málico (g/L) [2], ácido tartárico (g/L) y potasio (mg/L) [3]. Después de la caída de la hoja se pesó la madera de poda (kg/cepa).

A los datos obtenidos se les ha realizado un análisis de la varianza (ANOVA), para una fuente de variación (manejo de suelo). La significación de dichos análisis se determinó para los niveles de probabilidad $p < 0.05$ (*), $p < 0.01$ (**) y $p < 0.001$ (***). El programa utilizado en todos los análisis estadísticos fue Statistix 9.0.

3. Resultados

El transcurso de la maduración fue más lento y gradual en 2016 que en 2015 y 2017, realizándose la vendimia de 2016 el 2 de septiembre, fecha que es habitual en la zona. Sin embargo, en 2015 y 2017 coincidieron ambas el 10 de agosto y además con un grado de maduración superior al alcanzado en 2016.

Respecto a la producción, el tratamiento L ha sido más productivo que los tratamientos con cubiertas vegetales (Figura 2) con diferencias significativas en 2016 y 2017. Las producciones se vieron afectadas por las precipitaciones de cada campaña, observándose un marcado descenso de estas en los tres tratamientos en el año 2017, cuyos resultados se ven además influidos por las elevadas temperaturas que se alcanzaron a finales de primavera y en verano.

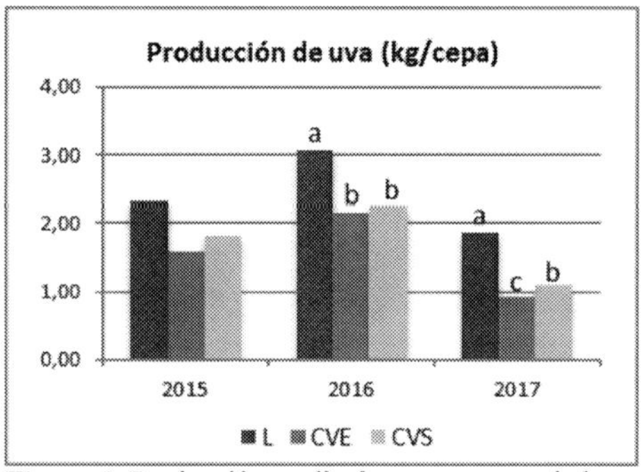

Figura 2. Producción media de uva por cepa de los tres tratamientos en los tres años de estudio.

En cuanto a los parámetros fisicoquímicos del mosto, se encontraron diferencias significativas en el contenido de sólidos solubles en los tres años evaluados. En 2015 el tratamiento CVS, fue el que presentó mayor concentración, diferenciándose significativamente de L, quedando CVE en posición intermedia. En 2016 hay una clara diferenciación entre el tratamiento L, que no alcanzó una concentración suficiente de sólidos solubles en la fecha que se vendimió, con los dos tratamientos de cubiertas vegetales. En 2017, el tratamiento de CVE fue el que presentó mayor contenido en sólidos solubles en la fecha de vendimia, proporcionando un grado alcohólico probable muy adecuado para la crianza biológica característica de los vinos finos, sin necesidad de adicionar alcohol de vino. (Figura 3).

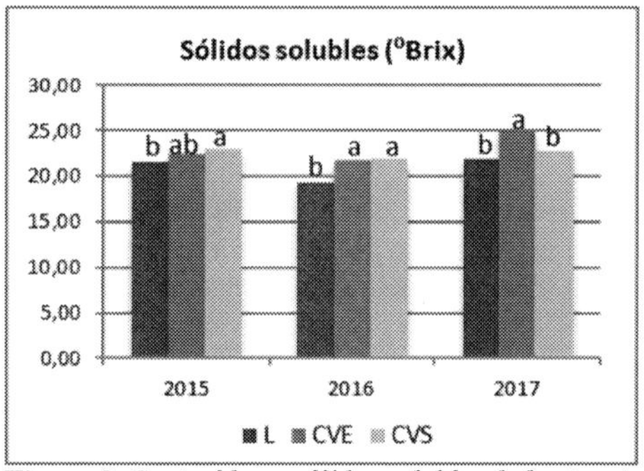

Figura 3. Contenido en sólidos solubles de los tres tratamientos en los tres años de estudio.

Para el resto de los parámetros analizados, no se encontraron diferencias significativas de relevancia. Tan solo en 2015 el contenido en ácido málico y nitrógeno fácilmente asimilable fueron ligeramente superiores en L que en CVE y CVS, aunque en magnitudes poco relevantes enológicamente hablando. (Tabla 1).

En los dos años en los que el tratamiento L presentó mayor producción, éste también presentó mayor LAI (Figura 4) y menor relación SA/LAI (Figura 5). El SA/LAI refleja el grado de solapamiento de la vegetación y da una idea de la exposición de las hojas a la radiación solar, siendo los valores óptimos cercanos a 1. En 2015, donde no se encontraron diferencias significativas respecto a este parámetro, fue cuando los valores se acercaron más al óptimo, mientras que en 2016 y 2017, los tratamientos de cubiertas presentaron valores mejores que L, aunque lejanos del óptimo.

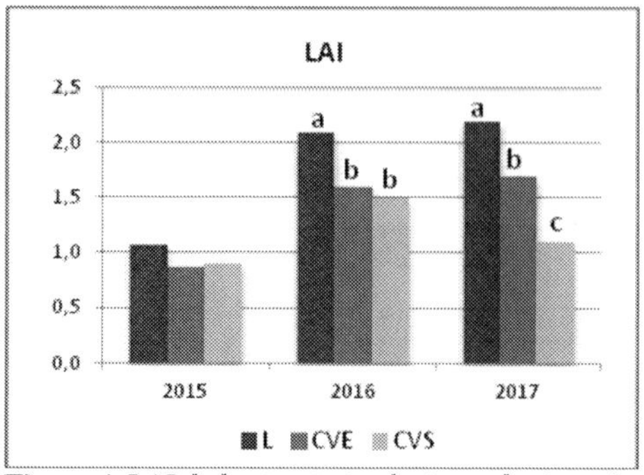

Figura 4. LAI de los tres tratamientos en los tres años de estudio.

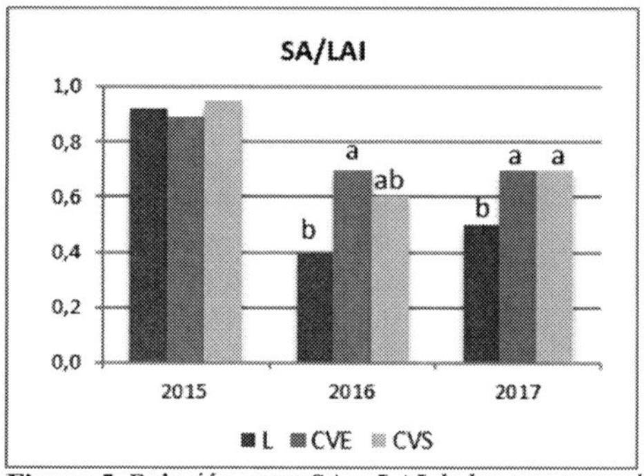

Figura 5. Relación entre SA y LAI de los tres tratamientos en los tres años de estudio.

En general, el tratamiento de laboreo presentó mayor desarrollo vegetativo de las plantas, con un mayor peso de madera de poda en los tres años evaluados (figura 6). Esto fue debido en 2015 y 2017 a un mayor peso de los sarmientos y a la mayor longitud de estos, y en 2016, por un mayor número de sarmientos. Las diferencias de LAI y madera del año 2015 con respecto a 2016 y 2017, son debidas también a que el primer año se realizó en el mes de mayo un despampano, sin embargo, en los años posteriores se decidió no repetir esta operación con la intención de aumentar la masa foliar en las plantas, lo que constata también los resultados obtenidos en el numero de sarmientos por cepa (Tabla 2).

https://doi.org/10.1051/e3sconf/20185001008

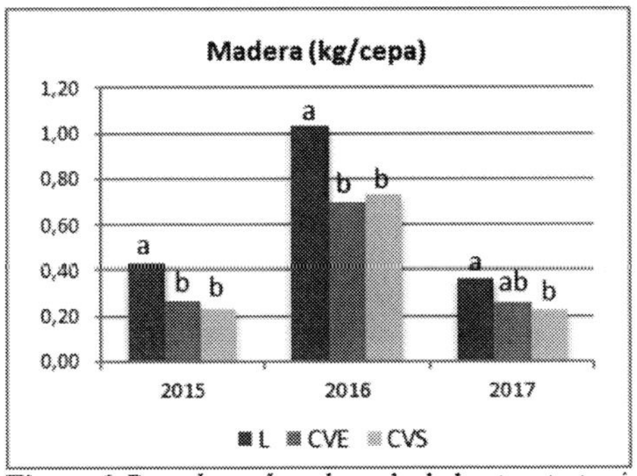

Figura 6. Peso de madera de poda de los tres tratamientos en los tres años de estudio.

Respecto a la relación entre la superficie foliar externa y la producción por cepa (SA/kg uva), los tres tratamientos presentaron en 2015 valores cercanos a los considerados como normales. En 2016 y 2017 se encontraron diferencias significativas, presentando L niveles inferiores a los tratamientos con cubiertas vegetales (Figura 7).

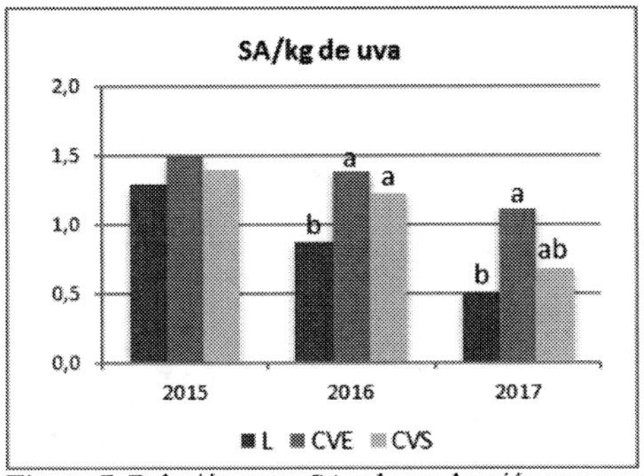

Figura 7. Relación entre SA y la producción por cepa de los tres tratamientos en los tres años de estudio.

En las Figuras 8, 9 y 10, se representan las integrales de estrés de los tres años evaluados. Durante el verano de 2015 fue cuando menos estrés acumularon los distintos tratamientos, diferenciándose en las primeras semanas el tratamiento CVS, como ligeramente más estresado para igualarse todos los tratamientos hasta vendimia.

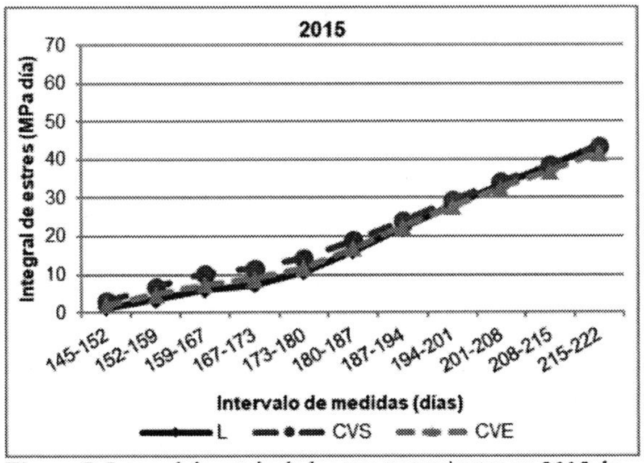

Figura 8. Integral de estrés de los tres tratamientos en 2015 de cuajado a vendimia.

En 2016 la vendimia fue veintidós días más tarde que en los otros dos años, por lo que las plantas acumularon más estrés hídrico al final de la maduración. En este caso, los tratamientos de cubiertas vegetales presentaron más estrés que L hasta el envero, momento a partir del cual, CVS se igualó a L.

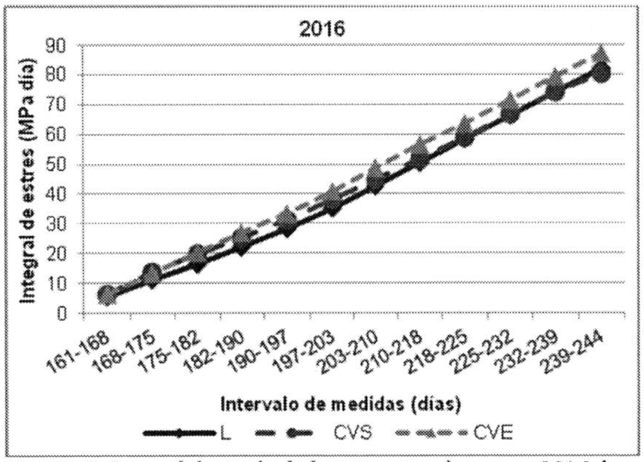

Figura 9. Integral de estrés de los tres tratamientos en 2016 de cuajado a vendimia.

En el año 2017, se encontraron diferencias del tratamiento L con respecto a los dos tratamientos de cubierta vegetal, tan solo en la última semana de la maduración.

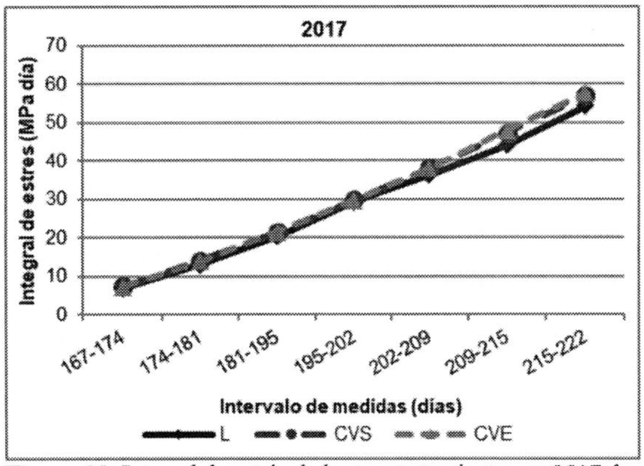

Figura 10. Integral de estrés de los tres tratamientos en 2017 de cuajado a vendimia.

4. Conclusiones

El tratamiento de laboreo fue más productivo que los tratamientos de cubierta vegetal. Sin embargo, en las fechas que se realizaron las vendimias, fue el que alcanzó el menor contenido de sólidos solubles. No encontrando diferencias claras ni en el rendimiento ni en los parámetros fisicoquímicos del mosto entre los dos tratamientos de cubierta vegetal.

El uso de cubiertas vegetales temporales, tanto espontánea como sembrada, ha influido en el vigor de las plantas, tal como se evidencia en la reducción del peso de la madera de poda y en el LAI.

Es necesario seguir con este estudio para evaluar el comportamiento de las cepas a largo plazo frente a los tres tipos de manejo del suelo en ecológico, para así contrastar los resultados en campañas de meteorología más normal en la zona.

4. Bibliografía

[1] P. Ramírez, J. Lasheras, J.M. León, J. Morales y I. López. Efectos del uso de una cubierta vegetal en el rendimiento, vigor y calidad de la uva. Consejería de Agricultura, Pesca y Desarrollo Rural. Instituto de Investigación y Formación Agraria y Pesquera. SERVIFAPA. Junta de Andalucía. (2014).

[2] DOCE. Reglamento (CEE) 2676/90 de la Comisión, de 17 de septiembre, por el que se determinan los métodos de análisis comunitarios aplicables en el sector del vino. DOCE. 272 (1990).

[3] H. Rebelein. Precise routine determination of tartaric and lactic acid in wine and similar beverages. Chemie, Mikrobiologie, Technologie der Lebensmittel. 2, 112-121 (1973).

5. Agradecimientos

Los resultados presentados han sido obtenidos en el marco del proyecto Transforma vid y vino del Instituto de Investigación y Formación Agraria y Pesquera (IFAPA) y del proyecto de Transferencia de Tecnología e Innovación en vid y vino del IFAPA, cofinanciados al 80% por el Fondo Europeo de Desarrollo Regional, dentro del Programa Operativo FEDER de Andalucía 2007-2013 y FEDER de Andalucía 2014-2020 respectivamente.

Tabla 1. Valores de las producciones medias, número de racimos, peso de 100 bayas y parámetros físico-químicos de los mostos analizados en vendimia de los tres tratamientos estudiados en las tres campañas.

	2015				2016				2017			
	L	CVE	CVS	g.s.	L	CVE	CVS	g.s.	L	CVE	CVS	g.s.
Peso de 100 bayas (g)	203.8	192.9	193.5	n.s.	184.5	184.3	186.7	n.s.	169.5	172.5	157.5	n.s.
Grado alcohólico probable (% v/v)	12.48b	13.01ab	13.40a	*	10.99b	12.61a	12.79a	**	12.75b	14.81a	13.33b	*
pH (20°C)	3.50	3.44	3.44	n.s.	3.54	3.51	3.52	n.s.	3.67	3.64	3.66	n.s.
Acidez total (g ác. Tartárico/L)	4.5	4.9	4.8	n.s.	4.7	4.4	4.5	n.s.	4.6	4.6	4.4	n.s.
Ácido tartárico (g/L)	6.9	7.1	7.1	n.s.	7.0	6.5	6.6	n.s.	7.1	7.1	7.0	n.s.
Ácido málico (g/L)	0.4a	0.3b	0.3b	**	0.4	0.2	0.2	n.s.	0.3	0.3	0.3	n.s.
Potasio (mg/L)	1609	1585	1538	n.s.	1746	1533	1612	n.s.	2003	1849	1843	n.s.
NFA (mg/L)	71.7a	52.7c	66.0b	***	53.8	49.0	57.8	n.s.	86.5	62.5	78.2	n.s.

Cada dato corresponde a la media de los tres bloques. Dentro de cada fila, letras diferentes a continuación de las medias indican diferencias significativas a $P \leq 0.05$ según el test LSD. Grado de significación (g.s.) n.s.: no significativo; * significativo a $P \leq 0.05$; ** significativo a $P \leq 0.01$; *** significativo a $P \leq 0.001$.

Tabla 2. Valores de los parámetros de desarrollo vegetativo de los tres tratamientos estudiados en las tres campañas.

	2015				2016				2017			
	L	CVE	CVS	g.s.	L	CVE	CVS	g.s.	L	CVE	CVS	g.s.
Longitud del sarmiento (m)	0.89a	0.70b	0.74b	**	0.88	0.81	0.79	n.s.	0.69a	0.64ab	0.50b	*
Número de sarmientos/cepa	12	11	11	n.s.	20a	15b	16b	*	24	21	20	n.s.
Peso medio del sarmiento (g)	38.1a	24.2b	21.6b	*	52.6	45.3	47.2	n.s.	15.3a	12.3ab	11.4b	*
Nudos por sarmiento	17	16	17	n.s.	18	17	17	n.s.	16	15	13	n.s.
Nudos por metro lineal	20b	24a	24a	*	21	21	22	n.s.	24	23	27	n.s.
SA	1.0a	0.8b	0.9b	*	0.9	1.0	0.9	n.s.	0.9	0.8	0.7	n.s.

Cada dato corresponde a la media de los tres bloques. Dentro de cada fila, letras diferentes a continuación de las medias indican diferencias significativas a $P \leq 0.05$ según el test LSD. Grado de significación (g.s.) n.s.: no significativo; * significativo a $P \leq 0.05$; ** significativo a $P \leq 0.01$; *** significativo a $P \leq 0.001$.

 https://doi.org/10.1051/e3sconf/20185001009

Caracterización del *terroir*: influencia en la composición de la uva y relación con el NDVI dentro de la D.O.Ca.Rioja / *Terroir* characterization: influence on grape composition and relationship with NDVI within D.O.Ca.Rioja

Álvaro Martínez[1,2,a] y Vicente D. Gómez-Miguel[2]
[1]Bodegas de los Herederos del Marqués de Riscal. Ctra. Nacional VI, km 172.600 Rueda (Spain)
[2]Departamento de Producción Agraria. Universidad Politécnica de Madrid (UPM), Spain

Resumen. Los factores naturales que definen el *terroir* son los principales responsables de la variabilidad observada entre plantas de una misma parcela (variabilidad intraparcela) o de distintas fincas (variabilidad interparcelas). En el presente trabajo se propone una metodología de zonificación del *terroir*, presentándose también su relación con algunas características vegeto-productivas del viñedo y con el Índice de Vegetación de Diferencia Normalizada (NDVI). La zonificación de los elementos del medio se ha llevado a cabo a partir de la fotointerpretación aérea mientras que la representación gráfica del NDVI se realiza partiendo de una imagen multiespectral capturada por el satélite Pleiades, de 0,5 metros de resolución espacial. Comparando los datos vegeto-productivos recogidos en las 4 parcelas de estudio, se ha observado que existen sectores, relacionados con zonas de deposición de material y asociados a suelos de carácter fluvéntico, con tendencia a ser más vigorosos, más productivos y con uvas de menor contenido fenólico y de menor pH. Por otro lado, analizando cada parcela de manera independiente, se observa que la distribución espacial del NDVI sigue un patrón semejante a la zonificación de los elementos del medio, algo que recomendaría hacer un uso de este índice por sectores previamente discriminados conforme a una cierta uniformidad del *terroir*.

Abstract. The natural factors that define *terroir* are main responsible for the variabililty observed between plants of the same plot (intra-plot variability) or of different farms (variability between plots). A methodology of terroir zoning is proposed, also is presented its relation with some vegeto-productive characteristics of vineyard and with the Normalized Difference Vegetation Index (NDVI). Zoning of natural factors has been carried out from aerial photointerpretation while graphic representation of NDVI was made starting from a multispectral image (0,5 meters of spatial resolution) captured by the Pleiades satellite. Comparing the vegetative and productive data collected in the four study plots, it has been observed that there are sectors related to deposition areas and associated with fluvent soils, with a tendency to be more vigorous, more productive and with grapes of lower phenolic content and lower pH. On the other hand, analyzing each plot independently, it is observed that NDVI spatial distribution follows a similar pattern to the natural factors zoning, something that would recommend making use of NDVI by sectors previously discriminated according to certain *terroir* uniformity.

1 Introducción

Los primeros estudios científicos relacionados con los elementos del medio vitícola y sus interacciones se llevan a cabo en el último cuarto del siglo XX, pudiendo considerarse la tesis doctoral del profesor René Morlat (1) uno de los estudios pioneros sobre la zonificación del *terroir* vitícola en el sentido moderno del término. El citado trabajo se desarrolla en la zona media del valle del Loira, siendo los países europeos los que tradicionalmente más importancia han otorgado a los elementos del medio en la caracterización de los vinos, protegiendo así el origen de los mismos. Dos ejemplos de esta tradición son la clasificación actual de vinos de Burdeos, que prácticamente no ha sufrido modificaciones desde su creación en 1855 y la clasificación de vinos de Oporto que tiene su origen en la delimitación de 1758 (actualmente esa zonificación se ha visto ampliada) llevada a cabo por la Companhia Geral da Agricultura das Vinhas do Alto Douro (empresa similar a los actuales Consejos Reguladores) a propuesta del Marqués de Pombal.

La clasificación de los vinos en el marco de la D.O.Ca.Rioja data de 1970 y está basada en el envejecimiento de los mismos, existiendo las clases Garantía de Origen (o Genérico), Crianza, Reserva y Gran Reserva. Recientemente (junio de 2017), el Consejo Regulador de la D.O.Ca. Rioja ha aprobado la figura de "Viñedo Singular", que, entre otros aspectos, protege viñedos que se caracterizan por poseer una delimitación natural técnicamente justificable, una edad de las plantas superior a 35 años, y un rendimiento más de un 20 % inferior al autorizado para el conjunto de la Denominación de Origen (D.O). Para una adecuada incorporación tanto de esta nueva protección como de hipotéticas nuevas figuras parece necesario el uso de metodologías de zonificación rigurosas y objetivas que permitan clasificar los diferentes *terroir* de la D.O., huyendo de potenciales métodos de caracterización subjetivos y poco exhaustivos que pondrían en riesgo su éxito.

1.1 *Terroir*

El concepto de *terroir* incluye, además de los elementos del medio vitícola, un factor enológico y un factor humano, presente en los dos anteriores (2). Después de la vendimia, el manejo de la cosecha queda subordinado a la enología, cuyos procesos han evolucionado notablemente en España desde los años 70 del siglo XX (3), mejorándose, entre otros aspectos, la higiene de las bodegas o el control de los procesos microbiológicos, térmicos, de envejecimiento, etc. Tomando, por lo tanto, el momento de la vendimia como punto de inflexión, podemos diferenciar entre un aspecto vitícola y otro vinícola dentro del *terroir* vitivinícola, sobre el primero gravita este trabajo.

El clima es un factor de gran importancia en la calificación de una región o zona en relación con la aptitud para producir uvas de calidad (4), siendo sus efectos notables en el nivel de maduración de una determinada variedad y en la tipicidad y la calidad del vino obtenido (4, 5). Para estudiar la influencia del clima en el viñedo es tradicional distinguir, en función de la escala, entre macroclima, mesoclima y microclima. El primero hace referencia al clima de una región y es el principal factor limitante para el cultivo de la vid (6), mientras que el mesoclima es característico de una ubicación topográfica y paisajística determinada y afecta a un conjunto de plantas por igual en una determinada unidad geomorfológica. Por último, el microclima se refiere al de la propia planta, rodea a hojas y racimos y tiene una gran influencia en el ciclo biológico (por ejemplo, es de gran importancia en el periodo de maduración de la uva), pudiéndose modificar, además, a través de las operaciones del cultivo.

Al igual que sucede con el clima, se producen vinos de alta calidad a partir de uvas cultivadas sobre geologías (estratigrafía y litología) diferentes. Según (7) la geología influye en tres factores que afectan a la calidad del vino: (i) modificando la temperatura alrededor de la planta y de los racimos; (ii) controlando el suministro de humedad y la velocidad en la toma de nutrientes; (iii) determinando la disponibilidad de nutrientes en el suelo. Respecto a este último punto, la geología es un factor importante en el control de la nutrición potásica de la vid (7). (8) apunta que la influencia directa de la geología sobre el viñedo es mínima en suelos profundos, afectando de forma indirecta en suelos superficiales, a través de la composición del suelo, la geomorfología y la capacidad de retención de agua. Por otro lado, el efecto de la geomorfología sobre el *terroir* es indirecto, a través de la altitud, la pendiente y la orientación o exposición (9, 10).

Una concepción tradicional del suelo permite considerarlo como el resultado de la interacción del clima y de los seres vivos, como factores activos, sobre el tipo de roca y el relieve, como factores pasivos, durante un tiempo de actuación determinado. Desde un punto de vista vitícola, el suelo es el sustentáculo de las plantas, el almacén de agua y nutrientes que estas utilizan y el regulador de los elementos del clima (temperatura y precipitación, por ejemplo) que actúan sobre ellas (6). De igual manera que sucedía con el clima, y lógicamente con la geología, se elaboran vinos de alta calidad a partir de uvas producidas sobre una gran diversidad de suelos, siendo imposible definir los mejores suelos en términos de pedregosidad, contenido en arcilla o limo, profundidad de suelo o contenido mineral (9). La descripción de suelos cultivados con viñedo ha estado durante mucho tiempo reducida a los tipos de textura, pero, por ejemplo, un mismo contenido en arcilla no tiene las mismas propiedades, ya que éstas dependen de la naturaleza mineralógica de la arcilla en cuestión. Así, las convicciones basadas en el empirismo popular se revelan a veces contradictorias debido a su carácter simplista (11).

La elección del portainjerto a utilizar y de la variedad a cultivar debe considerar las condiciones del medio en el que se implantará el viñedo, con el principal objetivo de conseguir una madurez adecuada de la uva. Varios autores (4, 12-14) discriminan entre zonas de maduración en función de distintos criterios.

Las distintas técnicas utilizadas para el cultivo de la vid y su influencia en la respuesta de la planta han sido, y son, objeto de numerosos estudios. Sin duda, las operaciones culturales (por ejemplo el sistema de conducción, el marco de plantación, la utilización y manejo del riego, la poda en verde, etc.) afectan, en distinta medida, al rendimiento de la cosecha y a la composición de la uva.

1.2 Variabilidad vitícola

Columela titula el capítulo primero del Libro III de De Re Rustica (15):"Que especie de vid convenga a cada terreno y a cada clima", donde habla de la versatilidad de la vid para adaptarse a distintos climas y suelos, y se dan indicaciones para elegir la variedad de vid adecuada según las características de estos dos factores. Las diferencias entre los factores naturales darán lugar a una variabilidad en las características del cultivo. Algunos de estos factores son el clima y el suelo (16, 17), la altitud, la pendiente y su orientación (18) y/o la geología (8, 16, 19). En un estudio llevado a cabo en el municipio de Oyón, dentro de la D.O.Ca. Rioja, (20) encuentran como elementos del medio responsables de la variabilidad interparcela la altitud, la geología, el suelo, la pendiente y la orientación.

Por otro lado, también es conocida la variabilidad existente en la maduración de las uvas dentro de la misma parcela (variabilidad intraparcela), así Cadet-de-Vaux en 1803 (21) señala: "...nunca se hará buen vino sin dos vendimias, porque nunca puede madurar igualmente el fruto". Más recientemente, (22) señalan que actualmente con ayuda de la Viticultura de Precisión (PV) se están realizando vendimias bien separando la uva vendimiada el mismo día en diferentes lotes, o bien vendimiando en días distintos cada zona de la parcela con el objetivo de obtener el mayor grado de calidad en cada bloque. (23) apunta que este tipo de vendimia es viable a distintas escalas de producción, desde pequeños hasta grandes elaboradores. Algunos de los factores del medio a los que puede deberse la variabilidad intraparcela son: (i) la topografía (9, 24); (ii) la profundidad del suelo (25); (iii) la capacidad de retención de agua del suelo (22, 26) y/o (iv) el contenido en nitrógeno del suelo (27, 28). Además, esta variabilidad tiene efectos tanto en el rendimiento (29) como en la calidad de la uva (30), siguiendo (en estos dos últimos ensayos) un patrón parecido año tras año.

En resumen, los factores naturales que definen el *terroir* son los principales responsables de la variabilidad observada entre plantas de una misma parcela (variabilidad intraparcela) o de distintas fincas (variabilidad interparcelas).

En el presente trabajo se propone una metodología de zonificación del *terroir*, así como su relación con algunas características vegeto-productivas del viñedo y con el Índice de Vegetación de Diferencia Normalizada (NDVI).

2 Material y Métodos

El ensayo se lleva a cabo en 4 parcelas (A, B, C, D) de viñedo, amparadas por la D.O.Ca. Rioja y certificadas en agricultura ecológica por el Consejo de Agricultura y Alimentación Ecológica de Euskadi, situadas en el municipio de Oyón.

Sobre estas parcelas se hace una zonificación bajo criterios vitícolas (variedad, sistema de conducción y edad del viñedo) y en las subparcelas resultantes se realiza una fotointerpretación aérea discriminando sectores (A1, A2, A3 para la parcela A y de forma análoga para el resto de parcelas). En cada sector se marcan 12 cepas, repartidas en 2 repeticiones de 6 plantas, sobre las que se realizarán análisis y medidas vegetativas (longitud del sarmiento, diámetro del sarmiento y peso de la madera de poda), productivas (número de racimos y peso de los mismos) y fisico-químicas de la uva vendimiada.

Según el Sistema Multicriterio de Clasificación Climática (MCCS) (5) el clima de la zona donde se enmarcan las parcelas es templado cálido (HI+1), de noches frescas (CI+1) y moderadamente seco (DI+1), situándose la mayoría de los viñedos de la zona sobre areniscas y lutitas de la Facies de Haro (Mioceno Medio-Superior). Algunos de los suelos encontrados sobre esta geología y su sistema cuaternario asociado son alfisoles (por ejemplo del subgrupo Haploxeralf Cálcico), entisoles (por ejemplo del subgrupo Xerofluvent Típico) o inceptisoles (por ejemplo del subgrupo Xerocrept Típico). Para más detalle de la zona de estudio ver (20).

Para el cálculo del índice de vegetación de diferencia normalizada (NDVI) se parte de una imagen multiespectral (incluyendo rojo e infrarrojo cercano) capturada por el satélite Pléiades el 19 de agosto de 2015, de 0,5 metros de resolución espacial. El cálculo y la representación gráfica del índice se lleva a cabo con el *software* ArcGIS 10.1 del Instituto de Investigación de Sistemas Medioambientales (ESRI). La definición de las clases (Muy baja, Baja, Media, Alta y Muy Alta) se realiza conforme a cinco cuantiles, correspondiéndose el primer cuantil con la clase Muy baja y el quinto cuantil con la clase Muy Alta. Con el objeto de conocer y poder operar con el valor del NDVI, se calculará el valor medio del índice dentro de cada sector.

El análisis estadístico de los datos se efectuó mediante Análisis de Componentes Principales (ACP), Análisis Factorial Discriminante (AFD) y ANOVA univariante, previa comprobación de normalidad y homogeneidad de varianzas de las variables. La significación de estos análisis se determinó para los niveles de probabilidad $p<0.05$ (*), $p<0.01$ (**) y $p<0.001$ (***). Las medias se comparan mediante el test de Duncan cuando hubo diferencias significativas en el análisis de varianza. Para los análisis ANOVA se utilizó el programa SPSS, versión 15.0 (SPSS Inc. Chicago, Illinois) y para el resto de cálculos estadísticos el complemento XLSTAT 2009.1.02 sobre Microsoft Excel 2007.

3 Resultados y Discusión

El clima en la zona de ensayo durante el año 2015, según la clasificación MCCS, fue templado cálido (HI+1), de noches muy frescas (CI+2) y sub-húmedo (DI-1)

En las Tablas 1 y 2 se presentan los resultados obtenidos en cada sector según aspectos vegeto-productivos y fisico-químicos, respectivamente.

Todos los parámetros relacionados con el crecimiento vegetativo y la producción (Tabla 1) han mostrado diferencias muy significativas. La parcela C (sectores C1 y C2) y el sector D2 aparecen como las zonas más productivas y de mayor peso de madera de poda, respecto a este último parámetro también el sector D3 es significativamente alto. Respecto a la variabilidad intraparcela, se observa que los sectores A2 y D2 (en mayor medida) presentan una tendencia más productiva y más vigorosa. Estos sectores se corresponden con zonas bajas de deposición de material, generalmente asociadas a suelos de carácter fluvéntico como los del grupo Xerofluvent.

Los parámetros relacionados con las características de la uva que presentan diferencias significativas (con distinto nivel de significación) son (Tabla 2): grado de alcohol probable, pH, Índice de Color (I.C), Índice de Polifenoles Totales (I.P.T.), Antocianos, Peso de 100 bayas y Potasio. Destacan de nuevo los sectores C1, C2 y D2 por ser los de menor contenido fenólico (I.C., I.P.T. y antocianos), así como por un menor pH, en particular el sector D2. Resultados similares han sido encontrados por

(31) dentro del ámbito de la D.O.Ca.Rioja, en el mismo municipio de Oyón.

Por otro lado, la representación gráfica del NDVI (Figura 1) sigue una distribución que se ajusta a la división realizada con ayuda de la fotointerpretación aérea. Por lo tanto se puede afirmar que tanto a una escala entre parcelas (20) como a nivel intraparcela, los elementos del medio (altitud, geología, geomorfología,...) tienen influencia sobre el NDVI. Además, algunas de las características vegeto-productivas del viñedo (datos no mostrados), como el peso de madera de poda, se correlacionan en gran medida con el NDVI. Fijándonos en la parcela D de la Figura 1, se puede apreciar como el sector D2 está asociado a valores muy altos del NDVI y, como se ha visto más arriba, este sector se caracteriza por ser vigoroso, productivo, de bajo contenido fenólico y bajo pH. Esta influencia de los elementos del medio sobre el NDVI recomendaría hacer un uso de este índice por sectores previamente discriminados conforme a una cierta uniformidad del *terroir*.

Para entender mejor la fuente de la variabilidad en el comportamiento de la vid se hace necesario realizar un estudio de suelos con el detalle adecuado en función del tamaño de la parcela. La caracterización de los suelos de cada sector, así como otras características relacionadas con el *terroir* se pondrán de manifiesto en la tesis doctoral titulada "Caracterización e influencia del *terroir* sobre la composición de la uva y los componentes vegeto-productivos de la vid (*Vitis vinifera* L.) en la D.O. Ca. Rioja", que próximamente verá la luz y donde también se podrá contrastar la presencia de estos suelos en el ámbito de la D.O.Ca. Rioja.

4 Conclusiones

En el presente trabajo se ha podido comprobar que la variabilidad intraparcela, presente en los viñedos del ensayo, sigue un patrón definido por los elementos del medio a través de la fotointerpretación aérea. Los sectores correspondientes a zonas de deposición de material se asocian con viñedos vigorosos, productivos, con bajo contenido fenólico y bajo pH.

A su vez, el NDVI sigue una distribución, dentro de cada una de las parcelas, conforme a la zonificación por fotointerpretación realizada, encontrando relación entre las zonas con valores muy altos del NDVI con los sectores más vigorosos y productivos. Al verse influenciado el valor del NDVI por los elementos del medio, sería recomendable hacer uso del citado índice sobre sectores de las parcelas cuyos factores del medio sean homogéneos.

Tabla 1. Valores de comparación de medias y sus agrupaciones mediante el test de Duncan, así como su significación (sg) para un nivel de probabilidad $p<0{,}001$ (***) para: Racimos (Nº de Racimos), Producción (Racimos*PRacimo), PRacimo (Peso del racimo, gr), Longitud (Longitud del sarmiento, cm), Calibre (Calibre del sarmiento, cm), Pesomad (Peso de madera de poda, gr) y Ravaz (Índice de Ravaz: Pesomad/Producción)

SECTOR	RACIMOS	PRODUCCIÓN	PRACIMO	LONGITUD	CALIBRE	PESOMAD	RAVAZ
A1	4a	330,5a	86,25abc	75,5ab	0,862a	228,7a	1,64ab
A2	6,42ab	481,17a	78,63abc	101,25c	1,048cd	276,67ab	1,89ab
A3	6a	552,00a	98,10bc	77,5ab	0,916ab	297,5ab	1,83ab
B1	5,78a	356,78a	68,17ab	82,33ab	0,867a	348,44ab	0,91a
B2	5,6a	316,2a	56,15a	84ab	0,851a	442,8bc	0,76a
C1	9,75bc	1991,5d	198,25d	122,08d	1,041cd	570,33c	3,5c
C2	10,25c	1901,33cd	186,23d	107,5c	0,993bcd	534,67c	3,6c
D1	11,92cd	1300,75b	105,08c	71,25a	1,087d	362,67ab	3,57c
D2	18,5e	1708,25bcd	89,11bc	107,08c	0,967abc	845,5d	2,49bc
D3	14,67d	1339,33bc	91,27bc	88,33b	0,867a	540,17c	2,7bc
sg	***	***	***	***	***	***	***

Tabla 2. Valores de comparación de medias y sus agrupaciones mediante el test de Duncan, así como su significación (sg) para un nivel de probabilidad p<0,05 (*), p<0,01 (**) y p<0,001 (***) para: Grado (Grado de Alcohol Probable, % volúmen), A.Total (Acidez Total, gr/l de Ácido Tartárico), Málico (Ácido Málico, gr/l), I.C. (Índice de Color), I.P.T. (Índice de Polifenoles Totales), Antoc (Antocianos, mg/l), P100Bayas (peso de 100 bayas, gr), Potasio (Potasio, mg/l) y Tartárico (Ácido Tartárico, gr/l)

SECTOR	GRADO	A.TOTAL	pH	MALICO	I.C.	I.P.T.	ANTOC	P100BAYAS	POTASIO	TARTÁRICO
A1	15,84d	3,25a	3,93abcd	1,95a	20,0855cd	67,85ef	846,5c	144,5a	2140ab	4,9a
A2	15,63cd	3,7abc	3,8abc	1,35a	23,3105d	59,9de	835cd	161ab	1785a	4,3a
A3	16,04d	3,6ab	4,11d	2,4a	22,8640d	72,6f	1032,5e	130a	3240c	4,7a
B1	15,09bcd	4,25c	3,99cd	1,85a	19,2365bc	58,2cd	835,5cd	148a	2075ab	4,6a
B2	15,67cd	3,85abc	3,97bcd	2,25a	19,2740bc	59,9de	847,5d	150,5a	2615b	4,8a
C1	14,26abc	4,25c	3,9abc	2,75a	14,4460a	47,4ab	689abc	196b	2630b	4,8a
C2	14,52abcd	3,85abc	3,79ab	2,15a	14,8075a	43,25ab	620a	193b	1945a	4,4a
D1	13,52a	3,85abc	3,84abc	1,55a	16,2510ab	52,5bcd	777bcd	140a	2260ab	4,5a
D2	13,74ab	3,6ab	3,73a	2,35a	16,0960ab	40,9a	636ab	102ab	1990a	4,6a
D3	14,56abcd	3,9bc	3,8abc	1,75a	18,4295bc	49,4abc	747,5abcd	142a	2015a	4,8a
sig.	*	ns	*	ns	**	***	**	*	**	ns

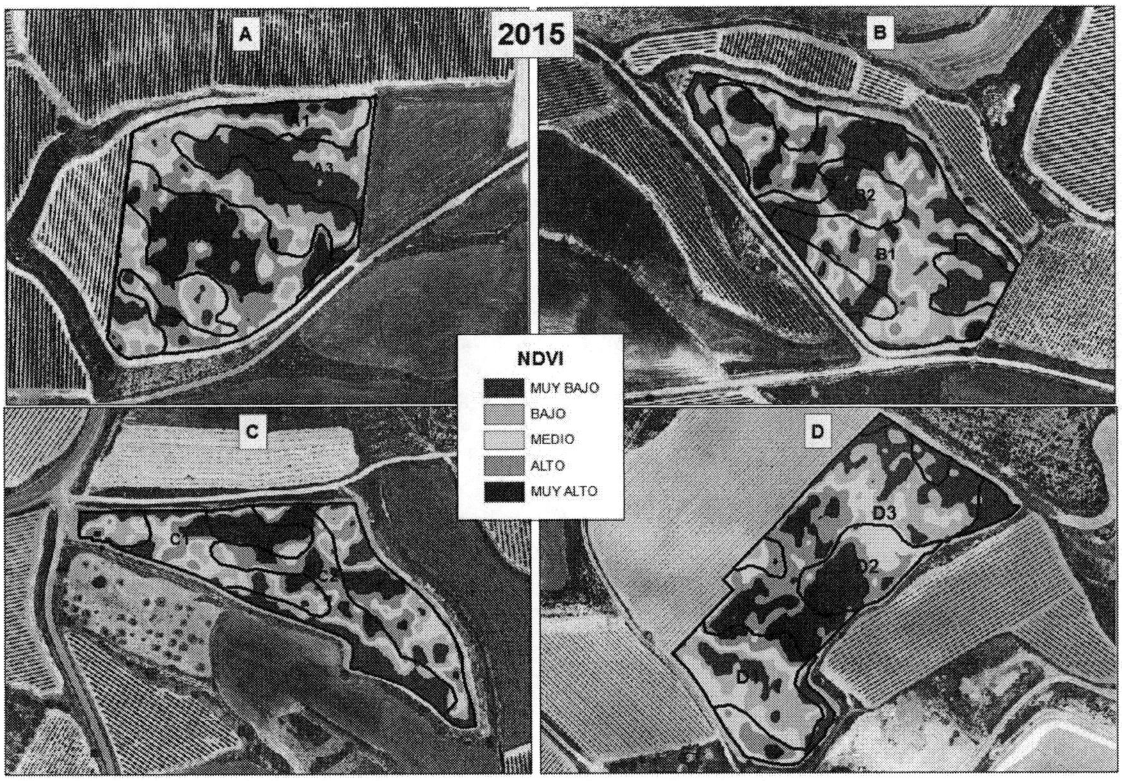

Figura 1. Representación gráfica del NDVI de las parcelas de estudio, calculado pixel a pixel. La clasificación se lleva a cabo conforme a 5 cuantiles y en cada parcela de forma independiente

Referencias

1. Morlat, R. in *Le Terroir viticole: contribution a l'etude de sa caracterisation et de son influence sur les vins, application aux vignobles rouges de moyenne Vallée de la Loire* (ed Université de Bordeaux II. Institut d'Oenologie), (fr), 1989).
2. OIV. Definición de *terroir* vitivinícola. (2010).
3. Pan-Montojo, J. in *Mediterráneo e Historia Económica* (eds Nadal, J. & Parejo, A.) (Caja Rural Intermediterránea. Cajamar, Almería, 2005).

4. Jones, G. V. in *Fine wine and terroir: the geoscience perspective* (eds Macqueen, R. W., Meinert, L. D. & Geological Association of Canada) (Geological Association of Canada, St. John's, Canada, 2006).
5. Tonietto, J. & Carbonneau, A. A multicriteria climatic classification system for grape-growing regions worldwide. *Agric. For. Meteorol.* **124**, 81-97 (2004).
6. Gómez-Miguel, V. in *Atlas das castas da Península Ibérica: história, terroir, ampelografia* (ed Böhm, J.) 104-153 (Dinalivro, Lisboa, Portugal, 2011).
7. Hancock, J. M. in *Encyclopedia of Geology* (ed Plimer, R. C. S. R. M. C. R.) 85-89 (Elsevier, Oxford, 2005).
8. Huggett, J. Geology and wine: a review. *Proc. Geol. Assoc.* **117**, 239-247 (2006).
9. Bramley, R. G. V. in *Managing wine quality* (ed Reynolds, A. G. (. G.) 445-479 (Woodhead; CRC Press, Oxford: Boca Raton: (uk), 2010).
10. van Leeuwen, C. in *Managing wine quality. V1: Viticulture and wine quality* (ed Reynolds, A. G.) 273-315 (Woodhead, Oxford, 2010).
11. Vaudour, E. in *Los terroirs vitícolas: definiciones, caracterización y protección* (Acribia, D.L., Zaragoza, 2010).
12. Jackson, D. & Lombard, P. Environmental and Management-Practices Affecting Grape Composition and Wine Quality - a Review. *Am. J. Enol. Vitic.* **44**, 409-430 (1993).
13. van Leeuwen, C. & Seguin, G. The concept of terroir in viticulture. *Journal of Wine Research* **17**, 1-10 (2006).
14. Gladstones, J. in *Wine, Terroir and Climate Change [Recurso electrónico].]* (Wakefield Press, Adelaide : (at), 2011).
15. Columela, L. J. M. in *Los doce libros de agricultura.* (ed Álvarez de Sotomayor y Rubio,Juan María) (Extramuros, Mairena de Aljarafe (Sevilla) : (sp), 2008).
16. Seguin, G. Terroirs and Pedology of Wine Growing. *Experientia* **42**, 861-873 (1986).
17. van Leeuwen, C. *et al.* Influence of climate, soil, and cultivar on terroir. *Am. J. Enol. Vitic.* **55**, 207-217 (2004).
18. Santesteban, L. G., Guillaume, S., Royo, J. B. & Tisseyre, B. Are precision agriculture tools and methods relevant at the whole-vineyard scale? *Precis. Agric.* **14**, 2-17 (2013).
19. Witze, A. The grapes of rock. *Nature* **438**, 1073-1074 (2005).
20. Martínez, A., Gomez-Miguel,V. Vegetation index cartography as a methodology complement to the terroir zoning for its use in precision viticulture. *OENO One* **51** (2017).
21. Cadet de Vaux, A. in *Arte de hacer el vino* (Maxtor, Valladolid, 1803).
22. Tisseyre, B., Ojeda, H. & Taylor, J. New technologies and methodologies for site-specific viticulture. *J. Int. Sci. Vigne Vin.* **41**, 63-76 (2007).
23. Bramley, R. G. V., Ouzman, J. & Boss, P. K. Variation in vine vigour, grape yield and vineyard soils and topography as indicators of variation in the chemical composition of grapes, wine and wine sensory attributes. *Aust. J. Grape Wine Res.* **17**, 217-229 (2011).
24. Bramley, R. G. V. & Hamilton, R. P. Terroir and precision viticulture: Are they compatible? *Journal International Des Sciences De La Vigne Et Du Vin* **41**, 1-8 (2007).
25. Lamb, D., Weedon, M. & Bramley, R. Using remote sensing to predict grape phenolics and colour at harvest in a Cabernet Sauvignon vineyard: Timing observations against vine phenology and optimising image resolution. *Australian Journal of Grape and Wine Research* **10**, 46-54 (2004).
26. Van Leeuwen, C. *et al.* Vine Water Status is a Key Factor in Grape Ripening and Vintage Quality for Red Bordeaux Wine. how can it be Assessed for Vineyard Management Purposes? *J. Int. Sci. Vigne Vin.* **43**, 121-134 (2009).
27. Chone, X., van Leeuwen, C., Chery, P. & Ribereau-Gayon, P. Terroir inluence on water status and nitrogen status of non-irrigated Cabernet Sauvignon (Vitis vinifera)Vegetative development, must and wine composition (example of a Medoc top Estate vineyard, Sain Julien Area, Bordeaux, 1997) . *South african journal for enology and viticulture* **22**, 8-15 (2001).
28. Chone, X. *et al.* Effect of vine nitrogen status on grape aromatic potential: Flavor precursors (S-cysteine conjugates), glutathione and phenolic content in Vitis vinifera L. cv. Sauvignon blanc grape juice. *J. Int. Sci. Vigne Vin.* **40**, 1-6 (2006).
29. Bramley, R. & Hamilton, R. Understanding variability in winegrape production systems 1. Within vineyard variation in yield over several vintages. *Australian Journal of Grape and Wine Research* **10**, 32-45 (2004).
30. Bramley, R. Understanding variability in winegrape production systems - 2. Within vineyard variation in quality over several vintages. *Aust. J. Grape Wine Res.* **11**, 33-42 (2005).
31. Unamunzaga, O. Zonificación de un viñedo de Rioja Alavesa según las propiedades del suelo y su influencia sobre producción y calidad. (2010).

Efficiency of wine grape growers at farm level: a case study

Micael Santos[1,a], Ana Marta-Costa[1], Cátia Santos[1] and Aníbal Galindro[1]

[1]University of Trás-os-Montes e Alto Douro (UTAD, www.utad.pt) and Centre for Transdisciplinary Development Studies (CETRAD), Portugal

[a]Corresponding author: micaels@utad.pt

Abstract. To determine the key variables of the vineyard efficiency is imperative to account the combined effects of the inputs interactions since they have implications on the overall final production. This paper estimates the productive efficiency of a wine-farm sample from the Douro Demarcated Region (DDR) while identifies economic, social and environmental indicators that characterizes the DDR grape production system. The data was collected by face-to-face surveys performed at farm level to build a pilot study. The majority of the sampled twenty farms are dedicated to mountain viticulture and mainly feature the cordon (simple and double) training system. Through the Data Envelopment Analysis (DEA) method, the productive efficiency of the sampled was performed and the results clinched different efficiency scores. The main explanation is related to the heterogeneity of the adopted production system. In addition, they revealed how grape producers could improve their productive efficiency by adopting particular practices and identifying the key factors of their system.

1 Introduction

Due to the relatively abundant vineyard area and the melancholic wine history, Portugal assumes itself as an international reference in the wine sector. However, that accomplishment is hauntingly relative since the country seems to be struggling to gain solid relevance in the highly competitive wine market. According to the OIV [1] worldwide data, Portugal is the 11th largest vineyard area (190 thousands of hectares in 2016), 11th wine producer and concerning the exports (both volume and value) Portugal caps the ninth place [1]. Nonetheless, this country watched those rankings deprecate during the period 2012-2016 since 18% of the total vineyard vanished and that fact caused an impact on the wine production and volume exports (15% and 18% respectively) [1]. On the other hand, the Portuguese wine exports increased 4% in value on the aforementioned period [1], suggesting that despite of the smaller volume of wine exported, the value of each bottle of Portuguese wine increased. However, that positive value evolution is still way far from the world's overall export value growth, which is roughly 14% [1].

Douro Demarcated Region (DDR) is the most important Portuguese wine region capping 22% and 21% of the total Portuguese wine production and vineyard area respectively [2,3]. This region is characterized by a steep mountain viticulture known by the coexistence of heterogeneous production systems. Their farms are assembled in several blocks where each block usually features a different variety of grapes and distinctive management endeavours. The steep slopes difficult the mechanization of the production process and obligates the producer to rely on huge amounts of seasonal workers, which are currently scarce due to the severe depopulation and aging of the region. Those challenges substantially increase the costs of producing grapes in DDR resulting in foregoing lower productivity [4]. One of the main costs of producing wine comes from viticulture, therefore, improving the efficiency at the grape-growing level can have an important impact on the commercial success of the Portuguese wine sector.
This paper also aims the recognition of economic, social and environmental indicators, which are able to characterize the grape production system of DDR farms and estimate their productive efficiency afterwards.

The article is divided in five sections. Besides the overview of the main objective, the following section introduces the notion of productive efficiency, with a brief literature revision. The used methodology is described in section three and the gathered results in section four. Finally, section five summarizes the main conclusions and suggestions for further research.

2 Productive Efficiency

Pioneering studies of productive efficiency literature starts with the works of Koopmans [5] and Debreu [6]. The first introduced the concept of technical efficiency (TE) and Debreu [6] developed the first measure of technical efficiency designed as "coefficient of resource utilization". Afterwards, Farrel [7] announced the frontier model to measure the economic efficiency (EE) broken down in TE and allocative efficiency (AE). The first (TE) states that when a certain level of inputs is given and the Decision Making Unit (DMU) is able to produce the maximum level of outputs or, fixing a certain level of output, the DMU is able to minimize the level of input [5,8,9]. The AE reflects the ability of a firm to use the inputs in their optimal proportions (given their respective prices) to minimize the cost or maximize the revenue [8,10]. There is also another type, which is the scale efficiency that tells us if a DMU is operating on an optimal scale [11].

Several studies have already estimated the productive efficiency in wine sector. Marta-Costa et al. [12], Freitas [13], Coelli and Sanders [14], Moreira et al. [15] and Henriques et al. [16] put their efforts into studies that aim the efficiency calculation of the grape producers, while Sellers-Rubio et al. [17], Sellers-Rubio & Más-Ruiz [18] and Aparicio et al. [10] studied the efficiency at the wineries level. Urso et al. [19] acquaint the efficiency of grape and wine producers in the same paper. Tóth & Gál [20] research the efficiency of wine producers at a country level and Vidal et al. [21] study the efficiency of the different denomination of origin (DO) existing in Spain.

The most popular variables among several studies are the grapes (quantity) and the general wine production (quantity or value). Tóth & Gal [20], Coelli & Sanders [14], Moreira [15] and Freitas [13] studied the quantity of output produced, while Marta-Costa et al. [12], Urso et al. [19], Sellers-Rubio et al. [17], Sellers-Rubio & Más-Ruiz [18], Aparicio et al. [10], Vidal et al. [21] and Henriques et al. [16] studied the value of output that was produced. Regarding the input variables, there are three variables mostly used in the precedent articles: labour, capital and land. In addition, some studies considered other inputs such as the intermediate consumption, total specific costs and irrigation [12-14] even though irrigation is only important for dryland agriculture [14].

The variables used to determine the causes of inefficiency in the wine sector are not consensual. Some are related to inner characteristics, such as the size of the farm and the farmer's gender, age education [16,19,22]. Meanwhile other efforts pointed out other features such as the diversification/specialization of products [14, 16], irrigation [14,16,22], credit access, quality of human capital and per capita wine consumption [23].

3 Methodology

This work uses the Data Envelopment Analysis (DEA) method to estimate the productive efficiency in twenty vineyard farms from DDR. In order to identify the best practices in grapes production, we assumed that the farmers were following a cost minimizing (input-orientation) strategy.

The input oriented model considers that a company will be efficient if given certain availability of outputs that company is able to minimize the available resources (inputs). Bearing in mind that the input and output oriented models estimate exactly the same frontiers, in the end they will be able to label the same companies as the efficient ones [18]. The problem of the farmer is the selection of a minimal but admissible input combination without jeopardizing the yield [24].

We assume that any farm j ($j = 1, 2, 3\ldots n$) produced a single output y_r (grape yield) using a combination of inputs x_{ij}. Following the constant returns to scale (CRS) input-oriented DEA model used by Cooper et al. [25], the efficiency of a farm relative to other farms is calculated using the following model (equation 1):

$$minimize\ \theta - \left(\sum_r s_r^+ + \sum_i s_i^-\right) \qquad (1)$$

Subject to following constraints,

$$\sum_{j=1}^{n} x_{ij}\lambda_j + s_i^- = \theta_0 x_{io}, \qquad i = 1,2,\ldots,m$$

$$\sum_{j=1}^{n} y_{rj}\lambda_j - s_r^+ = y_{ro}, \qquad r = 1,2,\ldots,s$$

$$\lambda_j, s_r^-, s_r^+ \geq 0;\ \forall_i, j, r$$

where x_{ij} and y_{rj} are inputs and outputs respectively, which were defined earlier; λ_j is the vector of weights of the efficient farms helping in projection of inefficient farms to an efficient frontier (i.e. the distance of

inefficient farms from the frontier); θ is an index of farm's technical efficiency.
However, a typical variable-return to-scale (VRS) is considered in reality farming activity because of the potential economies of scale [26]. Banker et al. [27] extended the earlier work of Charnes et al. [28] including an extra constraint $\sum_j \lambda_j = 1$ in equation (1) leading to a VRS frontier. The TE obtained in CRS DEA model is divided into two components obtained by VRS DEA model (equation 2).

$$TE_{CRS} = TE_{VRS} \times SE \qquad (2)$$

The second step studies the determinants of the inefficiency on farms alongside a brief econometric analysis with a Tobit model that can be expressed by equation (3) [19].

$$\theta_j = \alpha + X_j\vec{\beta} + \varepsilon_j \;\; j = 1, 2, \ldots, N \qquad (3)$$

where θ_j is the efficiency level of the *j-th* farm obtained by DEA, α is the estimated coefficient for the intercept, X_j is a matrix assembled by the explanatory variables of the respective vectors, β is a vector of the estimated coefficients, and ε_j is the stochastic error.

The data was gathered from a sample of 20 Portuguese grape producers by a structured face-to-face survey that was conducted with the owners or managers of the DDR holdings. This data is a preliminary sub-sample of the database that is being developed based on criteria such distinct spatial distribution in DDR and diversity of farms dimension and corresponding to the agricultural season of 2016-2017.
The used output variable was the grape production revenue in euros and the used inputs variables were the land (ha), labour (days), capital (euros) and intermediate consumption (euros). The determinants of the efficiency of farms used in tobit model were the vineyard area (ha), farmers' age, the grape production as the main source of income, the training systems adopted (cordon and guyot), and the vineyard landscaping (walled terraces/socalcos, patamares, vertical planting and plan).
The studied grape farms have an overall average size of 14,75 ha but accounting only the space dedicated singularly to grapevines that value drops to 9,42 ha. In addition, there is a substantial disparity among the size of farms, since the smallest farm has only 1,96 ha and largest one farm caps a total area of 24,43 ha of vineyard. As stated before the farms on DDR are usually divided in several blocks, that fact is confirmed by the sample mean value of 4,75 blocks within vineyard area.
Most of the grapes are produced using the cordon training system (75%), while only 25% selected the alternative guyot training system. According to the vineyard landscaping, the most are in levelled upland ("Patamares", 55%), followed by vertical planting (23%), plan (11%), walled terraces (5%) and mix situations (5%).

4 Results

The average values of the indicators from economic, social and environmental dimensions for the 20 studied farms are presented in Table 1, 2 and 3, respectively. The grapes production average is 4985 kg per ha but there are distinct productivities among farms, due to the production cycle of the vineyard and the productive system adopted. In addition, the corresponding revenue levels (without subsidies) are different due to the distinct prices payed to the high-quality grapes for Port and DO wines production. Summarily, the results show greater benefit/costs ratio for farms that adopt different training systems in their blocks (mixt), followed by the farms with only cordon system.

Table 1. Economic indicators of the studied grape farms (2017)

Training system	Cordon	Guyot	Mix	Total
Grape production (kg/Ha)	4873	4042	6057	4985
Revenue (€/Ha)	3251	2798	4015	3336
Capital (€/Ha)	51234	29360	47167	47139
Intermediate Consumption (€/Ha)	1078	1470	1426	1206
Real costs (€/Ha)	3301	3256	2937	3221
Benefit/cost Ratio	1,22	0,89	1,54	1,23
Grape production as main source of income (%)	46	33	75	50

Regarding the social domain (Table 2), the selected indicators bundled the farmers, education and age. The farmers have an average age of 50 years old and 55% of them have high education. In addition, the farmers with total dedication to vineyard, prospect of increasing or

https://doi.org/10.1051/e3sconf/20185001010

maintaining vineyard area and willingness to continue the viticulture were considered. Moreover, 95% of the farmers intend to continue the viticulture activity and there are only 65% that fully dedicate themselves to this sector (without parallel professional activities). The farmers majority (80%) intend to at least maintain or even increase the vineyard area of the farm in the upcoming years.

Table 2. Social indicators of the studied grape farms (2017)

Training system	Cordon	Guyot	Mix	Total
Labour (days/Ha)	70	74	81	73
Farmers with high school (%)	62	67	25	55
Farmers age	47	47	62	50
Farmers with total dedication to vineyard (%)	54	67	100	65
Prospect of increasing or maintaining vineyard area (%)	77	67	100	80
Continuity on viticulture (%)	100	100	75	95
Activity heirs (%)	46	67	75	55

The results from the studied farms within the environmental dimension (Table 3) show that the Patamares and vertical planting dominate the vineyard landscaping due to their slopes, as showed by the physiographic index. All farms with guyot system adopted the integrated farming and the crop protection assumes lower values per ha.

Table 3. Environmental indicators of the studied grape farms (2017)

Training system	Cordon	Guyot	Mix	Total
Fertilizations (€/Ha)	157	308	161	181
Crop protection (€/Ha)	422	402	510	437
Integrated farming (%)	85	100	75	85
Organic farming (%)	8	0	0	5
Soil analysis (%)	85	100	100	90
Favourable waste destination (%)	92	67	100	90
Physiographic index of the landscape (IQFP)	2,7	0,6	1,8	2,2
Walled Terraces (*Socalcos*) (%)	8	0	4	5
Patamares (%)	53	81	34	55
Vertical planting (%)	23	5	44	23
Traction (H/Ha)	30	34	38	32

Table 4 shows some descriptive statistics of the sample of the 20 studied grape farms and Table 5 shows the results of the productive efficiency estimation trough DEA using the STATA 14 software [29].

Table 4. Descriptive statistics of variables for the studied grape farms

Variables	Mean	Std. dev.	Min	Max
Grapes production (€)	32702,5	30293,1	5875	128819
Vineyard area (ha)	9,42	6,81	1,96	24,43
Labour (days)	586,39	375,93	213,25	1575
Intermediate costs (€)	17623,35	28470,25	1288	123467
Capital (€)	404144,5	344361,1	64795	1259041

The average of overall TE (TE_{CRS}) of grape production is 75,65%, and it ranged from 35% to 100%. The most efficiency farms are described as DMU 1, 6, 14, 17, 18 and 20. In the VRS analysis, as expected, there are higher levels of pure TE (TE_{VRS}), where ten farms have a score of 100%. The average of TE_{VRS} is 91,38% and it ranges from 42,39% to 100%. The presence of scale efficiency (SE) was proved by larger average value of TE_{VRS} than TE_{CRS} and the average of SE is 83,02%.
The results shows that the performance of the farms is uneven due to the high levels of inefficiency of some farms compared to the best practices of the farms with the highest levels of the efficiency. The low efficient levels of farms (12 and 19) is due to the pure TE and to the SE, respectively. Thirteen farms exhibit increasing returns to scale, six exhibit constant returns to scale and only one exhibit decreasing returns to scale.

Table 5. Efficiency scores of grape farms

DMU	TE_{CRS}	TE_{VRS}	SE	RTS
1	1,000000	1,000000	1,000000	-
2	0,790624	0,925318	0,854435	irs
3	0,735250	1,000000	0,735250	irs
4	0,763829	0,957323	0,797881	irs
5	0,536637	1,000000	0,536637	irs
6	1,000000	1,000000	1,000000	-
7	0,557135	0,895958	0,621832	irs
8	0,997661	1,000000	0,997661	irs
9	0,573671	0,753590	0,761251	irs
10	0,620784	0,978312	0,634545	irs
11	0,909096	0,909248	0,999833	irs
12	0,363374	0,423914	0,857187	irs
13	0,544799	1,000000	0,544799	irs
14	1,000000	1,000000	1,000000	-
15	0,692430	0,743937	0,930765	irs
16	0,694269	0,714644	0,971489	drs
17	1,000000	1,000000	1,000000	-
18	1,000000	1,000000	1,000000	-
19	0,350069	0,973369	0,359647	irs
20	1,000000	1,000000	1,000000	-
Mean	0,7564814	0,9137807	0,8301606	

Irs: increasing returns to scale; drs: decreasing returns to scale.

Table 6 shows the tobit model results for TE_{VRS}. The results reveals that size has a directly effect on the pure TE, therefore larger vineyard area originates more efficient farms. The age has a negative relationship with the efficiency performance of the farm, revealing that the farmer aging lowers the performance of the farm. In addition, farms that have grape production as the main source of income have lowers levels of pure TE. Regarding to the training systems, the results reveal that guyot and cordon have a negative influence on the efficiency, however only the coefficient of cordon variable is statistically significant at 5%. For the vineyards landscaping, all coefficients show lowers levels of TE_{VRS}, but only the vertical planting is statistically significant at 1%. It should be detached, that farms that used a mix of training systems and several vineyard landscaping had a positive effect on the farm efficiency.

Table 6. Tobit model results for TE_{VRS}

Variables	Coefficient
Constant	2,137537***
Vineyard area	0,0056197
Farmers' age	-0,0064585
Grape as main source of income	-0,167565
Guyot training system	-0,3896096
Cordon training system	-0,403977**
Walled terraces/Socalcos	-0,0602633
Patamares	-0,4470291
Vertical planting	-0,6721773*
Plan	-0,4935155

Statistically variables: *=10%; **=5%; ***=1%

5 Conclusion

The challenge of maintaining or improving agricultural productivity in DDR is enormous. Researchers and policy-makers are constantly looking for technologies that are economically and environmentally attractive to improve the efficiency of the farms. In this experimental work, the TE estimation elucidates that the majority of the sampled farmers are not efficiently using the available technology.

The mean of overall TE of the 20 farms from DDR is 75,65%, revealing that farms could reduce 24,45% of the used inputs in average. However, this efficiency is very unequal between farms, some with high levels while less than 40% have low efficiency. In average, these levels are mainly due to the scale efficiency, therefore farms are not operating at the efficient size.

About the determinants of the efficiency, some production practices proved to be more efficiency. For the 20 farms of the sample, the ones that use a mix of guyot and cordon training systems have better efficiency performances and cordon system seems to be the most inefficient. The same occurs in the vineyard landscaping, the farms that feature several systems appears to have better efficient performances and the vertical planting has a stronger negative relationship with efficiency. In addition, the farmers' age and their full dedication to the vineyard activity have negative influence on pure TE, while the size of vineyard area increase the efficiency scores.

To highlight that all results and conclusions of this work were obtained and studied for the twenty farms sample and cannot be extrapolated for DDR.

References

1. OIV (International Organisation of Vine and Wine), *OIV Statistical Report on World Vitiviniculture: World Vitiviniculture Situation*. Available at http://www,oiv,int/public/medias/5479/oiv-en-bilan-2017,pdf (accessed at 28/02/2017) (2017)
2. IVV (Instituto do Vinho e da Vinha), *Evolução da Área de Vinha*. Available at www,ivv,min-agricultura,pt/np4/35/ (accessed at 27/02/17) (2017)
3. IVV (Instituto do Vinho e da Vinha), *Evolução da Produção Nacional de Vinho por Região Vitivinícola*. Available at www,ivv,min-agricultura,pt/np4/35/ (accessed at 27/02/17) (2017)
4. J. Rebelo, *Rumo Estratégico para o setor dos vinhos do Porto e Douro*, 117-120 (2017)
5. T.C. Koopmans, *Analysis of Production and Allocation*, 33-97 (1951)
6. G. Debreu, *Econometrica* **22**, 14-22 (1951)
7. M.J. Farrell, *Journal of the Royal Statistical Society*, Series A (General) **120** (3), 253-290 (1957)
8. B.E. Bravo-Ureta, D. Solís, V.H.M. López, J.F. Maripani, A. Thiam; T. Rivas, Journal of productivity Analysis **27** (1), 57-72 (2007)
9. E. Fleming, S. Mounter, B. Grant, G. Griffith, R. Villano, *Wine Economics and Policy* **3** (2), 115-126 (2014)
10. J. Aparicio, F. Borras, J.T. Pastor, F. Vidal, *European Journal of Operational Research* **231** (2), 443-451 (2013)
11. B. Anang & A. Rezitis, 2016, Economics Bulletin, **36**(4), 2275-2290 (2016)
12. A. Marta-Costa, V. Martinho, M. Santos, Regional Science Inquiry **9** (2), 97-107 (2017)

13. R. Freitas, *Sobre a eficiência dos países produtores de uvas para vinho na União Europeia: uma aproximação DEA em duas etapas* (2014)
14. T. Coelli, O. Sanders, *Wine Economics, Quantitative Studies and Empirical Applications*, 231-249 (2013)
15. V.H. Moreira, J.L. Troncoso, B.E. Bravo-Ureta, Ciencia e Investigación Agraria **38** (3), 321-329, (2011)
16. P.D.S. Henriques, M.L.S. Carvalho, R.M.S. Fragoso, New Medit **8** (1), 4-9, (2009)
17. R. Sellers-Rubio, V. Alampi Sottini, S. Menghini, International Journal of Wine Business Research, **28** (1), 59-75 (2016)
18. R. Sellers-Rubio, F.J. Más-Ruiz, Review of Managerial Science **9** (1), 175-196 (2015)
19. A. Urso, G. Timpanaro, F. Caracciolo, L. Cembalo, Wine Economics and Policy (2017)
20. J. Tóth, P. Gál, Studies in Agricultural Economics **116** (2), 95-99 (2014)
21. F. Vidal, J.T. Pastor, F. Borras, D. Pastor, Spanish Journal of Agricultural Research **11** (2), 294-304 (2013)
22. B. Conradie, G. Cookson, C. Thirtle, South African journal of economics **74**(2), 334-343 (2006)
23. T. József, G. Péter, Studies in Agricultural Economics **116**, 95-99 (2014)
24. G. Murtaza, G.B. Thapa, Journal of Mountain Science **14** (4), 782-794 (2017)
25. W.W. Cooper, L.M. Seiford, K. *Tone, Introduction to data envelopment analysis and its uses: with DEA-solver software and references* (2006)
26. A. Ullah, S.R. Perret, Environmental management **54** (2), 288-300 (2014)
27. R.D. Banker, A. Charnes, W.W. Cooper, Management science **30** (9), 1078-1092 (1984)
28. A. Charnes, W.W. Cooper, E. Rhodes, European journal of operational research **2** (6), 429-444 (1978)
29. Y.B. Ji, C. Lee, Data envelopment analysis, The Stata Journal **10** (2), 267-280 (2010)

Acknowledgements

This work was supported by the R&D Project INNOVINE & WINE – Vineyard and Wine Innovation Platform - Operation NORTE -01-0145-FEDER-000038, co-funded by the European and Structural Investment Funds (FEDER) and by Norte 2020 (Programa Operacional Regional do Norte 2014/2020),

Soil typologies in the wine-growing areas of Mallorca, with special emphasis on available water content.

Joan ROSSELLO, José M. ESCALONA, Josep CIFRE, Jaume VADELL, Hipólito MEDRANO
Department of Biology. University of the Balearic Islands 07122 Palma Spain

ABSTRACT Under the conditions of the semi-arid Mediterranean climate, soil depth and water retention capacity are the most important characteristics of the soil related to the quality of the wines. The availability of water will mark the development of the vines and the development and maturity of the grape. The characterization and agronomic evaluation of the soil has been carried out from the excavation of 38 soil pits in the main wine growing areas of Mallorca. Cambisols are the most predominant soil typology followed by Cambisols, Regosols and Luvisols. The water classification of the soils has been made according to the maximum water availability (useful water) in 2 m of soil (mm water / 2 m soil). The proposed levels are: very low (<120 mm), low (120-180 mm), Medium (180-240 mm), High (240-300 mm) and Very high> 300 (mm). The most abundant soil type we find that 47% have high soil water availability, 27% very high, 14% normal and 14% low. Establishing an area based on available water content is a tool that allows us to adjust the selection of the most suitable grape varieties in each soil, thus taking advantage more efficiently of the potential of each type of grape.

1 INTRODUCTION

Grapevines are well adapted to semi-arid climate such as the Mediterranean, due to the large and deep root system and physiological drought avoidance mechanisms, such as an efficient stomatal control of transpiration and of xylem embolism, and/or the ability to adjust osmotically. However, the combined effect of drought, high air temperature and high evaporative demand during summer in these areas are known to limit grapevine yield, berry, and wine quality. The use of irrigation in these environments arises as a solution to prevent excessive canopy temperature, maintain quality in wine production and, in more extreme cases, guarantee plant survival. Nevertheless, irrigation is subject of social debate because the scarcity of water resources [1]. On the other hand in deep, rich soils, vines are vigorous and highly productive, but better wines are generally produced when the vines are cultivated on poor soils [2] . Growing certain grape varieties on un-irrigated fertile soils, promoting cultural practices that favor high yield, have been the cause of the poor quality of wines [3].

Traditionally, to overcome the limitations of poor soils, Mallorcan farmers developed different agronomic works carried out for the conservation of groundwater. Prior to planting, deep work is done to gain groundbreaking volume from the roots and facilitate penetration. With the crop already planted, surface tillage is done to eliminate competition with the weeds and maintain soil moisture [4].

Thus, when the availability of water in the soil is limited, if irrigation is not available, it is a matter of fine-tuning the water conservation tasks. On the other hand, when the objective is to limit the vigor, tillage can be reduced, favoring competition for water from spontaneous vegetation and direct evaporation from the soil [5].

Each viticultural variety has particular requirements, which makes it more suitable for specific conditions. Taking as reference two varieties of Mallorca: Prensal Blanc (white variety) and Callet (red variety), the first reaches its enological and productive potential on land with good water availability. On the other hand, in lands where water is a limiting factor, its production is reduced and quality is not improved significantly. Callet has very different behavior pattern; also, on land with good water availability is where it obtains the highest productions but it is well recognized that, in these conditions, the oenological quality is low. In thinner soils it reaches a reduced vigor, with more limited productions and a better oenological quality[4].

The soil acts on the functioning of the vine essentially through water supply and the nitrogen nutrition [6]. These effects are often in relation to the soil depth: in a deep soil, water and nitrogen supply are most of the time not limiting or not very limiting [7]. In order to gain a more synthetic approach than pedology, while continuing to use the concept of pedogenesis, Bodin and Morlat [8] put forward a terrain model based on the depth of soil and the degree of profile evolution [9].

2 MATERIALS AND METHODS

The characterization and agronomic evaluation of the soil has been carried out from the excavation of 38 soil pits in vineyard growing areas of Mallorca. Once made the opening of the soil pit, the description of the profile and the collection of soil samples from each area were taken.

The measurement of the water retained at 33 kPa was determined on soil samples, minimally altered, maintining the original structure of the soil and the coarse elements. The remaining determinations were made on air-dried soil and sieved at 2 mm. The determination of the water retained at 1.5 MPa (permanent wilting point) has been made on fine earth. The water retained by the coarse elements (gravel and stones) at 1.5 MPa was determined independently, correcting the results according to the proportion of fine earth and coarse elements. The plant available water capacity of the soil was calculated for each layer from the difference between the water retained at field

 https://doi.org/10.1051/e3sconf/20185001011

capacity and the permanent wilting point. The maximum available water of each of the soils was calculated from the maximum reserve of useful water of the upper layers that constitute the solum plus an estimate of the water available in the parent rock or original material up to a depth of 2 m. This depth give us a realistic idea of the reserves that the soil has, taken into account the depth of the roots as well as the rise by capillary action of the water. The Soil classification proposed according to its maximum useful water reserve agrees with the appreciations of the vine-growers in the area [3]. (Table 1).

Soil classification has been carried out according to the criteria of the World Reference Base of Soil Resources [10].

Table 1. Soil classification proposed according to its maximum useful water reserve

	Maximum useful water reserve (mm water in 2m depth)
Very low	< 120
Low	120-180
Medium	180-240
High	240-300
Very high	>300

3 RESULTS AND DISCUSSION

The most relevant differences between the soil types are found in the contents of total carbonates and active limestone with minimum values in the Luvisols and maximums in Regosols and Calcisols. The presence of carbonates affects the pH of the soil, being basic in all cases (values between 8.2 and 8.4). The soils texture oscillates between loam and clay soil with high clay values varying between 36% and 54%. Organic matter presents moderate values and similar levels in all soils (organic carbon contents between 1.1% and 1.4%). Parallel occurs with the contents of total nitrogen, reaching a balanced C/N ratio with values between 8.1 and 9.6. The coarse elements constitute a widespread component in the studied soils, affecting the volume of soil available for the roots. Regarding the capacity of water retention the values of field capacity are high; similarly the permanent wilting point is reflecting the high content of the fraction of clays (Table 2).

Most of the studied soils are Calcisol Haplic Calcisol (Aric, Clayic) (58%) followed by Calcisol Skeletic Calcisol (Aric, Clayic) 13% and Cambisol Calcaric Cambisol (Aric, Clayic) 8% (Table 3).

According to soil water availability 47% have high soil water availability, 27% very high, 14% normal and 14% low only 3% have very low. The Calcisols have variable water retention capacity, it depends on the thickness. In these soils, when the calcium carbonate is cemented and hard (petrocalcic horizon) it is an insurmountable barrier for roots limiting plant development. On the other hand, when the accumulation of carbonates has a soft consistency, the water reserve on this level may be important, greatly increasing the water reserve.

The water retention capacity of the soil is mainly due to the depth of the soils, the texture, the presence of coarse elements and degree of weathering of the parent rock. The parameters described above can give us an idea of the water content, but in order to have a realistic idea, a specific analysis must be made for this purpose. In certain cases, lands with high clay content are associated with high rock content and have limited depth which makes the availability of water moderate, while soils with the highest water retention capacity had a moderate content of clay but they are deep soils with few coarse elements.

These results agree with Van Leeuwen 2010 where he states that despite the closeness of pedology to agronomy, it is not possible however to transform a soil map into a map of potential wine quality. Sometimes, small differences between two soil profiles (differences in the amount of organic matter or in soil depth, for example) can modify the quality potential for wine production, without modifying the soil type in the pedological classification [8].

Table 2: Physical and chemical data of the arable layer of the representative soils from the Mallorca.

	Calcisols	Cambisol	Regosol	Luvisol
Coarse fragments (g/kg)	414	308	77.4	64
Sand (g/kg)	274	289	183.4	56
Silt (g/kg)	364	307	357.4	404
Clay (g/kg)	362	403	459.2	540
$CaCO_3$ (g/kg)	321	99.6	206.1	38
Fine $CaCO_3$ (g/kg)	57	14.5	76.5	8
Organic carbon (g/kg)	13.5	13.4	10.8	11.9
N (g/kg)	1.4	1.39	1.34	1.4
C/N	9.6	9.6	8.1	8.3
pH (H_2O; 1:2.5)	8.2	8.4	8.4	8.3
Bulk density (t/m^3)	1.62	1.62	1.07	1.31
FC (-33 kPa; mm/m))	224	256.7	287.9	352
WP (-1,5 MPa; mm/m)	128	143.9	162.5	207

FC: Field capacity; WP: Wilting Point.

Table 3: Distribution of Soil typologies in the wine-growing areas of Mallorca

		Soil water availability					
Unit	Soil types (IUSS Working Group WRB, 2014	Very low	Low	Medium	High	Very high	Total
Calcisol	Endopetric Calcisol (Aric, Chromic, Siltic)		1				1
Calcisol	Haplic Calcisol (Aric, Chromic, Siltic)			1			1
Calcisol	Haplic Calcisol (Aric, Clayic)		3	3	10	6	22
Calcisol	Skeletic Calcisol (Aric, Clayic)		5				5
Calcisol	Skeletic Calcisol (Aric, Loamic)		1				1
Luvisol	Calcic Rhodic Luvisol (Aric, Clayic)			1			1
Luvisol	Rhodic Luvisol(Aric, Clayic, Profondic)					1	1
Cambis ol	Calcaric Cambisol (Aric, Clayic)		2	1			3
Regosol	Calcaric Epileptic Regosol (Aric, Loamic)	1					1
Regosol	Calcaric Regosol (Aric, Loamic)					2	2
Total		1	12	6	10	9	38

https://doi.org/10.1051/e3sconf/20185001011

4 CONCLUSIONS

The vineyards in Mallorca are cultivated on different types of soil. Calcisols (78%) are the most predominant ones followed by Cambisols (8%), Regosols (8%) and Luvisols (6%). According to soil water availability 47% have high soil water availability, 27% very high, 14% normal and 14% low only 3% have very low. It is not possible to establish a direct relationship between these two classifications

The soil typology provides partial information about the agronomic aptitude. For example, the Calcisol, which is the dominant typology, covers practically the entire range of water availability.

The soil depth, the volume of fine earth, the texture and the type of parent material, are key parameters in the availability of water but a specific analysis must be done to have a real value.

To obtain quality grape production it is necessary to know in addition to the chemical fertility conditions the capacity of water retention and to apply the management of suitable work and irrigation. Even sometimes for some varieties, deep soils with high water retention can be conflicting to achieve optimal grape quality.

Having a classification of the soils according to the water capacity can be a very useful resource for choosing the right varieties for each soil and establish the cultivation tasks for the conservation of water and, also, the irrigation needs.

REFERENCES

[1] M. M. Chaves *et al.*, "Grapevine under deficit irrigation: hints from physiological and molecular data.," *Ann. Bot.*, vol. 105, no. 5, pp. 661–676, 2010.

[2] C. Van Leeuwen, "Soils and terroir expression in wines," in *Soil and Culture*, 2009, pp. 453–465.

[3] J. Rosselló, "Selección clonal e influencia de los factores ambientales y prácticas culturales en la producción y calidad de las variedades de vid Manto Negro, Callet y Prensal Blanc," Universitat de les Illes Balears, 2012.

[4] J. Vadell and J. Rosselló, "Tipologies de sòls i la seva influència sobre la producció i qualitat del raïm de vinificació a Felanitx," in *III Jornades d'Estudis Locals de Felanitx*, 2015.

[5] F. Martínez de Toda, *Claves de la viticultura de calidad: nuevas técnicas de estimación y control de la calidad de la uva en el viñedo*, 1st ed. S.A. Mundi-Prensa libros, 2008.

[6] O. Tregoat, C. Van Leeuwen, X. Choné, and J. P. Gaudillere, "Etude du régime hydrique et de la nutrition azotée de la vigne par des indicateurs physiologiques," *J. Int. des Sci. la Vigne du Vin*, vol. 36, no. 3, pp. 133–142, 2000.

[7] J. Coipel, B. Rodriguez Lovelle, C. Sipp, and C. Van Leeuwen, "«Terroir» effect, as a result of environmental stress, depends more on soil depth than on soil type (Vitis vinifera L. cv. Grenache noir, Côtes du Rhône, France, 2000)," *J. Int. des Sci. la Vigne du Vin*, vol. 40, no. 4, pp. 177–185, 2006.

[8] F. Bodin and R. Morlat, "Characterization of viticultural terroirs using a simple field model based on soil depth I. Validation of the water supply regime, phenology and vine vigour, in the Anjou vineyard (France)," *Plant Soil*, vol. 281, no. 1–2, pp. 37–54, 2006.

[9] C. Van Leeuwen, J. Roby, D. Pernet, and B. Bois, "Methodology of soil-based zoning for viticultural terroirs," *Bull. l'OIV*, vol. 83, no. 947-948–949, pp. 13–29, 2010.

[10] IUSS Working Group WRB. 2014, *World reference base for soil resources 2014 International soil classification system for naming soils and creating legends for soil maps.* World Soil Resources.Reports No. 106. Rome: FAO, 2014.

Fungi resistant grape variety diffusion: a multi-scale, multi-factor approach

Bruno Striffler[1], Julie Wohlfahrt[1] and Marie Thiollet-Scholtus[1]

[1]INRA - SAD - UR-0055-ASTER, 28, rue de Herrlisheim, 68000 Colmar, France

Abstract. Farm dependency to inputs is an issue concerning both farms and society. Viticulture is an input-intensive sector of agriculture. Some innovations in agriculture have been designed to reduce the use of inputs, such as fungi-resistant vine varieties. As the diffusion of innovation is uneven, we will compare the characteristics of innovations in both sides of the upper Rhine River. In this way we will interview farmers and analyze the geographical context. We will also compare the farms characteristics and their contexts at various scales. We expect to get a typology of innovations, with characteristics such as the impact on various scales (e.g. cropping system, farming system) and the radicalness of the changes. We also expect to get some information about the eventual levers and incentives to innovations' diffusion, adoption or abandonment. And finally, French and German innovations will be compared to others vineyards innovations. It is possible to sharply describe several innovations reducing vineyard inputs, located on both sides of Upper Rhine River.

1. Introduction

Viticulture is part of the French heritage, this crop represents only 3,7% of the French agricultural acreage but represent a turnover of more than 8 billions euros [1]. Viticulture consumes 20% of the national pesticides' use, mostly fungicides (about 80% of the pesticides) [2].

Fungi resistant vine varieties seem to be a promising alternative to drastically reduce pesticides uses in vine growing. At a time when some fungi resistant vine varieties are just registered at the French and European catalogue, it seems necessary to investigate why and how these varieties could be adopted by winegrowers. The choice of vine variety is one of the key practices in viticulture as it determines, in combination with biophysical properties of the vineyard, the type and quality of the vine. Thus the introduction of fungi resistant vine varieties (as known as PiWis) must be approached considering the whole system linking biophysical environment (soil, climate), winemakers' practices and economic environment (value chains, quality marks, certification, etc.). The choice of specific varieties must meet the requirements for certifications, such as Protected Designation of Origin and Protected Geographical Indications. In this way, growing a non-certified variety can be a hazard for the winemaker's business.

The aim of our work is to understand the incentives and locks for input-sparing innovations, especially the fungi resistant vine varieties' adoption by winemakers. We consider that the adoption process is dependent on factors at different scales, the cropping system (e.g.: agronomical factors), the farming system (e.g.: organizational factors, winegrower strategy) as well as factors at the landscape scale (e.g.: winemakers strategy, wine chain structure). To conduct the study, we will use a multi-scale survey framework, applied to the specific case of fungi resistant vine varieties.

This work is included in Interreg Vitifutur project, a European research project conducted at three scales dealing with the type of information we need to target the goals. (i) viticulture, winemaking and ageing practices will be characterized at cropping system's scale; (ii) business strategy will be described at farming system's scale; (iii) agro-pedo-climatic, the legislative set of constraints and the socio-technical environment (e.g. professional social network, commercialization network and infrastructures). In that research case, we will do a comparative study of the vineyards in both sides of the Rhine River: Alsace region on the French side and Baden-Württemberg on the German side. To conduct this study, we will use a classical method of qualitative and quantitative inquiry.

2. Material and methods

2.1. Data collection

2.1.1. Study areas

The survey will be conducted during March-April 2018 in the Upper Rhine region, in both France (Alsace region) and Germany (Bade-Württemberg land). On the French side, the study area is located on 17 municipalities between the cities of Colmar and Mulhouse. About 30 winegrowers will be interviewed.

On the German side, the study area is located in the Bade-Württemberg land, between the cities of Lahr/Schwarzwald and Bad-Krozingen. About 20 winegrowers will be interviewed.

 https://doi.org/10.1051/e3sconf/20185001012

In order to understand the incentives and the obstacles of the PiWis' and others innovations' diffusion at a regional scale, studied areas will be described in details.

2.1.2. Farmers' sampling

In France, the farmers have been chosen using the recommendation of stakeholders from the wine sector such as chamber of agriculture, Alsatian wine interprofessional council and local winegrowers' associations. They were asked for contacts of winegrowers, in the study region, using innovative technics and work organization to reduce the use of inputs. Among the list of the innovative farmers, we tried to choose a balanced number for each municipality.

In Germany, the farmers have been chosen using the recommendation of our project's partner of Landau University and the Baden-Württemberg winegrowers' association. They were asked for contacts of winegrowers, in the study region, using innovative practices and organizations to reduce the use of inputs, with a special focus on the PiWis.

Farmers have been chosen regardless of their certifications: they can be conventional or organic or biodynamic.

2.1.3. Survey's conduct

The whole questionnaire is conducted with the starting question: "What kind of innovation do you set up to reduce the inputs' use?". The survey will be conducted as a semi-structured interview using a questionnaire at field and farm scales. The questionnaire was divided on 12 themes such as viticulture practices, winemaking practices and business and management of the farm.

The focus on PiWis is more important in Germany than in France, as PiWis are already implanted in Germany contrary to France. Since innovation is an iterative concept, the input reduction can be both an innovation and its consequences, as shown in Figure 1. For example: reducing fertilizers can impact the cropping system's management, but the cropping system's management can also lead to a reduction of the fertilizers. We asked (i) if some change of the management has led to an input reduction and (ii) if the reduction of this input has led to a change of the management. These questions will ensure not to miss any side of an innovation. These questions will deal at any scale, i.e. cropping system, farming system. This approach can also help to understand the interactions between two innovations located in the same field or in the same farm, especially innovations including PiWis.

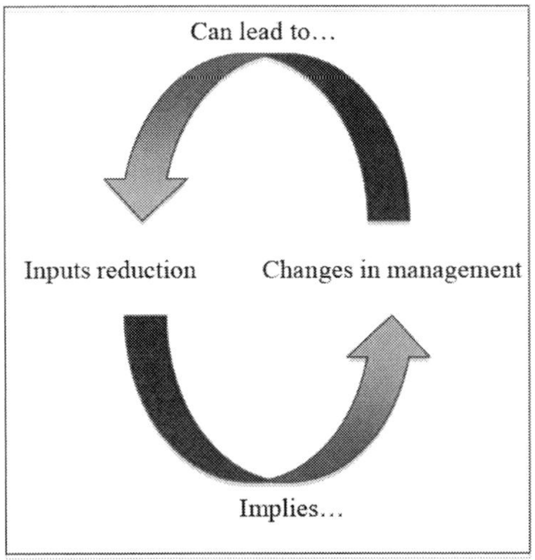

Figure 1. Iteration of the innovation process

2.2. Interviews analysis

2.2.1. Characterization of highlighted innovations

An analysis of each innovation will be conducted according to the theory of strategy analysis [3]. The characterization of highlighted innovation will be based on the questions in the surveys. The method of creating innovations typology is a constructivist agricultural method, i.e. types are defined by assumptions based on expert knowledge enriched by survey analysis [4]. For each innovation, its characteristics will be described and sorted according to their influence on the various dimension of the farm, i.e. organization or management.

The influence of an innovation is not necessarily the same in every farm. For each dimension of an innovation, its frequency of occurrence in each farm will be noted.

2.2.2. Typology of highlighted innovations

The output of the survey is a set of data, both quantitative and qualitative. To analyze all together, we did a Factor Analysis of Mixed Data (FAMD). The FAMD is used to see the influence and the correlation within the variables of a set. Thus this approach allows the selection of the variables explaining the most the set's diversity. This method mixes the Principal Component Analysis (PCA) and the Multiple

 https://doi.org/10.1051/e3sconf/20185001012

Component Analysis (MCA) using *factominer* package with the R software [5]. PCA and MCA used respectively for qualitative and quantitative variables. The interest of this method is the possibility of using both qualitative and quantitative variables from the surveys without transforming the qualitative variables in quantitative variable. The aim of the innovation FAMD is to represent simultaneously the quantitative variables, as a PCA, and the qualitative variables, as a MCA, on the same plane. Both quantitative and qualitative variables are active variables.

2.2.3. Geographical contextualization of highlighted innovations

We will interview study areas experts, to understand the regional context, the socio-technical environment and the legislative set of constraint. And to understand the importance of biophysical factors, we will also cross the location of highlighted innovations with climate, topography and soil databases. Innovations location will be mapped on the GIS software QGIS according to these databases.

3. Results and discussion

3.1. *Characterization and typology of actual highlighted innovations*

We expect this multi-scale analysis approach to pool the highlighted innovations that have common characteristics. We are also expecting to have a sharp description of the different dimension of innovations across the studied territories. The characteristics of each innovation will be its influences on the different dimensions of the farm; its characteristics will be as well it degree of evolution from the business as usual: improvement of the efficiency, product substitution or systemic redesign (ESR) [6].

3.2. *Characterization and typology of abandoned highlighted innovations*

We expect to describe and characterize the innovations that have been abandoned by the farmers. We will describe and organize it the same way as the one that have been adopted. We will also study the reason of the abandonment and the duration of the test.

3.3. *Geographical contextualization of highlighted innovations*

Innovations characterized in Alsace and Germany study areas will be compared together and with innovations characterized in others vineyards, i.e. Bordeaux, Burgundy and Mediterranean areas [7]. We will compare the diffusion of innovations and its eventual lock-in and incentives, especially in the PiWis' case at various scales.

4. Conclusion

Several innovations to reduce inputs in vineyard exist on both side of the upper Rhine River. Some of innovations are not long lasting according to geographical, political or business context. Some innovations seem to be implemented in several vineyards. Are they still innovations?

This approach, applied in study areas, could then be used for various purposes. The interest for the winemaker is to determine the utility of changing the farming practices, in the studied about fungi resistant vine varieties (i.e. PIWI). Integrated in a model, the results can be used to contextualize the farmers' behavior, in order to determine the effect of policies, for example adaptations in the Protected Designation of Origin regulation.

5. Acknowledgments

The authors thank Interreg VITIFUTUR for funding the research project.

References

[1] Fédération des Exportateurs de Vins et Spiritueux de France, Dossier de Presse, *Exportation des Vins et Spiriteux en 2017*. (2018)

[2] J.-P. Butault, C.-A. Dedryver, C. Gary, L. Guichard, F. Jacquet, J.-M. Meynard, P. Nicot, M. Pitrat, R. Reau, B. Sauphanor, I. Savini, T. Volay. Rapport de recherche. Ministère de l'Ecologie, de l'Energie, du Développement Durable et de la Mer. *Synthèse du rapport d'étude Ecophyto R&D : quelles voies pour réduire l'usage des pesticides?* (2010)

[3] M. Crozier and E. Friedberg. Edition du seuil. *L'acteur et le système*. (1977)

[4] E. Landais. Le Cour. de l'Env. de l'INRA, Paris. *Agriculture durable : les fondements d'un nouveau contrat social ?* **33**. (1998)

[5] S. Lê, J. Josse and F. Husson. J. Stat. Sof.t *FactoMineR : An R Packagefor Multivariate Analysis*. **25**. (2008)

[6] S.B. Hill, R.J. Macrae J. Sust. Agr.. *Conceptual framework for the transition from conventional to sustainable agriculture*. **7(1)**, 81-87. (1995).

[7] A. Petit, C. Cresson, M. Chovelon, J.-M. Barbier, J.-M. Meynard and M. Thiollet-Scholtus. RO **161**. *Recherche et évaluation de pratiques innovantes chez des viticulteurs. Offrir des pistes pour des solutions alternatives*. 1-3. (2016)

Assessment of biodiversity and agronomic parameters in two Agroforestry vineyards

Joséphа Guenser[1], Emilie Bourgade[2], Marc Vergnes[3] Thierry Dufourcq[4], and Séverine Mary[1]

[1] Univ. Bordeaux, Vitinnov, Bordeaux Sciences Agro, ISVV, 1 cours du Général de Gaulle, 33170 Gradignan, France
[2] Arbre et Paysage 32, 93 route de Pessan, 32000 Auch, France
[3] Institut Français de Vigne et du vin, pôle Bordeaux-Aquitaine, 33290 Blanquefort, France
[4] Institut Français de Vigne et du vin, Pôle Sud-ouest, Château de Mons, 32100 Caussens, France.
Corresponding author : Joséphа Guenser – josepha.guenser@agro-bordeaux.fr

Abstract. Sustainability of agricultural production systems is nowadays considered as a major challenge to face. Viticulture is particularly affected by environmental issues, especially because of its consumption in pesticides. Besides, the social demand in environment-friendly products is increasing, and the reputation of wines produced under Protected Denomination of Origin (PDO) is also built on the specificity of natural characteristics and resources such as soil, which has to be preserved. Biodiversity loss is largely admitted among the scientific community, and landscape simplification is known as a major driver in this process. Agroforestry, which combines trees with crops, could be a seducing response to biodiversity loss in agro-ecosystems, but the possibility of negative interactions between trees and vines (competition for water, nutrients, light) has to be considered. The Vitiforest project aims to assess south west of France agroforestry vineyard plots by spatializing different parameters in the domain of agronomy, ecology, micro-climate and economy. Field measurements were undertaken at different distances from the intercropped lines of trees in two agroforestry vineyard plots, in order to test potential effects of the trees. Arthropods were collected periodically with pitfall traps (for ground-dwelling individuals) and with D-vac system (aspiration of vine leaves). Pest insects (Empoasca vitis) were collected by specific yellow sticky traps. Data collected to describe biodiversity in these agroforestry systems was total arthropods abundances, abundances per order, carabidae richness and abundances. Nitrogen status was assessed by using Greenseeker® device, through the NDVI index. Our observations show that intercropped trees have no direct effects on vine nutrition. Slight effects on pests insects repartition were found, but these effects are inconstant according to experimental site and year of observation. The same trend is observed for arthropods abundances in the plots.

1 Introduction

Wine sector has currently several issues to face, particularly for products under DOC which are claiming their typicity and strong link to terroir specificities. Indeed, some parameters are changing, such as climate, biodiversity, and social demand of consumers. Climate change is currently known as a major factor which will impact the wine sector in the future [1]. Biodiversity has declined dramatically during the five past decades. The loss of natural habitats for the benefit of agricultural or urban land, and the pollution are identified as the main drivers of this global loss [2]. This worrying trend is still maintained nowadays, and jeopardizes the capacity of ecosystems to provide services which are necessary to human wellbeing (food production, pollination, climate regulation, etc.). The agricultural sector is particularly sensitive to this issue since it both provides and depends on EcoSystem Services (ESS) [3]. Moreover, general public (including wine consumers) has been more and more informed about these issues for the past two decades, and is demanding food and drink produced in the respect of environment [4]. Consequently, the wine sector has to find how to conciliate the production of grapes either in quality and quantity, considering the environmental issues of biodiversity, use of inputs, and anticipating the climate change consequences. In this context, a discipline known as agroecology has emerged in the 80's, and has been translated in terms of political line in France with the "Loi d'avenir pour l'Agriculture" in 2014. Agro-ecology is defined as "a way of thinking production systems which take advantage of ESS" (French Ministry of Agriculture and Food). Among the diverse aims of agro-ecology, such as the preservation of natural resources or the limitation of the use of inputs, the diversification of production systems is a major challenge [5]. Agroforestry (AF), defined by ICRAF as a land use system involving trees combined with crops and/or animals on the same unit of land, seems to satisfy at least this last point (diversification of production systems and landscape). But risks linked to the introduction of

 https://doi.org/10.1051/e3sconf/20185001013

trees in agricultural plots have also to be considered. The LUCAS land use and land cover survey indicates that AF represents about 15,4 million hectares in Europe, the largest area being occupied by livestock AF with 15,1 million hectares [6]. Other kinds of agroforestry remain therefore poorly represented, and references or feedbacks in French vineyards are very scattered. The Vitiforest project aims to describe the effects of introducing trees in vine plots in a south-west viticulture context. It allowed to explore parameters such as agronomy, ecology, micro-climate and economy. The results obtained for biodiversity and agronomic measurements are presented here. They were obtained in two AF vineyards in the Bordeaux and Côtes de Gascogne areas in 2015 and 2016.

2 Material and method

2.1 Experimental sites

Two vineyards were tested in this study. Vineyard 1 is located in the Bordeaux area (Lapouyade, Gironde), on a 3.5 hectares plot, planted in 2002 with *Vitis vinifera* cv Merlot, 6600 vines/ha. Fruit trees have been planted in 2009 in 2 of the rows to replace missing vines. The distance between trees is 5 to 9 meters. The two tree lines are about 50 meters apart from each other. This farm operates in certified organic practices. Vineyard 2 is located in Gascony area (Lagardère, Gers), on a 2.2 hectares plot, planted in 2008 with *Vitis vinifera* cv Sauvignon Gris, 4000 vines/ha. Trees, *Sorbus domestica, Sorbus pyralis* and *Pyrus pyraster* have been planted in 2008 on 3 specific inter rows.

The biodiversity monitoring occurred in these two plots during the years 2015 and 2016. Figure 1 shows their configuration.

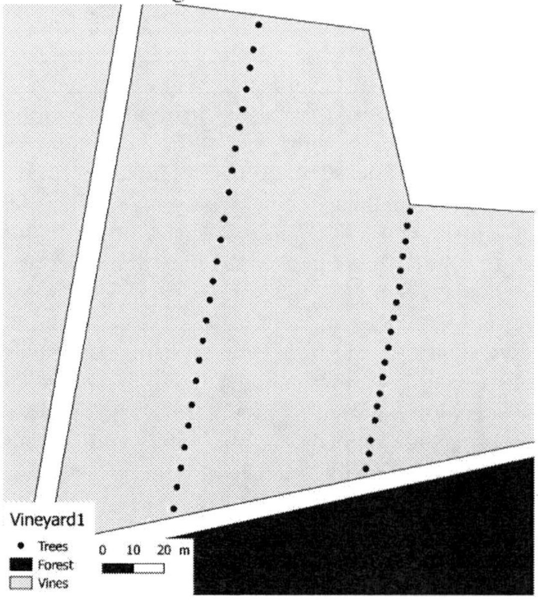

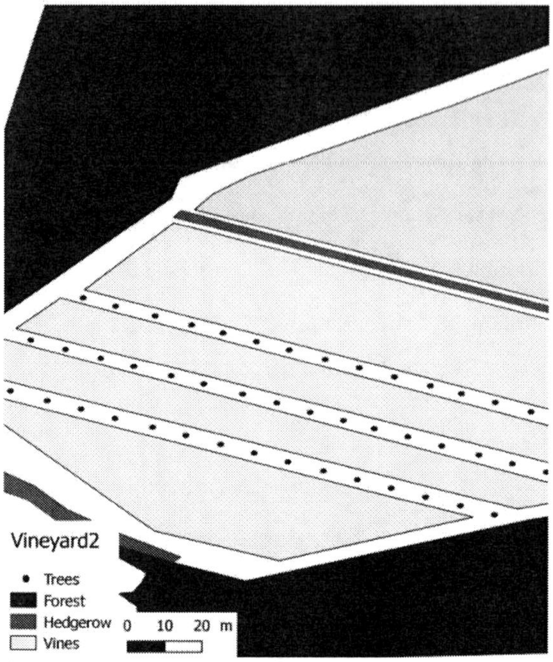

Figure 1 : Configuration of the two plots

These two vineyards were described by photo-interpretation in order to geo-localize AF trees (lines inside the vineyard plots) and surrounding trees (forests edges). Sampling points have been located and also geo-localized at different distances from either side of the trees lines, in order to test the effect of these distances on biodiversity parameters. It has previously been shown that trees seem to not influence NDVI index on these two plots [7]. Then, the potential effect of vigour on E. vitis repartition has been tested, as it has been demonstrated that E. vitis aggregates in areas with the most vigourous vine plants [8].

2.2 Biodiversity assessment

Different taxa were selected to describe biodiversity and biological communities associated with vines. Green leafhopper (*Empoasca vitis*) populations were observed and quantified all along the season thanks to yellow sticky traps (YST) [8]. Dimensions of YST were 5cm x 10 cm. These traps were settled in early may each year, between two vines, attached to the lower wire. YST were changed and adults of E. vitis counted every second week, until beginning of September.

Biodiversity was estimated using pitfall traps, specialized in ground-dwelling arthropods. Three sessions of trapping were done in 2015 and 2016, in May, June, and July. Pitfall traps were installed between two vines at each sampling point, and left for minimum three days before being taken away. Their content was transferred in jars containing 70% ethanol, in order to conserve arthropods trapped until identification at laboratory. Two sessions of aspiration (to catch arthropods in the vine leaves) were done each year, one in June and one in July. Aspiration of vine leaves was performed on four consecutive vines with a D-vac system, arthropods caught (called "foliage arthropods" in the following lines) were immediately transferred in plastic zipped bags and stored in a freezer until identification. For both samples got from pitfall traps and aspiration, identification consisted in sorting out and quantifying arthropods up to the order level. Among coleoptera order, carabids (family: carabidae) encompass a wide range of ecological functions, included predation. They were set apart, and specimens were sorted out up to the species level and counted.

2.3 Agronomic characterization

GreenSeeker® (N-Tech, USA) is an integrated optical sensing system that measures the Normalized Difference Vegetative Index (NDVI). It is mounted on a quad bike. GreenSeeker sensor uses light emitting diodes (LED) to generate red and near infrared (NIR) light. The light generated is reflected off the crop and measured by a photodiode located at the front of the sensor head. NDVI measured on a plant can vary between 0 and 1. A value of 1 represents a maximum coverage of surface by vegetation. The value 0 represents an absence of vegetation. We use a white board (NDVI value around 0) positioned behind the row measured. Thus, the GreenSeeker values vary between 0 and 1 depending on foliage porosity and leaves chlorophyll content. One session of measurements was performed per plot : in August 2015 in vineyard 1, and in August 2016 in vineyard 2.

3 Data processing and analyses

Distances between sampling points and the nearest trees (AF trees or forest edges) were calculated. Extraction of values for the plant vigor index produced by GreenSeeker sensor in a buffer zone of 2,5m radius around each sampling point was performed. These data were obtained by using the ArcGIS 10.0 software. We tested this main question : do the trees, inside the plot or forest border, have an effect on arthropods populations ? In arthropods populations, we include the following variables : E. vitis abundances, total foliage arthropods abundances, total ground-dwelling arthropods abundances, generalist predators (arachnids) abundances, carabids diversity and abundances. Number of arthropods were counts data and they were analysed using a GLM with a log link and Poisson errors. Overdispersion was checked by comparing residual deviance and residual degrees of freedom (R software; R Development Core Team 2010).

4 Results and discussion

4.1 Global results

The table 1 shows the total abundances trapped per site and per year for the five biodiversity variables monitored.

Table 1 : General abundances obtained for E. vitis, ground-dwelling arthropods and foliage arthropods

Mean +/- standard deviation per sampling point		Empoasca vitis abundance	Foliage arthropods abundance	Ground-dwelling arthropods abundances
Vineyard 1	2015	111,7 (±35,9)	174,8 (± 65)	117,6 (± 61,8)
	2016	78,1 (± 24,9)	134,5 (± 76,3)	111 (± 53,1)
Vineyard 2	2015	307,2 (± 88,4)	105,4 (± 40,0)	63,3 (± 19,2)
	2016	60,1 (± 24,9)	122,2 (± 37,7)	202 (± 60,8)

Arthropods trapped in the vines foliage belong to the following orders : Hemiptera (leafhoppers, true bugs, aphids, ...) are the most abundant with 48,9% of the individuals caught, Diptera (flies, mosquitoes), Arachnids (spiders), collembola and hymenoptera (parasitoids, wasps, bees, ...) represented respectively 15,8%, 14,5%, 7,8% and 7,3% of the total abundance. Ground-dwelling arthropods trapped mainly belong to hymenoptera (ants in majority) (36,2%), arachnids (19,2%), diptera (10,2%), Acarii (mites) (8,9%).

Focusing on carabids, their abundances and richness were very low (figure 2).

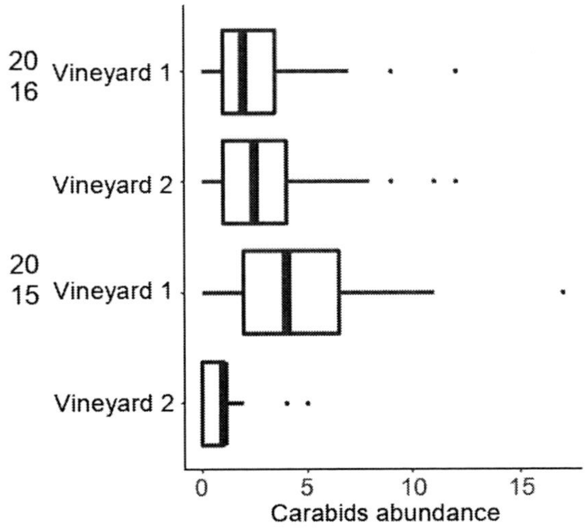

 https://doi.org/10.1051/e3sconf/20185001013

Figure 2 : Abundance of carabids trapped in the two vineyards in 2015 and in 2016.

Moreover, 80% of samples collected in the two vineyards in each sampling session of 2015 and 2016 contained 0 or 1 individual. For this reason, abundances only were taken into account for the study of carabids repartition according to the distance from the trees.

4.2 E. vitis repartition

When grouping total E. vitis abundances overall years of observation, no significant effect of AF trees or forest edges is noticed. A slight effect of AF trees

and forest edge is measured on Vineyard 1 in 2015, which is balanced by an opposite effect found in 2016 (table 2). No significant effects of trees on E. vitis repartition were found in the vineyard 2.

Table 2 : Effects of distance from the nearest tree (AF, forest and/or hedgerow) on E. vitis abundances

	Distance from AF trees		Distance from forest edge		Distance from hedgerow	
	t value	p value	t value	p value	t value	p value
2015 Vineyard 1	**3,73**	**<0,001**	**2,91**	**0,007**	/	/
2016 Vineyard 1	1,22	0,23	**-2,54**	**0,02**	/	/
2015 Vineyard 2	-0,23	0,82	-0,67	0,51	-0,22	0,83
2016 Vineyard 2	1,12	0,27	-0,56	0,58	-0,15	0,88

E. vitis abundances were positively linked to NDVI index in Vineyard 1 in 2015 (t = 2.60, p-value = 0.02). No significant effect was found in vineyard 2 in 2016.

4.3 Ground-dwelling and foliage arthropods

No effects of trees on ground-dwelling arthropods abundances was noted in the vineyard 1. Foliage arthropods were more abundant when the distance from the line of trees was big in 2015, but an opposite result was found in 2016 (Figure 3).

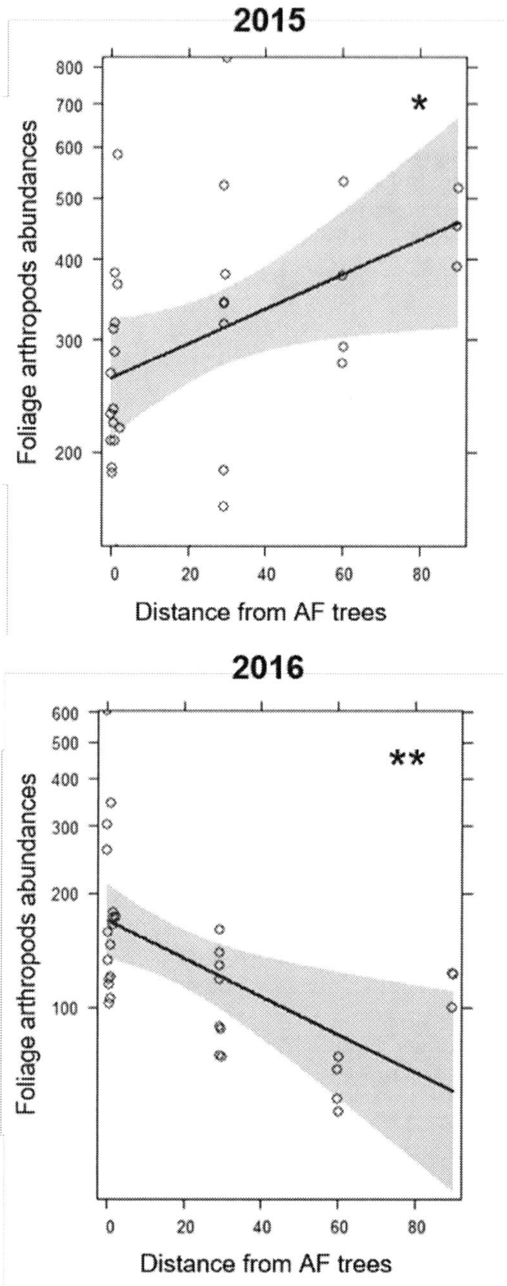

Figure 3 : Effects of distance from AF trees on foliage arthropods trapped in vineyard 1

In vineyard 2, a positive effect of the distance from the hedgerow on ground-dwelling arthropods (less abundances near the hedgerow) was found in 2015 and in 2016. Similar effects of distances from other trees have been partly observed: trees of forest edge in 2015, and intra-plot trees in 2016. Regarding foliage arthropods, except for a negative effect of the distance from the hedgerow in 2016 (i.e., number of arthropods decreasing when the distance from the hedgerow increased), the intra-plot trees and forest edge had no effect (table 3).

Table 3 : Effects of distance from the nearest trees (AF, forest and/or hedgerow) on arthropods trapped in vineyard 2

	Distance from AF trees		Distance from forest edge		Distance from hedgerow	
	t value	p value	t value	p value	t value	p value
2015 Ground-dwelling arthropods	1,91	0,06	**2,42**	**0,02**	**2,09**	**0,04**
2016 Ground-dwelling arthropods	**2,24**	**0,03**	0,95	0,34	**2,77**	**0,009**
2015 Foliage arthropods	-1,51	0,14	-0,37	0,71	-0,88	0,37
2016 Foliage arthropods	-1,50	0,14	-1,31	0,20	**-3,18**	**<0,01**

Arachnids abundances were analysed separately, only one significant effect is noted in vineyard 2 in 2015 : abundances are bigger near the forest edge than inside the vineyard (t= -2.02, p-value = 0.05). Regarding carabids, a slight effect exists in vineyard 2 in 2016 : carabids seem to be more numerous near the AF trees and their abundance decreases with the distance from the AF trees (t = -2,00, p-value = 0.05).

5 Conclusion

Some effects of trees on biodiversity (arthropods) have been observed, but they remain punctual and inconstant, so they cannot be generalized, even at the small scale of this trial. Our results show that AF practices only (introduction of trees in or around agricultural plots) seem not to be sufficient to explain arthropods repartition inside a vineyard. Other factors, like ground cover in the inter-rows for instance, could interfere. This potential effect could explain some contrasting results found in vineyard 1 between 2015 and 2016. In fact, ground coverage was established by the farmer in 2016 (the plot - excepted the two rows from either side of the trees lines - was kept under soil tillage until 2015). The combination of biodiversity-friendly practices (such as ground cover in the inter-rows associated with AF trees for instance) could therefore be considered as a promising solution to slow biodiversity loss in agricultural lands. If our study, run on two eight-years-old French AF vineyards, did not allow us to present AF as a key factor to solve the biodiversity issue, it does not seem to impact negatively grape production so far [7].

6 Aknowledgement

This study has been carried out within the frame of the CASDAR Vitiforest project, funded by French Ministry of Agriculture and Agence de l'Eau Adour Garonne. We thank the Emile Grelier estate, owned by Delphine and Benoit Vinet in Lapouyade, and Patrick Dubos in Lagardère for making this work possible. We also thank the master students who participated to the field work for their precious help.

References

[1] N.Ollat, J. Touzard, C. Van Leeuwen, Journal of Wine Economics, 11(1), 139-149 (2016)

[2] Millennium Ecosystem Assessment, Island Press, Washington, DC. (2005).

[3] W. Zhang, T.H. Ricketts, C. Kremen, K. Carney, S.M. Swinton, EcolEcon 64, 253–260 (2007)

[4] H. Assael, Consumer Behavior and Marketing Action (Kent Pub. Co., 1984)

[5] M. Duru, O. Therond, G. Martin, R. Martin-Clouaire, M. A. Magne, E. Justes, E. P. Journet, J. N. Aubertot, S. Savary, J. E. Bergez, J. P. Sarthou, Agron. Sustain. Dev. 35, 1259–1281, (2015)

[6] M. den Herder, G. Moreno, M.R. Mosquera-Losada, J.H.N. Palma,.A. Sidiropoulou, J.J. Santiago Freijanes, J. Crous-Duran, J. Paulo, M. Tomé, A. Pantera., V. Papanastasis, K. Mantzanas, P. Pachana, A. Papadopoulos, T. Plieninger, P.J. Burgess, Current extent and trends of agroforestry in the EU27. Deliverable Report 1.2 for EU FP7 Research Project: AGFORWARD 613520. (2016)

[7] T. Dufourcq, F. Lopez, M. Vergnes, L. Gontier. , *20th International GiESCO Symposium, Mendoza, Argentina*, 868-872. (2017)

[8] D. Decante, C. van Leeuwen, M. van Helden, Agricultural and Forest Entomology 11, 377–387 (2009)

Influence of irrigation on yield and quality of cv. Touriga Franca in the Douro Region

Inês Cabral[1], Tiago Nogueira[2], Anabela Carneiro[1,3] and Jorge Queiroz[1,3]

[1]DGAOT, Faculty of Sciences, University of Porto, Campus Agrário de Vairão, Portugal
[2] Quinta do Crasto S.A., Gouvinhas, Sabrosa, Portugal
[3]GreenUP/CITAB-UP, DGAOT, Faculty of Sciences, University of Porto, Campus Agrário de Vairão, Portugal

Abstract. The Douro Region is characterized by a typically Mediterranean climate, with extreme temperatures and irregular rainfall throughout the year. This experimental trial was carried out on a vineyard plot of Touriga Franca, grafted in 110R, at Quinta da Cabreira, property of Quinta do Crasto, S.A., located in the Douro Region. Four irrigation modalities were established, in three blocks, according to the percentage of cultural evapotranspiration (ETc): R_0, the control, without irrigation; R_{25}, irrigated with 25% ETc; R_{50}, with 50% ETc and R_{75}, with 75% ETc. Irrigation had repercussions on total leaf area, with statistically significant lower value for R_0 and greater for R_{75}. Concerning the ecophysiological parameters, the water leaf potential was lower in R_0 than at R_{50} and R_{75}. On the yield components, it was observed a higher number of clusters in R_{75} and a smaller quantity in R_0 and the bunch weight was also affected by the amount of water provided, with the R_{50} or R_{75} having a higher value. Even if the berry weight has not presented differences, those factors affect yield, that was higher in the R_{50} and R_{75}. Concerning the quality, no significant differences had been found on the various components tested.

1 Introduction

The vineyard in Portugal occupies a prominent place not only in terms of area – it is the third culture with the largest area in the country – but also because of its strong cultural and historical connection with the first Demarcated and Regulated region in the world (Douro), having an important socio-economic role.

Generically, the Douro Demarcated Region and in specific the Douro Superior Sub-region has some unique characteristics, either because of their soils and climate, being subject to strong water deficit during the Summer, motivated by low values of rainfall and high temperatures [1]. Despite this, the vine is a resistant plant and highly adapted to this type of conditions, having both anatomical and physiological characteristics that allow it to resist to stress conditions. Although these adaptive characteristics, in long drought periods some qualitative and quantitative changes might occur.

Nowadays, this problem is even more important due to the high inter-annual variability mainly justified by the climate change effect which brings new challenges for winegrowers [2].

The study of the irrigation effect on grape and wine quality and yield is an important problematic, although this practice is often considered a factor that has a negative influence on the quality of wines [3], persisting the idea that the water supply reduces the alcohol content and the phenolic extraction along with the increase of acidity [4].

In Portugal, until recently, irrigation was not practiced by winemakers because it was considered a derogatory factor, and it still has legal restrictions on its practice [4, 5]. However, due to the intensification of drought and the consequent increase in water stress, producers have been adopting this procedure in order to increase production, in a way that this practice does not affect quality. The increase of the irrigated vineyard area, attests that using a rational irrigation system it is possible to increase the yield and eventually improve the potential quality of the musts. Nevertheless, it is a theme that requires a case-by-case study and adjustment according to the region, place, variety, soil, slope or climatic factors. Some of the challenges are to determine the quantity and timing of its application, the effect of irrigation and which are the best parameters for monitoring the water supply, relating them to the plant response to irrigation [6].

One of the methods used in vineyard irrigation and particularly in drip irrigation is the "Regulated Deficit Irrigation" (DRI) method, which establishes different levels of water restriction throughout the various stages of the vine cycle, thus using the plant evapotranspiration as a methodology for estimating the amount of water to be applied.

The present essay emerges as a way of testing different irrigation modalities at Quinta da Cabreira, in the Douro Superior Sub-region, near Vila Nova de Foz Côa, with the aim to test how different treatments can affect both quality parameters and yield in one of the most widespread grape variety in Portugal, Touriga Franca.

 https://doi.org/10.1051/e3sconf/20185001014

2 Materials and Methods

2.1 Characterization of the plot

The trial was set up in 2015 (being presented in this paper the results of 2017), in a vineyard on a hillside with southern exposure. It had been selected three blocks, each one in a terrace with two rows. Each block was further divided into four irrigation modalities, corresponding each modality to twenty vines in each row, with data being collected in the central 10 vines. This way, the results of the trial had been collected in a total of 240 plants (10 vines × 2 rows × 4 irrigation modalities × 3 blocks).

The vineyard was planted in 2007, grafted on 110 Richter, with a distance between plants of 1 m x 2.2 m. The training system is a vertical shoot positioned canopy, pruned in a single cordon Royat. The average altitude of the field trial was 240 m and the rows were oriented Northwest-Southwest.

Four irrigation modalities were tested: R_0, the control treatment, which corresponds to non-irrigated vines; R_{25} corresponding to an irrigation of 25% of Cultural Evapotranspiration (ETc); R_{50} irrigated with 50% of ETc and R_{75} irrigated with 75% of ETc.

Water had been supplied six times, each 15 days, at night, in order to allow better soil water absorption and less evapotranspiration. It started in June 8th, according to the data provided by the probe existent in the plot and when the values of $\Psi_{predawn}$ fell below - 0.40 MPa.

2.2 Field Determinations

The foliar area was estimated using the model developed by Lopes & Pinto [7]. One shoot of 2 plants per treatment was tagged in order to monitor the foliar area throughout the vegetative cycle. The data were collected on four different dates, corresponding to Baggiolini states of fruitset (J), beginning of veraison (M), maturity (N) and harvesting (N).

Point Quadrat, following the method of Smart and Robinson [8], was assessed during the veraison/maturity (M/N) and at harvest (N) stages, at two different levels of canopy: at the vegetative and at the clusters level. Simultaneously, the exposed leaf area was also determined [9].

To estimate the evapotranspiration (ETc), the evapotranspiration values of the plant (ET_0) were collected from the meteorological station at Quinta da Cabreira and a value of Kc of 0.8 was selected having in account the period of irrigation and the characteristics of the soil [4, 10].

2.3 Measurements of physiological activity

The leaf water potential (Ψ_l) was measured using a pressure chamber according to the methodology described by Scholander [11]. After irrigation began, $\Psi_{predawn}$ measurements were performed every morning before irrigation, every two weeks.

2.4 Yield and fruit quality

The maturity control began at August 3th. At harvest, 200 berries were collected per modality, in the exterior and interior row, in a total of 24 samples. The weight of the sample, the weight of the berry, the probable alcohol (%), the pH, the total acidity (g/L), the anthocyanin (mg/L), polyphenols (mg/L)) and malic acid (g/L) content were assessed.

The harvest was settled according to the results of the samplings carried out, as well as the weather forecast. At harvest the number of clusters and yield per plant were registered in the 20 vines selected per modality.

2.5 Statistical Analysis

The data throughout the trial were analyzed using the SPSS® program version 23.0, and subjected to analysis of variance (ANOVA). The parameters were recorded as non-significant, with the abbreviation "n.s." and significant with "*" for $p < 0.05$, "**" for $p < 0.01$ and "***" for $p < 0.001$. When the data analysis showed statistical significance, the means were compared using the least significant difference (LSD) test.

3 Results and Discussion

3.1 Climatic characterization of the region

Comparing with Climatic Normal (1931-1960) [12], the rainfall was generally lower for the agricultural year (from November 2016 to October 2017) particularly in the months of January, March, April, June, July, September and October, with values in average 50 to 60% inferior to the Normal (534,9 mm for the reference period against 316,0 mm for the year of 2017).

The temperature values for the year of 2017 were higher (16.5°C) when compared to Climate Normal (16.1°C). That way, 2017 could be considered as an extremely hot and dry year [13].

3.2 Fertility components

The number of buds retained at winter pruning was of about 10 buds/vine and the percentage of budburst of about 96%, with no differences between modalities. However, the fertility of the modality R_0 (7.59 inflorescences/vine) was lower than the fertility of R_{50} (9.19) and R_{75} (9.37), presenting R_{25} an intermediate value of 8.51. The differences in fertility may have occurred due to accumulation of effect of irrigation provided in the previous years of 2015 and 2016.

3.3 Characterization of the vegetation

The results of the total leaf area, as shown in table 1, reflect an increase from the first to the second measurement, followed by a reduction on the next measurement in all modalities. As expected, R_0 and R_{25} had both the lowest total leaf area, since basal leaf senescence (caused by lack of water) occurs over time. In fact, in the first measurement, performed on May 16th, the total leaf area was similar in the four modalities, decreasing then on the following measurements. At the end of maturation, the value for the non-irrigated modality R_0, was less than half of the values of R_{50} and R_{75}.

Table 1: Average Total Leaf Area (m^2).

Modalities / day of measurement	R_0	R_{25}	R_{50}	R_{75}	Sig. (1)
May 16	2.08	1.81	1.98	2.27	n.s.
June 7	3.29	3.12	3.75	4.25	n.s.
July 20	3.27	2.66	3.49	3.56	n.s.
September 9	1.42	2.27	2.84	2.98	**
	a	ab	b	b	

(1) Significance of the test ANOVA. "n.s." – not significant; "**" – p<0,01.

The leaf layer number, presented in table 2, were similar in the first measurement on July 6th. However, in the measurement of September 9th, R_0 presented a value significantly smaller than R_{75}.

Table 2: Leaf Layer Number.

Modalities / day of measurement	R_0	R_{25}	R_{50}	R_{75}	Sig. (1)
July 6	2.54	2.48	2.40	2.60	n.s.
September 9	1.48	1.98	2.52	2.65	*
	b	ab	ab	a	

(1) Significance of the test ANOVA. "n.s." – not significant; "*" – p<0,05.

If the value observed for R_0 is similar to those proposed by Smart and Robinson [8], this value is considered very low for Douro Region by Queiroz [14], having in mind the risk of sunburn. This risk can be also assessed by the percentage of external clusters (Table 3), that show an increase in the percentage of exposed clusters to the direct radiation of the sun on the R_0 and R_{25} modalities.

Table 3: Percentage of Exposed Clusters.

Modalities / day of measurement	R_0	R_{25}	R_{50}	R_{75}
July 6	22.54	5.56	30.56	12.96
September 9	58.33	45.93	5.56	15.00

(1) Data not statistically analyzed.

3.4 Evapotranspiration and irrigation calculation

The value of Et_0 varies throughout the day and depends on several climatic factors, such as temperature, precipitation or wind speed.

According to the data of the meteorological station, as well as the volume of water to be applied and the flow rate of the drippers, it was possible to calculate the irrigation for each modality on the different days as shown in table 4. The rainfall reported during the fifteen days before the first three irrigations, was deducted to the crop evapotranspiration value (ETc).

3.5 Ecophysiological measurements

3.5.1 Predawn Water Potential

Table 5 exhibits the values of $\Psi_{predawn}$ on the different days of measurement, measured in the night before irrigation. The values are similar in the first measurements, between May 25th and July 20th, showing later disparities from the sixth measurement, corresponding to August 8th. Thus, although the irrigation started on June 8th, the differences between the modalities only became evident after August 8th, that is, two months later. These results may be related to the extremely dry and hot weather conditions that were observed throughout the grapevine cycle, posing us the question if the frequency irrigation of 15 days was adapted to the water supply of the plant.

Table 5: Predawn Water Potential (MPa).

Modality / Day	R_0	R_{25}	R_{50}	R_{75}	Sig. (1)
May 25	-0.33	-0.35	-0.37	-0.34	n.s.
June 8	-0.36	-0.43	-0.40	-0.40	n.s.
June 23	-0.41	-0.37	-0.40	-0.38	n.s.
July 7	-0.50	-0.49	-0.50	-0.48	n.s.
July 21	-0.57	-0.53	-0.50	-056	n.s.
August 9	-0.82 a	-0.71 ab	-0.63 ab	-0.52 b	*
August 18	-1.05 a	-0.81 bc	-0.88 b	-0.76 c	***

(1) Significance of the test ANOVA. "n.s." – not significant; "*" – p <0, 05; "***" – p <0,001.

Over time, analysing table 5, it can be noticed that there is a decrease in water potential in all modalities, reaching, according to Carbonneau [9], on August 18th a severe level of stress, but with greater significance in the non-irrigated modality R_0. It is also verified that the R_{75}

4: Date of irrigation, Rainfall in the period (mm), ET_C, ET_0, Volume of water applied (L), Watering time (m)

Irrigation Number	Day	Rainfall (mm)	ETc (mm)	ET_0 (mm)	Volume of water applied (L)			Watering time (m)
					R_{25}	R_{50}	R_{75}	
1	June 8	5,0	60,70	75,8	14,16	28,32	42,48	7 h 05
2	June 23	3,8	73,20	91,5	17,54	35,08	52,62	8 h46
3	July 7	3,2	64,08	80,1	15,38	30,76	46,14	7 h 41
4	July 21	0	75,04	93,8	18,76	37,52	56,28	9 h 23
5	August 9	0	73,20	91,5	18,30	36,60	54,90	9 h 09
6	August 18	0	69,68	87,1	17,42	34,84	52,26	8 h43 min
Total		11,0			101,56	203,12	304,68	

modality, despite being also in severe stress is the one with the lowest base water potential.

3.5.2 Daily Water Leaf Potential

As shown in table 6 and as verified in other studies, the increase of both temperature and solar radiation, led to an increase in water losses due to transpiration, with a decrease in water leaf potential from early morning (10h) until mid-afternoon (14h). After this measurement there is a slight increase in these values, as verified by the late afternoon measurement (around 17h30) and then a noticeable recovery in the pre-dawn measurement.

Table 6: Daily Water Potential (MPa).

Day	Hour	R_0	R_{25}	R_{50}	R_{75}	Sig. (1)
Jul 7	4h	-0.44 a	-0.32 b	-0,28 bc	-0,24 c	***
	10h	-1.43 a	-0.95 b	-0.73 c	-0.54 d	***
Jul 21	4h	-0.63 a	-0.53 ab	-0.44 b	-0.41 b	**
	10h	-1.42 a	-1.23 b	-0.96 b	-0.79 b	***
	14h	-1.55 a	-1.38 b	-1.24 b	-1.00 c	***
	17h30m	-1.28 a	-0.79 a	-0.63 b	-0.57 b	***
Aug 4-5	10h	-1.32 a	-0.93 b	-0.68 c	-0.63 c	***
	14h	-1.50 a	-1.46 a	-1.40 a	-1.15 b	***
	17h30m	-1.49 a	-1.45 a	-1.41 a	-1.14 b	**
	4h	-0.68 a	-0.38 b	-0.35 b	-0.26 b	***
Aug 18-19	10h	-8.53 a	-5.23 b	-4.80 b	-4.68 b	***
	14h	-9.78 a	-8.35 ab	-7.23 b	-6.95 b	***
	17h30m	-9.33 a	-6.50 b	-6.05 b	-5.58 b	***
	4h	-6.10 a	-3.75 b	-3.25 bc	-2.80 c	***

(1) Significance of the test ANOVA. "**" – p <0, 01; "***" – p <0,001.

There is a moderate to severe level of water stress in all modalities, which shows that this stress increases not only along the day but that is higher in the none irrigated or the less irrigated modalities. It is also verified that in the four measurements carried out, the differences between the modalities were statistically different, and it was also observed that, in general, the higher stress levels were in the non-irrigated modalities.

3.6 Fruit quality and Yield

3.6.1 Quality

Table 7 shows the evolution of the probable alcohol content (%) in the different modalities, during the maturation period. It was observed that there is an increase along the maturation period, with final values that are considered high for this grape variety, but no differences between the treatments.

Table 7: Evolution of Probable Alcohol (% v/v).

Day / Modality	Aug 16th	Aug 24th	Sept 9th
R_0	11.80	13.60	14.68
R_{25}	12.00	13.60	15.87
R_{50}	11.90	13.70	15.23
R_{75}	12.00	13.60	15.48
Sig. (1)	n.s.	n.s.	n.s.

(1) Significance of the test to Analysis of Variance of Means (ANOVA). "n.s." – not significant.

Although it was expected a dilution of sugar content in modalities with higher irrigation, this was not observed this year.

Table 8: Evolution of pH.

Day / Modality	Aug 16th	Aug 24th	Sept 9th
R_0	3.92	3.86	4.19
R_{25}	4.02	3.97	4.27
R_{50}	4.01	3.90	4.22
R_{75}	4.03	3.91	4.24
Sig. (1)	n.s.	n.s.	n.s.

(1) Significance of the test ANOVA. "n.s." – not significant.

Regarding the pH values (Table 8), it was observed a reduction of this parameter from the first measurement to the second, and an increase in the third measurement, which corresponds to the harvest period.

All the treatments showed a similar behaviour and, in addition, it was verified that the increase is not constant between the different treatments. In the last measurement, the modalities that have a lower pH were R_0 and R_{50}. Nevertheless, all the modalities presented very high values of pH.

Concerning the titrable acidity there were also no statistically differences between the modalities, as shown in table 9.

Table 9: Evolution of Titrable Acidity (g/L of Tartaric Acid).

Day / Modality	Aug 16th	Aug 24th	Sept 9th
R_0	3.50	4.23	4.56
R_{25}	3.72	4.34	4.24
R_{50}	3.68	4.17	4.60
R_{75}	3.81	4.19	5.44
Sig. (1)	n.s.	n.s.	n.s.

(1) Significance of the test ANOVA. "n.s." – not significant.

It was noted, however, that in all the treatments there was an increase of this parameter which was not expected.

Malic acid has generally a higher concentration at the beginning of maturation. The data in table 10 notice a decrease in all modalities over time, being this decrease more marked between the first and second measurements. Still, there was no evident relationship between the malic acid content (g / L) and the irrigation level. Despite the non-statistically significant differences, the R_0 and R_{50} values were lower than the remaining.

Table 10: Evolution of Malic Acid (g/L).

Day / Modality	Aug 16th	Aug 24th	Sept 9th
R_0	2.12	1.57	1.54
R_{25}	2.45	1.93	1.91
R_{50}	2.15	1.66	1.62
R_{75}	2.40	1.73	1.73
Sig. (1)	n.s.	n.s.	n.s.

(1) Significance of the test ANOVA. "n.s." – not significant.

Observing the anthocyanins, there is a decrease in their values in all modalities, being most evident in the measurements from August 16th to 24th, but also recorded from August 24th to September 9th. There is also a higher anthocyanin content, although the values were not statistically significant, in the R_0 and R_{50} modalities, the latter being the one that obtained the best results in the last measurement. This decreased was even more evident in the R_0 and R_{75} modalities, as evidenced in table 11.

Table 11: Evolution of Anthocyanins (mg/L).

Day	Aug 16	Aug 24	Sept 9
R_0	336	158	146
R_{25}	501	186	128
R_{50}	421	158	150
R_{75}	478	164	144
Sig. (1)	n.s.	n.s.	n.s.

(1) Significance of the test ANOVA. "n.s." – not significant.

Water stress may have a beneficial effect on the synthesis of anthocyanins, but when pronounced, it will lead to a decrease in the photosynthetic activity and to the synthesis of sugars, thus negatively affecting the synthesis of colouring substances [4].

Another analysed component was the polyphenols content. From the analysis of table 12, a decrease of this compound is observed in all modalities. Due to a problem of the reagents used in this analysis, it was not possible to have the result from the modality R_{75} modality on August 24th.

From August 24th to September 9th there is a statistically non-significant increase in the polyphenols content, with a concentration of this compound converging. It is observed a higher concentration of polyphenols in the modalities with higher irrigation supply, namely in R_{50} and R_{75}. As in most of the previously analysed parameters, the R_{50} modality was the one that exhibited a higher content of this constituent.

Table 12: Evolution of Polyphenols content (mg/L).

Day / Modality	Aug 16th	Aug 24th	Sept 9th
R_0	1392.2	572.9	836.4
R_{25}	1803.7	583.9	822.2
R_{50}	1795.8	259.7	877.8
R_{75}	1741.4	-	865.7
Sig. (1)	n.s.	n.s.	n.s.

(1) Significance of the test ANOVA. "n.s." – not significant.

3.6.2 Yield

Regarding the clusters per vine (Table 13) there are statistically significant differences, particularly between the R_0 and R_{75} modalities, which may, as referred, reflect the cumulative effect of irrigation, along the three years of study. Concerning the weight of the cluster, it was verified that, as expected, the water availability positively influenced the weight of the clusters, presenting R_0 and R_{25} lower values than the more irrigated modalities R_{50} and R_{75}. The berry weight, did not presented significant differences in any of the treatments.

In the analysis of the volume of the berries, as presented in table 13, a correlation between the weight of the berry and its volume is observed, as expected. Therefore, it is verified that the variations in weight are mainly due to the absorption of water or dehydration. It can be noticed that, although the differences were not significant between the modalities, the water availability could have contributed to the increase of the amount of pulp produced. In which case it was verified that the modalities R_{50} and R_{75}, had higher values of berry weight, thus demonstrating a tendency to an increase of this parameter. On the other hand, the increase of the berry volume will lead to an increase in yield along with a decrease in the skin / pulp ratio.

Table 13: Number of clusters/vine, cluster weight (g), berry weight (g), volume of the berry (mL) and yield/vine (kg) registered at harvest.

Modality	R_0	R_{25}	R_{50}	R_{75}	Sig. (1)
N° clusters/vine	7.60 a	8.80 ab	9.20 ab	9.27 b	*
Cluster weight (g)	150 a	172 ab	183 b	179 b	**
Berry weight (g)	1.47	1.36	1.53	1.56	n.s.
Berry volume (mL)	0.94	0.82	1.04	1.09	n.s.
Yield/vine (kg)	1.21 a	1.60 ab	1.61 b	1.71 b	**

(1) Significance of the test to ANOVA. "n.s." – not significant; "*" – $p<0,05$; "**" – $p<0,01$.

At harvest, apart from counting the number of clusters per vine, they were also weighed. Table 13 shows the yield per modality. In this table, it can be noticed the correlation between the amount of irrigation to which the vines were subjected and the yield. The results show statistically significant differences between the non-irrigated mode and the two modalities with higher irrigation supply.

4 Conclusions

The results obtained in the present study allowed to draw some conclusions about the effect of irrigation. If the decision is to irrigate, the level of irrigation must be adequate according to the goals of the viticulturist.

Although this study is a follow-up of other studies carried out since 2015, the results obtained this year are very influenced by the climatic conditions data of the year. Thus, it appears to be a cumulative effect on the vines according to the amount of applied water since the beginning of the essay.

The results obtained show a strong influence of irrigation on the vegetative growth of vines, leading to canopies with higher leaf layer number and to a lower bunch exposure, and consequently to a reduced sunburn sensibility. Contrariwise, non-irrigated vines presented a lower foliar surface due to the basal leaves senescence, motivated by a strong hydric and thermal stress.

Regarding the ecophysiological measurements, namely the predawn leaf water potential, the values obtained showed that with 15 days between irrigations, all the modalities, including the more watered, present at end of the period a strong water deficit. In fact, after the first 7 – 8 days all the vines presented visual signs of water stress, what was also confirmed by the values of the soil probe present in the plot. Therefore, it should be taken into account the hypothesis of irrigate more frequently. Thus, a weekly basis should be considered, so that the results could evidence the irrigation treatment instead of climatic factors.

Analysing the quality parameters, generally, no statistically significant differences between modalities were found, namely alcohol, pH or phenolic composition.

Finally, in the yield components, the number of clusters was lower in R_0 than R_{75}, which can be interpreted as a cumulative effect of irrigation along the three years. Regarding the weight of clusters there were also differences between R_0 and the two modalities with higher irrigation content. At harvest, there were statistically significant differences between the yield of the two treatments with the highest irrigation, when compared to the control treatment.

Being presented in this paper the results of the third year of the trial, it would important to continue this study, since the results showed large disparities among the three years, mainly because of the very different climatic conditions during this period.

5 References

[1] C. Almeida. *História do Douro e do Vinho do Porto*. Edições Afrontamento, 2006.
[2] G. Jones. *Uma Avaliação do Clima para a Região Demarcada do Douro: Uma análise das condições climáticas do passado, presente e futuro para a produção de vinho*. Vila Real, ADVID, 2013.
[3] A. Carbonneau, A. Deloire, B. Jaillard. *La Vigne: Physiologie, terroir, culture*. Paris: Dunod. 2007.
[4] N. Magalhães. *Tratado de Viticultura - A Videira, A Vinha e o "Terroir"*. Lisboa: Chaves Ferreira-Publicações S.A. 2008.
[5] J.-L, Simon, J. Schwarzenbach, M. Mischler, W. Eggenberger, W. Koblet. *Viticulture*. Editions Payot Lausanne: Paris. 1978.
[6] B. Pilar, P. Sánchez-de-Miguel, A.Centeno, P. Junquera, R. Linares, J. Lissarrague. Water relations between leaf water potential, photosynthesis agronomic vine response as a tool for establishing thresholds in irrigation scheduling. Sci. Hortic., **114**, 151-158. 2007.

[7] C. Lopes, P. Pinto. Easy and accurate estimation of grapevine leaf area with simple mathematical models. Vitis, 44(2), 55-61. 2005.
[8] R. Smart, M. Robinson. *Sunlight into wine - A handbook for winegrape canopy management* (8th ed.). New Zealand: Winetitles. 1991.
[9] A. Carbonneau. General Relationship within the whole-plant: Examples of influence of vigour status, crop load and canopy exposure on the sink "berry maturation" for the grapevine. Proc. Workshop Strategies to Optimize Wine Grape Quality. Acta Hortic. **427**, 99-118. 1996.
[10] T. Prichard, B. Hanson, L. Schwankl, P. Verdegal, R. Smith. Deficit Irrigation of Quality Winegrapes Using Micro-Irrigation Techniques. California: University of California Cooperative Extension. 2004.
[11] P. Scholander, E. Bradstreet, E. Hemmingson, H. Hammel. Sap Pressure In Vascular Plants: Negative Hydrostatic Pressure Can Be Measured In Plants. Science **148**(3668), 339-46. 1965.
[12] IPMA. Normais Climatológicas - Classificação Climática de Koppen. IPMA. 2017.
[13] B. Teixeira, C. Val, C. Carlos, I. Gonçalves, A. Nave, S. Martins, F. Morais. Boletim 14-17 Ano Vitícola 2017 - Balanço Final. Vila Real: ADVID. 2017.
[14] J. Queiroz. Condução e Relações Rendimento Qualidade de Castas Nobres do Douro. Porto: Faculdade de Ciências, Secção Autónoma das Ciências Agrárias, Universidade do Porto. 2002.

Soil type and soil preparation influence vine development and grape composition through its impact on vine water and nitrogen status

van Leeuwen Cornelis[1], de Rességuier Laure[1], Mary Séverine[2], Laveau Coralie[2], Mousset-Libeau Etienne[1,3], Marguerit Elisa[1], Roby Jean-Philippe[1] and Quiquerez Amélie[4]

[1]UMR EGFV, Bordeaux Sciences Agro, INRA, Univ. Bordeaux, ISVV, 33883 Villenave d'Ornon, France
[2]Vitinnov, Bordeaux Sciences Agro, ISVV, 33175 Gradignan cedex, France
[3]Château Fombrauge, 33330 Saint-Christophe des Bardes, France
[4]UMR CNRS 6298, ARTeHIS, Université de Bourgogne, 6, Boulevard Gabriel, 21000 Dijon, France

Abstract. The influence of soil type and preparation on vine development and grape composition was investigated in a 50 ha estate located in Saint-Emilion (Bordeaux, France) and planted predominantly with Merlot. Part of the vineyard was planted down the slopes and another part of the vineyard was planted on terraces, where soils were profoundly modified through soil preparation. Grape composition (berry weight, sugar, total acidity, malic acid and pH), vigor (pruning weight), vine nitrogen status (Yeast Available Nitrogen (YAN) in grapes) and vine water status ($\delta^{13}C$) was measured at a very high density grid of 10 data points per hectare. Water deficit was globally weak over the estate because of high soil water holding capacity whereas vine nitrogen status was highly variable. Vine vigor and grape composition were predominantly driven by vine nitrogen status. On terraces, where soils were deep, due to invasive soil preparation, water deficits were particularly small or non-existent and vine nitrogen status was highly variable. Grape quality potential was medium to low, except in places with low nitrogen status, but at the expense of low yields. On parcels planted down the slopes water deficits were recorded because vine rooting was limited by compact subsoils. Vine nitrogen status was homogeneous. Grape quality and yield were medium to high and relatively homogeneous. When possible, downhill plantations are to be preferred over terraces because in the latter vine yield and quality parameters are highly variable because of massive soil movements prior to plantation.

1 Introduction

Wine quality and style are influenced by environmental factors like climate, topography and soil type [1]. These influences are mediated through the vine and may be impacted by management practices [2]. On hills, soil erosion has an impact on soil distribution [3]. To quantify soil impact on vine vigor and grape composition, it has to be broken down in measurable effects. Among these, vine water and nitrogen status are of major importance [4]. In this study, the soil effect on vine vigor and grape composition was investigated in a 50 ha estate in Saint-Emilion (Bordeaux, France), planted predominantly with Merlot. Soils were developped either on Tertiairy massive limestone bedrock (located on the plateaux) or on Tertiairy unconsolidated limestone bedrock (located on the slopes beneath the plateaux). Soils were predominantly silty-clay or clayey-silt textured. Part of the soils contained limestone, part of the soils did not. Relations between vine water status, vine nitrogen status, vine vigor, berry size and grape composition were investigated through measurements implemented on a regularly grid of 10 samples per hectare. On a subset of parcels, the effect of pre-planting soil preparation was studied. Part of these parcels were located on terraces created prior to plantation (invasive soil preparation) and part of these parcels were planted down the slopes (non-invasive soil preparation). Soil pits were excavated to measure physical and chemical soil composition and to relate soil preparation to soil depth. Around each soil pit, vine vigor was measured, as well as yield parameters, vine nitrogen status, vine water status, and grape composition, including phenolic compounds. This approach allowed to investigate the effect of soil preparation, vine water and nitrogen status and their subsequent effect on vine vigor and grape composition.

2 Methods

2.1 Investigation of relations between vine water status, vine nitrogen status, vine vigor and grape composition

A regular grid of 10 locations per hectare (500 over the whole estate) was sampled prior to harvest. Each sample was composed of 60 berries taken randomly on 20 adjacent vines. Berries were weighted, pressed, and analyzed for $\delta^{13}C$ in order to determine vine water status [5] and Yeast Available Nitrogen (YAN) to determine vine nitrogen status [6]. Reducing Sugars, Total Acitidy (TA), Malate (MAL) and pH were assessed through classical methods. Shoots were counted and pruning wood was weighed on three vines per location and pruning weights were averaged per vine and per shoot at all the locations.

2.2 Investigation of the effect of soil preparation

The effect of soil preparation on soil depth, vine nitrogen and water status, vine vigor and grape composition was investigated on a subset of seven hectares. In this area, pre-planting soil preparation had been invasive, including the creation of terraces in sector A (Figure 1). Vines were planted down the slopes with limited soil preparation in sector B.

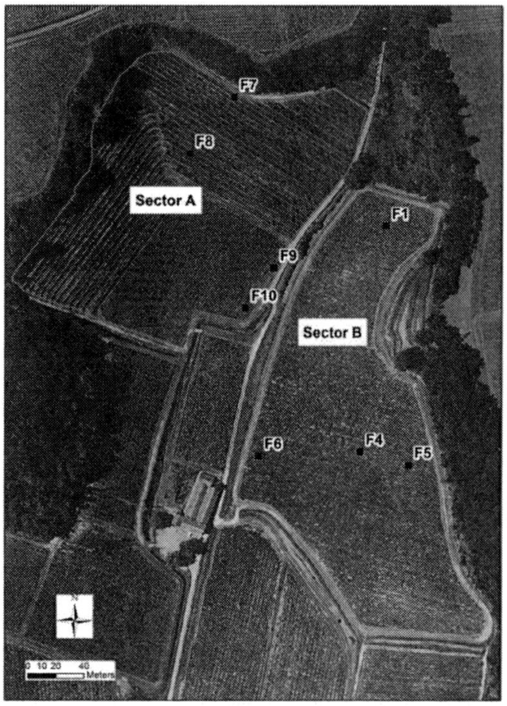

Figure 1. Location of the sector with invasive soil preparation (A) and the sector with conservative soil preparation (B).

2.2 Soil pits

Four soil pits were excavated in sectors A and B respectively (Figure 1) in order to determine soil type and soil depth (i.e. the depth at which the unaltered bedrock could be identified). In hillside context, soil depth is controlled by natural factors. It is the result of soil formation processes (duration and intensity) and of soil distribution along the slope driven by erosion processes. On cultivated areas, soil depth can be modified (generally increased, sometimes reduced) through pre-planting soil preparation operations, or the creation of terraces. Around each soil pit yield parameters were recorded at harvest (berry weight, number of bunches per vine, total yield per vine, bunch weight). Vine water status was assessed by means of stem water potential measurements (SWP, three measurements over the season) and $\delta^{13}C$. Only the results from the last SWP measurement (9 September 2016), when water deficits were most intense, are presented. Vine nitrogen status was assessed over the whole 7 ha section through N-tester measurements over a grid of 142 locations at two dates, at bloom and at veraison [6]. Only vine nitrogen status at veraison is presented. Grape composition was measured prior to harvest on four replicate samples per soil pit: YAN, Reducing Sugars, TA, MAL, pH, total anthocyanins and total phenolics.

3 Results

3.1 Relations between vine water status, vine nitrogen status, vine vigor and grape composition

A Principal Component Analysis (PCA) was carried out to investigate the relations between vine water status, vine nitrogen status, vine vigor and grape composition, on the data collected on the 500 locations over the whole estate (Figure 2). The first two dimensions (Dim 1 and Dim 2) explained 41% of the total variance of the dataset. The first Dim (24% of the total variance) separated the individuals according to the level of ripeness of the grapes, with individuals with high sugar on the left of the map and variables with high acidity on the right of the map. The second Dim (17% of the total variance) was driven by slope, U (North-South aspect) and Y (longitude). $\delta^{13}C$ was located close to the center of the Dim 1 - Dim 2 PCA mapping. The results of the PCA show that the total variance of the data set is better explained by YAN than by $\delta^{13}C$.

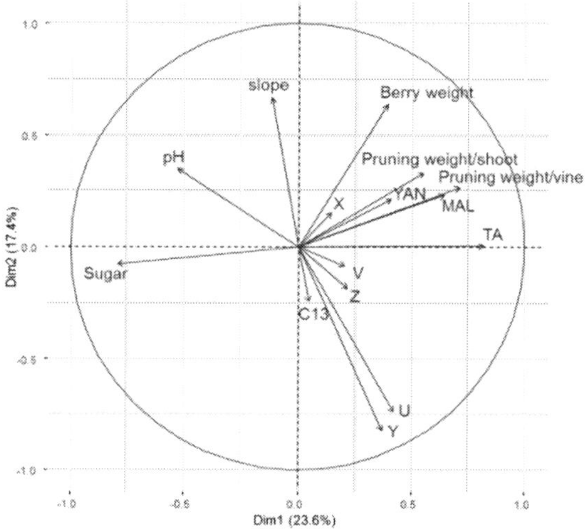

Figure 2. Projection of the first two dimensions of a PCA analysis carried out on the total data set.

Grape sugar content decreases with YAN (Figure 3a), while MAL increases with YAN (Figure 3b) and so does TA (data not shown). Hence, grapes from low nitrogen status vines produce grapes with greater ripeness at a given date. This effect is partially mediated through berry weight, which increases with YAN (linear regression significant $t = 2.34$; $p = 0.02$).

 https://doi.org/10.1051/e3sconf/20185001015

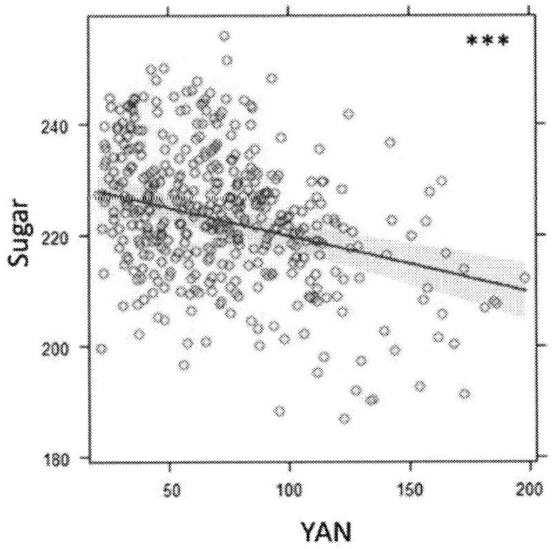

Figure 3a. Linear regression between grape sugar and YAN (t=5.66; p<0.001).

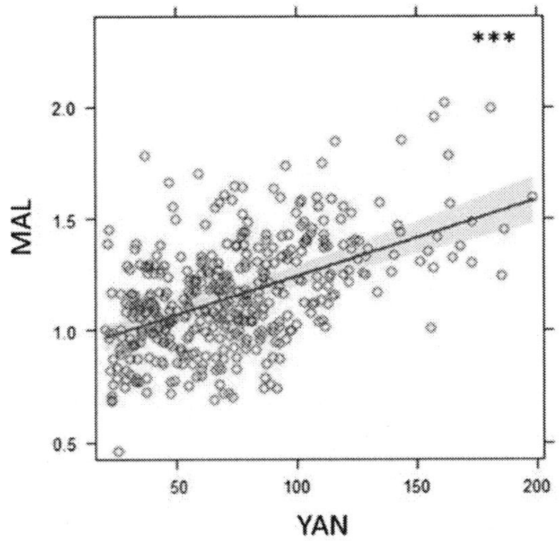

Figure 3b. Linear regression between malic acid and YAN (t=8.96; p<0.001..

Vine vigor is also driven by vine nitrogen status, as shown by the highly significant relation between YAN and pruning weight per vine (Figure 4a) and per shoot (Figure 4b).

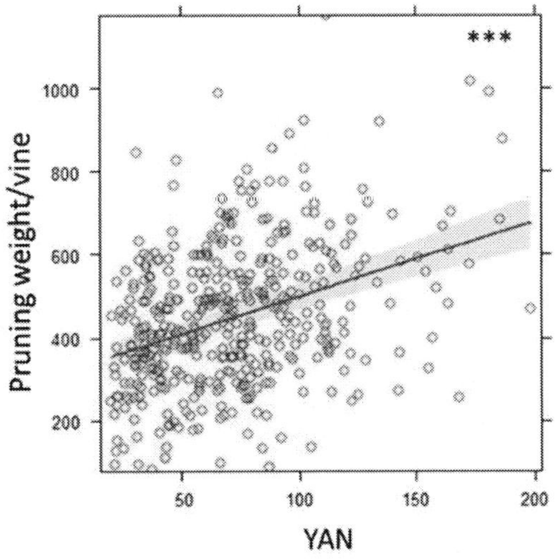

Figure 4a. Linear regression between pruning weight per vine and YAN (t=8.17; p<0.001).

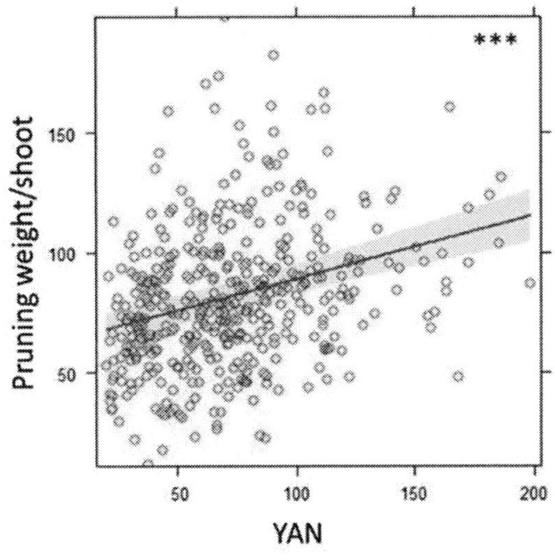

Figure 4b. Linear regression between pruning weight per shoot and YAN (t=7.65; p<0.001).

3.2 Vigor, yield and grape composition in relation to soil preparation before planting

Soils were deeper than 110 cm in all parcels with invasive soil preparation (sector A). Soils were shallow (55 to 80 cm deep before reaching the undisturbed unconsolidated limestone bedrock) in three of the four soil pits located in sector B (non-invasive soil preparation).

Stem water potential readings were taken around the location of the soil pits. At the end of the season, stem water potential readings were between 0.7 and 0.9 MPa in sector A, indicating weak water deficits. Three out of

 https://doi.org/10.1051/e3sconf/20185001015

the four readings were lower than -1.0 MPa in sector B, indicating moderate water deficits.

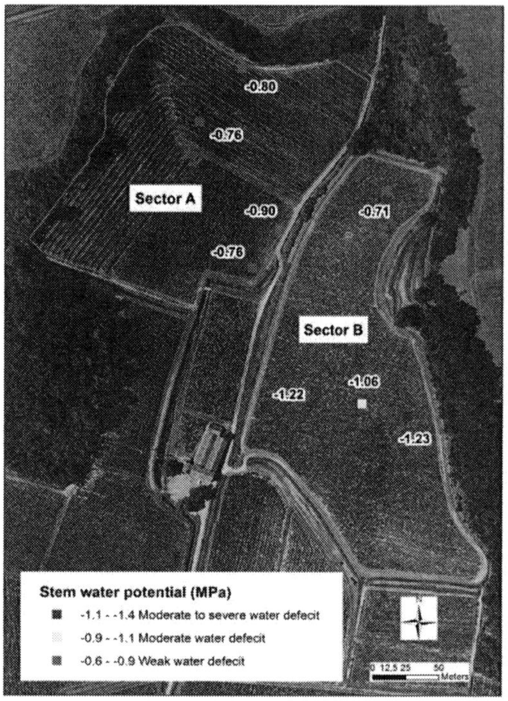

Figure 5. Stem water potential values at the location of the soil pits measured at the end of the season (21 September 2016).

Vine nitrogen status, as assessed by N-tester readings, was very heterogeneous in sector A, where soil was moved by bulldozers prior to planting. Vine nitrogen status was rather homogeneous in sector B where pre-planting soil preparation had been less invasive.

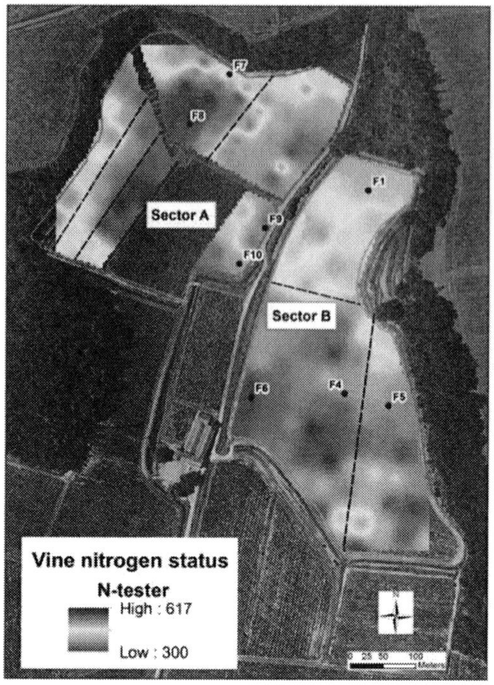

Figure 6. Vine nitrogen status map of sectors A and B, created by N-tester measurements at veraison over a regular grid of 142 data points.

Berry weight (Figure 7) and yield components (Table 1) were highly variable on the blocks with intense soil preparation (sector A). Berry weight and yield where low where vine nitrogen status was low (F7 and F9) and high where vine nitrogen status was high (F8 and F10; Figure 7). Homogeneous nitrogen status in sector B (F1, F4, F5 and F6) resulted in little variation in berry weight (Figure 7) and bunch weight (Table 1). However, bunch number and yield were high in F6, which was located down the slope and where vine N-status was slightly higher compared to other locations in sector B.

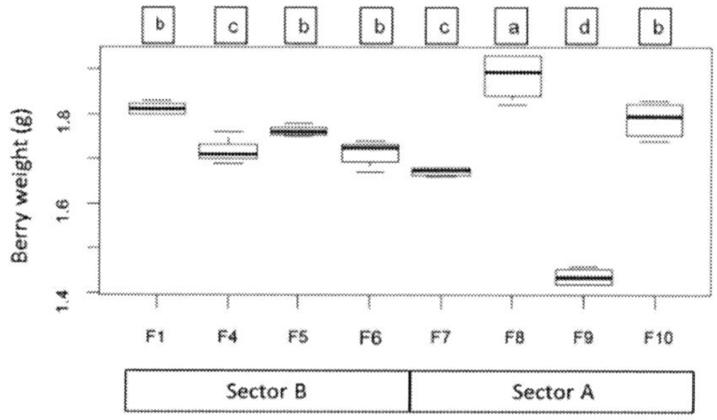

Figure 7. Berry weight at the location of the soil pits.

Table 1. Yield parameters at the location of the soil pits.

	Sector B				Sector A			
	F1	F4	F5	F6	F7	F8	F9	F10
Bunches/vine	9	8	9	13	6	11	6	11
Yield/vine (kg)	1.93	1.51	2.28	3.45	0.50	2.88	0.73	2.38
Bunch weight (g)	223	187	243	242	93	243	126	218

Grape composition was more heterogeneous in sector A compared to sector B (Table 2). In sector A, grape quality potential was very high on the locations with low vine nitrogen status (F7, F9), showing particularly high total anthocyanin concentrations, but low on locations with high vine nitrogen status (F8, F10). In low nitrogen locations, the high quality potential was at the expense of particularly low yields (0.5 kg/vine on F7, 0.73 kg/vine on F9).

Table 2. Grape composition at ripeness at the location of the soil pits.

		N-tester veraison	Grape sugar	TA	MAL	Total anthocyanin	Total phenolics index
Sector B	F1	average	220	3,9	1,9	1195	42
	F4	average	241	3,4	1,3	1537	51
	F5	average	241	3,4	1,4	1519	48
	F6	slightly high	225	3,9	1,5	1449	44
Sector A	F7	low	237	3,0	1,2	1804	49
	F8	high	215	3,8	2,2	1308	39
	F9	very low	240	2,8	1,1	1934	53
	F10	slightly high	221	3,7	1,9	1563	43

4 Discussion and conclusion

In the geological context of this study (limestone soils on Tertiairy bedrock with moderately high soil water holding capacity), grape quality potential and vine vigor were predominantly driven by vine nitrogen status. Only weak water deficits were recorded around the soil pits where the soil was profoundly modified before plantation, probably because increased soil depth allowed vines to have access to greater soil water reserves through deeper rooting. Moderate water deficits were recorded around three of the four soil pits located in downhill planted parcels and vine nitrogen status was moderate to low in all four. Vine nitrogen status was highly variable in parcels where the soil was moved before plantation (including terraces): in locations where the topsoil was removed vine nitrogen status was very low and where topsoil was accumulated, vine nitrogen status was very high. In those parcels with highly variable vine nitrogen status, grape quality potential was high in places with low nitrogen status (small berries, high sugar and anthocyanins), but vigor and yield were particularly low. In the same parcels, at locations where vine nitrogen status was high, grape quality potential was low (large berries, low sugar, anthocyanins and total phenolics, high malate) but vigor and yield were high. This study shows that in locations where soil water is not much limiting because of medium to high soil water holding capacity, vine nitrogen status is a major driver of the soil effect on yield, vigor and grape composition. Downhill plantations with minimal soil preparation do better respect soils and provide more homogeneous performances in terms of vigor, yield and grape quality potential. When the soil is moved during soil preparation prior to plantation (which is the case when creating terraces), vine water status is less limited because soil depth is increased and vine nitrogen status is highly variable, resulting in very heterogeneous performances in terms of grape quality potential, yield and vigor. The study also emphasizes that vine nitrogen and water status are relevant parameters for the interpretation of vine behavior in response to soil type, including anthropogenic modifications.

5 Acknowledgements

This study has been carried out with financial support from the French National Research Agency (ANR) in the frame of the Investments for the future Programme, within the cluster of excellence COTE (ANR-10-LABX-45). We are grateful to the staff of château Fombrauge F-33330 Saint-Christophe des Bardes and the Research and Development unit from *vignobles Bernard Magrez* for technical and financial support.

References

1. G. Seguin G., Experientia, **42** 861-873 (1986).
2. C. van Leeuwen, J.-P. Roby, L. de Resseguier, Practical Winery and Vineyard, **52** August 2017, 52-55 (2017).
3. J. Brenot, A. Quiquerez, C. Petit, J.-P. Garcia, Geomorphology, **100** 345-355 (2008).
4. C. van Leeuwen C., *In: Managing wine quality, volume 1: Viticulture and wine quality, Reynolds A. Ed., Woodhead Publishing Ltd.,* Oxford, UK. ISBN 978-1-84569-484-5, 587p (2010).
5. C. van Leeuwen, O. Trégoat, X. Choné, B. Bois, D. Pernet J.-P. Gaudillère, J. Int. Sci. Vigne Vin, **43** 121-134 (2009).
6. C. van Leeuwen, P. Friant, J.-P. Soyer, C. Molot, X. Chone, D. Dubourdieu, J. Int. Sci. Vigne Vin., **34** 75-82 (2000).
7. E. Chevigny E., 2014. Cartographie de la diversité des sols de versant par imagerie à haute résolution : contribution à la connaissance des terroirs. Thèse de doctorat de l'Université de Bourgogne.

Variability of climate, water and nitrogen status and its influence on vine phenology and grape composition inside a small winegrowing estate

Laure de Rességuier[1], Renan Le Roux[2], Théo Petitjean[1], Séverine Mary[3], Hervé Quénol[2] and Cornelis van Leeuwen[1]

[1]EGFV, Bordeaux Sciences Agro, INRA, Univ. Bordeaux, ISVV, 33883 Villenave d'Ornon, France
[2]LETG-RENNES, UMR 6554 CNRS, Université de Rennes 2, 35000 Rennes, France
[3]Vitinnov, Bordeaux Sciences Agro, ISVV, 33175 Gradignan cedex, France

Abstract. Climate is a major terroir factor in viticulture. In winegrowing regions, climate is studied at an increasingly refined scale. Results from the Life ADVICLIM project show substantial spatial variability of temperatures inside the region of Saint-Emilion, Pomerol and surrounding appellations (Bordeaux, France). In this study we investigated climatic variability at an even more refined scale, inside an 11 ha estate located in Saint-Emilion with significant topographic variability, planted with Merlot and Cabernet franc. Elevation ranges from 34 to 81 meters. 31 temperature sensors were set-up in 2013 inside the canopy, taking into account all parameters linked to the topography. Spatial temperature variability and its influence on vine phenology and grape composition were investigated. Vine water and nitrogen status were also taken into account through $\delta^{13}C$ and N-tester measurements. Over the growing season (April 1 through September 30), spatial temperature variability was greater on minimum temperatures (1.6°C) compared to maximum temperatures (1.3°C). Spatial variability in minimum temperature was driven by elevation and slope. Further investigation is required for spatial variability in maximum temperature, which could not be explained by environmental co-variables. Temperature variability among vintages was driven by maximum temperature, while minimum temperature showed little variation from one year to another. The average Winkler Index measured in the canopy ranged from 1774 degree.days to 1978 degree.days. This spatial variability of 204 degree.days can induces potentially 20 days difference in maturity dates. The timing of flowering varied from one vintage to another but inside a given vintage spatial variability was small, and so was variability induced by the cultivar. Veraison dates were highly variable among vintages. Inside a given vintage, spatial variability and cultivar effect were significant. Berry weight was driven by vine water and nitrogen status. Berry malic acid content was impacted by temperature. Vine water and nitrogen status also influenced berry malic acid content. Berry sugar content depended mainly on vine nitrogen status. The grapevine variety influenced berry sugar and malic acid content, as well as berry mass.

1 Introduction

Environmental factors like climate, soil, topography, play an important role on vine development [1] and grape quality potential [2].

In this study, climate is analysed at very local scale in an estate of 11.2 ha located in a part of Bordeaux area defined by a pronounced relief and multiple exposures.

The distinctive feature of this study is the density of the temperature sensor network: 1 sensor for 0.4 ha which allows to characterize temperatures at the winegrowing estate scale. On each parcel equipped with a sensor, phenological stages were monitored as well as grape composition and vine water and nitrogen status.

The first objective of this study is to characterize the temperature variability inside a winegrowing estate and to define the environmental parameters that contribute to this spatial distribution. The other major objective is to analyse the parameters which impact vine development and grape composition. Beside temperatures, the combined effect of soil and climate was investigated through vine water and nitrogen status.

The result of this study allows winegrowers to better adapt plant material and vineyard management practices to terroir components, especially in a context of climate change.

2 Materials and methods

2.1 Temperature network

31 temperature sensors (Tinytag Talk2, Gemini Data Loggers, UK) have been set up at the beginning of 2013, to monitor temperatures on this estate of 11.2 ha with an altitudinal gradient of 50 meters (Figure 1). To choose the location of the sensors, a Digital Elevation Model was created and used to cover the environmental parameters: elevation, slope and exposures.

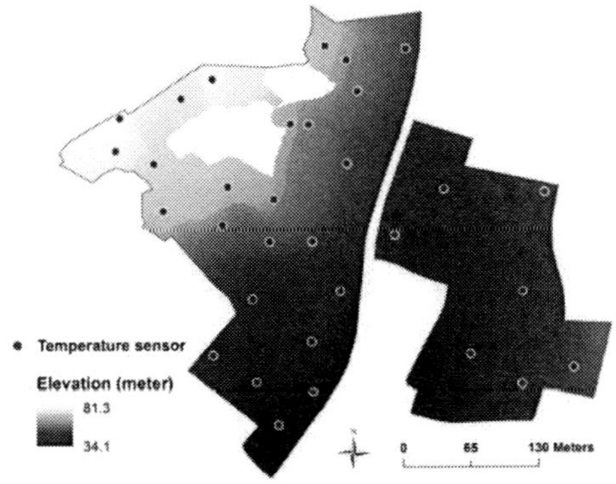

Figure 1. Location of the temperature sensors projected on a Digital Elevation Model

The temperature sensors were installed on vine posts inside vineyard parcels and record both minimum and maximum hourly temperatures. Temperature sum base of 10°C was computed from 1 April through 31 October. Because temperatures were measure in the canopy and not in a classical weather station, this temperature sum is referred to as "Canopy Winkler Index".

2.2 Ecophysiological measurements

Ecophysiological measurements have been carried out to monitor vine development and grape composition on 31 blocks located near the temperature sensors. This property is mainly planted with Merlot and Cabernet franc. One block is composed of 20 successive vines. In this estate, some parcels contain mixed plantations of Merlot and Cabernet franc. When those two grape varieties are present in the same block, 10 vines per variety are observed.

Phenological stages (flowering and veraison) were monitored, and the specific day when 50 percent of vine organs reach stage "I" for flowering and stage "M" for veraison [3] are recorded.

Once a year, a maturity control is carried out for all the blocks at the same date. Each sample is composed of 60 berries. Berries were weighted and Reducing Sugars, Total Acitidy (TA), Malate (MAL) and pH were measured. In order to determine vine water status and vine nitrogen status, $\delta^{13}C$ and Yeast Available Nitrogen (YAN) was measured [4,5].

Vine nitrogen status was also assessed through N-tester measurements at veraison stage with 3 replicates per block [5].

2.3 Statistical analyses

The effects of environmental parameters (elevation, slope, exposure) on mean, minimum and maximum average daily temperatures and on Canopy Winkler Index were analyzed by using multiple linear regressions. The maturity parameters have been analysed by a linear mixed model [6] including minimum temperature, YAN, $\delta^{13}C$ and grape variety as fixed effects. We considered "vintage" as a random effect to account for the repeated measurements conducted each year.

For creating the map of Canopy Winkler Index, a model based on a Support Vector Regression (SVR) was used [7].

3 Results and discussion

3.1 Growing season temperature variability

Figure 2 represents the daily mean, minimum and maximum temperatures during the growing season (April 1st to September 30th) over the 5 years studied.

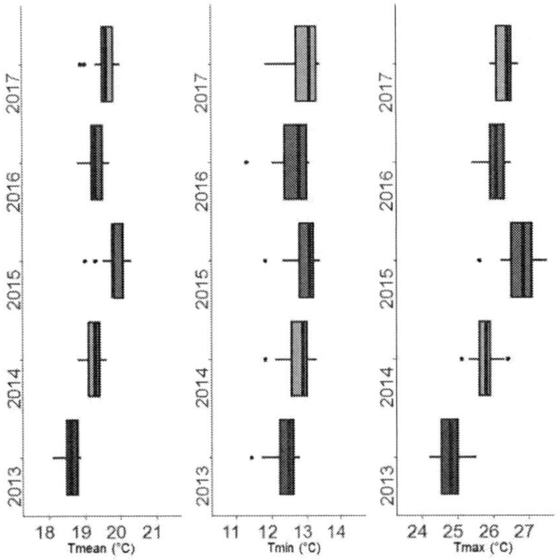

Figure 2. Boxplots on mean, minimum and maximum average daily temperatures over the growing season (from April 1st to September 30th)

A clear vintage effect is observed: 2015 is the warmest vintage and 2013 the coldest.

Inter annual variability is greater on mean and maximum temperatures than on minimum temperature which is more or less similar and constant each year. So the vintage effect is due to variations in maximum temperature which impact the mean temperature.

The range of temperatures between the coldest and the warmest sensor is greater for minimum temperatures (1.6°C in average on the 5 vintages) than for maximum temperatures (1.3°C in average). For mean temperature there is a range of 1°C (5 years average) between the coldest and the warmest sensor.

Statistical analysis (Table 1) show relationships between temperatures and topographic parameters. Minimum temperatures are driven by elevation and slope. At this local scale, the minimum temperature increases with elevation and slope. In 2017, there is also a relation with North/South exposure. South exposure is warmer than North exposure.

 https://doi.org/10.1051/e3sconf/20185001016

For maximum temperature, no linear relationships were found between temperature and topographic variables. The variations of maximum temperature at this micro scale is more complexe than for minimum temperature, with certainly more thermic variations during the day but also from day to day.

As for minimum temperature the spatial distribution of mean temperature is explained by elevation and slope.

Table 1. Relationships between average temperatures during vegetative season and environmental covariables

Year	Temp growing season	variance (%) explain by the model	Altitude	Slope	N/S exposure	E/W exposure
2013	Tmin	75	***	***	ns	ns
	Tmax	11	ns	ns	ns	ns
	Tmean	39	*	*	ns	ns
2014	Tmin	82	***	***	ns	ns
	Tmax	22	*	ns	ns	ns
	Tmean	34	*	ns	ns	ns
2015	Tmin	83	***	***	ns	ns
	Tmax	6	ns	ns	ns	ns
	Tmean	37	**	*	ns	ns
2016	Tmin	78	***	**	ns	ns
	Tmax	27	ns	ns	ns	ns
	Tmean	51	***	**	ns	ns
2017	Tmin	87	***	***	*	ns
	Tmax	30	ns	ns	ns	ns
	Tmean	69	***	***	**	ns

*ns: not significant; *** significant at $p<0.001$; ** significant at $p<0.01$; * significant at $p<0.05$*

3.2 Canopy Winkler Index

In order to improve the characterization of climate variability, the Winkler degree day summation [8], which is well adapted to study the influence of temperature on vine development, is used here. This index is based on the sum of temperatures above 10°C, from April 1st to October 31st. Because temperatures were measured inside the canopy, this index is referred to as Canopy Winkler Index.

Result shows that at this local scale, there is a variability of 204 degree.days on average over the 5 years studied between the coolest and the warmest sensor (Figure 3). Given this wide temperature range, maturity could be delayed by 20 days in the latest ripening parcels, compared to early ripening parcels.

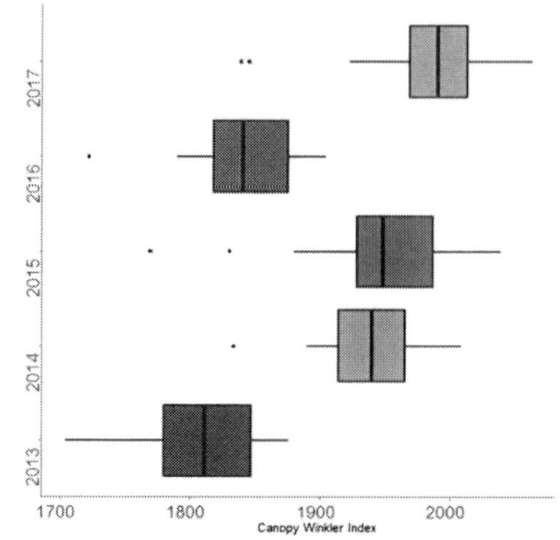

Figure 3. Boxplots of Canopy Winkler Index from 2013 to 2017

The elevation, slope and North/South exposure are covariables that explain the distribution of Canopy Winkler Index (Figures 4). Canopy Winkler Index increase by 50 degree.days with an increase in elevation of 40 meters (Figure 4a) and with steep slopes (Figure 4b). South exposure is warmer than North exposure (Figure 4c) and there is no linear relationship with East/West exposure.

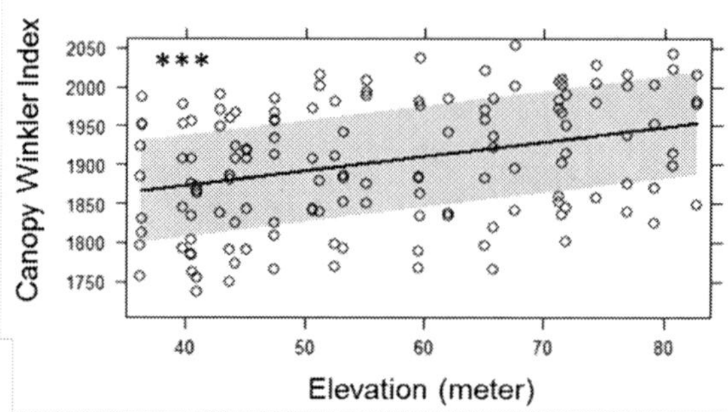

Figure 4a. Relationship between Canopy Winkler Index and elevation

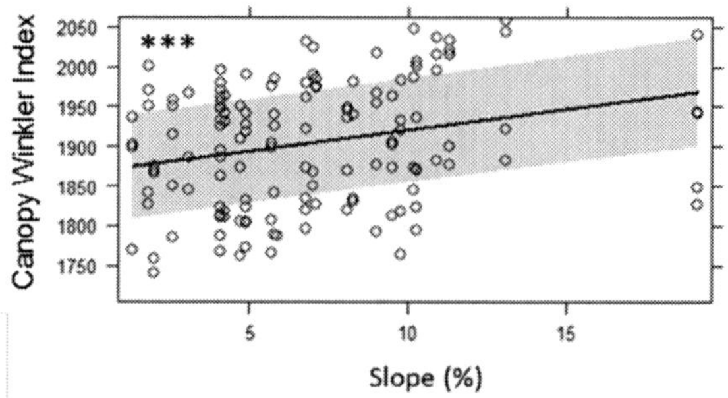

Figure 4b. Relationship between Canopy Winkler Index and slope

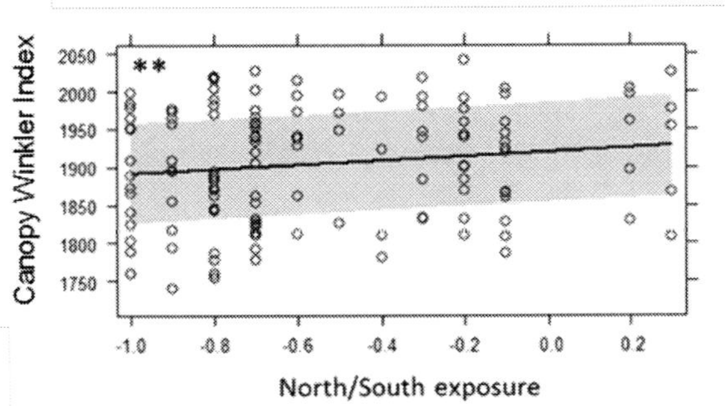

Figure 4c. Relationship between Canopy Winkler Index and North/South exposure

The map of average Canopy Winkler Index from 2013 to 2017 (Figure 5) shows that spatial distribution of the index is linked to the topography. The limestone plateau located at the top of the hill on the Western part of the estate and the East facing slopes are the warmest areas. The bottom of the valley is the coldest area. Cooler areas are also located on the West facing slopes.

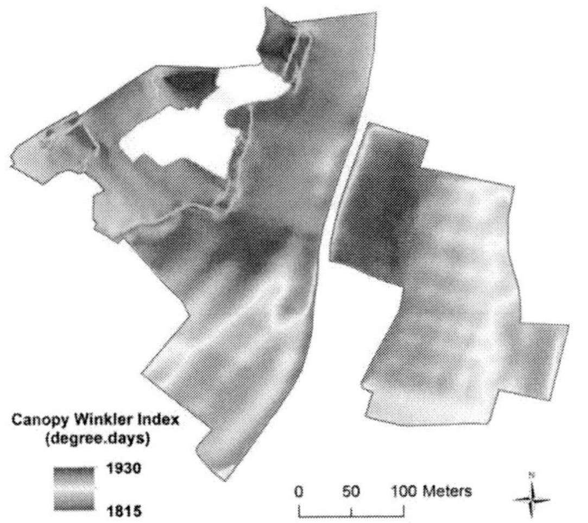

Figure 5. Spatial distribution of Canopy Winkler Index (average 2013-2017)

3.2 Phenology observations

The phenology recordings of Merlot and Cabernet franc did not reveal significant differences between the grape varieties at flowering (Figure 6). The timing of flowering varied from one vintage to another but inside a given vintage variability was small.

For veraison a grape variety effect is observed, Merlot is more precocious. Amplitude of veraison dates vary with the vintage. It is on average greater for Cabernet franc.

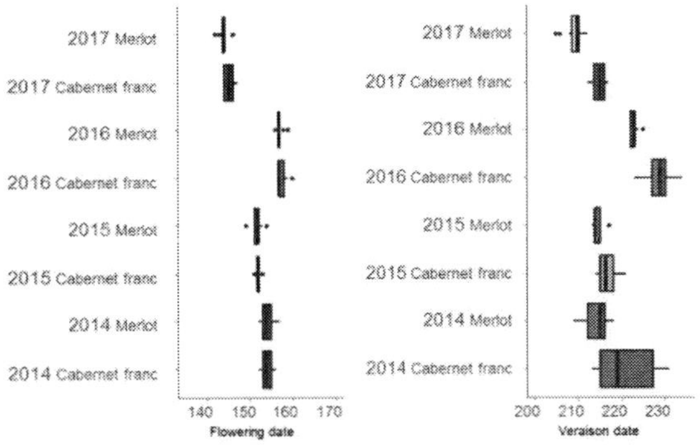

Figure 6. Boxplots of phenology observations from 2013 to 2017

3.3 Relations between temperature, vine water status, vine nitrogen status and grape composition

In order to investigate the relations between temperature, vine water status, vine nitrogen status, berry weight and grape composition, statistical analyses were carried out (Table 2).

For all parameters a grape variety effect is shown: Merlot has bigger berries and is earlier than Cabernet franc.

Berry weight is well explained by YAN, $\delta^{13}C$ and minimum temperature over growing season (Table 2) which is also linked with elevation and slope (Table 1). Berry weight decreases with the elevation of temperature and water deficit, and increases with YAN.

For grape sugar content there is a strong opposed link with YAN and a small link with temperature. Gape sugar content decreases with YAN, while Malate increases with YAN. Malate is also linked with $\delta^{13}C$ and minimum temperature. The increase of water deficit and minimum temperature decrease berries malate concentration.

For the maturity indicator Sugar/TA, there is a strong effect of YAN. This indicator decreases with an increase of YAN. There is also a small relationship with $\delta^{13}C$, the level of maturity increases with water deficit.

Table 2. Relationships between berry weight, Sugar and Malate with minimum temperature during the growing season, $\delta^{13}C$, YAN and grape variety

	Tmin Growing season	$\delta^{13}C$	YAN	Grape Variety
Berry Weight	*** Berry weight↘ when Tmin↗	*** Berry weight↘ when $\delta^{13}C$ ↗	*** Berry weight↗ when YAN ↗	* M > CF
Sugar	* Sugar↗ when Tmin↗	ns	*** Sugar↗ When YAN↘	*** M > CF
Malate	*** Malate↘ when Tmin↗	*** Malate↘ when $\delta^{13}C$ ↗	** Malate↘ When YAN↘	*** M < CF
Sugar/TA	ns	* Sugar/TA↗ when $\delta^{13}C$ ↗	*** Sugar/TA↘ when YAN ↗	* M > CF

*ns: not significant; *** significant at p<0.001; ** significant at p<0.01 ; * significant at p<0.05*

https://doi.org/10.1051/e3sconf/20185001016

4 Conclusion

This innovative study at this very local scale highlights the great temperature range that can be found in such a small territory with more than 200 degree.days of amplitude for Canopy Winkler Index between the coldest and the warmest sensors. For comparison, inside the region of Saint-Emilion, Pomerol and surrounding appellations which extends over a larger area (12200ha), 280 degree.days of amplitude is found for the same period [9].

Temperature variability among vintages was driven by maximum temperature, while minimum temperature showed small variation from one year to another. The great spatial variability of minimum temperature is driven by elevation and slope. But further investigation is required for spatial variability of maximum temperature, which could not be explained by these environmental variables.

The timing of flowering varied from one vintage to another but inside a given vintage spatial variability is small. Veraison dates were highly variable among vintages and inside a given vintage, spatial variability and cultivar effect were significant.

For maturity, berry weight is driven by vine water and nitrogen status and minimum temperature. There is a strong significant effect of nitrogen status and a smaller effect of minimum temperature on berry sugar content. Berry malic acid content is related to vine water status and minimum temperature with lower impact of vine nitrogen status.

The results of this study will allow winegrowers to better adapt plant material and vineyard management practices to terroir components which is of particular importance in a context of climate change.

Acknowledgements

This study has been carried out with financial support from the Conseil Interprofessionnel du Vin de Bordeaux, the European Union Life Adviclim project (LIFE 13 ENV/FR/001512), the Conseil des Vins de Saint-Emilion.

References

1. A. Parker, I. Garcia de Cortazar Atauri, C. van Leeuwen and I. Chuine. Aust. J. Grape Wine Res., **17**, n°2, 206-216 (2011).
2. C. van Leeuwen, P. Friant, X. Choné, O. Tregoat, S. Koundouras, D. Dubourdieu. Am. J. Enol. Vitic. **55**, 207–217 (2004).
3. M. Baggiolini, *Rev. Rom. Agr. Vitic.* **8**:4-6 (1952).
4. C. van Leeuwen, O. Trégoat, X. Choné, B. Bois, D. Pernert and J-P. Gaudillère. J. Int. Sci. Vigne Vin, **43**, n°3, 121-134 (2009).
5. C. van Leeuwen, Ph. Friant, J-P. Soyer, C. Molot, X. Chone, and D. Dubourdieu. J. Int. Sci. Vigne Vin. **34**, n°2, 75-82 (2000).
6. J.C. Pinheiro, D.M. Bates, Springer. New York (2000).
7. R. Le Roux, L. de Rességuier, T. Corpetti, N. Jégou, M. Madelin, C. van Leeuwen, H. Quénol, Agricultural and Forest Meteorology, **247**, 159-169 (2017).
8. A.J. Winkler, J.A. Cook, W.M. Kliewer, et al., *General Viticulture, 2nd ed, University of California Press,* 710p (1974).
9. L. de Rességuier, R. Le Roux, H. Quénol and C. van Leeuwen. *Proceedings of Climwine,* 66-72 (2016).

Composición aminoacídica del mosto de los cv. Cigüente, Macabeo, Moscatel de Alejandría y Verdejo cultivados en diversas regiones de España. Efecto del régimen hídrico a lo largo de tres campañas

M.E. Valdés[1]*, D. Moreno[1], D. Uriarte[1], A. Montoro[2], J.R. Castel[†3] y J. Yuste[4]

[1]Instituto Tecnológico Agroalimentario de Extremadura (CICYTEX), 06071 Badajoz (Spain)
[2]Instituto Técnico Agronómico Provincial, 02007 Albacete (Spain)
[3]Instituto Valenciano de Investigaciones Agrarias, Moncada 46113 Valencia (Spain)
[4]Instituto Tecnológico Agrario de Castilla y León, 47071 Valladolid (Spain)

Resumen. Este trabajo analiza el efecto del riego sobre la composición aminoacídica del mosto de 4 variedades blancas cultivadas en diferentes DD.OO. españolas durante el periodo 2012-2014. Se estudió la concentración de aminoácidos precursores de aromas de fermentación: Aspártico (Asp), Isoleucina (Ile), Leucina (Leu), Treonina (Thr), Tirosina (Tyr) y Valina (Val), y la suma de ellos (APN), del aminoácido azufrado Metionina (Met) y del total de aminoácidos libres (FAN). Las variedades, su localización y los tratamientos aplicados fueron: Cigüente (Ci) en Badajoz, secano y riego al 100% de ETc; Macabeo (Ma) en Albacete, riego a 25% y 33% aprox. de ETc; Moscatel de Alejandría (Mo) en Valencia, riego a 50% y 100% de ETc; y Verdejo (Ve) en Valladolid, secano y riego a 100% aprox. de ETc. El mosto analizado tenía un valor de 22-24 °Brix. Los análisis de aminoácidos se efectuaron mediante HPLC. Los tratamientos hídricos aplicados modificaron los valores de los aminoácidos en tendencia, significación y extensión según la variedad. En Mo apenas variaron. En Ci y, sobre todo, en Ma el efecto en amplitud y significación dependió del año. En Ve el riego provocó todos los años del estudio idéntica tendencia al descenso en los valores de los aminoácidos. Por tanto, la respuesta de la composición nitrogenada del mosto al régimen hídrico y las condiciones meteorológicas anuales fue diferente en cada variedad cultivada en diferente lugar.

1 Introducción y objetivos

El viñedo español está mayoritariamente localizado en zonas de clima árido y semiárido. Este hecho, unido a la situación climatológica actual, hace que el riego sea una práctica vitícola cada vez más usual. La cantidad de agua aplicada y el momento de aplicación determinan la composición físico química de la uva y, por tanto, las características del vino elaborado [1] [2]. Además, su efecto también depende en gran medida de la interacción de esta técnica con otros factores, entre los que destacan el cultivar y la situación edafoclimática en la que se encuentre el viñedo, así como el nivel de carga y el equilibrio vegetativo-productivo de las vides, esto es, del denominado efecto "terroir" [3] [4] [5]. Existe una amplia bibliografía en relación con el efecto del estado hídrico de la cepa en variedades tintas [6] [7]. Sin embargo, la información disponible sobre el impacto del riego en variedades blancas es aún escasa. Debido al creciente consumo de vino blanco, el riego se está comenzando a implantar en variedades blancas en la mayoría de zonas vitícolas españolas, siendo necesario el estudio de sus efectos en las variedades típicas y más extendidas en cada zona.

En cuanto a los metabolitos primarios del mosto, la cantidad y la distribución de aminoácidos tienen una importancia primordial. En lo referente a la cantidad global, porque es necesario un nivel mínimo del conjunto de todos ellos, excepto de la prolina (Pro), para el desarrollo y la finalización completa de la fermentación alcohólica. En cuanto a la distribución, porque determinados aminoácidos son precursores de sustancias volátiles, alcoholes, esteres y aldehídos, que se forman durante dicho proceso y que son responsables del aroma del vino y, por tanto, de su calidad organoléptica [8]. Frente a éstos, los aminoácidos azufrados, cistina, cisteína y metionina, son precursores de sustancias que causan mal olor y, por tanto, devalúan la calidad sensorial [9]. La distribución y la cantidad de aminoácidos de un mosto varía en función del cultivar, las condiciones edafoclimaticas, las técnicas de cultivo e incluso con el estado de maduración de la baya [10] [11]. En vista de todo lo anterior, es imprescindible determinar la respuesta, en cuanto a composición aminoacídica de las diferentes variedades en sus diferentes terroirs y conocer el efecto que las prácticas agronómicas tienen en dicha composición.

El objetivo de este trabajo es doble: analizar la composición aminoacídica del mosto procedente de variedades blancas cultivadas en diferentes zonas vitícolas españolas e investigar la variación de dicha composición en función del régimen hídrico y de la campaña. El estudio se ha efectuado en mostos que presentan similar concentración de azúcares, pues se trata de observar la incidencia del estado hídrico en la composición amínica de mostos que producirían vinos de similar grado alcohólico.

https://doi.org/10.1051/e3sconf/20185001017

2 Materiales y métodos

2.1 Características de los viñedos

El experimento se realizó durante las campañas 2012 a 2014 en tres zonas vitícolas de España, con las siguientes variedades: "Cigüente" (Ci, sinónimo Doña Blanca, en Lobón, Badajoz), "Moscatel de Alejandría" (Mo, en Vilamarxant, Valencia), y Verdejo (Ve, en Medina del Campo, Valladolid) y en 2013-2014 con la variedad "Macabeo" (Ma en Albacete). En la tabla 1 se muestran las principales características edafoclimatológicas de los viñedos y los respectivos climas vitícolas [12].

En todos los casos, se trata de viñedos conducidos en espaldera, mediante posicionamiento vertical de la vegetación, podados en cordon Royat biltaeral y sometidos a similares prácticas de cultivo. Los suelos, en todos los casos, son de tipo calcáreo, de reacción alcalina y de mediana a alta capacidad de almacenamiento de agua en general.

Tabla 1. Características de los viñedos en las diferentes localizaciones.

	Cigüente	Macabeo	Moscatel	Verdejo
Coordenadas	38°03'N, 2°06'W	39°14' N, 2°5' W,	39°33'N, 0°42'W	42°21'N, 4°56'W
Altitud, m	200	695 m	197	785
Patrón	110R	110R	161-49	110R
Año de plantación	2005	1999	1996	2006
Marco, m	2,95 x 1,35	3,00 X 1,5	2,75 x 1,8	2,6 x 1,25
N° de cepas/repetición	20	18	24	20
N° de repeticiones	4	3	4	4
Textura del suelo	Franco Arcilloso	Franco-Arcilloso-Arenoso	Franco Arcilloso	Franco-Arcillo-Arenoso
Profundidad del suelo, m	0,75	0,5-1,0	>1,5	>1
Capacidad de retención de agua, mm/m	130	130	200	120
Clima vitícola	Cálido, noches frías	Muy cálido, noches muy frías	Cálido, noches templadas	Cálido, noches muy frías
	Muy seco	Mod. Seco	Mod. seco	Mod. Seco

2.2 Tratamientos de riego y diseños experimentales

En la Tabla 2 aparecen descritos los tratamientos ensayados en las diferentes localizaciones. Las variables climáticas se midieron en estaciones meteorológicas automáticas ubicadas en cada parcela experimental o en su vecindad, y con ellas se calculó la evapotranspiración de referencia (ETo) mediante la ecuación de Penman-Monteith, [13]. La evapotranspiración del cultivo (ETc) se estimó como producto de ETo y del coeficiente de cultivo (Kc). El riego se aplicó a lo largo del ciclo vegetativo en función de la ETc calculada.

El riego se aplicó por goteo con 1-2 emisores auto-compensados por cepa. No hubo problemas de salinidad, dadas las características del suelo y la calidad del agua, que era de relativamente bajo contenido en sal en todos los sitios. El estado hídrico de las plantas se evaluó mediante la medida del potencial hídrico de tallo (Ψ) con cámara de presión [14] [15]. Las mediciones se realizaron a mediodía (11:30 a 12:30 horas solares) a intervalos semanales o quincenales entre Mayo y Octubre, sobre 1-2 hojas por planta en al menos 6 plantas por tratamiento.

Tabla 2. Descripción de los tratamientos aplicados.

Variedad	Tratamiento	Descripción
Cigüente	Ci-S	No regado
	Ci-100R	Riego al 100 % de la ETc potencial
Macabeo	Ma-25R	Riego al 25% de la ETc potencial
	Ma-33R	Riego al 33% de la ETc potencial
Moscatel	Mo-50R	Riego al 50% de la ETc potencial
	Mo-100R	Riego al 100% de la Etc potencial
Verdejo	Ve-S	No regado
	Ve-100R	Riego al 100% de la Etc potencial

2.3 Metodología analítica

El contenido en sólidos solubles se determinó por refractometría (refractómetro digital ATR ST plus Schmidt+Haznsch, Berlin, Germany). El mosto de todos los tratamientos de cada variedad presentó, en el momento del análisis, un valor de Sólidos Solubles Totales de 22 ± 2 °Brix. El análisis de la composición aminoacídica de la uva de todas las variedades se llevó a cabo en el INTAEX según Valdés et al. [16]. Se efectuó mediante HPLC en un autoanalizador de aminoácidos (Biochrom 30, Pharmacia) empleando una columna intercambiadora de cationes. Como eluyentes se emplearon tampones de citrato de litio de diferente pH. Una vez separados, los aminoácidos reaccionan con ninhidrina y se detectan por espectrofotometría a 570 nm, excepto la prolina (Pro), que se detecta a 440 nm. La cuantificación se efectuó mediante calibración externa, empleando una recta de calibrado para cada aminoácido. A partir de las concentraciones de todos los aminoácidos se evaluó el valor de la concentración total de aminoácidos libres (*Free Aminoacids Nitrogen*, FAN) y de aminoácidos asimilables por las levaduras (*Aminoacid Assimilable Nitrogen*, AAA-N), restándole a FAN el valor de Pro. El cálculo de la concentración total de aminoácidos precursores de aromas de fermentación (*Precursors Aminoacid Nitrogen*) se efectuó como sumatorio de los aminoácidos Aspártico (Asp), Isoleucina (Ile), Leucina (Leu), Fenilalanina (Phe), Treonina (Thr), Tirosina (Tyr) y Valina (Val).

Análisis estadístico

En cada variedad, para determinar el efecto del régimen hídrico cada año, los datos se sometieron a un test de Student. Para establecer la significación del efecto año, en cada variedad y régimen hídrico se efectuó un ANOVA y el posterior test de separación de medias. La significación del factor variedad, se determinó mediante un ANOVA y el posterior test de medias al que se sometieron los datos procedentes de todos los tratamientos y años de todas las variedades. Con los datos medios procedentes de todos los tratamientos y años se realizó un ANOVA y el posterior test de medias y un análisis de componentes principales (ACP). El análisis estadístico fue ejecutado mediante el software informático XLstat-Pro (Addinsoft, París 2011).

3 RESULTADOS Y DISCUSIÓN

3.1 Estado hídrico de las cepas

La tabla 3 muestra las lluvias registradas durante los años del estudio y el estado hídrico de las cepas, a través del potencial hídrico de tallo de los diferentes tratamientos en las diferentes localizaciones.

Los datos expuestos en la tabla 3 muestran que la campaña 2013 fue la más lluviosa en todas las zonas vitícolas y que la más seca fue la 2012 en Ci y Ve y la 2014 en Mo. Respecto a Ma, las dos campañas en estudio, (2013 y 2014) fueron climatológicamente muy diferentes, mucho más húmeda la de 2013.

Tabla 3. Precipitaciones anuales, agua de riego aportada y estado hídrico (potencial de tallo) de las cepas durante el período vegetativo de las diferentes campañas

Variedad	Precipitaciones anuales registradas (mm)		
	Año		
	2012	2013	2014
Cigüente	268	585	514
Macabeo		480	187
Moscatel	280	352	197
Verdejo	122	407	329
Tratamiento	**Agua aportada (mm)**		
	Año		
	2012	2013	2014
Ci-S	0,00	0,00	0,00
Ci-100R	356,46	359,77	294,27
Ma-25R		70,00	123,61
Ma-33R		84,00	151,61
Mo-50R	156,05	120,00	131,09
Mo-100R	280,64	233,82	261,32
Ve-S	0,00	0,00	0,00
Ve-100R	250,71	208,28	216,05
Tratamiento	**ψtallo medio periodo vegetativo (MPa)**		
	Año		
	2012	2013	2014
Ci-S	-1,29	-1,18	-1,09
Ci-100R	-0,56	-0,62	-0,65
Ma-25R		-0,88	-1,20
Ma-33R		-0,61	-0,90
Mo-50R	-0,81	-0,69	-0,75
Mo-100R	-0,78	-0,61	-0,66
Ve-S	-1,18	-1,00	-1,07
Ve-100R	-0,77	-0,73	-0,81

En general, el riego disminuyó el estrés hídrico de las cepas de todas las variedades (valores de Ψ menos negativos). La reducción dependió en gran medida de la variedad considerada y para cada una de ellas, de la campaña considerada. Según los datos mostrados en la tabla 3, durante las tres campañas del estudio las cepas Ci-S y las Ve-S mostraron un grado de estrés [17] catalogado como *moderado a fuerte*, y el de las Ci-100R y Ve-100R fue *leve* según. La aplicación de riego en Mo fue la que menor efecto provocó en el estado hídrico de las cepas que mantuvieron siempre un grado de estrés clasificado como *leve a moderado*. Finalmente, en Ma la situación fue muy diferente en cada una de las dos campañas analizadas: en 2013 el grado de estrés de las Ma-25R y las Ma-33R fue *moderado a fuerte* y *leve a moderado* respectivamente; sin embargo, en 2014 las cepas Ma-25R sufrieron un estrés *severo* y las Ma-33R *fuerte a severo*.

3.2 Composición aminoacídica del mosto

3.2.1. Caracterización varietal

La tabla 4 muestra los valores medios interanuales de los aminoácidos analizados y del PAN, FAN y AAA-N de los de las variedades en estudio

De entre los aminoácidos precursores aromáticos Thr y Va, fueron en este orden, los que alcanzaron los mayores valores medios interanuales en todas las variedades estudiadas. Frente a ellos, Met fue el de menor concentración en todas ellas.

En todas las variedades, los valores medios de AAA-N se hallan en los intervalos normales para este parámetro. En todos los casos, la concentración es suficiente para asegurar una buena marcha de la fermentación alcohólica y una correcta finalización de la misma. La variedad más rica en aminoácidos fue Mo, cuyos valores medios de FAN fue superior a los del resto de variedades. Los mostos Ci presentaron los valores más bajos de este parámetro debido, entre otras causas, a los altos valores de Prolina, aminoácido no asimilable por las levaduras, hallados en esta variedad (valores no mostrados).

Tabla 4. Composición aminoacídica media interanual (2012-2014) de las variedades Cigüente, Macabeo, Moscatel y Verdejo.

Aminoácidos (mg/L)	Cigüente	Macabeo	Moscatel	Verdejo
Asp	20,04a[1]	18,67ab	14,67b	24,68a
Ile	11,56b	8,59b	11,86ab	15,17a
Leu	22,87a	14,65b	25,19a	20,68a
Phe	19,23bc	17,46c	23,73ab	25,12a
Thr	53,33b	50,97b	39,71c	87,39a
Tyr	7,34c	19,04a	16,35ab	10,68bc
Val	33,01a	29,88a	28,43a	32,59a
Met	4,65a	1,59b	4,41a	4,28a
Parámetros aminoacídicos (mg/L)				
PAN	167,39b	159,27b	159,94b	216,32a
FAN	1366,22b	1638,91a	1756,67a	1373,47b
AAA-N	983,03b	1029,59b	1263,91a	1094,94b

[1]Para un mismo aminoácido o parámetro, letras diferentes indican diferencias entre variedades ($p<0,05$)

En Ve se registraron los mayores valores medios interanuales de Asp, Ile, Phe y especialmente Thr. Estos aminoácidos son importantes precursores de aromas de fermentación [18]. En consecuencia, el valor medio de PAN fue superior a las restantes. El valor medio del PAN representa en esta variedad un 16% del FAN, superior al 12%, 10% y 9% en Ci, Ma y Mo. De estos resultados aquí expuestos puede deducirse que en función de su composición aminoácidica, Ve es la variedad con el potencial de aromas de fermentación más elevado de entre las analizadas.

3.2.2. Efecto de la campaña y del régimen hídrico

La tabla 5 muestra el efecto del año, el régimen hídrico y la interacción de ambos factores en la concentración de aminoácidos y en ambos parámetros analizados. En todas las variedades, y en ambos regímenes hídricos establecidos en cada una de ellas, la mayoría de aminoácidos y parámetros analizados, se vieron afectados de forma significativa por el factor año. Similares resultados hallaron Vilanova et al. en la variedad Mencía cultivada en diferentes zonas de la Ribeira Sacra (España) [19].

En la variedad Cigüente, los mayores valores de sustancias nitrogenadas se observaron en ambos regímenes hídricos, en el año más seco, el 2012, y en consecuencia el valor de FAN de ese año fue significativamente superior al de las otras dos. También en los dos tratamientos aplicados en Mo, las mayores concentraciones de aminoácidos también se registraron en la campaña más seca, que en esa localización fue la 2014. Recíprocamente, en Ve los valores más pequeños correspondieron al 2013 que fue la campaña más húmeda.

La amplitud y la significación del efecto de la campaña efecto fueron superiores a los del régimen hídrico, conclusión a la que también llegaron los estudios de Ortega et al. realizados en la variedad Verdejo en la D.O. Rueda, así como Bouzas et al. en el cultivar Albariño procedentes de Rías Baixas y Ribeiro [20] [21]. Hay que destacar que la variación debida a la campaña fue muy similar en cuanto a significación y extensión en los dos regímenes hídricos impuestos en cada variedad, de manera que, mediante la aplicación de agua en el caso de Ci y Ve o la aplicación de agua en mayor cantidad en los casos de Mo y Ma, no se evitaron las variaciones debidas al efecto campaña. Además, se observó interacción significativa tratamiento*año en un número variable de aminoácidos y parámetros según la variedad: 5 en Ci, 2 en Mo, 2 en Ve y en todos excepto Ile en Ma, variedad en la que el efecto del riego sobre la concentración de las sustancias nitrogenadas varió en función del año considerado.

En ninguna de las campañas en estudio de Ci, la aplicación de riego (Ci-S *vs* Ci-100R) modificó significativamente la concentración total de aminoácidos libres (FAN). En cambio, sí tuvo efecto sobre la de algunos aminoácidos, en concreto los valores de Phe de Ci-100R son menores que los de Ci-S en las tres campañas. Este resultado puede tener una repercusión negativa en el aroma de los vinos ya que este aminoácido es precursor del 2-fenil etanol, sustancia formada durante la fermentación alcohólica que aporta aroma a rosa a los vinos. En la campaña 2014, se observó un descenso generalizado en los valores de los aminoácidos mostrados en la tabla 4, que provocó una disminución del valor del PAN que del 15% en Ci-S pasó a suponer únicamente un 10% en Ci-100R. Únicamente en esta campaña, a diferencia de las otras dos anteriores, el valor de Asp en Ci-100R fue menor ($p<0,05$) que en Ci-S.

La respuesta a la aplicación de una mayor cantidad de agua de riego en la variedad Macabeo (Ma-25R *vs* Ma-33R) sobre la concentración de las sustancias en estudio, fue muy diferente en los dos años analizados. El primer año de la aplicación del riego no se observó ningún cambio significativo, en tanto que, en 2014, año más seco y en el que la aplicación de agua en Ma-33R fue superior al año anterior, se registró un aumento significativo de todos los aminoácidos y en consecuencia de los parámetros nitrogenados.

En la variedad Mo el estado hídrico de la cepa en las campañas 2012 y 2013 no tuvo repercusión en las concentraciones individualizadas de ninguno de los aminoácidos analizados En 2014, en Mo-100R respecto de Mo-50R, excepto en Asp, se hallaron descensos generalizados, de escasa amplitud en todos los casos, en los valores de todos los aminoácidos que fueron significativos en Phe, Thr. Estos descensos provocaron un descenso significativo del 8% del PAN. También se registraron disminuciones ($p<0,05$) del 7% de AAA-N y del 13% del FAN.

Finalmente, en la variedad Ve, la aplicación de agua (Ve-S vs Ve-100R) provocó en las tres campañas una similar tendencia al descenso en los valores de prácticamente todos los aminoácidos, excepto de Asp. En el año 2012, se hallaron disminuciones significativas de Ile, Leu, Phe, en 2013 de Ile, Leu, Thr, y Val, y en el 2014 únicamente de Ile. Estos resultados coinciden parcialmente con los de Ortega et al. [19]. En sus investigaciones, al igual que en este trabajo, el riego provocó descensos de los valores de Thr, Sin embargo, a diferencia de los resultados hallados en este trabajo, estos investigadores hallaron que, en uvas maduras, el riego en esta variedad provocó un aumento de la concentración de aminoácidos. Por otra parte, es interesante destacar que el sentido de la variación provocada por el riego dependió en gran medida de la madurez de las uvas. A este respecto hay que señalar que, en este estudio, el análisis se ha efectuado siempre en uvas maduras.

De lo anterior podría deducirse que la respuesta de la composición nitrogenada al estado hídrico de la cepa es diferente para cada cultivar en su medio tradicional. En algunas variedades, la repuesta es independiente de la climatología anual (Mo y Ve), frente a otras en la que respuesta dependió de ese factor (Ci y Ma). Centrándonos en estas últimas, en el caso de Ci, la respuesta tuvo siempre igual sentido pero diferente amplitud y en el caso de esta última varió en amplitud y signo.

Tabla 5. Efecto de la campaña y el tratamiento en la composición aminoacídica del mosto de las variedades Cigüente, Macabeo, Moscatel y Verdejo.

Aminoácidos (mg/L)	Año	Ciguente			Macabeo			Moscatel			Verdejo		
		CiS	Ci100R	Tr*año[3]	Ma25R	Ma33R	Tr*año	Mo50R	Mo100R	Tr*año	VeS	VeR	Tr*año
Asp	2012	25,54a[1]	40,76a***[2]					16,53a	16,82a		28,66a	39,62a***	
	2013	13,55b	18,59b*	**	23,00	16,23	***	15,52a	15,56a	ns	12,96c	11,00b	*
	2014	14,06b	7,75c***		10,74	25,20[4]***		11,37b	12,24b		22,71b	33,15a*	
Ile	2012	16,93	11,46a					8,06c	8,84b		29,75	9,73*	
	2013	10,88	10,27a	ns	9,34	10,40	*	12,91b	12,28a	ns	14,31	8,79***	ns
	2014	14,02	5,83b*		5,80	9,46**		15,20a	13,85a		19,98	8,48*	
Leu	2012	29,31	30,88a					17,50c	19,94b		38,36a	16,83a*	
	2013	20,79	20,16b	ns	16,25	16,94	***	27,32b	26,28a	ns	19,36b	10,99b***	ns
	2014	21,78	14,31c*		9,20	17,19****		31,21a	28,85a		24,63ab	13,92ab	
Phe	2012	30,32a	21,66a*					17,11c	18,32c		35,69a	18,71*	
	2013	16,26b	12,99b*	*	25,08	22,86	***	25,18b	23,38b	ns	22,67a	16,72	ns
	2014	26,19ab	7,96c*		6,85	18,32****		30,36a	28,01a*		35,43a	21,49	
Thr	2012	76,49a	66,70a					33,80a	35,94b		95,78ab	95,99	
	2013	34,45c	43,44b	**	46,70	34,11	***	32,27b	30,59c	ns	78,53b	52,11***	ns
	2014	58,34b	40,55b*		46,35	71,43****		55,48a	50,16a*		105,88a	96,03	
Tyr	2012	5,38b	5,48					10,72b	11,31b		7,47b	7,43b	
	2013	4,72b	5,22	ns	4,38	4,66	***	9,11b	7,97c	*	8,17b	6,96b	*
	2014	11,09a	12,17*		13,30	46,56**		31,77a	27,19a		21,54a	12,53a	
Val	2012	40,07a	42,18a					22,14c	23,99b		48,61a	29,69a	
	2013	27,28b	26,33b	ns	23,57	21,49	***	25,76b	25,01b	ns	25,68b	18,05b**	ns
	2014	34,81ab	27,41b*		27,64	43,14***		38,17a	35,51a		44,11ab	29,40a	
Met	2012	9,68a	7,75a					4,21b	4,62		7,85	4,31	
	2013	2,82b	4,13b	ns	nd	nd	***	4,49ab	4,42	ns	3,09	0,85	ns
	2014	2,64b	0,86c*		nd	nd		4,69a	4,02		6,63	2,94	
Parámetros aminoacídicos													
PAN	2012	224,04a	219,12a					125,95c	135,16b		284,33a	218,01a	
	2013	127,94b	137,01b	*	148,32	126,68	***	148,09b	141,08b	ns	181,68b	124,63b	ns
	2014	180,29ab	115,98b*		119,88	231,31****		213,56a	195,81a*		274,28ab	215,01a	
AAA-N	2012	1185,89a	1291,31a					1105,90b	1153,18b		1207,54ab	1062,10a	
	2013	618,07b	922,05b***	**	866,55	870,00	***	1093,87b	1061,84b	ns	963,34b	756,29b	ns
	2014	958,98a	921,86b		971,46	1329,69****		1643,14a	1525,48a*		1333,29a	1247,11a	
FAN	2012	1732,62a	1798,95a					1515,97b	1600,71b		1434,48	1351,66a	
	2013	1147,68b	1132,35b	ns	1330,25	1127,61	***	1581,86b	1488,10b	*	1330,76	910,71b	ns
	2014	1208,01b	1177,73b		1616,49	2276,29****		2330,81a	2022,57a**		1678,61	1534,61a	

[1] Diferentes letras para un mismo aminoácido y tratamiento indican diferencias significativas entre distintas campañas (p<0,05).

[2] *,** y *** indican para un determinado aminoácido y campaña, diferencias significativas (p<0,05, p<0,01 y p<0,001) debidas al tratamiento.

[3] *, ** y *** indican interacción significativa tratamiento x año (p<0,05, 0,01 y 0,001 respectivamente).

[4] *,** y *** en un mismo aminoácido y tratamiento en la variedad Macabeo, indican diferencias significativas (p<0,05, p<0,01 y p<0,001) entre las dos campañas analizadas.

Con el objetivo de facilitar la interpretación de los resultados correspondientes a la incidencia de los diferentes tratamientos en la composición aminoacídica interanual de los mostos, se sometieron los valores medios interanuales de cada tratamiento de Ci, Mo y Ve a un análisis de componentes principales. La razón de no incluir Ma se debe a que únicamente se dispone de dos años de esta variedad, y el comportamiento fue de diferente signo en cada uno de ellos. La figura 1 muestra los resultados de la aplicación del Análisis de Componentes Principales (PCA). El componente principal F1, que explica el 45,46% de la varianza total, permite diferenciar los tratamientos VeS y CiS de Ve100R y Ci100R respectivamente. Los primeros están situados a la derecha del eje de modo que poseen mayores valores de Ile, Met, Phe y PAN. Sin embargo, los valores de estos aminoácidos apenas diferencian Mo-50R de Mo-100R.

La componente principal F2, asociada principalmente a los valores de Thr en el eje positivo y a FAN en el negativo, permite diferenciar las muestras procedentes de los tratamientos con mejor estado hídrico procedentes de las diferentes variedades en función de los valores de Thr, Asp, FAN y AAA-N, de manera que las muestras con mayores valores de Thr son las Ve-100R, mientras que las de Mo-50R son las de mayor FAN. También este eje permite constatar la poca diferencia exixtente entre la composición de las muestras M-100R y la delas M-50R

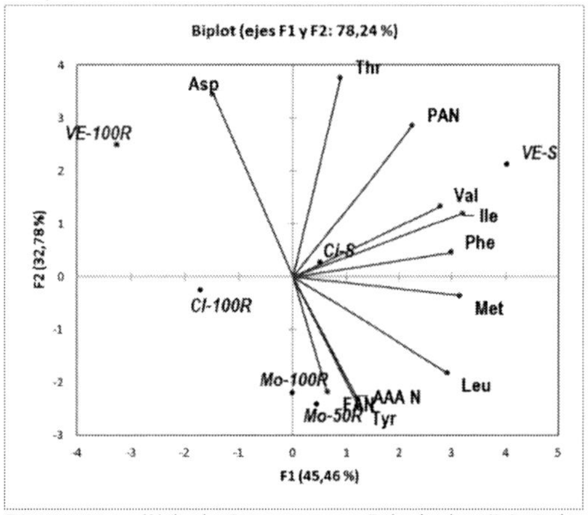

Figura 1. Análisis de Componentes Principales (PCA) de los valores medios interanuales de los aminoácidos de los diferentes tratamientos.

Por tanto, de lo anterior, puede concluirse que no se puede establecer un efecto generalizado del riego sobre la concentración de aminoácidos precursores de aromas de fermentación. Los resultados aquí expuestos indican que la respuesta depende en gran medida del binomio variedad-terroir y dentro de cada uno de ellos, incluso de la campaña en estudio.

Son necesarios más estudios de estas variedades en otras localizaciones para determinar respectivamente el grado de contribución de la variedad y del terroir a la respuesta del viñedo observada.

4 BIBLIOGRAFÍA

1. E. Gamero, D. Moreno, M. Vilanova, D. Uriarte, M.H. Prieto, M. E. Valdés, Aust. J. Grape Wine Res. **20(3)**, 394–400 (2014)
2. I. Talaverano, E. Valdés, D. Moreno, E. Gamero, L. Mancha, M. Vilanova, J. Sci. Food Agric. **97 (5)**, 1533-1542 (2017)
3. E. Vadour, *Les terroirs viticoles. Definitions, characterisation, protection* (EDP Dunod, Paris, 2003)
4. M. E. Valdés, M. H. Prieto, D. Moreno, E. Gamero, A. Montoro, F. Mañas, R. López-Urrea, P. López-Fuster, J. Yuste, M.V. Alburquerque, J.R. Yuste, E. Barajas, A. Yeves, D. Pérez, D.S. Intrigliolo, J.R. Castel, *IXe International Terroirs Congress* (Bourgogne-Dijon, Champagne-Reims, 2012)
5. J.R. Castel, M. E. Valdés, M. H. Prieto, D. Uriarte, L. Mancha, A. Montoro, F. Mañas, R. López-Urrea, P. López-Fuster, J. Yuste, M.V. Alburquerque, J.R. Yuste, E. Barajas, A. Yeves, D. Pérez, D.S. Intrigliolo, *IXe International Terroirs Congress* (Bourgogne-Dijon, Champagne-Reims, 2012)
6. P. Romero, R. Gil-Muñoz, F. M. Amor, M. E. Valdes, J. I. Fernández, A. Martinez-Cutillas, Agr. Water Manag. **121,** 85-101 (2013)
7. I. Garrido, J. Llerena, M. E. Valdés, L. Mancha, D. Uriarte, M. H. Prieto, F. Espinosa, J. Int. Sci. Vigne Vin. **48**, 1-32 (2014)
8. P. Hernández-Orte, J. F. Cacho, V. Ferreira,. J. Agric. Food Chem. **50**, 2891-2899 (2002)
9. N. Moreira, F. Mendes, O. Pereira, P. Guedes de Pinho, T. Hogg, I. Vasconcelos,.Anal. Chim. Acta. **458**,157-167 (2002)
10. M.L Asensio, M.E. Valdes, F. Cabello, Characterisation of some Spanish white grapevine cultivars by morphology and amino acid analysis. Sci. Hortic. **93**, 289-299 (2002)
11. T. Garde-Cerdán, C. Lorenzo, J.F. Lara, F. Pardo, C. Ancín-Azpilicueta, M.R. Salinas, J. Agric. Food Chem. **57**:2410-2419 (2009)
12. J. Tonietto, A. Cabonneau, Agric. Forest Meteorol. **124**, 81-97 (2004)
13. R.G. Allen, W.O. Pruitt, M.E Jensen, *Lysimeters for evapotranspiration and environmental measurements*. (EDP Allen RG, Howell TA, Pruitt WO, Walter IA, Jensen MEASCE, New York, 1991)
14. X. Chone, C. Van Leeuwen, D. Dubourdieu, J. P. Gaudillère, Ann. Bot. **87**, 477-483. (2001)
15. P.F. Scholander, H.T. Hammel, E.D. Bradstreet, E.A. Hemmingsen, Science **148**, 339-346 (1965)
16. M. E. Valdés, M. Vilanova, E. Sabio, M. J. Bernalte, Food Chem. **125**, 430-437 (2011)
17. A. Deloire, H. Ojeda, O. Zebic, N. Bernard, J.J. Hunter , A. Carbonneau. www.infowine.com, revue internet de viticulture et oenologie **19** (2006)
18. Guitart, A., Hernandez-Orte, P., Ferreira, V., Pena, C., Cacho, J. Am. J. Enology Vitic. **50**, 253-258. (1999)
19. M. Vilanova, I. Rodríguez, P. Canosa, I. Otero, E.

 https://doi.org/10.1051/e3sconf/20185001017

Gamero, D. Moreno, I. Talaverano, M. E. Valdés, Food Chem. **169**, 187-196 (2015)

20. M. Ortega-Heras, S. Pérez-Magariño, V. Del-Villar-Garrachón, C., González-Huerta, L.C. Moro, A. Guadarrama, S. R., Villanueva Gallo, S. Martín de la Helguera, J. Sci. Food Agric. **94**, 2073-2082 (2014).
21. Y. Bouzas-Cid, E. Díaz-Losada, E. Trigo-Córdoba, E. Falqué, I. Orriols, T. Garde-Cerdán, J. M. Mirás-Avalos, Sci. Hortic, **227**, 313-325 (2018)

 https://doi.org/10.1051/e3sconf/20185001018

Spatial variability in climate, phenology, and fruit composition across a reference vineyard network in Southern Oregon

Gregory V. Jones

Center for Wine Education and Department of Environmental Studies
Linfield College, McMinnville, Oregon, USA

Abstract. This paper provides a summary of observational data from a reference vineyard network in the Southern Oregon American Viticultural Area (AVA). The network was established to collect baseline data on spatial variations in temperature, grapevine phenology, and composition characteristics across a collection of varieties grown in the region. The results from the period 2003 through 2017 show a wide diversity of climate types for winegrape production ranging from relatively cool to relatively warm conditions (GST 14.8 to 17.5°C and GDD 1109 to 1657). One unique climate aspect is that the region experiences higher diurnal temperature ranges during the growing season than regions in Europe and many others in the western US. Grapevine phenological timing and intervals between events are similar to other wine regions with budbreak averaging 14-Apr, flowering 14-Jun, véraison 19-Aug. However, harvest dates are up to ten days to two weeks later than many regions in Europe. Ripening period and harvest composition observations reveal the diversity of potential varieties that the region can ripen. Over all varieties, sites and vintages observed harvest composition averages 23.7 °Brix, 6.4 g/L total acidity, a pH of 3.44, and yields of 2.7 tons/acre.

1 Introduction

Within region weather and climate controls on site plant growth and fruit composition are evident in every wine region worldwide [1]. To understand spatial variations in climate, grapevine phenology and fruit characteristics numerous reference vineyard networks have been developed in regions worldwide (e.g., Bordeaux, Italy, etc.) [2,3,4]. The aims of reference vineyard networks are typically to document baseline characteristics in within region growth and production characteristics, the controls that landscape variations have on these characteristics, and provide growers comparative measures in a small sample of vineyards within the region.

This research describes observations from a reference vineyard network in the Southern Oregon American Viticultural Area (AVA) (Figure 1) [5]

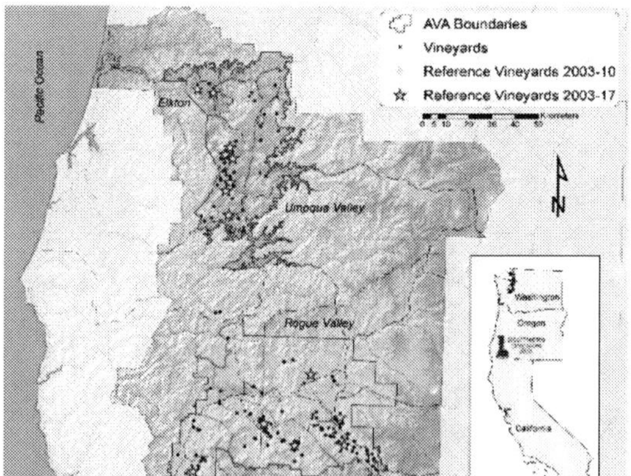

Figure 1. Location of the Southern Oregon AVA and sub-AVAs along with the reference vineyards used to monitor weather and climate, phenology, and composition described in this paper.

one of 18 regions in the state recognized and defined by the federal government based on geographic, historical, and cultural identity along with climate, soil, geology, and terrain uniqueness compared to surrounding areas. The Southern Oregon AVA has five sub-regions and represents approximately 24% of Oregon's production, acreage, number of vineyards, and number of wineries. The region grows over 70 different types of winegrapes ranging from cool climate varieties (e.g., Pinot Noir, Pinot Gris, Chardonnay, etc.) to warm climate varieties (e.g., Syrah, Malbec, Tempranillo), with a mix of approximately 70% red and 30% white varieties being grown. The region currently has nearly 300 growers, with nearly 3300 planted hectares, producing 25,000 tons that are crushed at 125 wineries [5].

The Southern Oregon (AVA) is one of the many winegrape producing regions found within the intermountain valleys along the west coast of the United States. The landscape of the Southern Oregon AVA is extremely diverse, coming from the joining of three mountain ranges of varying ages and structure: the Klamath and Siskiyou Mountains to the southwest to southeast, the Coastal Range to the west, and the Cascades to the east and north [5]. The Klamath Mountains extend through the south and southwestern portion of the AVA and consist of complex folded and faulted igneous and metamorphic rocks that are the oldest in the region. The Cascade Mountains to the east consist of the younger High Cascades and the older, more deeply eroded Western Cascades that make up the eastern boundary of the AVA. The region is protected from the ocean largely by the Coastal Mountains, which are composed of mostly oceanic sedimentary rocks and volcanic islands that were accreted to the landscape over the last 50 million years [5]. The Rogue Valley AVA is drained mainly by the Rogue River and its major tributaries; the Applegate River, the Illinois River, and Bear Creek, while the Umpqua Valley AVA is drained by hundreds of smaller tributaries of the North and South Umpqua Rivers.

The agricultural landscape of the Southern Oregon AVA is mostly comprised of valley lowlands with some isolated hills, stream terraces or benches, and footslopes of alluvial fans scattered by hilltops and ridges. Vineyards in the region are found on flat to very steep slopes (up to 40% or more) that are distributed along isolated hills, stream terraces or benches, and at the foot of alluvial fans. Elevations of existing vineyard sites range from approximately 60-90 m in the northwestern portion of the Umpqua Valley AVA to 750 m in the Bear Creek Valley of the Rogue Valley AVA. Most of the current vineyards are planted to ~180-200 m in the Umpqua Valley AVA and ~450-550 m in the Rogue Valley AVA [5].

From the diverse geology of the region comes a widely varying mix of metamorphic, sedimentary, and volcanic derived soils [5]. The lower elevations of the valleys are mostly deep alluvial material or heavy clays while the hillside and bench locations have mixed alluvial, silt, or clay structures. Complex faulting, especially in the western portion of the Rogue Valley AVA and southern portion of the Umpqua Valley AVA, can produce large variations in soil types over areas the size of a vineyard. Drainage and moisture-holding capacity vary greatly by soil type, and while most soils in the region do retain water into the growing season, available water for irrigation during mid to late summer growth is generally needed [5]. Soil fertility is generally sufficient for winegrape production but varies greatly over the region with issues generally related to either imbalances of nitrogen, calcium, potassium, phosphorous, magnesium, boron, or zinc. Soil pH also varies from region to region (roughly from 4.5 to 7.0) and is mostly due to differences in climate and parent rock material. In general, the soils in the northern and western portions of the Southern Oregon AVA are slightly more acidic than those of the south as a result of more rainfall and greater leaching potential [5].

From a climate perspective the Southern Oregon AVA offers the most diverse growing conditions in Oregon and arguably in the western United States [6]. Heat accumulation varies from cool climate suitability in the northern Umpqua Valley and Illinois Valley (1200-1350 GDD) to intermediate values in the central Umpqua Valley and Applegate Valley (1350-1500 GDD) to warm climate suitability in the Bear Creek and Rogue River region (1500-1700 GDD). The Umpqua Valley AVA in general has longer frost-free periods (180-220 days) and milder growing seasons, experiencing annual precipitation values that average 750 to 1500 mm from south to north [5]. The Rogue Valley AVA has the higher elevations, that along with their general north-south tending valleys, and their proximity to the Pacific Ocean and intervening topographical barriers create a climate transect of wetter and cooler conditions in the western parts of the region to the warmer and drier eastern areas. Precipitation varies from 300-650 mm in most of the vineyard areas in the Rogue, declining in amount from west to east (all of the Southern Oregon AVA experiences less than 15 percent of the total precipitation occurring during the growing season of April through October). The frost-free growing season is shorter in the Rogue Valley AVA (145-185) due to higher elevations that bring later and earlier frost potential in the spring and fall, respectively [5].

The reference vineyard network described in this research was established in 2003 within the

Rogue Valley AVA and in 2004 in the Umpqua Valley AVA, both sub-regions of the Southern Oregon AVA (Figure 1). The purpose of the reference vineyard network is to provide an in depth look at spatial variations in important characteristics of temperature, phenology, composition, and yields in the region. This paper will provide a description of the observations from this network during 2003-2017.

2 Data and Methods

The network was developed initially to include 29 vineyards from which temperature, phenology, and sample and harvest composition data were collected during 2003-2010. During 2011-2017 the network was reduced to 14 locations with temperature, phenology, and harvest composition data collected (no sampling was done). All participating vineyards were chosen based upon their spatial and elevational representation in the Southern Oregon AVA and willingness to participate and observe phenology in a consistent manner (using the Modified E-L System) [2]. The reference vineyards are spread out over a 110 km east-west and 240 km north-south area across a range of landscapes with a mean elevation of 391 m, ranging from 103 m to 611 m (Figure 1). At each reference vineyard site temperature devices were installed at 1.5 m height in a standard housing and set to record at 15 minute intervals during both the dormant (Nov 1-Mar 31) and growing seasons (Apr 1-Oct 31). Each year the data was downloaded and the observations were then aggregated to hourly and daily average, maximum, and minimum values, along with standard growing degree-days (GDD) and summarized by site, sub-region, and region. For phenological observations, the four main events of budbreak, bloom, véraison, and harvest for a collection of varieties (i.e., Cabernet Franc, Chardonnay, Merlot, Pinot Gris, Pinot Noir, Sangiovese, Syrah, Tempranillo, Viognier, Grenache, Malbec, and Riesling) commonly grown across the region were submitted by growers (not all varieties are grown at every site). The phenological data were then examined for average dates and intervals between dates for the entire region, sub-regions, and by variety. During the 2004-2010 period a "snapshot of ripening" 100 berry sample was collected from the same vines during 13-15 September for each variety and site. The samples were then analyzed for °Brix, titratable acidity, pH, and berry weights using standard industry methods. Finally, harvest composition values for °Brix, titratable acidity, pH, and yields were submitted by growers at the end of the season and summarized by region and variety for the entire time period of 2003-2017. The results will detail both the temporal and spatial variability in the weather/climate, phenology, fruit composition and harvest characteristics across the Southern Oregon AVA. Below the general climate, phenology and harvest composition characteristics are presented for the 14 locations for the entire time period (2003-2017) and the sampling characteristics for the entire 29 locations during 2003-2010.

3 Results

For the dormant season from November through March, the 14 locations exhibit an average temperature of 5.7°C ranging 2 degrees across the locations (Table 1). Maximum temperatures during the winter average 11.6°C across the region however they did occur as high as 18.7 to 33.2°C during warm fall periods. Minimum temperatures across the network during the winter drop below freezing on average 53 times with the warmer sites experiencing as few as 27 and the colder sites experiencing as many as 87. However, while average minimum temperatures remain above freezing (1.5°C) absolute minimums dropped as low as -3.4 to -21.8°C during the 2003-2017 period.

Table 1. Dormant season (November through March) for 14 locations with data during 2004-2017

Variable	Mean	STD	Max	Min
DTavg (°C)	5.7	0.7	6.7	4.7
DTmax (°C)	11.6	0.7	13.0	10.4
DAbsTmax (°C)			33.2	18.7
DTmin (°C)	1.5	1.3	3.1	-0.4
DAbsTmin ((°C)			-3.4	-21.8
# Days < 0°C	53	22	87	27

For the April through October growing season the 14 locations have a climate maturity type that ranges from a cool to warm climate [6], averaging 16.3°C or an intermediate climate (Table 2). GDD averages 1412 for the locations, ranging from 1109 to 1657 or a Region Ia to nearly a Region III on the Winkler Index [6]. Maximum temperatures during the vintage average 26.1°C but the absolute maximums were observed to range from 35.3 to 47.0°C across the locations. The number of days above 35°C average as low as 8 per year to as many as 48 per year. While minimum temperatures across the region average 8.0°C during the growing season, temperatures dropped as low as -6.3°C in early April during one vintage. Most sites experience frost days during the growing season with an average of 7 per year, ranging from a low of 0 to a high of 19 at the coolest location. The last spring frost (0°C) averages 14-Apr across the sites with a standard deviation of +/- 25 days during the 2004 through 2017 vintages. The first fall frost (0°C) averages 27-Oct for the 14 sites varying +/- 18 days during the observation period. The result is an average frost-free period of 197 days with a range of 149 to 236 days across these 14 locations. Finally, the growing season diurnal temperature range (DTR) averages 18.4°C across these sites. Other work examining only the August and

September months finds that the region has a 22.4°C average DTR during ripening [5].

Table 2. Growing season (April through October) for 14 locations with data during 2004-2017

Variable	Mean	STD	Max	Min
GSTavg (°C)	16.3	0.7	17.5	14.8
GSTmax (°C)	26.1	1.4	28.2	23.1
GSAbsTmax (°C)			47.0	35.3
GSTmin (°C)	8.0	1.2	9.4	5.7
GSAbsTmin (°C)			3.4	-6.3
GSDTR (°C)	18.4	0.9	19.1	17.5
GDD (C°)	1412	138	1657	1109
# Days > 35°C	27	12	48	8
# Days < 0°C	7	6	19	0
Last Spring (0°C)	14 Apr	25	22 May	3 Jan
First Fall (0°C)	27 Oct	18	7 Dec	24 Sept
Frost-Free (Days)	197	28	236	149

Table 3 details the phenological characteristics across the entire 2003-2017 time period including all sites (29) and varieties (12) with observations. Budbreak occurs on average 14-Apr, but has been as early as 8-Mar and as late as 1-Jun. The earliest variety in the region to break bud is typically Chardonnay (12-Apr), while the latest is commonly Tempranillo (18-Apr). Flowering has occurred as early as 30-Apr and as late as 14-Jul, averaging 14-Jun. On average the earliest flowering varieties are Chardonnay and Pinot Noir (8-Jun) and the latest is Grenache (21-Jun). Véraison averages 19-Aug across the network but has been observed as early as 13-Jul and as late as 23-Sep. The earliest varieties to experience véraison on average are Pinot Gris and Tempranillo (13-Aug) and the latest on average is Grenache (28-Aug). Harvest dates average 8-Oct across the sites, varieties and vintages (Table 3) however harvest has occurred as early as 23-Aug (early vintage and variety) and as late as 30-Nov. Chardonnay is on average the first to harvest (2-Oct) along with Pinot Noir and Tempranillo (6-Oct) while Cabernet Franc and Sangiovese are on average the latest to harvest (17-Oct). During the 2003-2017 vintages the budbreak to flowering interval averaged 61 days, budbreak to véraison averaged 126 days, and budbreak to harvest 172 days (Table 3). Flowering to véraison averaged 64 days while flowering to harvest averaged 112 days. Véraison to harvest averages 49 days across these locations. Each of these intervals vary +/- 8-13 days across sites, vintages and varieties.

Table 4 displays summary statistics for a 'snapshot of ripening' 100 berry sample from the same vines at each location, across twelve varieties, during seven years (2004-2010), and on the same days (13-15 September). Ripening tends to progress to 18.4 °Brix on average by mid-September, with +/- 2.4 °Brix across sites and varieties. °Brix has ranged from as low as 11.3 (cool year and site) to 22.8 (warm year and site) with Pinot Gris and Tempranillo having higher °Brix while Grenache and Riesling were the lowest. Titratable acidity from the sampling averages 11.9 g/L varying +/- 4.2 g/L across sites and varieties. Merlot and Pinot have the lowest average TA at this time while Riesling and Grenache are the highest. By mid-September pH across varieties and sites averages 3.09 varying +/- 0.17 (Table 4). Tempranillo has the highest average pH by this time with Sangiovese, Grenache and Riesling having the lowest average pH. Samples of 100 berries exhibit an average of 130.4 g, ranging 31.7 g across sites, years and varieties. The smallest berry weights are consistently from Merlot and Cabernet Franc while Tempranillo and Grenache have the highest berry weights.

Table 3. Phenological statistics across all sites and varieties during 2003-2017. The number of observations per event each year averages 70-80. All dates are reported by reference vineyard growers.

Variable	Median	STD	Max	Min
Budbreak (BB)	4/14	9	6/1	3/8
Flowering (F)	6/14	8	7/14	4/30
Véraison (V)	8/19	8	9/23	7/13
Harvest (H)	10/8	11	11/30	8/23
Budbreak to Flowering (BB-F)	61	9	81	40
Budbreak to Véraison (BB-V)	126	10	155	105
Flowering to Véraison (F-V)	64	8	87	50
Budbreak to Harvest (BB-H)	172	13	206	148
Flowering to Harvest (F-H)	112	12	146	91
Véraison to Harvest (V-H)	49	11	81	27

Table 4. Sampling composition statistics during 13-15 September each year for 2004-2010 for 29 locations in the Southern Oregon AVA. Observations come from approximately 100 samples per year

Variable	Median	STD	Max	Min
°Brix	18.4	2.4	22.8	11.3
TA (g/L)	11.8	4.2	25.3	6.6
pH	3.09	0.17	3.51	2.71
Weight 100 berries (g)	130.4	31.7	226.9	75.3

Harvest composition across the reference vineyards show an average °Brix of 23.7 that tends to range +/- 1.3 across vintages, sites, and varieties (Table 5). No one variety stands out with higher average °Brix however Chardonnay and Riesling tend to have the lowest average °Brix at harvest. Titratable acid levels at harvest average 6.4 g/L, ranging +/- 1.4 g/L for individual varieties, sites or vintages. The higher average TA at harvest is generally found with Riesling and Sangiovese while the lowest is typically seen with Tempranillo. Harvest pH values average 3.44, varying +/- 0.22 across sites, varieties and vintages. Given its lower average TA, Tempranillo tends to have the highest average pH at harvest while Riesling has the lowest average pH. Yields during 2004-2017 averaged 2.7 tons/acre with a fairly large vintage, site and variety variation of +/- 1.3 tons/acres. Yields were reported to be on average the highest for Chardonnay and Pinot Gris and lowest for Sangiovese.

Table 5. Harvest composition statistics for 2003-2017 for all locations in the Southern Oregon AVA. Observations come from approximately 125-150 samples per year.

Variable	Median	STD	Max	Min
°Brix	23.7	1.3	27.3	20.2
TA (g/L)	6.4	1.4	10.2	3.6
pH	3.44	0.22	4.21	3.02
Yield (tons/acre)	2.7	1.3	7.2	0.7

4 Discussion and Conclusions

The reference vineyard network in the Southern Oregon AVA was established to provide an in depth look at spatial variations in important characteristics of temperature, phenology, composition, and yields in the region. During 2003-2017 numerous aspects of winegrape growing and production have been documented providing growers baseline information for the region.

The results detail a region that has a cool winter with low to moderate winter extreme temperature risk. The growing season has average temperatures that are intermediate for winegrape production (16.3°C) and growing degree-days that range from a Region Ia to a Region III on the Winkler Index (average 1412, Region II). The GDD values compare to Valtellina Superiore in Italy, Bordeaux in France, and Rioja in Spain [6]. The region experiences an average last spring frost on 14-Apr and a first fall frost on 27-Oct resulting in an average frost-free period of 197 days (ranging from 149-236 days over the sites). Furthermore, the growing season diurnal temperature range averages 18.4°C across the Southern Oregon reference vineyard network which is higher than a nine region study conducted in France, Spain, Germany, Italy, and Slovakia (18.4°C vs. 16.1°C) and is the result of both warmer daytime maximum temperatures (26.1°C vs. 21.8°C) and lower nighttime minimum temperatures (8.0°C vs. 10.4°C) across the growing season [7].

Phenological observations from the network show an average regional budbreak on 14-Apr, bloom on 14-Jun, véraison on 19-Aug, and harvest on 8-Oct. These events vary 16-22 days across sites, varieties and vintages with harvest dates varying the most. The phenological timing in Southern Oregon is very similar to the averages from a 17 variety, nine region study conducted in France, Spain, Germany, Italy, and Slovakia (13-Apr, 13-Jun, 17-Aug), except for harvest which averages 10 days earlier (28-Sep) in the European regions [7]. A similar comparison can be made with an 18 variety average in Conegliano, Italy (+/- 4 days), except for harvest dates which occur two weeks earlier in the Italian region [8]. Similar bloom and véraison dates are also observed in a reference vineyard network in Bordeaux while harvest is typically one week later in Southern Oregon [9]. The interval from budbreak to harvest averages 172 days across the region, which is within the average frost-free period although the average budbreak and average last spring frost are both on 14-Apr indicating some frost risk, especially in the cooler areas in the southern portion of the AVA. Average phenological interval lengths in Southern Oregon are overall similar to average interval links from a European study across multiple varieties, regions, and countries [7]. The only differences were for bloom to harvest (108 days in Europe vs. 112 days in Southern Oregon), véraison to harvest (43 days vs. 49 days), and the overall budbreak to harvest interval (168 days vs. 172 days) [7].

Fruit composition characteristics during ripening exhibit a wide range due to vintage variations with °Brix levels reaching 11.3 to 22.8 across sites and varieties. Acid levels in mid-September on average are still relatively high (11.8 g/L) but in the warmer years have been below 8 g/L by mid-month. pH values during this period average near 3.00, although have ranged from 2.71 to 3.51 during warm or cool vintages, respectively. Berry weights in mid-September range from very low weights of 100 g or less (per 100 berries) from Merlot and Cabernet Franc to over 200 g (per 100 berries) for Tempranillo and Grenache.

Growers and producers ripen a wide range of varieties in the region, achieving 23.7 °Brix, 6.4 g/L, 3.44 pH, and yields of 2.7 tons/acre on average at harvest. From the mid-ripening sampling to harvest, which averages 22 days, the region typically sees increases of 5-7 °Brix, a reduction of 4-6 g/L of titratable acidity, and an increase in pH of 0.40 to 0.60.

The network is an ongoing research project that provides growers with information on spatial variations in climate, grapevine phenology, and composition characteristics of a young wine region in Oregon. This project summary highlights the large climatic diversity in the region, which in turn allows the region to grow over 70 different varieties of winegrapes.

5 References

1. G.V . Jones, M. Moriondo, B. Bois, A. Hall, A. Duff, Analysis of the spatial climate structure in viticulture regions worldwide. Le Bulletin de l'OIV 82(944,945,946):507-518 (2009).
2. G.V . Jones. Winegrape Phenology in Phenology: An Integrative Environmental Science 2nd Edition, Mark D. Schwartz, Editor. Springer. 610 pp. (2013).
3. P. Catania, M. Vallone, G. Lo Re, M. Ortolani, A wireless sensor network for vineyard management in Sicily (Italy). Agric Eng Int: CIGR Journal Open access at http://www.cigrjournal.org Vol. 15, No.4 139-146, (2013).
4. R. Le Roux, L. de Rességuier, T. Corpetti, N. Jégou, M. Madelin, C. van Leeuwen, H. Quénol, Comparison of two fine scale spatial models for mapping temperatures inside winegrowing areas. Agri & F Meteo. 247. 159-169. 10.1016/j.agrformet.2017.07.020, (2017).
5. G.V . Jones, Southern Oregon AVA landscape and climate for wine production. Proceedings of the 11th International Terroir Congress, Linfield College, McMinnville, Oregon, pp 33-34 (2016).
6. G.V . Jones, A.A. Duff, A. Hall, J. Myers, Spatial analysis of climate in winegrape growing regions in the western United States. Am J Enol Viti, 61:313-326 (2010).
7. G.V . Jones, E. Duchene, D. Tomasi, J. Yuste, O. Braslavksa, H. Schultz, C. Martinez, S. Boso, F. Langellier, C. Perruchot, G. Guimberteau, Changes in European Winegrape Phenology and Relationships with Climate, Proceedings of the 14th International Meeting GiESCO (2005).
8. D. Tomasi, G.V. Jones, M. Giust, L. Lovat, F. Gaiotti, Grapevine Phenology and Climate Change: Relationships and Trends in the Veneto Region of Italy for 1964-2009. Am J Enol Viti, 62:3(329-339) (2011).
9. G.V . Jones, R. E. Davis, Climate Influences on Grapevine Phenology, Grape Composition, and Wine Production and Quality for Bordeaux, France, Am J Enol Viti, 51(3):249-261, (2000).

Monitoring a commercial vineyard in New Zealand to maintain soil health and achieve long-term sustainability

Robert E White[1] and Daniel Watson[2]

[1]Faculty of Veterinary and Agricultural Sciences, The University of Melbourne, Australia; email robertew@unimelb.edu.au

[2]National Vineyard Manager, Craggy Range Vineyards, Hawke's Bay, New Zealand; email Daniel.Watson@craggyrange.com

Abstract

This paper illustrates the benefits of long-term monitoring of soil and vine health in Craggy Range's Te Muna vineyard in the Martinborough region of New Zealand. The soils at Te Muna are formed on two river terraces. The higher and older terrace, planted to Pinot Noir, is very gravelly; the lower and younger terrace, planted to Sauvignon Blanc, contains more silt and very fine sand. Both terraces are freely drained. Soil pH, organic C and available P were monitored in selected blocks on both terraces. Under current management that includes cover crops and sheep grazing in winter, soil organic C has been stable for 10 years. Because of the regular application of RPR and intermittent liming, soil pH (water) has risen from *c*.5.5 to 6.5–7. Values for soil available P, measured by Olsen bicarbonate and Mehlich-3 extractions, diverged over time. The Olsen test indicated satisfactory P levels: conversely, the Mehlich test showed P values rising to very high levels, confirming Western Australian experience that this acid extraction is unsuitable for soils treated with RPR. Petiole and leaf blade P concentrations confirmed these results, but they showed greater interannual variation than the soil P measurements. Overall, the long-term sustainability of the vineyard should be assured.

1. Introduction

Craggy Range Vineyards has a reputation for distinctive wines of high quality, being named as 'New World Winery of the Year' in 2014. The company has a long-term commitment to maintaining healthy soils by continuous monitoring of key soil properties. This paper describes some of the practical issues that the vineyard manager must confront in monitoring the nutritional status of soil and vines to ensure the productive capacity of the vineyard and its long-term sustainability.

2. Background information

Craggy Range's Te Muna vineyard in the Martinborough region of the southern North Island, New Zealand, is used as a case study of the company's strategy. The annual rainfall is 783 mm with January-February being the driest months and June-July the wettest. The soils have formed on two river terraces. The higher and older terrace was laid down 50–60 000 years ago and is very gravelly; the lower younger terrace was laid down in Recent times and is also gravelly, but has more silt and very fine sand (Figure 1).

Of the two main varieties grown at Te Muna, Pinot Noir is planted on the upper terrace; Sauvignon Blanc mainly on the lower. Both varieties are on rootstocks. The soils on both terraces are freely drained except for some blocks on the lower terrace where there is an underlying clay layer. Drip irrigation, but no fertigation, is applied at the rate of 0.5 to 0.75 ML/ha. Volunteer cover crops are grown in the mid-rows and sheep are grazed for four months in winter. There is a regular program of fertilizing with an N, P, K mix, supplemented with magnesite, ulexite, S prills, reactive phosphate rock (RPR) and potassium sulfate (K_2SO_4), as deemed necessary from soil and plant testing. The RPR input varies from 100–150 kg/ha, K_2SO_4 from 50–100 kg/ha and S prills at 40 kg/ha. Lime is also applied infrequently to some blocks.

Figure 1 Soil profile on the lower terrace at Te Muna; the scale is 10 cm

 https://doi.org/10.1051/e3sconf/20185001019

2.1 Methods

The vineyard of 93 ha has been divided into 40 blocks of varying size according to soil variation and vine performance. Since the early 2000s, soil samples (0–15 cm) have been taken from specific, under-vine locations in 'indicator' blocks that are representative of a zone, based on soil differences and vine performance. Each year, normally in June-July, 20 soil samples are taken with a 2.5 cm corer and bulked for each indicator block. The sampling locations are the same each year. The fine earth fraction (<2 mm) of all samples is analysed for soil pH, organic carbon (C) and the macro- and micronutrients. Because the soils are so gravelly, this fraction comprises only 40–50% of the soil volume. Leaf petioles and blades are also sampled, usually at flowering but sometimes also at veraison.

The analyses are carried out by an internationally accredited laboratory in New Zealand. Soil pH is measured at a 1:2 soil:solution ratio in water, organic C by an NIR calibration of total C by Dumas combustion, and available phosphorus (P) by Olsen bicarbonate [1] and Mehlich no. 3 [2] extractions. Plant analyses are done by ICP on acid digests of the dried, ground samples.

3. Outcomes of the long-term monitoring

3.1 Soil organic carbon

As shown in Figure 2, the long-term trend in soil organic carbon (SOC) is remarkably stable. The upper terrace soil, represented by Block 1, is between 5 and 6% C; the lower terrace represented by Block 42 is about 3% C. Although the lower terrace soil has more silt and fine sand than the upper terrace soil, it is a younger soil and probably has not developed to its equilibrium C content yet. This difference was consistent for other blocks located in the upper and lower terraces. The vineyard management regime involving cover crops and sheep grazing for four months in winter has maintained SOC levels and this should benefit the structure of the topsoil.

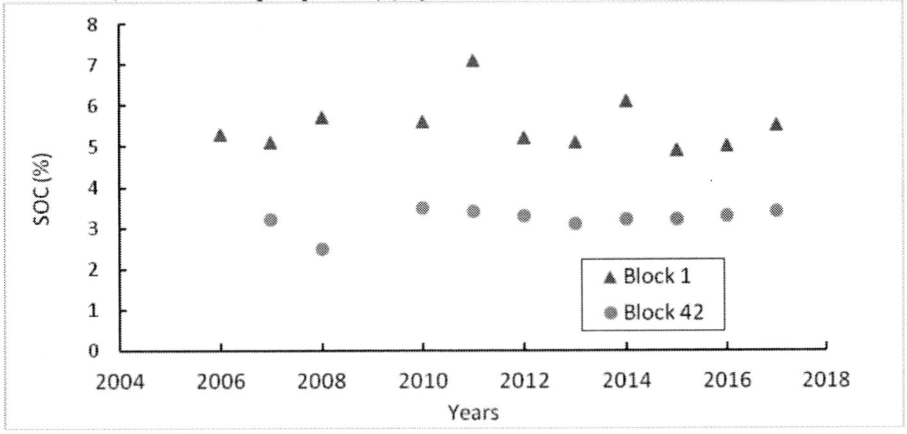

Figure 2 Long-term trend in SOC in the upper (Block 1) and lower terrace (Block 42) soils

3.2 Soil pH

The graphs in Figure 3, again for Blocks 1 and 42, show that soil pH has been increasing steadily over the monitoring period, approaching a value between 6.5 and 7. This trend is primarily the result of regular applications of RPR, which has a liming effect equivalent to about 40% of its equivalent weight in agricultural lime. Ideally, the soil pH (measured in water) should not be raised above 6.5–7 (corresponding to about 5.8–6.3 in 0.01 M $CaCl_2$), otherwise soil micronutrients such as Fe and Zn may become less available.

3.3 Available soil phosphorus

The graphs in Figures 4 and 5 show the long-term trends in available soil P for Blocks 1 and 42, respectively, by both the Olsen and Mehlich tests. The trends are very similar for topsoil on the upper and lower terraces: Olsen values stay relatively constant, between 20 and 40 mg P/kg soil, over time. On the other hand, Mehlich values increase to high levels, up to 160 mg P/kg in the case of Block 1 (upper terrace).

These trends are not unexpected when RPR has been used as a P fertilizer. First, previous research in New Zealand [3] has shown that the alkaline Olsen extraction did not adequately reflect the increase in P availability following RPR application to soils of varying P sorption capacity. Second, the Mehlich test, being an acid extraction, overestimates the P availability of soils to which RPR has been applied. Extensive appraisal of trials in Western Australia [4] concluded that the Mehlich no. 3 test was not suitable for soils to which RPR had been applied. These authors gave a range of <10 (very low) to 40–70 (optimum) to >70 (high) to assess the results of the Mehlich test for soil P availability. Thus, the Mehlich results in Figures 4 and 5 apparently show very high available P levels. However, Olsen P values show little change over time and are mainly in the desired range of 15-40 mg P/kg, as identified by the New Zealand soil testing laboratory.

On a few occasions early in the monitoring period, subsoils were analysed for available P by the Olsen test. Table 1 gives examples. These figures show

Table 1 A comparison of Olsen P values in the topsoils and subsoils at Te Muna

Year	Block	Olsen P topsoil (mg/kg)	Olsen P subsoil (mg/kg)
2006	1	49	22
2006	1	35	22
2012	42	39	26
2006	Mean for all subsoils		19

that subsoil P values are 22–26 mg P/kg, with an average value for the whole vineyard of 19 mg P/kg. This appears to be adequate, except that these are values for the fine earth fraction, which comprises only about 50% of the soil volume.

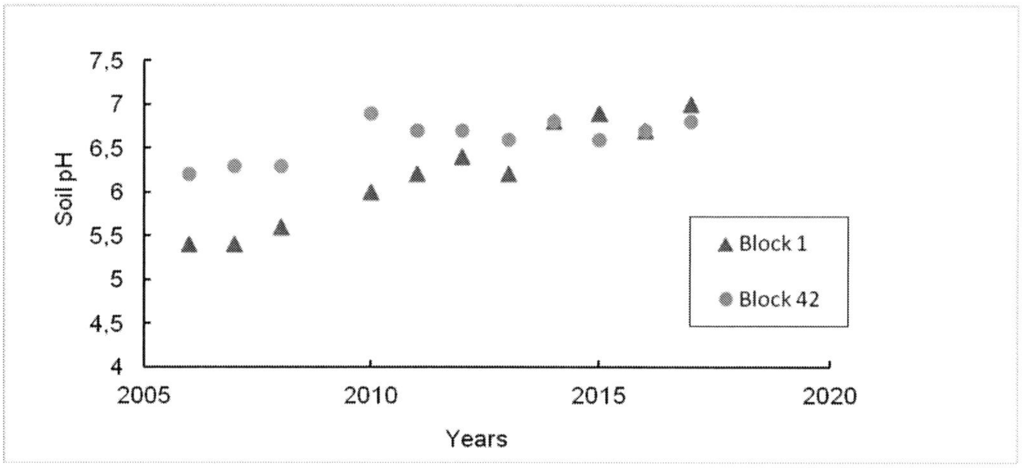

Figure 3 Long-term trend in soil pH (water) for the upper (Block 1) and lower terrace (Block 42) soils

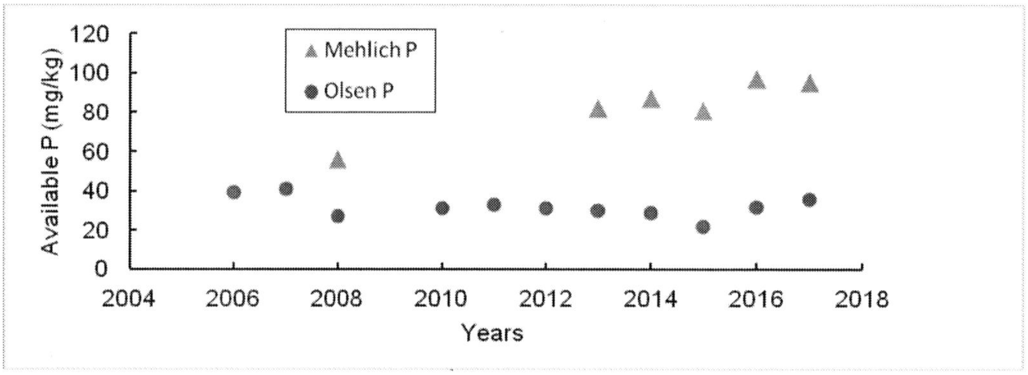

Figure 4 Long-term trend in available soil P in the upper terrace soil (Block 1)

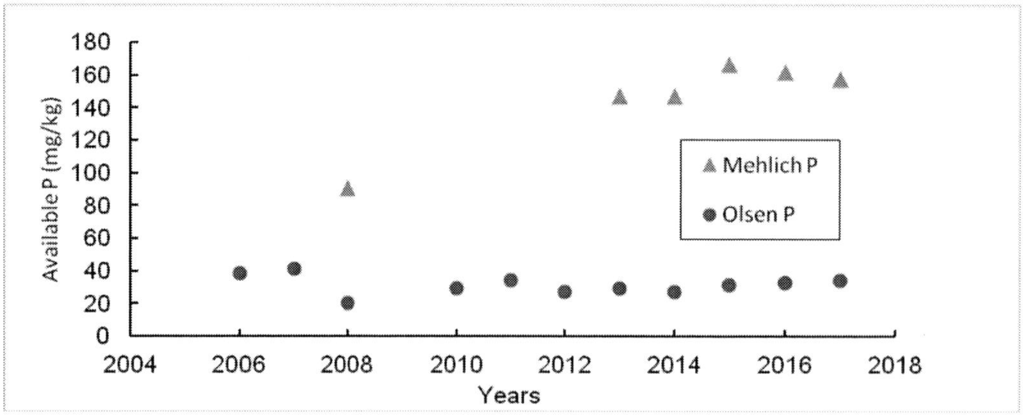

Figure 5 Long-term trend in available soil P in the lower terrace soil (Block 42

https://doi.org/10.1051/e3sconf/20185001019

Effectively, therefore, these subsoil values are equivalent to only 10–13 mg P/kg, which is deficient. Unfortunately, no data are available for subsoil Mehlich values. Although topsoil Olsen P values should be similarly discounted for the presence of inert gravel, they still fall mainly in the desired range of 15-40 mg P/kg. Hence one concludes that the application of RPR and some soluble P fertilizer to these soils has maintained available P in the desired range. Even after discounting the high Mehlich P values for the presence of gravel, it appears that the Mehlich test is overestimating P availability.

3.4 Plant analyses for P

Given the uncertainties associated with the interpretation of the soil tests for available P, plant analysis may give a more reliable indication of the P status of the vines. However, there are two problems with plant analysis for a mobile element such as P. First, concentrations can change rapidly with phenological development of the vine: for example, a decrease in concentration occurs from flowering through fruit set to veraison. The data in Table 2 illustrate this change for Sauvignon Blanc vines in Block 35 (lower terrace) from flowering to early veraison.

Table 2 A comparison of leaf blade and petiole P concentrations at flowering and veraison at Te Muna

Block	Date of sampling	Leaf P (%)	Petiole P (%)
35	25.11.2008	0.55	
	27.01.2009	0.29	
	24.11.2006	0.35	0.57
	18.01.2007	0.23	0.23

Second, the plant P concentration tends to be lower in vines under moisture stress, irrespective of the soil P status. Generally speaking, petiole values reflect the current P status of the vine better than blade values which reflect the accumulation of P in the vine up to the time of sampling.

Figures 6 and 7 show the trends in petiole and leaf P concentrations for Pinot Noir in Block 1 (upper terrace) and Sauvignon Blanc in Block 42 (lower terrace), sampled at flowering. Based on New Zealand experience, the desired ranges for petiole P and leaf P at flowering are 0.18–0.45% and 0.16–0.25%, respectively. Figure 6 shows that leaf P is relatively stable over time and in the desired range, which is consistent with the Olsen soil P results (Figure 4). Petiole P is more variable over time, ranging from 0.1% to 0.36%. The lower values in 2010, 2011, 2012 and again in 2014 and may well reflect the later sampling times in these years.

Figure 7 (Sauvignon Blanc, Block 42) shows a generally consistent story with Figure 6 – petiole and leaf P concentrations are generally in the desired range, except for petiole P in 2010 and 2011 when the vines were sampled late. However, leaf P values were also low in 2010 and 2011 so that vine moisture stress could have affected P concentrations in those years. Apart from the interannual variability introduced by time of sampling and moisture stress, it is important to note that the plant analyses indicate the vines are in the desired P range, with a trend towards higher values over time in the Sauvignon Blanc in Block 42, a more vigorous site on the lower terrace.

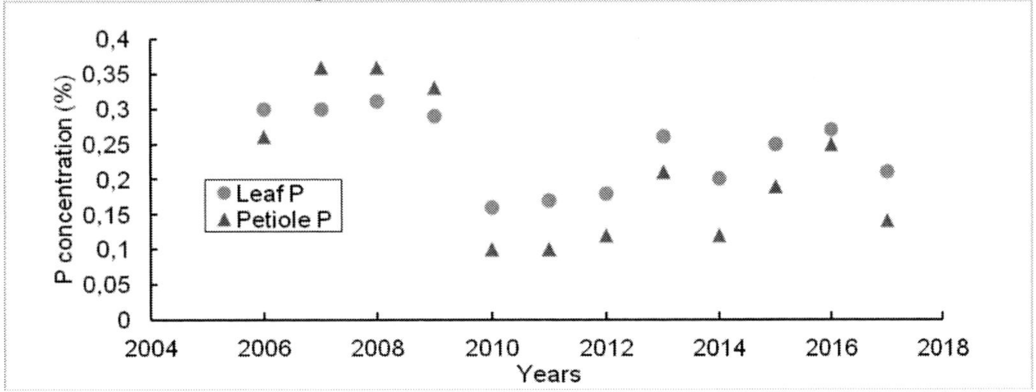

Figure 6 Trends in leaf blade and petiole P concentrations for Pinot Noir on the upper terrace (Block 1)

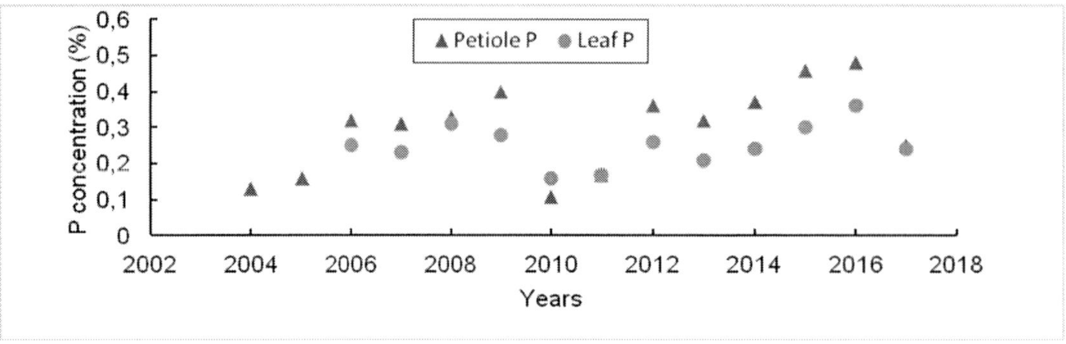

Figure 7 Trends in leaf blade and petiole P concentrations for Sauvignon Blanc on the lower terrace (Block 42)

4. Conclusions

The examples of soil and plant analysis given here are representative of soil conditions in the Te Muna vineyard since the early 2000s. Clearly the prevailing soil management system is maintaining or improving soil health. Soil organic C has remained constant at 5-6% for the older upper terrace soil and around 3% for the younger lower terrace soil. Some nitrogen (N) is infrequently supplied in an N:P:K mix, but the main source of N is from soil organic N. For example, the C:N ratio of the soil organic matter in the topsoil is consistently in the range 10–12. The potentially available N from mineralization in the topsoil is between 60 and 120 kg N/ha, with the higher values generally occurring in the lower terrace soils.

Soil pHs have been increasing steadily into the 6.5–7 range through the use of occasional lime applications, but more consistently through the use of RPR. They are now at the level where the rate of RPR dissolution will be extremely slow. As the Mehlich soil P test shows, there is a substantial reserve of RPR residues in the topsoil so that future P fertilizer additions can be scaled back. For example, in 2017 P inputs were reduced to 6.5–13 kg/ha, with higher values in this range being applied to the more gravelly, upper terrace soils.

Soil and plant analyses show that the P status of the soils is generally in the desired range. The use of RPR as a P fertilizer is justified on these freely drained, gravelly soils, because soluble P fertilizer can be leached. Furthermore, for a perennial plant such as the grapevine a slow-release source of P is satisfactory for the plant's needs and minimizes any impact of leached P on water quality in the nearby Huangarua River.

Overall, the management of the Te Muna vineyard is consistent with Craggy Range's commitment to the excellence and long-term sustainability of its wine business.

Acknowledgements

One of us (REW) gratefully acknowledges the opportunity to visit the Te Muna vineyard offered by Matt Stafford, Chief Winemaker, Craggy Range Pty Ltd.

References

1. S.R. Olsen, C.V. Cole, F.S. Watanabe, L.A. Dean, USDA Circular no. 939

2. A. Mehlich, Comm. Soil Sci. Plant Anal. **15**, 1409 (1984)

3. S.R. Saggar, M.J. Hedley, R.E. White, Fert. Res. **33**, 81 (1992)

4. K. Walton, D. Alle, in Supersoil 2004 (ed. B. Singh) Proc. 3rd Australian New Zealand Soils Conference, Sydney, Australia, 5-9 Dec. (2004)

Adaptation to climate change of the French wine industry: a systemic approach – Main outcomes of the project LACCAVE

Nathalie Ollat [1], Hervé Quénol[2], Gérard Barbeau[3], Cornelis van Leeuwen[1], Philippe Darriet[4], Inaki Garcia de Cortazar Atauri [5], Benjamin Bois[6],Hernan Ojeda[7], Eric Duchêne[8], Eric Lebon[9], Philippe Vivin[1], Laurent Torregrosa[10], Jean-Marie Sablayrolles[11], Geneviève Teil[12], Philippe Lagacherie[13], Eric Giraud-Héraud[4], Patrick Aigrain[14] and Jean-Marc Touzard[15]

1 : EGFV, ISVV-Bordeaux Sciences Agro-INRA- Bordeaux Univ, ISVV33140 Villenave d'Ornon, France,
2 : LEGT, CNRS-Univ de Rennes, Place du Recteur Henri Le Moal, 35043 Rennes, France
3 : UVV, INRA, 42 rue Georges Morel BP 60057, 49 071 Beaucouzé, France
4 : Laboratoire Œnologie, ISVV-Univ Bordeaux, 33140 Villenave d'Ornon, France
5 : AgroClim, INRA, Domaine Saint Paul, Site Agroparc, CS 40 509 – 84914 Avignon, France
6 : Institut Universitaire de la Vigne et du Vin - Université de Bourgogne, 21000 Dijon, France
7 : Experimental Unit of Pech Rouge, INRA, 11430 Gruissan, France
8 : SVQV, INRA-Univ Strasbourg, 28, rue de Herrlisheim, 68000 Colmar, France
9: LEPSE, INRA-Supagro, 2, place Viala, 34060 Montpellier, France
10 : AGAP, Univ. Montpellier, CIRAD, INRA,Supagro-, 2, place Viala, 34060 Montpellier, France
11 : SPO, INRA-Supagro, 2, place Viala, 34060 Montpellier, France
12 : SAD APT, INRA-Agroparitech, 16 rue Claude Bernard, 75005 Paris, France
13 : LISAH, INRA-IRD-SupAgro, 2, place Pierre Viala, 34060 Montpellier, France
14 : FranceAgriMer, Service Etude et Prospective, 93555, Montreuil, France
15 : Innovation, INRA-Supagro, 2, place Viala, 34060 Montpellier, France

Abstract: Taking into account the major economical role and specificities of the French wine industry, adaptation to climate change is a very challenging issue. In 2011, 23 research teams launched a systemic and multidisciplinary program to analyze the impacts from the vine to the region, to define adaptation strategies combining technical, spatial and organizational options and to evaluate the perception by the actors and consumers of climate change issues. Thermal variability was studied at local scale to develop high resolution atmospheric models which better simulate future climate trends. Impacts on growth/developmental conditions and vine responses were estimated from the calculation of eco-climatic indices and a combination of functional models. Genetic and physiological bases of grapevine adaptation to high temperature and drought were analyzed. Improving oenological and cultural practices as well as plant material innovation have been investigated as major technical adaptations. How these options could be implemented at the plot level was examined to elaborate decision tools. Multi-agent modelling was developed for this purpose. Surveys were performed to evaluate the perception of the main actors regarding climate change and their level of acceptability towards technical changes. Consumer acceptability of new types of wines was also investigated with an experimental economy approach. Finally a foresight exercise was conducted to design four potential adaptation strategies: conservative, innovative, nomad and liberal. Outcomes of this exercise are now used as a tool for wine industry members to develop their own strategic plan for adaptation.

1-General objectives

The general objectives of the LACCAVE project are to further characterize the long-term impacts (2050) of climate change on the grapevine growing and wine production, particularly at regional scales, and to elaborate additional knowledge on innovations and possible adaptation strategies for the sector. The purpose is not to propose technical solutions ready to use. The goal is to improve our understanding of the effects of climate change, and then to explore various levers and their combinations to define adaptation strategies at the local or regional scale. It is also important to assess the economic, sociological and environmental consequences of these strategies. Finally it aims at coordinating the French research on these issues, at communicating the results at the national and international levels, in order to better meet the demand of the industry and society.

More specific objectives are i) to build a multidisciplinary scientific network, ii) to develop a shared vision and a collective questioning on the issue, iii) to develop methods and analysis tools, iv) to coordinate responses to other funding calls. The project has the ambition to support a set of targeted or integrative studies and achieve a prospective study on adaptation strategies. To prepare a more operational phase, the objective is also to share information and initiate actions with stakeholders.

2- Overview of the results

2.1- The construction of a national scientific network

LACCAVE brought together all the French teams working on the theme "vine, wine and climate change": 96 researchers and students, 20 INRA and 3 non INRA laboratories. The project improved the visibility of their work at national and international scale. This community exchanged very actively, taking into account the multidisciplinary dimension of the issue. Existing or new collaborations have been developed (leading in many cases to new formal projects). The LACCAVE members shared a systemic vision and analysis framework of adaptation (combination of adaptation levers at several scales, taking into account the perception by stakeholders and consumers). Workshops and thematic seminars, annual general assemblies of the participants, coordination board meetings, the foresight exercise and the organization of the international conference about sustainable grape and wine production in the context of climate change "ClimWine2016" managed to build this community of knowledge.

2.2 The improvement of climate simulations at local level

LACCAVE provided the framework for improving knowledge on climate variability and climate evolution at the local level. Climatologists and modelling specialists have worked on climate scenarios across a 8km x 8km grid developed by MétéoFrance and available on the DRIAS platform (http://www.drias-climat.fr/), in order i) to obtain useful information for the development of adaptation strategies in different French wine regions, and ii) to disseminate these outputs to stakeholders [1, 2, 3]. Furthermore, temperature variability at local level has been studied in a modelling perspective in the Loire Valley [4] and the region of St Emilion [5]. These results highlight that within a small wine region, the variability of temperature can be equivalent to i) that observed on average between two wine regions, and ii) the increase in average temperature expected for 2050, offering flexibillity/opportunity for adaptation at the local level [6, 7].

2.3. Overview of Climate Change impacts on vine and wine

LACCAVE provided an overview of the impacts of climate change on vine and wine [8, 9]. This synthesis has benefited from the collective ability to bring together results from a wide range of existing works, such as historical observations, experiments and simulations (in different disciplines). The advancement of phenological stages was confirmed and clarified [1, 3]; the accentuation of water deficit for the vines in the Mediterranean area has been clearly shown [10]; the possible impacts of climate change on grapes component (sugar, acids, aroma precursors, anthocyanins ...) and the wine quality have been reviewed [11, 12, 13], providing an illustration of how climate change can influence the quality of food, and thus its markets and regulations [14]. A survey has been launched at the international level, in order to better assess the correlations between disease incidence and climatic data, and to improve pest risk simulations [15].They have mostly been a common basis to start working on adaptation, defined as a set of actions to respond these impacts [16].

2.4 Advanced knowledge on the genetic basis of adaptation

In connection with other projects, the genetic determinants of vine response to water stress and to high temperatures were assessed. Phenology (methods, genetic variability analysis, modelling) was analyzed in collaboration with the ACCAF-PERPHECLIM project. Genetic variability was studied in clonal populations, intraspecific vinifera or between rootstock genotypes [17, 18]. The results highlight the importance of hydraulic conductivity and night transpiration as genetically controlled characters involved in the management of water losses of the vine [19, 20, 21]. Furthermore, the metabolism of ABA and the molecular determinants of the transport of water in the plant have varying characteristics depending on the genetic origin of rootstocks, and could be associated with adaptation to drought [22]. Interactions between rootstock and scion play also a key role in the response of plants. The effect of high temperatures on i) the vegetative growth of the vine, ii) the berries and iii) the composition of the grapes has been studied. Variable and complex responses have been demonstrated according to the organ or molecule, and to the phenological stage impacted by the stress [23, 24, 25]. Studies carried out in the context of joint projects revealed significant interaction between the effects of CO_2, temperature and UVB [26]. A major concerted effort was made on ecophysiological modelling in order to identify gaps and to better integrate the various existing models [27, 28].

2.5- Characterization of adaptation practices in the vineyards

Enological and viticultural practices, but also the potential of vine varieties (scion and rootstock) have been investigated as possible solutions for adaptation. In winemaking, yeast strains producing less alcohol have been selected in Montpellier and Bordeaux [29, 30]. Other technologies such as de-alcoholisation or acidification have also been experimented in the context of industrial partnerships. The management of the aromatic potential of wine in a context of climate change was also been analyzed and debated [31, 11]. As far as grapevine vine management is concerned, a wide range of practices has been assessed [32], including irrigation [33]. Research on highly innovative processes such as the use of recycled water for irrigation and coupling grapevine growing with the production of electricity by photovoltaic coverage have also been initiated. First results are promising and studies should continue.

The diversity of plant material (varieties and rootstocks) has been studied in different vineyards, using pre-existing experiments [34, 18]. A large experiment for rootstock assessment has been planted. These experiments have raised a significant amount of data on the phenology and ripening features of many varieties, but also on their agronomic and biochemical characteristics. Data have been stored in specific information systems for further analyses.

An overview of the possible changes of practices at different scales of space and time was conducted [32] and can be used as a decision making tool. Several surveys were also carried out among wine producers in different regions in order to assess their perceptions of both climate change and new practices [35, 36]. Sociological surveys show how the changes in practices for the climate challenge are influenced by those related to environmental and quality issues [36]. Combined approaches with climate modelling at local level, assessment of impacts on grape growing, analysis of practices and perceptions of wine growers draw a methodological basis for exploring the best practices of adaptation at vineyard level.

2.6. Land use management and modelling of landscapes evolution

LACCAVE considered the local level as a key issue for adaptation [37], leading to reconsider the organization / location of the vineyard plots according to the variability of climate impacts and the management of local public goods (soil, water, landscape, biodiversity ...). To support these local strategies several methods have been explored combining scenario building, exploration of coordination problems and co-construction of strategies with stakeholders. Multi-agent models connected with cartographic representations (on climate and land use) and models integrating hydrological, agronomical and ecological processes were elaborated [38, 39]. For example, the simulations of future landscapes of Banyuls vineyards (extreme south of France) have linked climate change, land management and possible evolution of the wine quality (which depends on the location of the plots). The outputs have been elaborated and discussed with the grape growers of the local cooperative in order to explore new collective rules for vine plantation and grape payment [38].

2.7. Economic analysis of conditions for adaptation to climate change.

Beyond the theoretical assessments on the economic dimensions of adaptation in the wine industry [16], surveys have been carried out in the Bordeaux region, linking the perceptions of climate change by the wine producers with their expected strategies, and their potential costs at the farm level. In this regional context, producers are mainly considering the risks of climate variability [40], adopting different strategies according to their economic resources and their (inherited/projected) trajectory. Studies in other regions confirmed that risk management is crucial, combined with different local issues (drought in Languedoc, diseases in Champagne [36]. Economic decisions (investment) result from uncertain cost/benefice calculations, taking into account the wine value, the expected costs and a scope of options allowed by each quality label [41].

The analysis of consumer's willingness to pay for wines affected by climate change is a key contribution of LACCAVE. Methods of experimental economics were developed to show that consumer's behavior is also a lever for adaptation. Wine tasting sessions were organized with large consumers panels in order to assess their judgment and willingness to pay for wines (still) impacted by climate change (more alcohol, less acidity, changes in aroma and color) and wines produced with oenological innovations in order to correct the wine composition. The influences of climate change on several sensory characteristics of the wines have been demonstrated, suggesting influences on the (future) wines markets [42]. This work is an important scientific verification of the instability of consumer's preferences according to the length of exposure to a product [43]The "wines of climate change" were more appreciated in the first tasting session, but were progressively depreciated (in comparison to current wines) when the consumer repeated the tasting along a week. Finally the results show that consumers evaluate differently (partly related to age and gender) the oenological practices used to correct the effects of climate change on quality (de-alcoholisation and acidification) [42].

2.8. The role of innovations and research for adaptation

Within LACCAVE, the role of agricultural research on the development and dissemination of innovations that contribute to adaptation strategies was investigated (new varieties, agronomic practices, wine-making technology...). A national survey, combined with a bibliometric study of scientific publications on "vine and wine", showed the importance of ongoing research (around the two poles of Montpellier and Bordeaux), but without clear relations with the regional vineyards where the research centers are located [44]. A double survey was thus carried out in 3 vineyards (Champagne, Bordeaux, Languedoc) with one hundred wine producers and 80 scientists in order to analyze the links between i) the perception of the climate challenge, ii) the implemented/expected actions coping with this challenge (in research or adoption of innovation) and iii) the advice networks built for these actions. If various wine producers profiles (according to their vision of innovations and climate change) are shared between the regions, their willingness to innovate largely depends on each "regional innovation system" [45] and the role played wine inter-professional organization, able to build a real "demand for research and innovation" [45]. These results call for developing multidisciplinary and participatory research, including the wine inter-professional partners [14]. They show that the organization of links between research and wine producers is also a lever for adaptation.

2.9. The construction of adaptation scenarios for the vine and wine sector

Finally LACCAVE launched a foresight exercise in order to explore potential adaptation strategies in the French wine industry, in the context of climate change. The exercise was led by a group of 12 LACCAVE participants (covering the full expertise range) and representatives of governmental bodies in charge of Appelation system (INAO) and Agriculture in general (FranceAgriMer). An original method was tested to describe the pathways leading to the implementation of different adaptation strategies [46]. Starting from a "median" climate scenario for 2050, 4 major adaptation strategies have been defined: "conservative" integrating marginal changes in the current vineyards; "Innovations to remain in place" opening the existing vineyards to radical innovations in order to maintain their current locations ; "Nomad vineyards" highlighting the consequences of potential vineyards relocation; "Liberal" testing a situation where "everything is possible everywhere." The group has collected, selected and combined a wide range of assumptions that could be related to these 4 strategies (influence-dependency matrix between 70 assumptions) and based on expert knowledge, previous foresight exercise and the survey described in 2.8. Four pathways were finally described by linking micro-scenarios (outputs of the matrix analysis), providing possible stories that lead to the 4 adaptation strategies in 2050. The 4 strategies were submitted to the industry in 6 different wine regions during participative workshops. Outcomes of this exercise and the workshops are currently used to initiate a new phase of co-construction of "climate strategies" at local and national scales.

3. Impacts and perspectives

LACCAVE allowed to raise the awareness of the grape and wine industry about climate change, but also of policy makers. A stakeholder group has now been launched at national level to define a national plan for adaptation. LACCAVE also contributed to build a shared and systemic vision of adaptation among the French scientific community working on vine and wine. This shared vision is crucial to elaborate proper innovations for the actors. Communication towards various publics was very active in order to provide a scientific and realistic message about the expected impacts of climate change on this national industry [47]. It can be summarized as following:

- Adaptation strategies depend first on the level of mitigation and could be reasonably implemented in all French vineyards if global warming stays in the range of the current simulations made by climatologists for 2050 [48].

- There is no single and simple solution. Different levers including technical innovations, geographical strategies, and organisational changes, and regulation policies should be combined.

-Perception and adaptation capacities from the different stakeholders of the industry should be taken into account.

- Solutions should be integrated, by considering the value chain for wines, including the evaluation of consumer preferences. Global issues should also be taken into account as short-term challenges for the industry (lowering the pesticide inputs), other environmental aspects and public health considerations.

- These adaptation strategies must be coordinated at local and regional levels where climate impacts are specific and where the use of resources can be optimized;

- Taking into account the large uncertainties, the best way to adapt will be based on the collaborative capacity between researchers and stakeholders, even at regional level.

Perspectives of this first step are to further investigate effective solutions operating at different scales. Based on the knowledge already collected, challenging issues of adaptation as water management, pest management, soil management, better adapted varieties, vineyard location, wine quality and economic issues should be considered in a complex and systemic way. Simulated eco-climatic indicators and vine responses should be delivered to the industry in order to better define adaptation strategies and solutions. Participative activities with stakeholders should be launched in order to elaborate from national to estate levels adaptation plans and to design adapted and resilient cropping systems. Information systems should also be delivered to collect existing initiatives and innovations. Finally working at international levels should allow to learn from the experience of other wine regions and to share with them our analytical approach of adaptation. These perspectives will be developed with the new project LACCAVE 2.21.

ACKNOWLEDGMENTS

The authors would like to thank all the participants of LACCAVE and the contribution of stakeholders all through the project. Without their work, this paper and many others would not have been possible. We dedicate this contribution to our estimated colleague, friend and co-author Eric Lebon who passed away in December 2016. This work was supported by the LACCAVE and PERPHECLIM projects funded by the meta-program Adaptation of Agriculture and Forests to Climate Change (AAFCC) of the French National Institute for Agricultural Research (INRA).

REFERENCES

1. B. Bois, I. Garcia de Cortazar-Atauri, H. Quénol, J. Caubel, Mid-term LACCAVE meeting Montpellier (2014)
2. E. Lebon, I. Garcia de Cortazar Atauri, Innov. Agron. INRA, Montpellier, pp 1-12 (2014)
3. I. García de Cortázar-Atauri, E. Duchêne, A. Destrac-Irvine, G. Barbeau, L. de Rességuier, T. Lacombe, A.K. Parker, N. Saurin, C.van Leeuwen, OenoOne, **51**: 115 (2017)
4. E. Neethling, T. Petitjean, H. Quénol, G. Barbeau, Mitig Adapt Strateg Glob Change. DOI 10.1007/s11027-015-9698-0 (2016)
5. L. de Resseguier, R. Le Roux, H. Quénol, C. van Leeuwen, *Proceedings of ClimWine2016*, 66 (2017)
6. C. Bonnefoy, M. Madelin, H. Quénol, Revue Internationale de Géomatique, **24/3**, 377 (2014)
7. E. Neethling, T. Petitjean, H. Quénol, G. Barbeau, Mitig Adapt Strateg Glob Change. DOI 10.1007/s11027-015-9698-0 (2016)
8. N. Ollat, J.M. Touzard, J. I. S. V. V LACCAVE special issue, 1-7 (2014)
9. N. Ollat, J.M. Touzard, I. Garcia de Cortazar-Atauri, H. Quénol, C. van Leeuwen, *Proceedings of ClimWine2016* (2017)
10. H. Ojeda, N. Saurin 2014. Innov. Agron., **38**, 97 (2014)
11. P. Darriet, A. Pons, L. Allamy, A. Schüttler, C. van Leeuwen, C. Thibon, *3ième assise des vins du Sud Ouest, Colloque IFV*, (2016)
12. C. van Leeuwen, A. Destrac, OENO One, **51**, 147 (2017)
13. E. Duchêne, G. Butterlin, P. Claudel, N. Jaegli https://www.researchgate.net/publication/306941071_Consequences_of_elevated_temperatures_during_ripening_on_the_biosynthesis_of_monoterpenols_in_grape_berries (2016)
14. N. Ollat, J.M. Touzard, C. van Leeuwen, J. Wine Economics **11**, 139 (2016)
15. B. Bois, S. Zito, A. Calonnec, OENO one, **51**,133 (2017)
16. V. Viguier, F. Lecocq, J.M. Touzard, J.I.S.V.V. Laccave Special issue, 55 (2014)
17. A. Bigard, D.T. Berhe, E. Maoddi, Y. Sire, J.M. Boursiquot, H. Ojeda, J.P. Péros, A.Doligez, C. Romieu, L. Torregrosa (2018) Vitis vinifera L. fruit diversity to breed varieties anticipating climate changes. Frontiers Plant Sci. doi: 10.3389/fpls.2018.00455.
18. E. M. Wolkovich, I. García de Cortázar-Atauri, I. Morales-Castilla, K. A. Nicholas, T. Lacombe. Nature Climate Change, **8**, 29 (2018)
19. A. Coupel-Ledru, E. Lebon, A. Christophe, A. Doligez, L. Cabrera-Bosquet, P. Péchier, P. Hamard, P. This, T. Simonneau, J. Exp. Bot. **65**: 6205 (2014)
20. A. Coupel-Ledru, E. Lebon, A. Christophe, A. Gallo, P. Gago, F. Pantin, A. Doligez, T. Simonneau, PNAS DOI: 10.1073/pnas.1600826113 (2016)
21. A. Coupel-Ledru, S. Tyerman, D. Masclef, E. Lebon, A. Christophe, E. Edwards, T. Simonneau Plant physiol. **175**, 1121 (2017)
22. L. Rossdeutsch, E. Edwards, S.J. Cookson, F. Barrieu, G.A. Gambetta, S. Delrot, N. Ollat(2016) BMC Plant Biology, **16**, 91 (2016)
23. L. Torregrosa L, A. Bigard, A. Doligez, D. Lecourieux, M. Rienth, N. Luchaire, P. Pieri, R. Chatbanyong, R. Shahood, M. Farnos, C. Roux, A. Adiveze, J. Pillet, Y. Sire, E. Zumstein, M. Veyret, L. Le Cunff, F. Lecourieux, N. Saurin, B. Muller, H. Ojeda, C. Houel, J-P Péros, P. This, A. Pellegrino, C. Romieu, OENO One, **51**, 155 (2017)
24. E. Duchêne, OENO One, **50**, DOI: 10.20870/oeno-one.2016.50.3.98 (2016)
25. F. Lecourieux, C. Kappel, P. Pieri, J. Charon, J. Pillet, G. Hilbert, C. Renaud, E. Gomès, S. Delrot, D. Lecourieux, Frontiers in Plant Science **8** (2017)
26. J. Martínez-Lüscher, F. Morales, M. Sánchez-Díaz, S. Delrot, J. Aguirreolea, E. Gomès, I. Pascual (2015) Plant Sci, **236**,168 (2015)
27. P. Vivin, E. Lebon, Z. Dai, E. Duchêne, M. Marguerit, I. García de Cortázar-Atauri, J. Zhu, T. Simonneau, C. van Leeuwe, S. Delrot, N. Ollat, OenoOne **51**, 181 (2017)
28. J. Zhu, Z. Dai, P. Vivin, G. A. Gambetta, M. Henke, A. Peccoux, N. Ollat, S. Delrot, Annals of Botany, DOI:10.1093/aob/mcx141 (2017)
29. V. Tilloy, A. Ortiz-Julien, S. Dequin, Applied and Env. Microbiology, **80** : 2623 (2014)
30. S. Dequin, J.L. Escudier, M. Bely, J. Noble, W. Albertin, I. Masneuf-Pomarède, P. Marullo, J.M. Salmon, J.M. Sablayrolles, OENO One, **51**, 205 (2017)
31. A. Pons, L. Allamy, A. Schüttler, C. Thibon, P. Darriet, OENO One, **51**, 141 (2017)
32. G. Barbeau, E. Goulet, E. Neethling, N. Ollat, J.M. Touzard, *Changement climatique et terroirs viticoles* (2014)

33. H. Ojeda, N. Saurin, S. Alvarez Gei, R. Symoneaux, C. Coulon-Leroy, *Proceedings of ClimWine2016,* 194 (2017)
34. A. Destrac, C. van Leeuwen, *Proceedings of ClimWine2016,* 166 (2017)
35. E. Neethling, T. Petitjean, H. Quénol, G. Barbeau, Mitig. Adapt. Strateg. Glob. Change, **22**, 777 (2017)
36. G. Teil, *Proceedings of ClimWine2016,* 229 (2017)
37. H. Quénol, *Pour une géographie de la conservation : Biodiversités, natures et sociétés* (2015)
38. E. Delay, N. Becu, *Proceedings of ClimWine2016,* 204 (2017)
39. F. Vinatier, A. González Arnaiz, P. Lagacherie, *Proceedings of ClimWine2016,* 213 (2017)
40. A. Alonso Ugaglia, E. Giraud-Heraud, S. Pérès, Y. Surry,https://www6.inra.fr/laccave/ClimWine2016/Download-oral-presentations/Session-Perception-and-adaptability-of-climate-change-by-the-actors (2016)
41. J. Boyer, J.-M. Touzard, Forum Innovation, Paris, Cité des Sciences et de l'Industrie (2016)
42. A. Fuentes-Espinoza, S. Pérès, A. Pons, S. Tempère, A. Samson, J.L. Escudier, P. Darriet, E. Giraud-Héraud, *Proceedings of ClimWine2016,* 235 (2017)
43. E. Giraud-Héraud, Enometrics XXIII, COLMAR - FR : May 25-28, (2016)
44. J. Boyer, J. M. Touzard, *Proceedings of ClimWine2016,* 235 (2017)
45. J. Boyer, Revue d'Economie et de Management de l'Innovation, **52** (2016)
46. P. Aigrain, B. Bois, F. Brugière, E. Duchêne, I. Garcia de Cortazar Atauri, J. Gautier, E. Giraud-Héraud, H. Hannin, N. Ollat, J.M.Touzard, *Proceedings of ClimWine2016,* 253 (2017)
47. J.M. Touzard, N. Ollat N., Revue Parlementaire, **543** (2015)
48. IPCC - The physical Science Basis - Summary for Policymakers (2013)

Diferenciación del vino a través del terroir: el caso de la DOP Campo de Borja (España)

Miguel LorenteBlasco[1] y Ernesto Franco Aladrén[2]
[1]Grupo de Estudios de Desarrollo Territorial, Universidad de Zaragoza (GEDETUZ). España. mlorenteb@yhoo.es
[2]Centro de Transferencia Agroalimentaria - Gobierno Aragón. Unidad de Enología. Avda. de Movera, s/n, 50194, Zaragoza (España). efranco@aragon.es

Abstracts: The Regulatory Council of the DOP Campo de Borja (Spain) promoted a zoning study aimed at knowing the geographical environment and its effect on the quality of wines, in order to delimit the space in Units of Terroir (UT) base to apply a model Productive and commercial reasoned following the concept terroir. It was to seek the specialization and recognition of wine values linked to the territory of origin. The study of the medium, reflected in a map of cartographic soil units (SMU) to scale 1:25.000, was accompanied by an oenological study carried out during three vintages to know the variability of the oenological capacities of the SMU. The results indicate that in the DOP have obtained different types of units of terroir depending on the characteristics of the mistelas used to classify and delimit territorially Units of Terroir.

1 Introducción

El término francés terroir, utilizado en todo el mundo, expresa el fenómeno geográfico que rige la Denominación de Origen (DO), arraigado en la idea de hacer inseparables la producción y el territorio de origen del vino. Del mismo se deriva la servidumbre de entender el vino como el resultado de una cultura emanada de las interacciones producidas en la Naturaleza entre el medio y la vid, y cuyo resultado se manifiesta en las características distintivas de los vinos. Se deduce, por tanto, que la aplicación coherente de la DO va unida a la idea del terroir, lo que lleva implícito conocer el medio y su potencial enológico a través de la zonificación, y utilizar este fenómeno en la producción y comercialización del vino.

En Europa la producción de vinos de calidad se centra en un sistema de distribución territorial basado en DOP e IGP. A este respecto Huglin (1978) consideraba que las DO se deben basar en la delimitación de zonas con condiciones ecológicas particularmente privilegiadas o simplemente originales.

Se considera zonificación, en un sentido amplio, a la división racional de un área geográfica en zonas homogéneas de acuerdo a ciertos criterios. Desde el punto de vista de la producción enológica la zonificación se concibe como la investigación del territorio para repartirlo en zonas relativamente homogéneas, como consecuencia de la interacción entre el viñedo y el medio ambiente (Fregoni et al., 1998). Por tanto, la aplicación coherente y razonada de la normativa sobre la DO implica la zonificación territorial, con el fin de determinar los terrenos de especial aptitud para el cultivo de la vid que deben constituir los límites de todos los niveles de las DOP.

Un estudio realizado en la DOP Campo de Borja con el objetivo de promover su desarrollo, indicaba que para mejorar su posicionamiento comercial se debía avanzar en especialización y en el reconocimiento de valores del vino vinculados a el territorio de origen. Esto llevó al Consejo Regulador a promover un trabajo de zonificación destinado a conocer el medio geográfico y su efecto en la calidad de los vinos con el fin de delimitar el espacio en Unidades de Terroir (UT), base para aplicar un modelo productivo y comercial razonado siguiendo el concepto terroir.

El trabajo pretendía proporcionar recursos técnicos a los agentes de la producción, al mismo tiempo que aportaba justificación ideológica (la DO y el terroir son una ideología en el modo de concebir el vino), pero se sabía de las limitaciones que surgen a la hora de aplicar las innovaciones en las zonas productoras, especialmente en el medio rural. Como el trabajo de la DOP Campo de Borja pretendía ser una herramienta para promover el desarrollo, y es sabido que no basta con que lo nuevo sea mejor que lo existente sino que debe ser integrable en la cultura de los que la van a adoptar, en paralelo al trabajo de zonificación se desarrollaron una serie de actividades de comunicación encaminadas a producir cambios en el *statu quo*, difundiendo los conocimientos mediante conferencias, salidas al campo y cursos para los técnicos sobre manejos de SIG, etc.

2 Material y métodos

El estudio se estructura tomando la vid como elemento vivo que se desarrolla afectado por el medio geográfico (con sus factores naturales y humanos), y que, como consecuencia de esta interacción, produce la materia prima del vino con características intrínsecas particulares. Este proceso biológico nos lleva a analizar el medio geográfico para delimitar las zonas homogéneas con incidencia en el desarrollo de la vid, y las características enológicas de estas unidades de suelo para determinar las Unidades de Terroir (UT) con posibilidad de ser utilizadas de un modo razonado en la producción de vino. Por tanto, se asocia el término de zonificación a subdivisión (clasificación) de un territorio en base a las características ecopedológicas y topográficas (Fregoni *et al*, 1998), y al resultado enológico fruto de la interacción vid-medio.

Finalmente, esta información se une al SIGPAC y al Registro Vitícola que recoge los datos relacionados con las parcelas cultivadas de vid, permitiendo abordar el análisis estadístico desde la perspectiva del medio geográfico, de la caracterización enológica y del viñedo, para establecer las UT. La interrelación de las tablas de atributos de las capas permite tener información, a escala parcelaria, de todos los factores que afectan al desarrollo del viñedo y a las características enológicas, permitiendo una gestión razonada de la producción de acuerdo a los criterios del terroir.

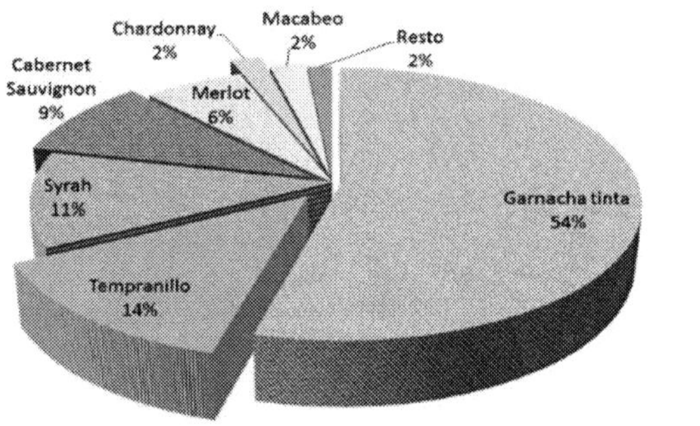

Figura 1. Ocupación de las variedades en la DOP Campo de Borja.

El territorio de la DOP tiene una superficie de 65.300 hectáreas y el viñedo ocupa 6.661 hectáreas repartidas en 10.074 parcelas, lo que representa una dimensión parcelaria media de las viñas de 0,6 ha y un índice de ocupación del viñedo en la zona de 9,5 %. La variedad garnacha es la tradicional de la DOP y ocupa el 55 % de la superficie, seguida a gran distancia de otras variedades de reciente implantación (Figura 1). La preponderancia e interés de la garnacha hace que el estudio enológico se realice sobre esta variedad.

2.1. Estudio del medio

El método utilizado en el estudio del medio se ha aplicado en distintas DOP de España y otros países por Sotés y Gómez-Miguel (1993-2015). Se trata de un análisis geográfico del territorio en el que el resultado final es un mapa cuyas unidades cartográficas (SMU) sintetizan las relaciones entre clima, litología, geoformas y serie de suelos. El manejo de la información generada en las capas tratadas por un Sistema de Información Geográfica (SIG) da como resultado la cuantificación de los contenidos y la posibilidad de su tratamiento estadístico.

En el estudio del medio se tuvo en cuenta el clima, la topografía, la geología y los suelos, según Soil Survey Manual y Soil Taxonomy USDA (1990-2014). Como base de reconocimiento se utilizaron fotogramas aéreos a escala 1:20.000, ortofotos, mapa topográfico nacional, mapas geológicos del IGME y mapa forestal español. En el trabajo de campo se describieron 519 perfiles y 1.560 observaciones (densidad 3,2/km^2). En el laboratorio se analizaron 2.309 muestras.

El mapa final a escala 1:25.000 está constituido por SMU cuyos contenidos reflejan los tipos de suelos en unidades taxonómicas (STU) o taxa (Figura 2).

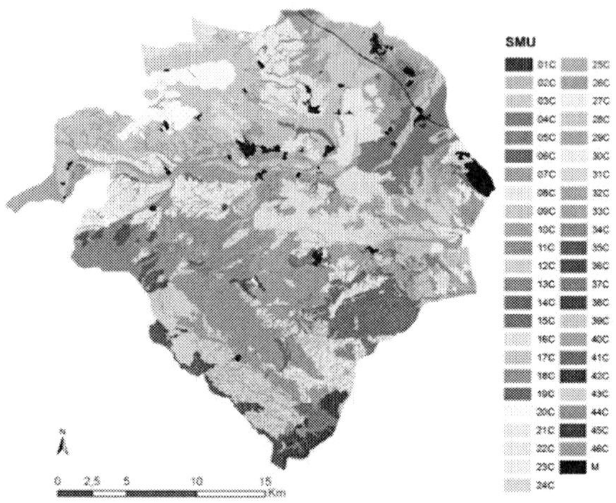

Figura 2. Mapa de suelos en SMU. Fuentes: Gomez-Miguel et al, 2015)

2.2. Los suelos de la DOP Campo de Borja

El mapa de suelo de la DOP está compuesto de 46 SMU aunque diez de ellas ocupan prácticamente el 50 % de la superficie. El 64 % del viñedo se encuentra en el 33 % de la superficie compuesta por ocho SMU. Tomando las unidades taxonómicas principales (STU) y secundarias de suelo, como representantes de mayor ocupación de las SMU, los resultados muestran que dos STU destacan sobre las demás, puesto que soportan un tercio del viñedo. Las seis STU siguientes en ocupación albergan al 45 % y otras siete el 17 % (Figura 3). Esto nos demuestra que los viticultores han seleccionado las tierras de forma empírica a la hora de plantar las viñas, Lorente (2017).

2.3. Estudio enológico

Definidas las SMU el siguiente paso fue verificar su potencial enológico para descubrir el vínculo entre el vino y el lugar donde se producen las uvas. Pero abordar un estudio de estas características supone manejar un número elevado de parcelas durante varios años y disponer de medios técnicos adecuados. En nuestro caso solo disponíamos de recipientes de poco volumen donde las fermentaciones resulta difícil controlar e introducen factores que no están directamente relacionados con el medio geográfico y pueden desvirtuar el vínculo entre el vino y el territorio, por lo que los resultados a veces resultan erróneos. Por ello se opta por una vinificación sin fermentación alcohólica, mistela, que se lleva acabo en depósitos de 25 litros, tal como se describe en el trabajo de Franco y Lorente (2018).

Para el estudio se tomaron 60 parcelas de la variedad garnacha tinta con el portainjerto R-110, en secano y plantadas antes del año 2000, de cuyas uvas se elaboraron 60 mistelas en depósitos de 25 litros y 12 vinos en depósitos de 1.000 litros de parcelas que también se obtenían mistelas. El estudio se realizó en los años 2012, 2013 y 2014. Este método se ajusta a los fines perseguidos puesto que se buscaba determinar la expresión de aromas primarios de la uva, y para ello no era necesario realizar la fermentación alcohólica.

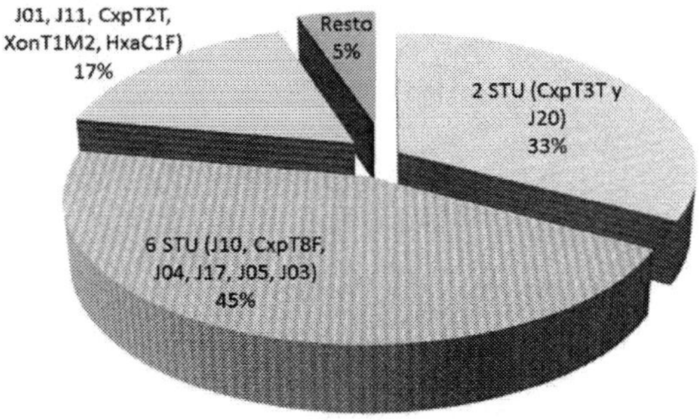

Figura 3. Ocupación del viñedo por STU

Las mistelas y vinos fueron sometidos a análisis fisico-químico y sensoriales, y los resultados de todas parcelas fueron tratados estadísticamente para clasificarlos y ver si se formaban grupos diferentes con sus características. Esta clasificación unida a la información del mapa de suelos permitiría conocer la tendencia enológica de cada SMU y determinar el vínculo entre el vino y el territorio.

Hay que destacar que a lo largo de los tres años de ensayos se produjeron una serie de incidencias por diversos motivos que impidieron realizar el trabajo de acuerdo a los protocolos previstos. En 2013 se produjo un fuerte *corrimiento* de la flor que mermó la cosecha en numerosas parcelas, hasta el punto de impedir realizar algunas vinificaciones. En consecuencia, 24 parcelas de mistela disponen de tres resultados, 27 dos resultados y 17 solamente de un año.

2.4. Estudio estadístico

Con los resultados analíticos obtenidos se buscaba comprobar, mediante análisis factorial, si las diferencias entre los valores de las parcelas eran significativas, ya que en caso contrario se produciría un caso de uniformidad. Igualmente se quería conocer si los valores de las parcelas podían ser clasificados en grupos homogéneos para ver posteriormente si existía correlación con las SMU. Se pretendía establecer un modelo que, a partir de las variables analizadas en las mistelas, permitiera definir sus características, comprobar su variabilidad espacial y relacionarlas con las unidades cartográficas de suelo. Para ello se realizó un tratamiento estadístico utilizando la técnica de análisis multivariante (Júdez 1989, Dolédec 1995), en el que se consideraban variables los valores de los elementos organolépticos y parámetros del análisis físico-químico obtenidos mediante la metodología descrita Franco y Lorente (2018).

En el estudio estadístico se utilizaron técnicas de análisis de regresión, análisis exploratorio de Tukey (estadística descriptiva) y análisis multivariante (análisis de componentes principales –ACP-, clasificación automática según diferentes criterios de distancia). Dichas técnicas se aplicaron utilizando el programa informático SPSS.

2.5. Relación entre las características enológicas de las mistelas y las SMU

Tras determinar a través de la clasificación automática los grupos en los que las mistelas se clasifican según sus características enológicas y la pertenencia de cada parcela a un grupo concreto, con la herramienta de geoprocesamiento de ArcGis se cruzan las capas en formato vectorial que recogen las unidades cartográficas de suelo (SMU) y las parcelas de ensayo extraídas de los recintos del SIGPAC. Esta operación de intersección permite añadir a las parcelas un nuevo campo con la SMU en la que se encontraban.

Las parcelas de cada SMU se agrupan y se puede observar su tendencia enológica a partir de los grupos clasificatorios. Esta operación permite comprobar si en cada SMU las parcelas se clasificaban mayoritariamente en uno o dos grupos concretos o si, por el contrario, esta dominancia no se produce, por lo que en el primer caso se podría entender que existe relación entre el SMU y el tipo de uvas que producen, mientras que en el segundo caso esto aún no quedaba demostrado. Se considera grupo dominante cuando más del 50 % de las observaciones se clasifican en ese grupo. En algunos casos esa dominancia la producen dos grupos contiguos con algunas características comunes y en otros, normalmente por falta de observaciones, no se aprecia ninguna dominancia.

3 Resultados

3.1. Analíticos y sensoriales de las mistelas

El análisis de componentes principales (ACP) de la concentración de azúcar, color e IPT, muestra un efecto de la añada muy moderado, a pesa que en las medias se observe diferencia importante que se atribuían a la diferencia de pluviometría, (Franco y Lorente, 2018). En la F-Q como en el sensorial se ve que las añadas de 2013 y 2014 son más parecidas entre sí, coincidiendo con una pluviometría superior a la de 2012. También se aprecia que algunas variables presentan pocas diferencias entre los años, mientras que en otras la variabilidad es mayor

Los resultados del ACP de todas las mistelas, con independencia del años y con las variables físico-químicas más explicativas, intensidad de color e índice de polifenoles totales en la componente 1, y azúcares en la componente 2, según el trabajo previo de Franco y Lorente (2018), se muestra en la Figura 4, observando en la zona positiva de la componente 2 mayor presencia de las mistelas de 2012, mientras que las de los años 2013 y 2014 están mayoritariamente distribuidas en la zona negativa de la componente 2, la menor lluvia de 2012 justificaría este comportamiento, sin embargo no podemos asignar diferencia significativas en la analítica de las mistelas en función del año.

El ACP del análisis sensorial indica que la componente 1 engloba los aromas a fruta negra, fruta pasa y tinta, y la componente 2 la fruta roja y el balsámico, según el trabajo previo de Franco y Lorente (2018), en este caso la distribución de las mistelas es más uniforme que con parámetros químicos y tampoco podemos asignar diferencia significativas en el análisis sensorial de las mistelas en función del año, Figura 5.

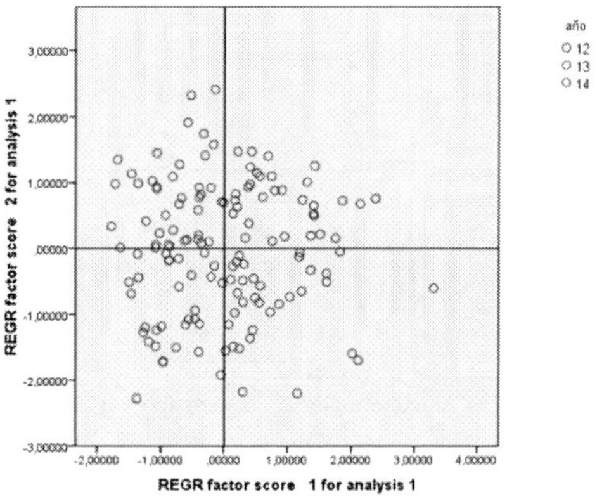

Figura 4. Distribución de las mistelas en ACP de los parámetros analíticos.

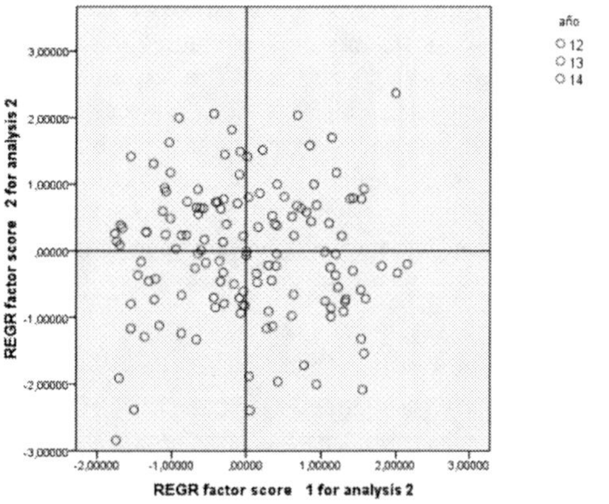

Figura 5. Distribución de las mistelas en ACP de los parámetros sensoriales.

3.2. Clasificación de las mistelas

Si bien estadísticamente no podemos hablar de un efecto añada nítido, sí que se observa una influencia de la añada, que condiciona la clasificación de las mistela.

La aplicación de la clasificación automática con la base de los parámetros físico-químicos y del análisis sensorial clasifican las mistelas en cuatro grupos ajustados a los cuadrantes formados por los dos ejes que representan los componentes principales. Ahora bien, la clasificación por cuadrantes es similar en los tres años y solo se representa la clasificación de las mistelas el año 2014.

La distribución de las mistelas en función de los tres parámetros químicos, se muestra en la figura 6, indica que las mistelas del primer cuadrante se caracterizan por moderado IPT y color, las del segundo cuadrante presentan los máximos valores de IPT y color, las del tercer cuadrante IPT y color son elevadas, inferiores a las del segundo y superiores a primero y cuarto y las del cuarto son las menores en los tres parámetros. Figura 7, remarcar que la concentración de azúcar es muy similar en los cuatro grupos y en consecuencia quien realmente clasifica las mistelas es el IPT y la intensidad colorante.

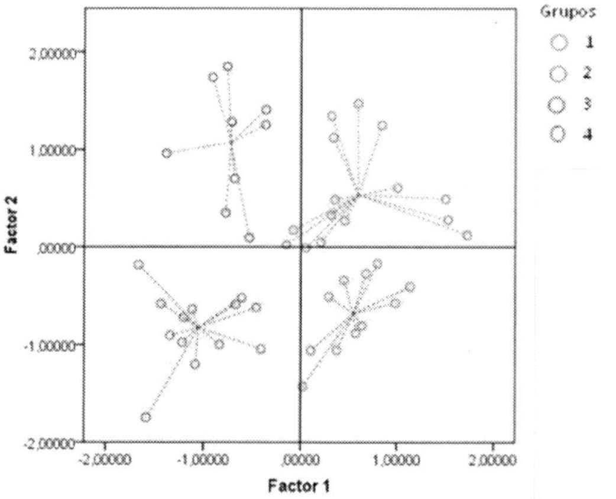

Figura 6. Clasificación en grupos de las mistelas en función de los parámetros químicos del año 2014

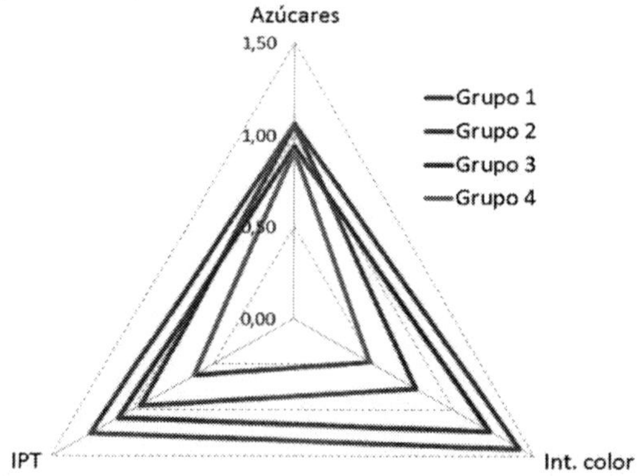

F

igura 7. Representación de los valores medios de los parámetros químicos de los grupos resultantes de la clasificación. Años 2012, 2013 y 2014.

En el caso del análisis sensorial el grupo uno, del primer cuadrante, se caracteriza por predominar los aromas a fruta roja y balsámico pero con menores valores de la fruta negra y tinta, el grupo dos presenta valores altos en todos los descriptores, el grupo tres los valores bajos son de fruta roja y balsámico, y el grupo cuatro todos los descriptores son bajos, figura 8.

En la figura 9, se representan los valores de los diferentes descriptores del aroma de las mistelas en función de la clasificación en cuatro grupos, el grupo cuatro es el de valores más bajos de los descriptores y en consecuencia el de menor calidad, mientras el grupo dos presenta los valores más elevados en todos los descriptores menos en la tinta, que el grupo tres es superior. Las diferencias importantes observadas entre los

 https://doi.org/10.1051/e3sconf/20185001021

valores de los cuatro grupos indican que esta metodología es adecuada para clasificar las mistelas en grupos en función de sus características aromáticas.

Una vez clasificadas las mistelas en cuatro grupos para cada uno de los tres años de estudio, la pregunta es ¿Cuántas mistelas de la misma parcela repiten en el mismo grupo? Con tres observaciones, es decir con tres mistelas de la misma parcela, la probabilidad de que se repita grupo dos o tres años es del 78% en el sensorial y del 56% en el químico.

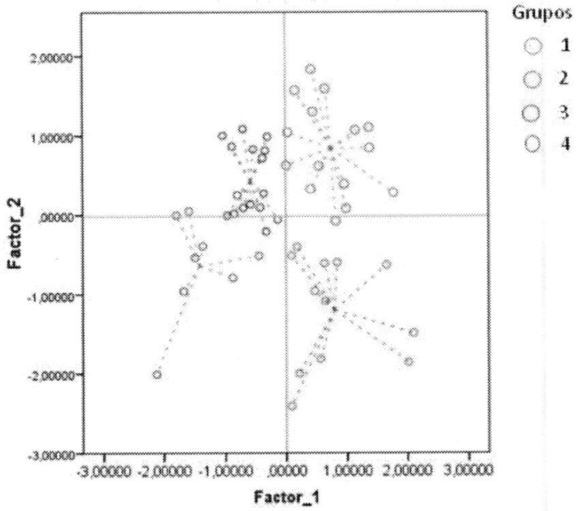

Figura 8. Clasificación en grupos de las mistelas en función de los descriptores sensoriales del año 2014

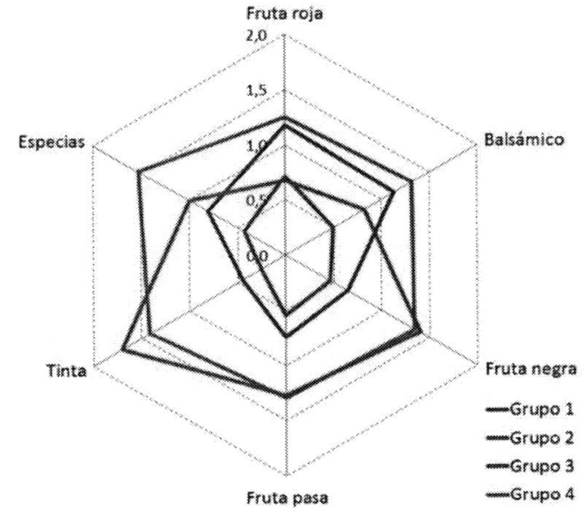

Figura 9. Representación de los valores medios de los descriptores del aroma de los grupos resultantes de la clasificación. Años 2012, 2013 y 2014.

3.3. Asignación de las parcelas a las SMU, en función de las características enológicas de las mistelas obtenidas.

Cuando las SMU se asocian a uno o varios grupos enológicos concretos, se puede decir que constituyen una unidad de terroir. Para este estudio se tomaron todas SMU con parcelas de ensayo excepto las que disponían menos de cuatro observaciones.

Los grupos 1 y 4 mayoritariamente se producen en unidades cartográficas del Cuaternario (fondos de valle y terrazas), mientras que las del Terciario y Triásico se asocian a los grupos 2 y 3. Esto hace que se produzca una gran similitud entre el mapa de características enológicas y el de Litología.

En la mayoría de los casos las unidades cartográficas tienen más probabilidades de clasificarse en uno o dos grupos concretos, lo que se podría considerar como potencial enológico de la SMU que sirve para definir las UT. Las observaciones de las parcelas cuyos valores del análisis estadístico (ACP) se representan próximas a los ejes, pequeñas variaciones por la añada u otros factores pueden influir en la clasificación, cambiando de grupo, de un año a otro, por tanto, en algunos casos el potencial de las SMU debe considerarse con cierta flexibilidad o considerar que la probabilidad para clasificarse en grupos diferentes es mayor.

Por ello, si en el mapa de SMU se agrupan las unidades que se relacionan con grupos enológicos concretos, se convierte en el mapa de unidades de terroir (UT), así postulamos seis unidades de terroir (UT) que se muestran en la figura 10, tres de estas unidades se corresponden con los grupos 2,3 y 4, mientras las otras dos comparten grupos, 1-4, y 2-3. En las SMU con menos de 4 observaciones es difícil determinar su potencial enológico por falta de información, son las indicaciones de pocas referencias.

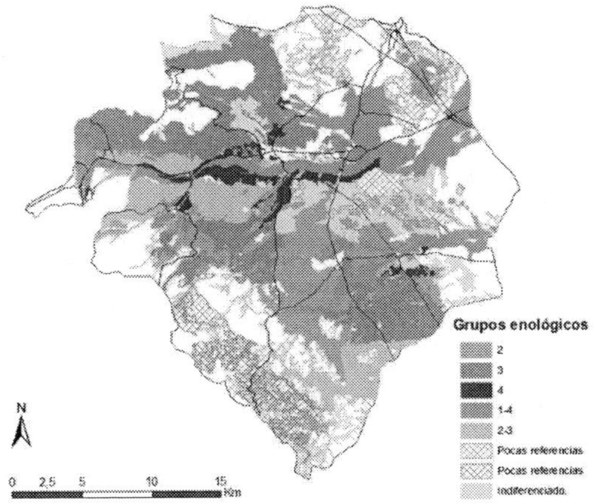

Figura 10. Mapa de Unidades de Terroir, según los grupos enológicos, SMU con al menos cuatro observaciones.

Se ha observado que los vinos de mayor expresión y valor se obtienen en UT cuyas mistelas se han clasificado en el grupo 2 y 3. A su vez se constata que las zonas clasificadas 1 y 4, se tiende a arrancar la garnacha por falta de color y calidad organoléptica.

4. Conclusiones

1ª La metodología desarrollada para evaluar la potencialidad enológica de los viñedos resulta válida y es sencilla de aplicar, vinificación de mistelas. Esta facilita el estudio de las aptitudes de los viñedos a una escala de mayor precisión. La línea de trabajo emprendida ha abierto un camino para continuar profundizando en el estudio del terroir.

2ª En este estudio concreto en la D.O.P. Campo de Borja y durante los años 2012, 2013 y 2014 y con la información obtenida, hemos clasificado las mistelas en cuatro grupos, que se corresponden con características enológicas bien diferenciadas entre los grupos.

3ª Se han postulado seis unidades de terroir (UT), tres se corresponden con los grupos 2,3 y4, mientras las otras dos comparten grupos, 1-4, y 2-3. Con estos datos se observa claramente el efecto terroir en la composición de las mistelas y por tanto en las uvas y en los vinos. Aunque el reducido número de observaciones de algunas SMU impide establecer un modelo predictivo estadísticamente concluyente en estas unidades, por lo que es necesario continuar con el estudio enológico durante un periodo más largo, con el fin de disponer de más información.

4ª No obstante y a pesar de la observación anterior, como conclusión final se puede afirmar que algunas SMU han mostrado su potencial enológico de un modo estable, bien asociándose a un grupo concreto o a dos grupos vecinos con características comunes, se pueden vincular a un tipo de uvas y, en consecuencia, a una unidad de terroir.

5ª La variabilidad en la composición de las uvas de las parcelas estudiadas, clasificadas en cuatro grupos según sus características enológicas y sensoriales, nos indica que el territorio de la DOP Campo de Borja, a pesar de su aparente uniformidad, presenta claras diferencias enológicas según las características del medio, posibilitando la implantación de un sistema productivo basado en concepto terroir.

5. Referencias bibliográficas

DOLÉDEC, A.-F. (1995). Recherche des composantes principales des terroirs vitico*les afin d'elaborer un outil d'aide à la gestión au moyen d'observatoires et de traitements statistiques de données spatialisées*. Application au vignoble champenois. Tesis INA P-G.

FRANCO, E, LORENTE,M. (2018). Making mistelas: New methodology for evaluating the oenological characterisctics of Terroir Units. *XII Congreso Terroir*. Zaragoza.

FREGONI, M. et al (1998). La zonazione viticola della Collina cesenate. *Vignevini,* 1/2: 39-57

HUGLIN, P. (1978). Nouveau mode d'évaluation des possibilités héliothermiques d'un milieu viticole. *Comptes Rendues de l'Académie d'Agriculture*. 1117-1126.

JÚDEZ, A. (1989). *Técnicas de análisis de datos multidimensionales*. MAPA. Madrid. 301pp.

LORENTE,M. (2017). Zonificación de la D.O.P. Campo de Borja con criterios de terroir. *Tesis Doctoral. Universidad de Zaragoza*.

SOTÉS,V., GÓMEZ-MIGUEL,V. (1992). Criterios de diferenciación y delimitación de comarcas y regiones vitícolas. Exámen de los factores naturales y humanos que concurren. *Vitivinicultura,* 3,5: 29-40.

Nueva metodología para evaluar las características enológicas de las unidades de un terroir. Vinificación de mistela.

Ernesto Franco[1], Miguel Lorente[2]

[1] Centro de Transferencia Agroalimentaria - Gobierno Aragón. Unidad de Enología. Avda. de Movera, s/n, 50194, Zaragoza (España). efranco@aragon.es
[2]Grupo de Estudios de Desarrollo Territorial, Universidad de Zaragoza (GEDETUZ). España. mlorenteb@yhoo.es

Abstract. To To characterize terroir units that express the interaction between the vine and the geographical environment it is necessary to study the oenological parameters of each terroir. Nevertheless, this study presents many difficulties if the number of areas to study it is huge as it was the case of the zoning of DOP Campo de Borja. Grape analyses during harvest are expensive and give limited information. At the same time, fermentations in small-volume containers are difficult to control and introduce factors that are not directly related to the geographical environment. These small-scale fermentations can distort the link between the wine and the soil giving inaccurate results. In the project for the zoning of the DOP Campo de Borja has been used a new simple method which consists in vinification mistela. This new method consisted on producing a wine spirit directly from the grapes but without alcoholic fermentation. Therefore, with this new methodology of making mistelas, an wine produced without fermentation, was possible to obtain the contribution of the terroir of the DOP Campo de Borja to the composition of the grape and, consequently to the wines

1 Introducción

Esta presentación es el resultado del trabajo desarrollado por el Centro de Transferencia Agroalimentaria (CTA) en el "Proyecto de valorización de la D.O. Campo de Borja a través de la vinculación del vino con el territorio" encaminado a dotar del máximo prestigio comercial al vino de la D.O. Campo de Borja, vinculándolo a su territorio de origen como vector de desarrollo territorial, a partir del conocimiento de las aptitudes del medio geográfico que permita una gestión razonada del cultivo de la vid, de la elaboración de los vinos y de la comunicación en la comercialización.

Con el proyecto se pretende identificar y delimitar las áreas homogéneas, denominadas Unidades de Terroir, para gestionar la producción de vino bajo el concepto de calidad, autenticidad y originalidad. En definitiva, se pretende encontrar el vínculo entre los diferentes terroir que se delimiten en la D.O. y las características de los vinos de Garnacha cultivados en dichos terroir.

Ahora bien, en la calidad y características de los vinos, influyen numerosos factores, unos relacionados con la uva, como la vinífera, el terroir, madurez y sanidad y otros relacionados con la vinificación y conservación del vino, como levadura, maceración, prácticas enológicas, crianzas, etc. Sin lugar a dudas estas últimas modifican sustancialmente las características del vino y pueden disminuir el peso de la vinífera y del terroir.

En base a lo anterior, y teniendo en cuenta que el análisis de uvas en la fecha de la vendimia es costoso y la información obtenida limitada y que la fermentación en recipientes de poco volumen es difícil de controlar e introduce factores que no están directamente relacionados con el medio geográfico y pueden desvirtuar el vínculo entre el vino y el territorio, por lo que los resultados a veces resultan erróneos, se opta por una vinificación sin fermentación alcohólica, mistela, un tipo de vino de licor, mosto o uva estrujada al que se adiciona alcohol de origen vínico y definida en el anexo IV del Real Decreto 1127/2003 y posteriores reglamentos CE. Se parte de la experiencia previa de vinificación de mistelas, línea de recuperación de vinos dulces tradicionales que desarrolla el CTA desde 1995.

En las mistelas al no producirse fermentación, tanto los aromas como polifenoles proceden solo y exclusivamente de la uva, y si bien el alcohol no es inodoro su papel es evitar la fermentación y ejercer acciones de disolvente y esterificación, favoreciendo la expresión aromática de la variedad por procesos de hidrólisis tanto química como enzimática.

2 Objetivo

Es postular la vinificación, de mistela tinta con uvas de Garnacha Tinta, como una herramienta para establecer el vínculo entre las características analíticas y sensoriales del vino y el terroir. Relacionando las características de las mistelas con los vinos tintos obtenidos de las mismas parcelas.

3 Material y métodos

Las vinificaciones de mistelas tintas y vinos tintos se han realizado durante los años 2012, 2013 y 2014, en el Proyecto se ha trabajado con 79 parcelas, si bien en este estudio solo se muestran las 23 parcelas que han sido objeto de vinificación en los tres años de trabajo. Un total de 23 parcelas de diferentes terroir de la D.O.P. Campo de Borja, se han vinificado cada año como mistela y además en seis de ellas se han vinificado vinos tintos.

3.1 Vinificación de mistela tinta

3.1.1 La uva de Garnacha Tinta de cada terroir se vendimia en el entorno de la madurez que marca la Bodega de la D.O.P. Campo de Borja receptora de la uva de esa parcela.

3.1.2 Se vendimia en cajas de 25 Kg, 50 Kg de uva de cada parcela.

3.1.3 En campo se controla la producción mediante el dato de Kg/cepa, trasladándola a la uva a la Bodega Cava de Almonacid de la Sierra del CTA.

3.1.4 La uva se despalilla y estruja y se encuba en depósitos de acero inoxidable de 25 l de capacidad, dosificando 40 mg/kg de anhídrido sulfuroso en forma de solución acuosa del 15%.

3.1.5 Transcurrido 30 minutos de la operación anterior, se procede a la adición de 140 cc/Kg de uva de alcohol de 96%, homogeneizando toda la masa.

3.1.6 Diariamente, y de forma manual, se homogeneiza toda la uva del depósito. La temperatura se mantiene constante a 12°C.

3.1.7 Después de siete días de maceración se prensa, y la mistela obtenida se corrige hasta 15% de alcohol si es necesario.

3.1.8 Transcurridos 15 días de la prensa se procede al trasiego para retirar lias, dosificando 30 mg/l de anhídrido sulfuroso en forma de solución acuosa al 15%

3.1.9 Transcurridos 30 días de la operación anterior se procede a un segundo trasiego para retirar lias, si fuera necesario se dosificará anhídrido sulfuroso, para mantener el nivel de sulfuroso libre en 20 mg/l.

3.2 Vinificación de vino tinto

De seis parcelas en las que se han vinificado mistelas, se realizan vinificaciones semi industriales de vino tinto, siguiendo los siguientes pasos:

3.2.1 La uva de Garnacha Tinta de cada terroir se vendimia a la vez que su correspondiente vendimia de mistela como se describe en 3.11.

3.2.2 Se vendimia en cajas de 25 Kg, 1200 Kg de uva de cada parcela.

3.2.3 En campo se controla la producción mediante el dato de Kg/cepa, trasladándola a la uva a la Bodega Piloto de Ainzón del CTA.

3.2.4 La uva se despalilla y estruja, se encuba en depósitos de acero inoxidable de 15 hl de capacidad, dosificando 40 mg/kg de anhídrido sulfuroso en forma de solución acuosa al 15%

3.2.5 Se inocula levadura seca activa, de acción mecánica, a razón de 20 g/Kg de uva y activadores de fermentación.

3.2.6 Diariamente se realizan dos remontados, la maceración se prolonga durante 10 días después de iniciada la fermentación alcohólica.

3.2.7 Transcurrido el periodo anterior la uva se prensa, favoreciendo posteriormente la degradación maloláctica.

3.2.8 Concluida de degradación maloláctica o transcurridos 30 días desde el final de la fermentación alcohólica, se procede al trasiego para retirar lias, dosificando 50 mg/l de anhídrido sulfuroso en forma de solución acuosa al 15%

3.2.9 Transcurridos 30 días de la operación anterior se procede al segundo trasiego para retirar lias, si fuera necesario se dosificará anhídrido sulfuroso, para mantener el nivel de sulfuroso libre en 25 mg/l.

3.3 Métodos analíticos

En las mistelas se han analizado los siguientes parámetros, densidad relativa, grado alcohólico, azúcares reductores, acidez total, pH, sulfuroso libre y total, ácido málico, potasio, intensidad de color e IPT. En los vinos se han analizado, densidad relativa, grado alcohólico, azúcares reductores, acidez volátil, acidez total, pH, sulfuroso libre y total, ácido málico, potasio, intensidad de color e IPT. En ambos casos se han empleado métodos habituales y oficiales de análisis.

3.4 Estadística

En el estudio estadístico se utilizan técnicas de análisis de regresión, análisis exploratorio de Tukey (estadística descriptiva) y análisis multivariante (análisis de componentes principales ACP). Dichas técnicas se aplicaron utilizando el programa informático SPSS

3.5 Metodología sensorial

El análisis sensorial descriptivo, se realiza conjuntamente entre jurados del CTA, LAAE (Universidad de Zaragoza) y Comité de cata de la D.O.P. Campo de Borja. El panel lo componen 11 jueces experimentados y para la descripción de las mistelas y vinos se utilizan las fichas de las Figuras 1 y 2. Se valora el color, aroma y gusto, el jurado debe puntuar entre 1 y 3 puntos cada uno de los descriptores. La puntuación de cada descriptor se obtiene ponderando su valor como se muestra en las fichas, la puntuación final es la suma del valor parcial de cada descriptor. En las fichas se consideran descriptores negativos en cuyo caso la ponderación es negativa y por tanto estos restan de la

https://doi.org/10.1051/e3sconf/20185001022

puntuación total. Este modelo permite trabajar con datos de organolépticos parciales, aroma, gusto y/o los totales, posibilitando a su vez relacionarlos con parámetros de viñedo y con la composición analítica de mistelas y vinos.

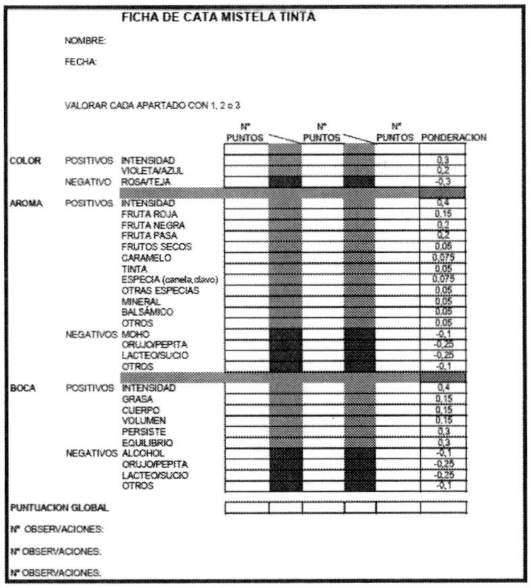

Figura 1. Ficha descriptiva de mistelas tintas.

Además, se pedía a los catadores una valoración global de cada mistela o vino que serviría para relacionarla con la puntuación total de los descriptores con el fin de conocer la bondad de la evaluación. En el análisis de regresión entre ambos parámetros se produce una correlación alta entre ambos valores (r= 0,91 en mistelas y 0,97 en vinos), lo que nos indica que el trabajo del jurado fue muy coherente en sus apreciaciones.

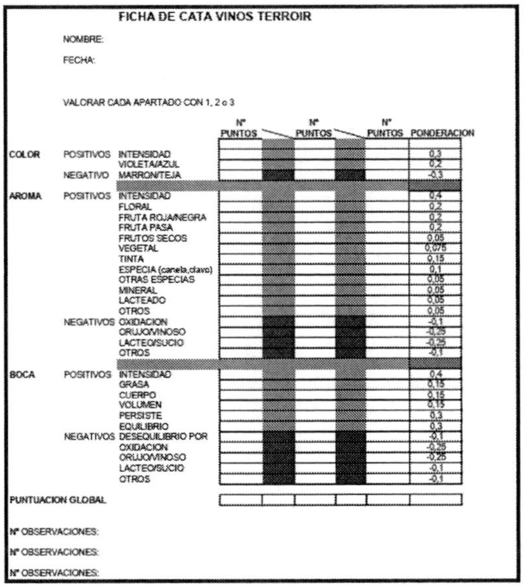

Figura 2. Ficha descriptiva de vinos tintos.

4 Resultados

4.1 Agronómicos

Se realiza un seguimiento de maduración de cada una de las parcelas conjuntamente con la Bodega que recepciona la uva, se observa si hay anomalías en la misma y se controla la producción. En la **tabla 1** se muestran datos estadísticos de la producción en los tres años de ensayo.

Tabla 1. Producción por años de las 23 parcelas objeto de vinificación.

	201 2	**20** 13	**20** 14
Med ia	1,8 2	1,4 5	3,4 9
Máx imo	4,7 6	3,1 3	6,2 5
Mín imo	0,7 4	0,4	0,9 8
Des v	0,9 8	0,7 4	1,5 5

El rendimiento medio de las parcelas en 2012 y 2013 es similar, mientras que en 2014 fue muy superior a la de los dos años anteriores, observando en los tres años una desviación muy elevada. Es de destacar que en 2013 se produjo un corrimiento fisiológico importante de la variedad Garnacha, que afecto con gran intensidad en todo Aragón.

4.2 Analíticos y sensoriales de las mistelas.

Teniendo en cuenta que la acidez y el pH se modifican sustancialmente por la adición de alcohol, ambos parámetros se descartan como elementos de valoración analítica de las mistelas. El análisis de componentes principales (ACP) muestra que el primer factor recoge la aportación principal de la intensidad de color y en menor medida de IPT y el segundo factor del azúcar, como se muestra en la **figura 3**. De acuerdo con lo anterior, los parámetros analíticos estudiados son azúcar, intensidad de color e IPT en mistelas. Los resultados analíticos de las mistelas de los años 2013 y 2014 son muy similares, en ambos años se observa precipitaciones similares, **tabla 2,** pero con diferente producción, la de 2014 es más de dos veces superior a la de 2013. En 2012 las mistelas son más concentradas en los tres parámetros, estas diferencias se plasman en particular en la intensidad de color e IPT, **tabla 3**, y que se podrían relacionar con una menor precipitación en 2012.

Tabla 2. Datos climáticos básicos en los tres años de estudio.

Observatorio Borja	2012	2013	2014
Pp anual (mm)	336,6	504,9	543,1
Pp Jun-Jul (mm)	38,8	57,3	42,2
°C T. media anual	14,7	13,9	14,8
°C T. media Junio	22,8	18,4	20,8

En el caso del análisis sensorial los años 2012 y 2014 son más parecidos, mientras que las mistelas del 2013 son las peor valoradas, **tabla 3**. La relación entre el aroma y gusto se muestra en la **figura 4**, la distribución de las valoraciones es uniforme y homogénea, es decir los criterios de cata utilizados en los tres años son idénticos, aspecto básico para una posterior clasificación de las mistelas. Dado que la puntuación del aroma y del gusto es la suma de los diferentes descriptores, a su vez estos también se convierten en una herramienta para relacionar la cualidad del aroma con el terroir.

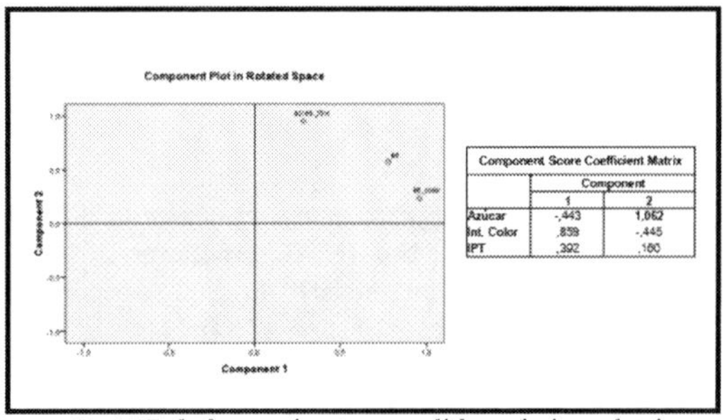

Figura 3. ACP de los parámetros analíticos de las mistelas. Años 2012, 2013 y 2014.

El análisis de componentes principales (ACP) de los descriptores del aroma, que son los relacionados en la **figura 1**, muestra que los que mayor capacidad de discriminación tienen son fruta roja, negra y pasa, junto con balsámico y tinta, **figura 5**. En la componente primera tienen más peso fruta negra, pasa y tinta, mientras en la componente segunda pesan más la fruta roja y balsámico.

Tabla 3. Valores analíticos y sensoriales de las mistelas por años

Año	Valores	Azúcar	Color	IPT	Aroma	Gusto	Total
	Media	274,00	9,92	71,67	1,89	3,04	6,23
2012	Máximo	311,00	20,77	105,00	2,71	3,91	8,06
	Mínimo	205,00	2,16	26,90	0,84	1,88	3,38
	Desv	20,68	4,22	19,44	0,52	0,61	1,33
	Media	249,56	6,79	38,70	1,60	2,75	5,43
2013	Máximo	282,50	17,30	57,30	2,41	3,62	7,45
	Mínimo	207,00	2,05	21,80	0,41	1,78	1,78
	Desv	19,14	3,55	9,44	0,56	0,50	1,54
	Media	246,57	7,93	47,88	1,89	3,05	6,06
2014	Máximo	320,00	15,71	79,80	2,56	4,04	7,94
	Mínimo	189,00	0,30	11,30	1,13	2,11	3,88
	Desv	26,96	3,84	16,29	0,41	0,51	1,06

Azúcar reductor en g/l; I. Color Abs 420+520+620 nm; IPT Abs 280nm.

Un aspecto importante es la producción del viñedo y cómo puede afectar a las características de las mistelas y los vinos, cabe preguntarse ¿Qué relación se puede observar entre los parámetros analíticos, sensoriales y la producción del viñedo?

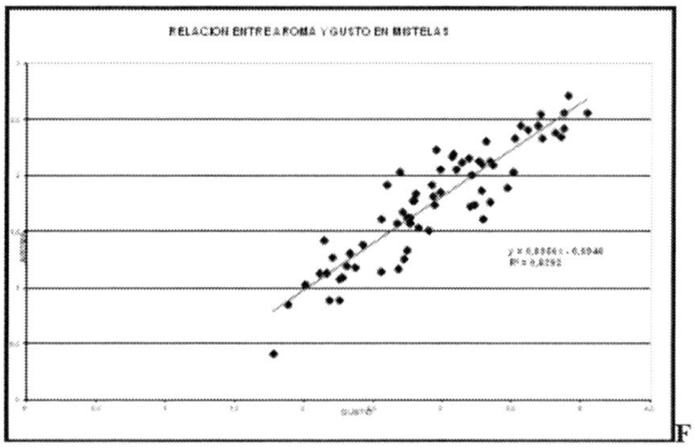

Figura 4. Relación entre el aroma y el gusto de las valoraciones de los diferentes descriptores del aroma y del gusto de las mistelas de los años 2012,2013 y 2014.

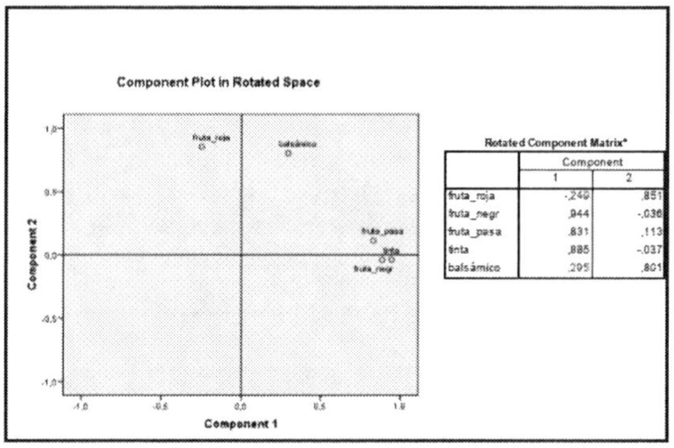

Figura 5. ACP de los descriptores del aroma de las mistelas. Años 2012, 2013 y 2014.

En la **tabla 4**; se muestra la relación entre los diferentes parámetros analítico, el análisis sensorial y la producción del viñedo, la tendencia es que a mayor producción los valores analíticos son menores y no observando relación entre el aroma y el gusto con la producción. Estas observaciones son muy importantes, dado que no ligan la analítica y los parámetros sensoriales de las mistelas con la producción, por tanto, las diferencias observadas entre las mistelas mayoritariamente se deben a otros factores diferentes a la producción, como los relacionados con el terroir.

También hay que destacar que los parámetros analíticos tampoco están relacionados con los sensoriales y solo el color tiene cierta relación con el sensorial total,

 https://doi.org/10.1051/e3sconf/20185001022

dado que este incluye la valoración sensorial del color de la mistela. La independencia entre los parámetros analíticos y los sensoriales, indica que el factor terroir tiene un peso importante en un entorno de madurez similar.

Se observa una cierta relación entre IPT y la intensidad de color. La relación entre los parámetros sensoriales es muy buena e indica que cualquiera de ellos puede ser usado para valorar la calidad global de las mistelas.

4.3 Analíticos y sensoriales de los vinos.

En el caso de vinos tintos se tienen en cuenta todos los parámetros relacionados directamente con las características del vino, grado de alcohol, acidez total, pH, intensidad de color e IPT. El análisis de componente principales, ACP, de los parámetros analíticos, **figura 6**, nos indica que la componente primera se debe a la intensidad de color, pH y acidez total y la segunda al grado e IPT, estos parámetros analíticos son muy similares a los que también se relacionan con las características de las mistelas.

Tabla 4. Coeficientes de correlación (r) de parámetros analíticos, sensoriales y producción de las mistelas

	Azúcar	Color	IPT	Aroma	Gusto	Total
Producción	-0,33	-0.02	-0,13	0,00	0,00	0,00
Azúcar		0,51	0,73	0,28	0,30	0,35
I. Color			0,82	0,54	0,62	0,70
IPT				0,44	0,48	0,56
Aroma					0,91	0,93
Gusto						0,94

En los vinos no se observan diferencias en cuanto al grado, como si se observaba en el azúcar de las mistelas y si en la intensidad de color e IPT, **tabla 5,** en el análisis sensorial los años 2013 y 2014 son más parecidos y los vinos del 2012 son los mejor valorados.

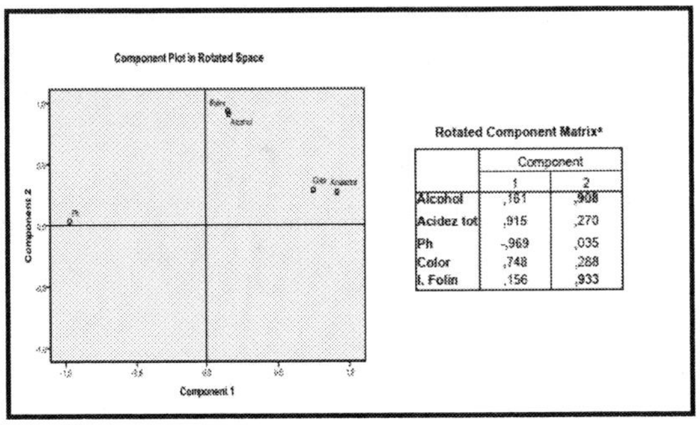

Figura 6. ACP de los parámetros analíticos de los vinos. Años 2012, 2013 y 2014

Tabla 5. Valores analíticos y sensoriales de los vinos años.

Año	Valores	Grado	Color	IPT	Aroma	Gusto	Total
	Media	16,47	13,54	77,43	2,25	3,17	6,79
2012	**Máximo**	0,85	2,25	7,12	0,28	0,40	0,67
	Mínimo	17,30	17,62	86,80	2,55	3,69	7,65
	Desv	15,50	11,07	66,30	1,81	2,71	5,94
	Media	16,30	15,40	74,97	2,02	2,73	5,93
2013	**Máximo**	0,89	3,53	8,81	0,54	0,86	1,58
	Mínimo	17,50	18,24	85,50	2,60	3,82	7,83
	Desv	14,95	10,60	63,00	1,38	1,56	3,66
	Media	16,66	12,52	67,72	1,72	2,93	5,91
2014	**Máximo**	1,53	4,47	15,58	0,35	0,47	1,03
	Mínimo	18,30	18,50	91,68	2,12	3,52	6,87
	Desv	14,35	6,72	48,36	1,19	2,42	4,37

La relación entre los parámetros analíticos, los sensoriales y entre todos ellos se muestra en la **tabla 6**, solo se observa una buena relación del IPT con la intensidad de color y el grado, y también con el aroma. Como también ocurre en las mistelas si hay alta relación del aroma y del gusto con la valoración total del vino.

Tabla 6. Coeficientes de correlación (r) de parámetros analíticos y sensoriales de los vinos

	I. Color	IPT	Aroma	Gusto	Total
Grado	0,52	0,81	0,49	0,47	0,53
I. Color		0,79	0,62	0,46	0,61
IPT			0,85	0,65	0,77
Aroma				0,79	0,93
Gusto					0,95

4.4 Relación entre los parámetros analíticos y sensoriales de vinos y mistelas

El efecto añada es mucho menor en los vinos que en las mistelas, como se observa al comparar los datos de las **tablas 3** y **5**. En el análisis sensorial tanto las mistelas como los vinos de 2012 fueron los mejor valorados, los vinos de los años 2013 y 2014 son más parecidos, a diferencia de las mistelas que eran 2012 y 2014.

La relación de los parámetros analíticos y sensoriales entre los vinos y las mistelas se muestra en la **tabla 7**, se observa mejor relación entre los parámetros analíticos que entre los sensoriales, siendo la relación del IPT y de azúcar/grado las que muestran la mayor correlación. No obstante, la correlación entre mistelas y vinos, para los seis parámetros estudiados, es suficientemente buena para poder relacionar las características de las mistelas con la de los vinos tintos obtenidos de las mismas parcelas.

Tabla 7. Coeficientes de correlación (r) de parámetros analíticos y sensoriales entre mistelas y vinos

VINO	Grado	Color	IPT	Aroma	Gusto	Total
MISTELA						

 https://doi.org/10.1051/e3sconf/20185001022

Azúcar	0,76					
Color		0,74				
IPT			0,76			
Aroma				0,71		
Gusto					0,73	
Total						0,71

5 Conclusiones

La acidez total y el pH no se pueden usar como parámetros de valoración analítica de las mistelas, mientras que azúcar, intensidad de color e IPT son los parámetros que discriminan. En el caso de los vinos son también estos parámetros los que diferencian.

La producción del viñedo no se relaciona con los parámetros analíticos y sensoriales de las mistelas. En 2012 las mistelas y los vinos son más concentrados en los tres parámetros analíticos y los mejor valorados sensorialmente, estas diferencias podrían relacionar con una menor precipitación en 2012. No se liga la analítica y los parámetros sensoriales de las mistelas con la producción, por tanto, las diferencias observadas entre las mistelas mayoritariamente se deben a otros factores diferentes a la producción, como los relacionados con el terroir.

De los descriptores del aroma de las mistelas la mayor capacidad de discriminación se debe a los descriptores fruta negra, pasa y tinta en la componente primera y fruta roja y balsámico en la segunda componente.

La independencia entre los parámetros analíticos y los sensoriales de las mistelas, muestran que el factor terroir tiene un peso importante en un entorno de madurez similar.

Se confirma el postulado que la vinificación, de mistela tinta con uvas de Garnacha Tinta, como herramienta para establecer el vínculo entre las características analíticas y sensoriales del vino y el terroir es adecuada, dado. La correlación de los seis parámetros estudiados, es suficientemente buena para poder relacionar las características de las mistelas con la de los vinos tintos obtenidos de las mismas parcelas.

Referencias

Real Decreto 1127/2003. BOE, núm. 228, de 23 de septiembre de 2003, pág 34823- 34828

Caracterización aromática de variedades minoritarias del Piedemonte Pirenaico

Ricardo López[1*], *Laura* Culleré[1], *Elisa* Gracia-Moreno[1], *Fanny* Prezman[2] y *Ernesto* Franco Aladrén[3]

[1]Laboratorio de Análisis del Aroma y Enología, Instituto Agroalimentario de Aragón (IA2), Departamento de Química Analítica, Facultad de Ciencias, Universidad de Zaragoza, E-50009 Zaragoza, España
[2]Institut de la Vigne et du Vin- Pôle Sud Ouest. Lisle Sur Tarn. Francia
[3]Unidad de Enología. Centro de Transferencia Agroalimentaria. Zaragoza. España

Resumen. Mediante los conservatorios de variedades y la colaboración ciudadana se han identificado más de un centenar de variedades minoritarias procedentes de las regiones que rodean los Pirineos. Con 78 de dichas variedades identificadas se han realizado más de 200 microvinificaciones que han permitido, por una parte, evaluar su potencial enológico y, por otra, caracterizar sensorial y químicamente dichas variedades.

1 Introducción

La globalización del mercado del vino durante los últimos treinta años ha provocado una disminución de las variedades de uva utilizadas para vinificación. En la actualidad de las más de 10000 variedades de uva conocidas, 13 cubren más de un tercio de la superficie plantada y 33 variedades cubren el 50% [1]. En España esta situación es también extrema, ya que sólo 10 variedades representan el 75% de la superficie plantada.

Este fenómeno de homogenización de las variedades empleadas para vinificación conlleva un empobrecimiento genético a causa de la selección clonal y una dificultad de los pequeños productores para conseguir competitividad en un mercado cada vez más global.

Una posible solución a estos problemas puede encontrarse en las variedades de vid minoritarias empleadas como herramienta de innovación. Las regiones francesas y españolas que rodean a los Pirineos han tenido una gran diversidad de variedades de vid. Esto es debido a su larga tradición como región vitícola, ya desde tiempos del Imperio Romano, pasando por los intercambios culturales del Camino de Santiago y, por supuesto, por su gran variedad geográfica. Gracias a prospecciones llevadas a cabo estos 10 últimos años en antiguas parcelas vitícolas, se han implantado diversos conservatorios o colecciones varietales en estas regiones. Sin embargo, el potencial agronómico y enológico de dichas variedades minoritarias aún está pendiente de ser evaluado.

Este es el objetivo del proyecto europeo "Valorización de las variedades de vid minoritarias del Piedemonte Pirenáico" (VALOVITIS). Dentro de este proyecto, se ha identificado de manera exhaustiva más de un centenar de variedades procedentes de las regiones que rodean los Pirineos. Con las variedades identificadas se han realizado más de 200 microvinificaciones. Los resultados de la caracterización de las diferencias varietales en lo que respecta a su composición en compuestos químicos volátiles se presenta en esta comunicación.

2 Material y métodos

2.1 Variedades

Un total de 78 variedades minoritarias se emplearon para elaborar los vinos. Las variedades utilizadas procedían del conservatorio de variedades del Centro de Transferencia Agroalimentaria del Gobierno de Aragón del conservatorio de variedades del Institut de la Vigne et du Vin- Pôle Sud Ouest en Francia.

2.2 Vinificaciones

Las fermentaciones y tratamientos de las variedades españolas se realizaron en la bodega experimental del Gobierno de Aragón en Cariñena, mientras que las de las variedades francesas se realizaron en las instalaciones del Institut de la Vigne et du Vin- Pôle Sud Ouest en Francia. Debido al escaso número de plantas disponibles para cada variedad los vinos se elaboraron a escala de microvinificación (5 L), siguiendo un protocolo estándar común para las vinificaciones en tinto, rosado y blanco en cada país. Las vinificaciones se realizaron con las cosechas de 2016 y 2017.

2.3 Análisis cuantitativo de compuestos volátiles

Un total de 82 compuestos volátiles implicados en el aroma del vino se analizaron en todas las muestras, Para ello se utilizaron cuatro métodos de análisis diferentes,

* Corresponding author: riclopez@unizar.es

https://doi.org/10.1051/e3sconf/20185001023

todos ellos publicados y validados. Los compuestos mayoritarios (aquellos con concentraciones en el rango de mg/l) se analizaron mediante microextracción líquido-líquido y cromatografía gas (GC) con detección FID [2]. Los compuestos minoritarios y traza (concentraciones por debajo de 1 mg/l hasta el μg/l) se analizaron mediante extracción en fase sólida (SPE) y GC con detección por espectrometría de masas (MS) [3]. Los tioles varietales (en concentraciones de ng/L) se determinaron mediante SPE y GC-MS con ionización química negativa [4]. Por último, las alquilmetoxipirazinas y rotundona se analizaron mediante SBSE y cromatografía gas multidimensional GC-GC-MS [5].

2.4 Análisis estadístico

Los análisis estadísticos se llevaron a cabo con el programa XLSTAT 19.6 (Addinsoft).

3 Resultados y discusión

Es de sobras conocido que existe una parte del aroma del vino cuyo origen está en gran parte determinado por la variedad de uva utilizada, es lo que se conoce como aroma varietal. A la hora de evaluar el potencial aromático de las variedades minoritarias, parece lógico analizar su capacidad de producción de las moléculas volátiles responsables de estos aromas varietales. Por esta razón se ha dividido el estudio de los resultados de composición química de acuerdo con las familias de compuestos varietales más interesantes. A continuación, se muestra un resumen de los hallazgos más relevantes.

3.1 Tioles varietales

Existen varios mercaptanos polifuncionales que pueden ser responsables de aromas positivos y cuya presencia en el vino está asociada a la variedad de uva utilizada. Los más destacados en este grupo son la 4-metil-4-mercapto-2-pentanona identificada como un compuesto clave en el aroma del Sauvignon blanc [6], el 3-mercaptohexanol y el acetato de mercaptohexilo también compuestos impacto en los vinos de Verdejo y de Sauvignon blanc entre otros [4]. Entre las variedades minoritarias estudiadas hay varias que destacan por su contenido en tioles varietales. En la figura 1 puede observarse como los vinos de 6 variedades minoritarias mostraron contenidos superiores a 10 veces el umbral sensorial del 3-mercaptohexanol. Estos niveles son comparables a los encontrados en los vinos de Sauvignon blanc y producen en el vino notas de fruta tropical [4]. El vino rosado elaborado con la variedad San Antoni también destacó por un nivel muy alto en acetato de mercaptohexilo.

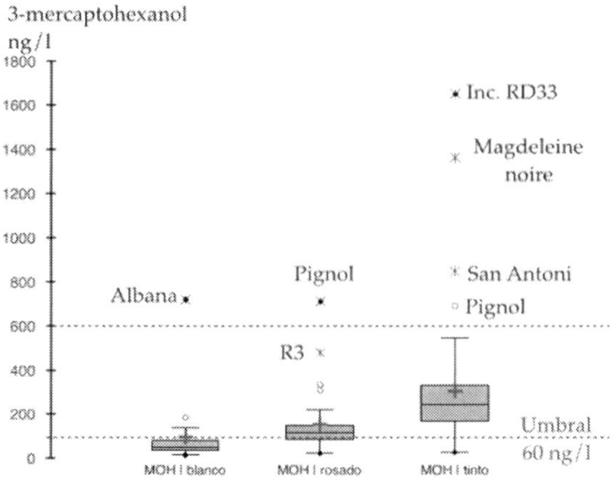

Fig. 1. Diagrama de cajas para el contenido en 3-mercaptohexanol en los vinos de variedades minoritarias

3.2 Rotundona

La rotundona es un sesquiterpeno de origen varietal y gran potencia sensorial, ya que a partir de los 16 ng/L imparte una característica y apreciada nota de pimienta negra [7]. La rotundona ha sido localizada en otras variedades originarias de la zona Pirenaica como el Duras [8]. Entre los vinos producidos en este trabajo hubo 7 variedades que destacaron por su contenido en rotundona, pero dos de ellas (Negral y Camarou) mostraron niveles claramente por encima del valor umbral (Figura 2). Estas mismas variedades elaboradas como vino rosado no mostraron niveles apreciables de rotundona, debido a que su precursor sólo se encuentra en los hollejos.

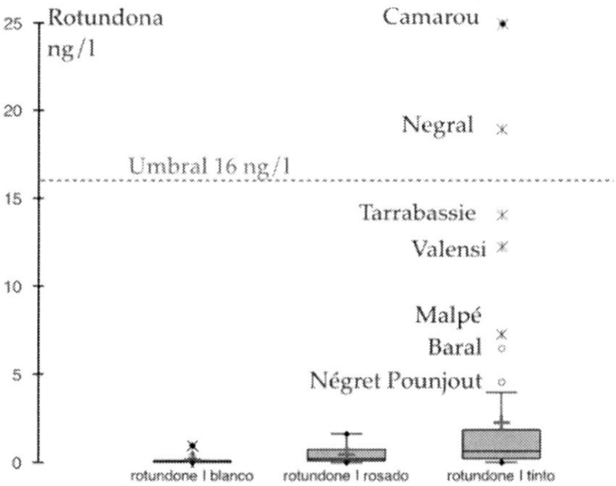

Fig. 2. Diagrama de cajas para el contenido en rotundona en los vinos de variedades minoritarias

3.3 Terpenoles y norisoprenoides

Los vinos producidos no destacaron por niveles fuera de lo normal de norisoprenoides con relevancia para el aroma como β-damascenona o β-ionona. Sin embargo, es destacable que los vinos españoles contenían un nivel medio mayor de estos compuestos, debido probablemente al mayor número de horas de insolación. No se encontraron niveles significativos de TDN. Respecto a los terpenoles, sólo una de las variedades mostró niveles por encima del valor umbral de linalool en la elaboración en rosado, confirmando su carácter terpenodependiente.

3.4 Alquilmetoxipirazinas

Sólo dos de las variedades produjeron vinos con niveles superiores o cercanos al valor umbral sensorial de 3-isobutil-2-metoxipirazina. En ambos casos se trataba de variedades francesas: Gros Cabernet con 12 ng/L y Bequignol con 42 ng/L. Especialmente en el último caso aparecieron notas vegetales debido a este compuesto.

3.5 Compuestos fermentativos

Aunque la producción de compuestos fermentativos está fundamentalmente controlada por la levadura utilizada, existe también una influencia varietal a través del perfil de aminoácidos característico de la variedad [9]. En el caso de las variedades minoritarias estudiadas se encontraron diferencias significativas en algunos compuestos fermentativos importantes desde el punto de vista del aroma. Entre otros destacaron acetato de isoamilo, acetato de feniletilo o γ-nonalactona. En la Figura 3 puede observarse el caso del acetato de isoamilo, que en la variedad Plant de Fronton mostró una cantidad muy llamativa, con una cantidad cercana a los 8 mg/L.

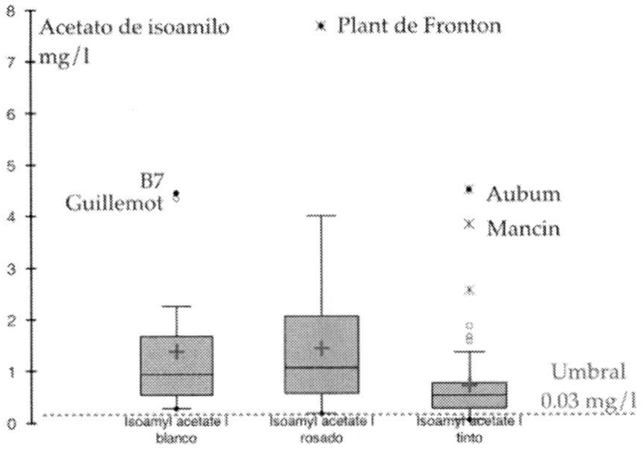

Fig. 2. Diagrama de cajas para el contenido de acetato de isoamilo en los vinos de variedades minoritarias

4 Conclusiones

Muchas de las variedades minoritarias localizadas presentan un potencial enológico, desde el punto de vista sensorial, altamente interesante. Algunos de los vinos elaborados contienen compuestos de gran impacto sensorial en cantidades similares a las de variedades muy apreciadas como Sauvignon blanc o Shiraz. El trabajo realizado dentro de este proyecto abre la puerta a una futura implantación de estas variedades en las regiones pirenaicas como una herramienta de mejora de la calidad y la competitividad.

VALOVITIS es un proyecto de colaboración transfronteriza entre España y Francia cofinanciado por el Fondo Europeo de Desarrollo Regional (FEDER) a través del Programa Interreg V-A España-Francia-Andorra (POCTEFA 2014-2020), destinado a reforzar la integración económica y social de la zona fronteriza España-Francia-Andorra.

Referencias

[1] Distribution of the world's grapevine varieties, International Organisation of Vine and Wine (OIV), 2017.
[2] C. Ortega, R. López, J. Cacho, V. Ferreira, J. Chromatogr. A., **923** (2001).
[3] R. Lopez, M. Aznar, J. Cacho, V. Ferreira, J. Chromatogr. A., **966** (2002).
[4] L. Mateo-Vivaracho, J. Zapata, J. Cacho, V. Ferreira, J. Agric. Food Chem., **58** (2010).
[5] Y. Wen, I. Ontañón, V. Ferreira, R. Lopez, Food Chem., **255** (2018).
[6] P. Darriet, T. Tominaga, V. Lavigne, J.N. Boidron, D. Dubourdieu, Flavour Fragrance J., **10** (1995).
[7] C. Wood, T.E. Siebert, M. Parker, D.L. Capone, G.M. Elsey, A.P. Pollnitz, et al., J. Agric. Food Chem., **56** (2008).
[8] O. Geffroy, T. Dufourcq, D. Carcenac, T. Siebert, M. Herderich, E. Serrano, Ais. J. Grape Wine Res., **3** (2014).
[9] P. Hernandez-Orte, J.F. Cacho, V. Ferreira, J. Agric. Food Chem. **50** (2002).

Geoquímica de los suelos de la D.O. Uclés

Vineyard soils geochemistry: D.O. Uclés.

Sandra Bravo[1,a], Francisco Jesús García-Navarro[1]José Ángel Amorós[1], Caridad Pérez-de-los-Reyes[1] y Raimundo Jiménez Ballesta[2]

[1]Universidad de Castilla- La Mancha, Escuela Técnica Superior de Ingenieros Agrónomos, Ronda de Calatrava, 7, 13071 Ciudad Real, España.

[2]Universidad Autónoma de Madrid, Departamento de Geología y Geoquímica, Facultad de Ciencias, 28049 Madrid, España.

Resumen. La D.O. Uclés es una de las más jóvenes de España. Se encuentra situada entre las provincias de Cuenca y Toledo, ocupando una superficie total de viñedo de 1.700 ha. Se han estudiado 17 perfiles de la zona occidental, al Oeste de la Sierra de Altomira, los cuales se muestrearon y clasificaron siguiendo las nomas FAO (1996).
Los suelos estudiados se caracterizan por tener: pH alcalino (pH = 8,24), porcentajes de arcilla de un 20%, valores de conductividad eléctrica elevados (0,44 dS/m) y contenidos altos en CaCO3 (22%) El elemento mayoritario con mayor abundancia relativa es Si (con un valor medio de 196,6 g/kg), seguido por Ca (120,22 g/kg) y Al (63,86 g/kg). Entre los elementos traza llama la atención los elevados contenido en Sr (445,75 mg/kg) Rb (81,64 mg/kg) y Ba (223,61 mg/kg). Se han podido diferenciar varios grupos de suelos utilizando los contenidos en los elementos químicos anteriores, las características físico-químicas y las elevadas correlaciones entre contenido totales en hierro y aluminio con ciertos elementos químicos del grupo de las tierras raras.

Abstract. D.O. Uclés is one of the youngest wine quality areas in Spain. It is located between the provinces of Cuenca and Toledo, occupying a total vineyard area of 1,700 ha. We have studied 17 profiles from the western zone, west of the Sierra de Altomira, which were sampled and classified according to FAO standards (1996). The studied soils characteristics are: alkaline pH (pH = 8,24), percentages of clay of 20%, high electrical conductivity values (0,44 dS / m) and high contents in CaCO3 (22%). The element with more concentration is Si (with a mean value of 196,6 g / kg), followed by Ca (120,22 g / kg) and Al (63,86 g / kg). Among the trace elements, Sr is the more abundant (445,75 mg / kg) followed by Rb (81,64 mg / kg) and Ba (223,61 mg / kg). It has been possible to differentiate several groups of soils using the contents in the previous referenced chemical elements, the physical-chemical characteristics and the high correlations between the total content of iron and aluminium with certain chemical elements of the Rare Earth group.

1 Introducción

El suelo es un recurso natural de gran valor. El contenido y la distribución de los elementos químicos dependen de la composición de la roca madre, del clima, de la topografía y del uso y tipo de suelo. El estudio de su composición geoquímica puede servir para evaluar su fertilidad, la posible disponibilidad de nutrientes y aportar información sobre el origen geoquímico del suelo [1] y problemas de toxicidad [2].

La importancia del viñedo en Castilla-La Mancha radica no sólo en su extensión (aproximadamente 500.000 ha) sino también en las tradiciones y el impacto económico del sector en la Región. Por ello, aparecen diferentes figuras de calidad que intentan acoger la gran diversidad de viñedos existentes en Castilla-La Mancha. Bajo esta premisa se crea, en 2006, la Denominación de Origen (D.O.) Uclés (una de las más jóvenes de España) que cuenta con 1.700 ha dedicadas al cultivo de la vid, ubicada entre las provincias de Cuenca y Toledo.

Una primera aproximación para realizar la zonificación de la D.O., es el estudio de los componentes del suelo se convierte en una herramienta básica para la gestión racional del territorio. Caracterizar los suelos existentes en una determinada zona vitícola nos permite, establecer

https://doi.org/10.1051/e3sconf/20185001024

un primer paso dentro de la trazabilidad de los productos y recalcar la importancia del "Terroir" para la valoración de los vinos [3, 4].

Con este estudio se pretende caracterizar los suelos estudiados y tener un conocimiento del marco edafológico en el que se desarrollan los viñedos de la D.O. Uclés. A su vez, pretendemos conocer los elementos químicos más abundantes en la zona y su relación con determinados parámetros físico-químicos, se intentará establecer entre ellos relaciones que nos permitan abordar en un paso próximo una zonificación de la D.O. Uclés. Para ello hemos seleccionado un total de 17 elementos químicos (9 elementos mayoritarios: Mg, Al, Si, P, S, K, Ca, Fe, Mn; y 8 elementos traza: Sr, Ba, Rb, Ce, Cs, Y, La y Nd.) y 4 parámetros físico-químicos (pH, conductividad eléctrica, contenido en arcilla y contenido en carbonato cálcico).

2 Materiales y métodos

Para la realización de este trabajo se han estudiado 17 suelos vitícolas de la zona occidental de la D.O. Uclés, representando aproximadamente un 30% de la superficie vitícola de la zona. Los perfiles fueron muestreados en 2015 y se describieron siguiendo la metodología descrita en las normas FAO [5].

El clima de la zona, siguiendo la clasificación de Papadakis, se considera "mediterráneo templado" [6]. En concreto, la temperatura media de la zona es de 13,3°C. Atendiendo a las precipitaciones la zona presenta unos valores medios de 497 mm anuales, con mayores precipitaciones en municipios situados en altitudes superiores a 900 metros [7]. La descripción geológica de la zona indica que los suelos se desarrollan sobre arcillas yesíferas o yesos sacaroideos [8], caliche, calizas y calizas margosas [9].

Los perfiles fueron abiertos con máquina retroexcavadora. Las muestras de cada horizonte fueron guardadas en bolsas de plástico y etiquetadas para su transporte al laboratorio, donde se dejaron secar durante una semana a temperatura ambiente, extendidas sobre papel secante. Una vez secas, las muestras fueron tamizadas (<2mm) y conservadas para su análisis posterior.

El pH, la conductividad eléctrica y el contenido en carbonato cálcico fueron determinados según los procedimientos descritos por Porta [10], mientras que la textura se obtuvo siguiendo el procedimiento descrito por Gee y Bauder [11]. Los contenidos en elementos mayoritarios y traza se determinaron mediante fluorescencia de Rayos X, utilizando un espectrofotómetro secuencial (Philips Magix Pro con ánodo de Rodio).

Todos los datos obtenidos se sometieron a diferentes análisis estadísticos mediante IBM SPSS 24 (bajo licencia de la Universidad de Castilla-La Mancha).

3 Resultados y discusión

A partir de los datos obtenidos podemos decir que los suelos de la D.O. Uclés tienen un pH alcalino, conductividad eléctrica media elevada (por la presencia, en algunas zonas, de sales de yeso), porcentajes elevados de carbonato cálcico y valores normales para la fracción arcilla [8]. Como puede verse en el gráfico 1, los valores de pH se agrupan en un rango muy pequeño siguiendo una distribución normal; por el contrario, podemos observar una asimetría positiva en la distribución de la conductividad eléctrica provocada por la presencia de suelos con unos valores superiores a la media.

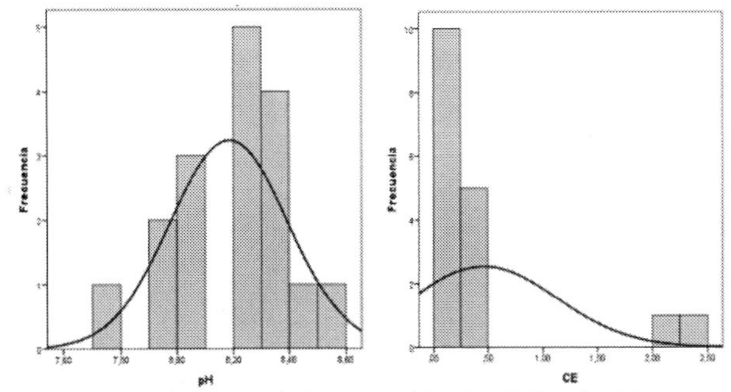

Figura 1. Histograma de los contenidos de pH (izquierda) y Conductividad Eléctrica (dS/m) (derecha) en los suelos estudiados.

Los contenidos en carbonato cálcico y arcilla en los suelos de la D.O. Uclés se comportan como distribuciones normales, aunque los rangos obtenidos para el carbonato cálcico son mayores que los obtenidos para los contenidos en arcilla (Figura 2). En general, los suelos estudiados se caracterizan por tener un contenido en arcilla entre medio y bajo, datos que resultan coherentes con el entorno geológico [8]. Aunque los datos medios obtenidos de $CaCO_3$ nos indican que son suelos con elevado contenido en este compuesto, los rangos de variación nos demuestran una gran diferencia de concentraciones, por lo que podremos diferenciar entre suelos de origen calizo y suelos que no lo son [12].

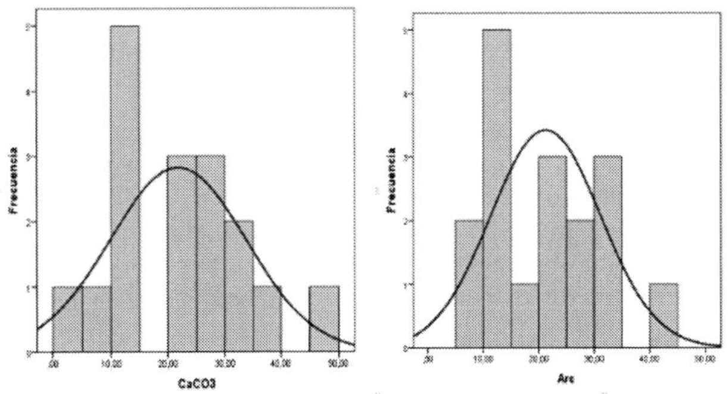

Figura 2. Histograma de los contenidos de carbonato cálcico (%) (Izquierda) y arcilla (%) (Derecha) en los suelos estudiados.

 https://doi.org/10.1051/e3sconf/20185001024

La Tabla 1 muestra una comparativa con los datos medios de concentración de elementos químicos obtenidos a nivel mundial [1, 13] y de Castilla-La Mancha [14]. Los elementos mayoritarios que presentan mayor concentración a nivel mundial en los suelos destacan Si (330,0 g/kg), Al (71,0 g/kg) y Fe (40,0 g/kg) [1, 13]. Datos que son similares a los obtenidos en este estudio, con la excepción de que en los suelos de la D.O. Uclés el Ca (120,2 g/kg) aparece en segundo lugar, por delante de Al y Fe (63,9 g/kg y 22,9 g/kg, respectivamente). Estos datos son acordes a los esperados debido al predominio de suelos calizos y a la presencia de yesos en la zona de estudio [15,16].

Tabla 1. Contenido medio en elementos químicos de los suelos estudiados (Mg a Mn en g/kg; de Ba a Nd en mg/kg)

Elemento	Media+DS	CLM[a]	Munndo[b]
Mg	10,2±6,2	7,3	5,0
Al	63,9±15,9	53,9	71,0
Si	196,6±56,0	237,4	330,0
P	0,8±0,3	0,6	0,8
S	4,2±12,0	2,7	0,7
K	17,1±4,4	16,4	14,0
Ca	120,2±57,7	108,7	15,0
Fe	22,9±5,4	22,1	40,0
Mn	0,4±0,1	0,4	1,0
Ba	222,7±41,3	191,8	513,0
Sr	445,8±6222,2	241,8	220,0
Rb	81,6±24,0	63,1	111,0
Cs	6,4±3,2	4,9	3,5
Ce	54,3±18,0	45,5	55,0
Y	21,5±5,8	13,8	29,0
La	31,4±11,3	25,4	35,5
Nd	25,4±8,5	9,9	30,5

[a] [14] [b] [1,13]

El carácter calcáreo de los suelos está diferenciado por la presencia de sulfato de calcio en la zona N-E, alcanzando valores muy superiores a las medias regional y mundial. El K y el P presentan valores muy similares a tanto a las medias mundiales [1, 13] como a los contenidos más frecuentes en Castilla-La Mancha [14].

Desde el punto de vista geoquímico, el elemento más relevante es el Sr, presenta valores que llegan a duplicar tanto la media regional [14] como la media mundial [1, 13], valor que está influenciado por los elevados contendidos en yesos [14].

El resto de elementos traza estudiados de la D.O. Uclés presentan algunas peculiaridades: el Ba muestra valores más bajos (222,7 mg/kg) que la media mundial de los suelos (513,0 mg/kg) pero mas altos que la media de los suelos de la región (191,8 mg/kg). El contenido en Rb depende en gran medida de los minerales heredados de la roca madre, de origen carbonatado y evaporítico, presentando valores medios mundiales de 111,0 mg/kg. Los suelos aquí estudiados poseen valores medios inferiores a la media mundial (81,6 mg/kg) y similares a los registrados por Amorós et al., [15].

Dentro de los elementos incluidos en el grupo de las Tierras Raras cabe destacar las concentraciones moderadas en Ce (54,3 mg/kg), Y (21,5 mg/kg), La (31,4 mg/kg) y Nd (25,4 mg/kg); valores que son similares a los mundiales (55,0 mg/kg; 29,0 mg/kg; 35,5 mg/kg y 30,5 mg/kg respectivamente) y superiores en todos los casos a los valores medios de la región [14, 15].

El análisis de componentes principales (ACP) revela como las variables Nd y Al son las que más peso positivo tienen en la componente 1 (0,96 y 0,95 respectivamente) mientras que el peso negativo recae en el Ca (-0,78) y el $CaCO_3$ (-0,66). Por el contrario, las variables que más influyen en la parte positiva de la componente 2 son Mg (0,64) y conductividad eléctrica (0,63) y en la parte negativa el Si (-0,46). Esto nos permite agrupar los suelos estudiados atendiendo a las concentraciones de Nd y Al y a los valores de conductividad eléctrica (Figura 3). Podemos ver cómo por encima de la línea se encontrarían los suelos de elevada conductividad.

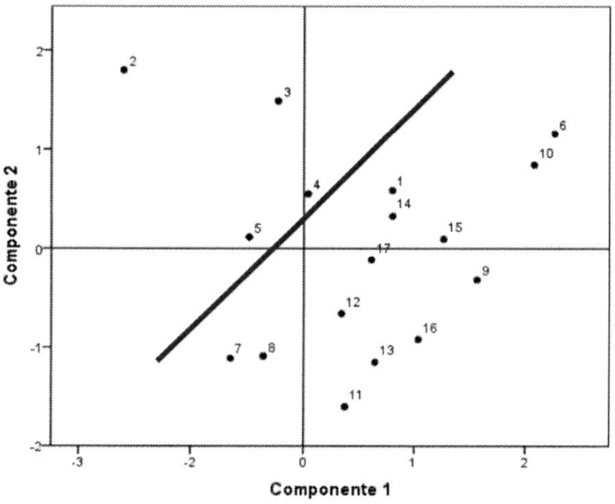

Figura 3. Distribución de los suelos estudiados en función del ACP.

Al estudiar la matriz de correlaciones observamos que las correlaciones más significativas y positivas se dan entre Al y Fe con los elementos del grupo de las Tierras Raras; destacando las relaciones del Al con Nd (0,94) y La (0,90) y las del Ce con Al y Fe (0,92 en ambos casos). Además, podemos destacar que el Ca tiene una correlación negativa con prácticamente todos los elementos, excepto con el Sr (0,48), aunque sólo son significativas las que establece con Ce (-0,87) y Al (-0,72).

Lantano (La) es un elemento que puede ser tenido en cuenta a la hora de realizar una caracterización geográfica [17], en nuestro caso cabe destacar las correlaciones significativas entre La y Cs (0,84) y entre La e Y (0,87); aunque también son elevadas las relaciones del La con Rb (0,8) o K (0,85) aunque éstas no son significativas.

4 Conclusiones

Las propiedades físico-químicas estudiadas ponen de manifiesto la riqueza y variedad de los suelos de esta D.O., indicando, con ello, la gran edafodiversidad de la zona.

El estudio de los elementos mayoritarios y traza nos permite diferenciar los suelos estudiados en base a sus concentraciones y su influencia en las propiedades físico-química.

Las relaciones obtenidas entre los diferentes elementos (especialmente aquellos que pertenece al grupo de las Tierras Raras) nos hacen pensar que podrían utilizarse para establecer la huella geoquímica de los suelos, y poder realizar, en un futuro, una trazabilidad del suelo a la uva y al vino.

Referencias

1. A. Kabata-Pendias, *Trace elements in soil and plants* (ED CRC, 2011)
2. P. Higueras, J.A. Amorós, J.M. Esbrí, F.J. García-Navarro, C. Pérez-de-los-reyes, G. Moreno, J. Geochem. Explor. **123**, 143-151 (2012)
3. J.M. Huggets, *Proceeding of the geologists associations*, **117**, 239-247 (2006)
4. C. Van Leewen, *Proceedings of clime and viticulture*, Zaragoza (2007)
5. FAO ISRIC ISSS. 2006. World soil resources report 103, FAO, Rome, 132 pp (2006)
6. MMA. Ministerio de Medio Ambiente, *Guía para la elaboración de estudios del medio físico.* (Centro de publicaciones del MMA. (2004)
7. F.J. García-Navarro, J.A. Amorós, C. Pérez-de-los-Reyes, S. Bravo y R. Jiménez-Ballesta, E.T.S.I.Agrónomos (Sin publicar)
8. IGME. Instituto geológico y minero de España. *Mapa geológico de España* **Hoja 607** (1975)
9. IGME. Instituto geológico y minero de España. *Mapa geológico de España* **Hoja 632** (1976)
10. J. Porta, *Técnicas y experimentos en Edafología.* (1986)
11. G.W. Gee, J.W. Bauder, *Physical and mineralogical methods* (ASA-SSSA, 1986)
12. S. Bravo, J.A. Amorós, C. Pérez-de-los-Reyes, F.J. García-Navarro, R. Ruedas, R. Jiménez-Ballesta, *J Chem* (2015)
13. D. L. Spark, *Environmental soil chemistry* (ED Elservier, 2003)
14. J.A. Amorós, C. Pérez-de-los-Reyes, F.J. García-Navarro, S. Bravo, J.L. Chacón, J. Martínez. R. Jiménez-Ballesta, *Atlas de suelos vitícolas de Castilla-La Mancha* (Universidad de Castilla-La Mancha, 2015)
15. J.A. Amorós, S. Bravo, F.J. García-Navarro, C. Pérez-de-los-Reyes, J.L. Chacón, J. Martínez. R. Jiménez-Ballesta, Vitis **51**, (2013)
16. S. Bravo, F.J. García-Navarro, J.A. Amorós, C. Pérez-de-los-Reyes, P.L. Higueras, *J Soil Sediments* **16** (2016)
17. A. Pisciotta, L. Tutone, F. Saiano, *Procedding of 18th International Symposium GIESCO,* 365 – 369 (2013).

Zonificación de suelos vitícolas en Villanueva de Alcardete (Toledo, La Mancha, España) utilizando elementos traza.

Zonification of vineyard soils in Villanueva de Alcardete (Toledo, La Mancha, Spain) using trace elements.

José Ángel Amorós[1,3,a], Sandra Bravo[1,3], Francisco Jesús García Navarro[1,3], Caridad Pérez-de-los-Reyes[1,3], M. Sánchez-Ormeño[1], Jesús Gracía-Pradas[1], Raimundo Jiménez Ballesta[2] y Pablo Higueras[3]

[1]Universidad de Castilla- La Mancha, Escuela Técnica Superior de Ingenieros Agrónomos, Ronda de Calatrava, 7, 13071 Ciudad Real, España

[2]Universidad Autónoma de Madrid, Departamento de geología y Geoquímica, Facultad de Ciencias, 28049 Madrid, España

[3]Universidad de Castilla- La Mancha, Instituto de Geología Aplicada, Plaza Manuel Meca, 1, 13400 Almadén, España

Resumen. Se realiza un estudio de suelos en el término municipal de Villanueva de Alcardete (Toledo, España) con la finalidad de señalar zonas homogéneas para el cultivo del viñedo. Se practicaron 15 calicatas en lugares representativos y se realizaron análisis físico químicos y clasificaciones edafológicas. También se consultaron fuentes clásicas de información pero se apunta como novedosa herramienta para la zonificación la composición geoquímica de los suelos. Se determinaron para cada perfil 11 elementos mayoritarios (Al, Ca, Fe, K, Na, Mg, Mn, P, S, Si, Ti) y 27 elementos traza (As, Ba, Ce, Co, Cr, Cs, Cu, Ga, Hf, La, Mo, Nb, Nd, Ni, Pb, Rb, Sc, Sn, Sr, Ta, Th, U, V, W, Y, Zn, Zr). Una vez estudiadas las fuentes de información y los datos obtenidos, el mapa de distribución del Estroncio (Sr) representa bastante bien las distintas zonas de suelos del área estudiada y puede ser utilizado (con las debidas comprobaciones y correcciones) para establecer lotes de suelos homogéneos.

Abstract. A soil survey is carried out in the municipality of Villanueva de Alcardete (Toledo, Spain) in order to identify homogeneous zones for the cultivation of the vineyard.15 pits were made in representative sites and Physico-Chemical analyses and Pedologic classifications were performed. Classic information sources were also consulted, although the geochemical composition of the soils is suggested as a novel tool for zoning. 11 major elements (Al, Ca, Fe, K, Na, Mg, Mn, P, S, Si, Ti) and 27 trace elements (As, Ba, Ce, Co, Cr, Cs, Cu, Ga, Hf, La, Mo, Nb, Nd, Ni, Pb, Rb, Sc, Sn, Sr, Ta, Th, U, V, W, Y, Zn, Zr) were determined for each profile. Once the sources of information and the data obtained have been studied, the distribution map of the Strontium (Sr) represents quite well the different soil zones of the studied area and can it be used (with the necessary checks and corrections) to establish pools of homogeneous soil.

1 Introducción

La aplicación del concepto "Terroir" [1] se está mostrando desde hace años como una herramienta muy útil para la valorización y autentificación de los vinos de calidad. De los distintos componentes del "Terroir", el estudio del suelo es el que mejores resultados está ofreciendo para los fines perseguidos, por lo que muchos trabajos de zonificación de suelos vitícolas se han emprendido en los últimos años a muchos niveles (Denominación de Origen, Cooperativa, bodegas particulares, etc...). La metodología de Zonificación clásica se basa en la integración de todos los datos disponibles (Mapas Geológicos, Edafológicos, datos altitudinales, fotografías aéreas, imágenes satelitales y obtenidas por drones, etc...). Dicha integración es facilitada hoy en día por el uso de los Sistemas de

 https://doi.org/10.1051/e3sconf/20185001025

Información Geográfica. En cualquier caso, sigue siendo inexcusable un buen conocimiento del terreno y la toma de datos guiados por expertos locales para la validación de los resultados. El estudio de la Geoquímica del suelo es otra herramienta que se está mostrando de gran utilidad para diversos fines: el contenido en elementos mayoritarios aporta información sobre componentes estructurales y macronutrientes, mientras que el contenido en elementos traza (concentración inferior a 0,01%) [2, 3] nos informa sobre el origen de los suelos, su huella geoquímica y la posible toxicidad o carencias en micronutrientes [4,5]. Concretamente, en el presente trabajo, se ha estudiado el contenido en elementos traza y su distribución como otra fuente de información a tener en cuenta en los estudios de zonificación vitícola.

Por todo lo explicado anteriormente, se acomete el estudio de los suelos vitícolas del término municipal de Villanueva de Alcardete (Toledo, España), de 14.700 has totales, de las que 8.660 has corresponden a viñedo.

2 Materiales y métodos

Durante el mes de mayo del año 2017 se seleccionaron quince perfiles representativos del término municipal de Villanueva de Alcardete (Toledo) y se describieron morfológicamente según los criterios de la FAO [6]. Se realizaron análisis convencionales (textura, pH, conductividad eléctrica, contenido en carbonato cálcico, densidad aparente, contenido en materia orgánica, nitrógeno, relación carbono/nitrógeno, fósforo, cationes de cambio, capacidad de intercambio catiónico y porcentaje de saturación por bases) [7, 8]. Así mismo se determinaron las concentraciones de 11 elementos mayoritarios y 27 elementos traza mediante la técnica de Fluorescencia de Rayos X utilizando un espectrofotómetro secuencial (Philips Magix-Pro con ánodo de Rodio en el tubo de rayos X).

Para la realización de los mapas de distribución de los elementos químicos presentes en el suelo se ha utilizado el software informático ARCGIS 10.3 ©ESRI 2018.

3 Resultados y discusión

Los suelos de Villanueva de Alcardete son, en general, poco evolucionados (Entisoles y, especialmente, Inceptisoles) con texturas de franco-arenosas, pH en torno a 8 y con un predominante carácter cálcico. Se han clasificado según la Soil Taxonomy (Soil Survey Staff, 2014) como Entisoles, en concreto, Lithic Xerorthent (perfiles 1, 3 y 13) y Typic Xerorthent (perfil 11) e Inceptisoles: Typic Haploxerept (perfiles 2, 5), TypicCalcixerept (perfiles 8, 10, 12 y 14), y Calcic Haploxerept (perfiles 6, 9 y 15). El perfil 7 corresponde a un Inceptisol muy influenciado por el ambiente fluvial (Fluvaquentic Endoaquent).

En cuanto a las propiedades químicas, el pH de los perfiles es moderadamente alcalino [4], la conductividad eléctrica es baja en la mayoría de los perfiles, destacando el perfil 7 (1,49 dS/m). El porcentaje en carbonato cálcico es alto en todos los perfiles (>20%) excepto el perfil 5, siendo el porcentaje de caliza activa alto (>9 %) para todos los perfiles excepto los perfiles 1, 8 y 15 que lo presentan medio (de 6-9 %). Los contenidos de materia orgánica de los horizontes superficiales son bajos (1,7% en el perfil 3), normales, altos o muy altos (5,4 % en el perfil 7). La relación C/N es de buena a mediana en todos los perfiles (entre 8 y 15). Por último, la capacidad de intercambio catiónico (C.I.C) está por encima de los 8-10 meq/100g por lo que los suelos se consideran aptos para el cultivo.

En la Tabla 1 aparecen los valores medios, con su dispersión, de concentración de elementos de los horizontes superficiales de los perfiles estudiados. Para facilitar su comparación y discusión se ha añadido otra columna con los valores medios de Castilla-La Mancha [9] y en los niveles medios mundiales [10,11].

Tabla 1. Contenido medio y desviación típica de los 15 perfiles estudiados en elementos mayoritarios ($g \cdot kg^{-1}$ hasta el Ti, en gris) y traza ($mg \cdot kg^{-1}$ a partir del Ti). a) [9] b) [10, 11]

Elemento	Media	des.Tip	CLM .a)	Mundo. b)
Na	0,87	0,37	1,3	5,0
Mg	13,24	11,75	17,7	5,0
Al	40,12	15,55	81,0	71,0
Si	140,03	71,87	292,0	330,0
P	1,03	0,59	1,9	0,8
S	2,17	5,57	2,4	0,7
K	12,73	5,30	15,4	15,0
Ca	215,49	78,74	301,0	15,0
Fe	14,96	4,63	25,6	40,0
Mn	0,32	0,12	0,4	1,0
Ti	2,32	0,79	4,4	5,0
Ba	202,80	52,21	389,0	513,0
Sr	421,98	570,54	380,0	220,0
V	30,51	10,49	50,0	95,0
Cr	26,91	5,74	54,8	68,5
Ce	30,27	14,78	57,7	55,0
Rb	48,31	18,72	86,2	111,5
Ni	13,64	3,85	16,9	37,0
Zn	39,83	13,35	35,7	78,5
Nd	16,43	6,30	21,6	30,5
Pb	16,24	3,65	19,9	32,0
Nb	9,40	1,86	8,0	13,0
Co	5,73	0,84	5,8	9,0
Cs	2,19	1,97	7,7	3,5
Ga	6,35	2,13	11,1	18,5
La	18,10	7,38	23,3	35,5
Y	14,32	4,05	17,9	29,0
Zr	159,34	69,00	167,0	353,0
Sc	7,44	2,20	8,0	8,0
Th	5,94	2,82	9,6	8,5
U	1,92	0,69	3,8	3,0
Cu	17,31	4,70	10,3	27,0
As	0,47	1,15	7,4	15,5
Mo	1,02	0,37	0,9	1,1
Ta	1,84	0,50	2,2	3,0
Hf	5,04	2,10	4,9	7,0
W	1,36	1,09	3,2	3,0

De la observación de los datos geoquímicos, merece la pena destacar el marcado carácter cálcico de los suelos estudiados (Cretáceo los más antiguos, Mioceno la mayor extensión y Cuaternario los más modernos) [12]. Con contenidos en Ca superiores a 200 $g{\cdot}kg^{-1}$, nos encontramos en niveles algo menores que la media de suelos vitícolas de Castilla-La Mancha (301 $g{\cdot}kg^{-1}$) y muy por encima de los niveles mundiales (15 $g{\cdot}kg^{-1}$). Valores relativamente altos de Mg (más de 13$g{\cdot}kg^{-1}$) respecto a los mundiales (5 $g{\cdot}kg^{-1}$) y muy similares a otras zonas de Castilla-la Mancha nos orientan sobre la procedencia geológica de las rocas del sustrato (dolomias) de algunos suelos de la parte oriental. También aparecen valores relativamente altos de S (más de 2$g{\cdot}kg^{-1}$) con mucha variabilidad, que nos señalan la presencia ocasional de yeso ($CaSO_4$) sobre todo en las terrazas cerca del cauce del Cigüela.

El Sr (422$mg{\cdot}kg^{-1}$) es un elemento muy característico de la zona y presenta grandes variaciones en contenido. Son en general contenidos superiores a los niveles generales de Castilla-La Mancha (380 $mg{\cdot}kg^{-1}$) y mundiales (220 $mg{\cdot}kg^{-1}$). Es un elemento clave para la posible determinación de relaciones isotópicas Sr^{86}/Sr^{87} que nos servirían para diferenciar nuestros productos y para mejorar su trazabilidad [4, 13]. Su distribución aparece en la Figura 1.

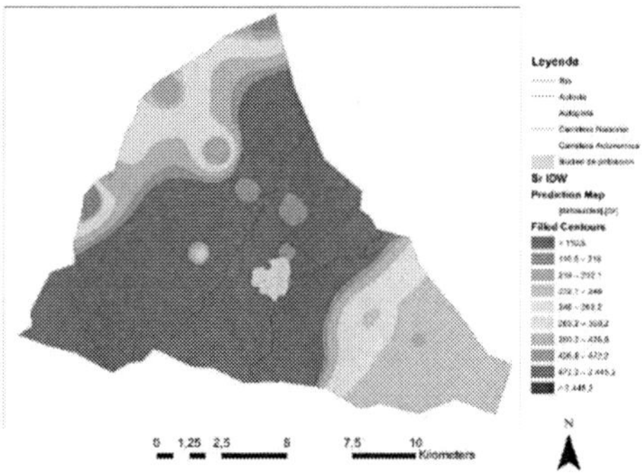

Figura 1. Mapa de distribución de estroncio (Sr) medido en $mg.kg^{-1}$ en el término municipal de Villanueva de Alcardete (Toledo, España).

El Ba (202 $mg{\cdot}kg^{-1}$) presenta, sin embargo, valores más bajos que los regionales (389 $mg{\cdot}kg^{-1}$) y los mundiales (513 $mg{\cdot}kg^{-1}$), siendo un elemento relativamente poco abundante en la zona. Su distribución se muestra en la Figura 2. Sucede lo mismo con el Rb (48 $mg{\cdot}kg^{-1}$) con valores medios más bajos que en Castilla-La Mancha (86 $mg{\cdot}kg^{-1}$) y en el Mundo (111$mg{\cdot}kg^{-1}$) ya que este elemento es más típico de zonas con sustrato de rocas graníticas [8]. Otros elementos traza de interés son los del grupo de las tierras raras (REE): Lantano (La) con concentraciones desde 8,7 a 33 $mg{\cdot}kg^{-1}$, con una media de 17,41 $mg{\cdot}kg^{-1}$ y una desviación típica de 7,79 y Cerio (Ce) desde 2,5 a 55,8 $mg{\cdot}kg^{-1}$, con una

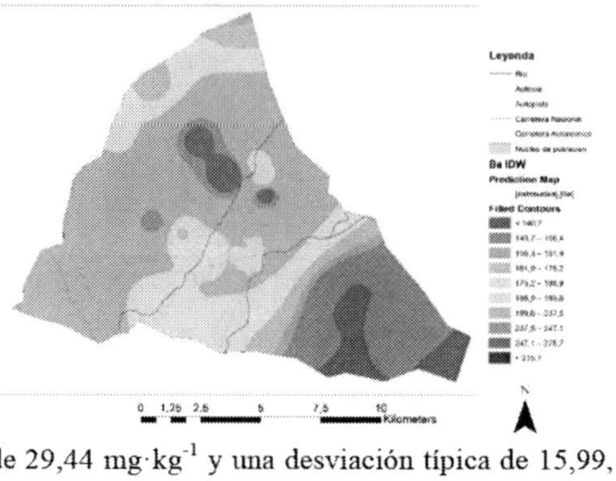

media de 29,44 $mg{\cdot}kg^{-1}$ y una desviación típica de 15,99, entre otros.

Figura 2.- Mapa de distribución de bario (Ba) medido en $mg.kg^{-1}$ en el término municipal de Villanueva de Alcardete (Toledo, España).

4 Conclusiones

Nos encontramos ante una amplia zona vitícola con suelos aparentemente uniformes y con una topografía poco diferenciada.

El desarrollo de los suelos sobre los materiales de partida, es escaso y similar por lo que resulta difícil decidir qué suelos son diferentes.

La concentración de Sr (mayor en suelos más modernos) nos sirve en este caso para establecer zonas diferenciales respecto a este elemento que coinciden sustancialmente con su antigüedad y situación topográfica.

La geoquímica puede ser de gran ayuda para la zonificación de suelos vitícolas, seleccionando los elementos adecuados y contrastando las diferencias y coincidencias con otras fuentes de información.

Referencias

1. C. Van Leewen, y col. *J. Wine Research*. **17**, nº 1, 1-10 (2006)
2. R. E. White, *Understanding Vineyard Soils*. Oxford University Press. (2009).
3. A. Wild, *Condiciones del suelo y el desarrollo de las plantas según Russel*. Ed. Mundiprensa. (1992).
4. J.A. Amorós, S. Bravo, C. Pérez-de-los-Reyes, F.J. García-Navarro, J.A. Campos, M. Sanchez-Ormeño, R. Jiménes-Ballesta, P. Higueras, App Geoschem, **88**, (2018).
5. S. Bravo, E. García-Ordiales, F.J. García-Navarro, J.A. Amorós, C. Pérez-de-los-Reyes, R. Jiménez-

 https://doi.org/10.1051/e3sconf/20185001025

Ballesta, J.M. Esbrí, E.M. García-Noguero, P. Higueras, *Environ Sci Pollut Res*, 2017

6. FAO ISRIC ISSS. 2006. World soil resources report 103, FAO, Rome, 132 pp (2006)
7. J. Porta, *Técnicas y experimentos en Edafología*. (1986)
8. G.W. Gee, J.W. Bauder, *Phisical and mineralogical methods* (ASA-SSSA, 1986)
9. J.A. Amorós, C. Pérez-de-los-Reyes, F.J. García-Navarro, S. Bravo, J.L. Chacón, J. Martínez. R. Jiménez-Ballesta, *Atlas de suelos vitícolas de Castilla-La Mancha* (Universidad de Castilla-La Mancha, 2015)
10. D. L. Spark, *Environmental soil chemistry* (ED Elservier, 2003)
11. A. Kabata-Pendias, *Trace elements in soil and plants* (ED CRC, 2011)
12. IGME. Instituto geológico y minero de España. *Atlas geológico de España*. (2012)
13. R. Petrini, L. Sansone, F.F. Slijko, A. Buccianti, P. Marcuzzo, D. Tomasi, Food Chem, **170** (2015).

E3S Web of Conferences 50, 01026 (2018) https://doi.org/10.1051/e3sconf/20185001026
XII Congreso Internacional Terroir

Edafodiversidad en viñedos de Castilla-La Mancha, España

Pedodiversity in Vineyards of Castilla-La Mancha, Spain

Jiménez Ballesta R.[1], Pérez-de-los-Reyes C.[2], Amorós A[2], Bravo S.[2] y García Navarro F. J[2]

[1]Universidad Autónoma de Madrid, Departamento de Geología y Geoquímica, Facultad de Ciencias, 28049 Madrid, España

[2]Universidad de Castilla- La Mancha, Escuela Técnica Superior de Ingenieros Agrónomos, Ronda de Calatrava, 7, 13071 Ciudad Real, España

Resumen. En el desarrollo de estrategias relativas al "Terroir", se requiere una comprensión del grado de edafodiversidad. En un intento de inventariar y zonificar los suelos vitícolas de Castilla La Mancha (CLM), desde 2008 venimos realizando un análisis comprensivo y armonizado de las distintas coberturas edáficas, de los edafotaxa y de su distribución espacial. Fruto de este trabajo se desprende que, a pesar de que CLM ocupa una superficie dominantemente plana, existe gran variedad de sustratos litológicos y posiciones geomorfológicas, lo que conlleva una notable edafodiversidad, incluso a escala detallada. En concreto, se han caracterizado, según Soil Taxonomy (2014), fundamentalmente: Entisoles, Inceptisoles, Alfisoles y Ultisoles; mientras que según FAO-ISRIC-ISSS (2006) existen fundamentalmente: Cambisoles, Calcisoles, Luvisoles, Arenosoles, Gleysoles, Gypsisoles, Fluvisoles, Leptosoles, Regosoles, Acrisoles y Alisoles. Esta tipología de suelos la atribuimos fundamentalmente a procesos tales como alteración mineral parcial, calcificación o acumulación de carbonatos o yesos, argiluviación, rubefacción, etc. También cabe señalar, entre otros, dos procesos singulares y de moderada extensión territorial, como son la salificación y la gleización. Entre los cualificadores más comunes cabe citar los relacionados con la naturaleza de los materiales originarios o con procesos incipientes edafogenéticos, tales como Calcaric, Haplic, Profondic, Skeletic, Leptic, etc.; otros están relacionados con procesos de formación, tales como Chromic, Calcic, Luvic, etc. Sin embargo, desde nuestro punto de vista, se necesitan nuevos calificadores, especialmente en el caso de los Calcisoles.

Abstract. In the development of strategies related to terroir is required understand the pedodiversity. In an attempt to inventory and zoning the vineyard soils of Castilla La Mancha (CLM), (Spain), since 2008 we have been carrying out a comprehensive and harmonized analysis of the different soil coverings, the pedotaxa and their spatial distribution. The result of this work is that although CLM occupies a predominantly flat surface, which might suggest a certain monotony or scarce pedodiversity, given the variety of lithological substrates and existing geomorphological positions, a remarkable pedodiversity can be appreciated (including at a detailed scale). In effect, the pedodiversity or diversity of genetic types is very appreciable in CLM. Specifically they have been characterized, according to Soil Taxonomy (2014) mainly: Entisols, Inceptisols, Alfisols and Ultisols. According to FAO-ISRIC-ISSS (2006) there are mainly: Cambisols, Calcisols, Luvisols, Arenosols, Gleysols, Gypsisoles, Fluvisols, Leptosols, Regosols, Acrisols and Alisols. This type of soils are attributed to various processes such as partial weathering, calcification or accumulation of carbonates, argiluviation, rubefaction, gypsum accumulation, etc. Finally, it is worth mentioning two singular processes of slight territorial importance, such as the salification and gleization. Among the qualifiers, the most common are those related to the nature of the original materials or to incipient pedogenetic processes, such as Calcaric, Haplic, Profondic, Skeletic, Leptic, etc; others are related to soil forming processes, as Chromic, Calcic, Luvic, etc. However, in our point of view, new qualifiers are needed, especially in the case of Calcisols.

1 Introducción

Suelo y clima son los dos pilares del término francés "Terroir", concepto que implica la existencia de estrechas relaciones entre la composición de la uva, las características del vino y el territorio de producción. De hecho, uno de los rasgos más frecuentemente discutidos, en relación a un gran vino, es el suelo.

El cultivo de la uva para vinificación es muy conocido desde hace siglos en Castilla-La Mancha (CLM), una región que cubre una superficie de 79,463 km^2. De hecho, CLM es la región de mayor concentración vitícola del mundo con algo más de 500.000 has.

El suelo y la historia vitícola del mismo, constituye el alma del argumento tradicional del "Terroir" [1]. En efecto, los suelos son componentes esenciales de los "Terroirs" vitivinícolas, junto con factores climáticos, vegetales y humanos [2] [3], ya que influyen en el rendimiento del viñedo y la composición de bayas y vinos [4] [5], Pero el terroir es difícil de estudiar sobre una base científica ya que intervienen varios factores [6],

https://doi.org/10.1051/e3sconf/20185001026

de tal modo que las cualidades del suelo de cultivo influencian la composición de las bayas de la vid, así como las características cualitativas a través de la composición química; pero la humedad edáfica, altitud, orientación etc., también deben tenerse en cuenta.

Con esta premisa, el principal objetivo de este trabajo es examinar el grado de edafodiversidad del territorio de CLM dedicado al cultivo de la vid, para entender su significado en relación al cultivo de la vid, dado que existe una estrecha relación entre edafodiversidad y el cultivo de la vid.

2 Rasgos del medio físico de Castilla La Mancha

Desde un punto de vista geológico, el territorio de Castilla-La Mancha es relativamente complejo, ya que incluye la mayoría de los dominios geológicos de España [7]. Lo que predomina en CLM son materiales carbonatados tales como margas (muy variadas), calizas, y a veces areniscas calcáreas o limos calcáreos. Todos ellos aparecen en formaciones geomorfológicas muy variadas (glacis, terrazas, conos, vertientes, etc.), ocupando proporcionalmente grandes extensiones de sedimentos terciarios o cuaternarios; en este último caso son de origen aluvial o coluvial; a estos materiales se asocian, en determinados entornos, yesos y margas yesíferas. Hacia el NW dominan materiales graníticos y/o sus depósitos sedimentarios asociados como son las arcosas; además, en esta última zona, abundan los materiales metamórficos (pizarras, esquistos, etc.), si bien estos se extienden de forma diseminada por todo el territorio. Por último, cabe resaltar la presencia de materiales volcánicos en una extensión considerable.

El clima de esta extensa región obviamente no es único, aunque si tiene la impronta del régimen mediterráneo, lo que va a conferir a los suelos su caracterización a nivel de clasificación bajo el régimen xérico. Se trata de un clima caracterizado por inviernos rigurosos fríos y relativamente húmedos, veranos calurosos y secos, con el sello de tener unas irregulares en la distribución de la lluvia, (puede establecerse un rango entre 375 mm en el centro a más de 1000 mm en las áreas de montaña); hay que añadir las fuertes oscilaciones térmicas diarias y, en cierto modo, una aridez generalizada.

Con estos rasgos aparecen paisajes con viñedo como los que aparecen en la figura 1.

Figura 1. Zonas dedicadas al cultivo de la vid en CLM: algunos paisajes característicos, resaltando los suelos rojos.

3 Materiales y métodos

Se han descrito y muestreado numerosos perfiles (más de 150), en los que se han definido los rasgos macromorfológicos, (con los criterios de FAO-ISRIC-IUSS, 2006), que incluyen desde la profundidad al drenaje, pasando por la estructura, consistencia, presencia y abundancia de fragmentos gruesos, textura etc.; así como toda una serie de parámetros edáficos tales como pH, conductividad eléctrica, contenido en materia orgánica, salinidad, capacidad de cambio y saturación en bases, contenido en carbonatos y caliza activa, además de N, P y K. Adicionalmente se ha llevado a cabo una determinación de los contenidos totales de un amplio número de elementos químicos, (tanto mayoritarios como minoritarios). La metodología utilizada para efectuar los análisis aparece en Amorós et al. [8].

4 Resultados y discusión

En el cultivo de la vid hay que tomar en consideración tanto los factores extrínsecos como los intrínsecos. Entre los extrínsecos cabe citar: pendiente, rocosidad y pedregosidad. Respecto de la pendiente, una mayoría de los suelos de CLM se sitúan en topografías planas o casi planas, y solo en ocasiones en topografías inclinadas, raramente abruptas. Ello conlleva que la mayoría de los suelos no tengan problemas de profundidad efectiva, lo que, unido a una moderada presencia de rocas duras, nos permite concluir que prácticamente no existen suelos con contacto lítico próximo a la superficie, si bien en ocasiones, por la pendiente se detecta cierta actividad erosiva.

En lo que se refiere a los fragmentos gruesos, hay que señalar que en determinadas zonas son abundante (figura 2); existen algunas áreas (por ejemplo, sobre la raña, o sobre calizas planas o casi planas) en donde es muy palpable la existencia de una fase pedregosa. Desde el punto de vista del tamaño y de la naturaleza mineral de los mismos (grava, gravilla, etc.) son muy variables. Su presencia puede constituir un factor de calidad en el viñedo, ya que aumenta el drenaje y la aireación, al tiempo que aumenta la capacidad de retener calor del sol para luego cederlo a la planta.

Figura 2. Intensa pedregosidad superficial. Foto superior: Cogolludo (cuarcitico); foto inferior Villanueva de Alcardete (calizo)

Entre los factores intrínsecos cabe resaltar el color del suelo que obedece originalmente al color de la roca de partida, pero que, conforme avanza la edafogénesis, los colores pueden ir cambiando; este es el caso de los abundantes y genuinos suelos rojos de CLM (Rhodoxeralfs). Otros muchos suelos de CLM tienen tendencia a ser claros, sobre todo si abundan los carbonatos; y como no son ricos en materia orgánica absorben moderadamente la radiación solar.

Normalmente son suelos en los que existe un adecuado equilibrio entre profundidad efectiva, buen drenaje y capacidad de retención de agua para que la vid no sufra demasiado en verano. No es por tanto habitual observar problemas de drenaje, salvo que se produzcan lluvias tardías que afecten el período de maduración, como puede suceder en algunos suelos sobre la raña (figura 3).

Figura 3 Suelos sobre sedimentos tipo raña (Cogolludo, Guadalajara), en los que pueden aparecer signos de gleización, ya que en periodos prolongados de lluvia se generan encharcamientos como el de la figura. Nótese el color rojo intenso al disponer de hierro y arcilla en suspensión.

En la tabla 1 aparecen algunas propiedades químicas de los suelos de CLM. Como se sabe es bien conocida la influencia de estas propiedades en la nutrición de la vid, pues afecta a los patrones de crecimiento específicos de la planta. Sin embargo, somos escépticos sobre el hecho de que la composición mineral del suelo afecte a las características organolépticas del vino, ya que no es posible para la vida la traslación de moléculas que afecten a las papilas gustativas desde el suelo a la baya y tampoco hacia los vinos.

Tabla 1. Resultados obtenidos

	pH (H_2O)	C.E. (dS/m)	$CaCO_3$ (%)	Arcilla (%)	N (%)	P (mg/kg)
Media	8,01	0,21	20,27	22,01	0,07	11,4
Mínimo	5,01	0,02	0,00	0,70	0,01	0,14
Máximo	8,85	2,57	61,70	46,93	4,04	975

Sobre materiales carbonatados y, por tanto, suelos carbonatados, pivota una parte importante de suelos de CLM. En efecto, la presencia del carbonato cálcico es muy común, de tal modo que una proporción significativa de viñedos en CLM crecen en sustratos edáficos carbonatados y en muy diversas proporciones, de modo que no es extraño encontrar horizontes de suelos con valores superiores al 60%, mientras que otros tienen valores inferiores al 10%. Al tiempo, los contenidos en caliza activa en numerosas ocasiones son elevados.

La acumulación de carbonatos puede aparecer en formas masivas, pero existen numerosos ejemplos de apariciones como masas concrecionadas (nódulos), como muñequillas, en forma laminar, acintadas; otras veces como micelios y pseudomicelios, calcanes, etc.... (Figura 4). Así, el carbonato cálcico puede aparecer en horizontes cálcicos o petrocálcicos, que pueden subyacer a un cámbico o a un argílico: otras veces aparece directamente

 https://doi.org/10.1051/e3sconf/20185001026

bajo el antrópico, prácticamente en superficie. Por esta circunstancia no es raro encontrar algunos horizontes petrocálcicos acumulados parcialmente en la superficie, en forma de fragmentos; esta circunstancia se produce tras ser arados los suelos, ya que necesitan ser roturados para permitir el desarrollo radicular.

Figura 4. Detalle de la acumulación de carbonato en fases diferentes, dentro de un horizonte petrocálcico en Tomelloso (Ciudad Real). Obsérvese en la base el carácter asalmonado de algunas láminas.

Castilla La Mancha es una región típicamente mediterránea, por lo que el contenido en materia orgánica es bajo; generalmente no se superan valores del 3%. En consecuencia, la estructura se genera por la presencia de otros coloides como la arcilla y sólo en los horizontes superficiales se asocia a la materia orgánica. Por el escaso contenido en materia orgánica la fertilidad es moderada, si bien la presencia de texturas francas o franco-arcillosas favorece la formación de estructuras poliédricas o prismáticas en horizontes subsuperficiales mejorando la fertilidad.

En cualquier caso, tanto sobre materiales carbonatados como yesíferos, los suelos se caracterizan por tener pHs básicos o moderadamente básicos; y generalmente con texturas finas. En efecto, aunque los valores de pH son muy variables, oscilando entre 5,0 y 8.5, una buena proporción tienden a la neutralidad o moderada basicidad. En efecto, una gran proporción de suelos (probablemente más del 80%) tienen valores de pH entre 6,5 y 8,0, con el consiguiente exceso de Ca (y/o Mg) y su influencia en otros elementos como el K y el Fe; por ello uno de los posibles problemas que pueden aparecer en viñedos de CLM es la clorosis férrica. Suelos con valores en el entorno de 5 pueden estar afectados por la presencia de Al o Mn en posiciones interlaminares; en estos casos pueden detectarse deficiencias en Ca, K, N, Mg, Mo, P y S, al tiempo que posibles excesos en Cu, Fe, Mn, Co y Zn.

Se han definido suelos dedicados a viñedo sobre yesos o margas yesíferas, observándose que son viñedos adaptados a suelos poco o moderadamente salinos. En estos casos son suelos moderadamente profundos, con una salinidad en el en torno a 2 dS/m; no son realmente abundantes en la Comunidad.

La textura es una propiedad edáfica muy importante en viticultura. En este sentido se detecta una significativa proporción de texturas equilibradas, si bien en ocasiones se encuentran áreas dominadas por texturas arenosas o bien arcillosas. Existen suelos arenosos en varios entornos dentro de CLM, como es el caso de los arenales de S Gregorio (Cuenca). Se trata de suelos que están bien drenados y retienen el calor, produciendo vinos "suaves" con menos color, acidez más ligera y tanino. Frente a este tipo de suelos existen otros arcillosos que retienen más fácilmente el agua, por lo que tienden a mantenerse más frescos, generando algunos de los vinos blancos más complejos del mundo. Apenas se han detectado suelos limosos, aunque existen en determinados entornos de CLM; en este caso muestran cierta habilidad para retener el agua y el calor. Lo ideal es disponer de texturas francas, equilibradas en arcilla, limo y arena, cosa que sucede con bastante frecuencia. Se trata de suelos fértiles que generalmente conllevan viñedos vigorosos, por lo que se generan vinos equilibrados. Es el caso de numerosos suelos formados sobre margas y otros materiales sedimentarios que comportan un gran potencial de elaboración de vinos (probablemente cerca del 50% de toda CLM).

Los suelos de viñedos de CLM son notoriamente bajos en fósforo (tabla 1), limitación que habitualmente se corrige fácilmente mediante la incorporación de cantidades adecuadas de fósforo durante las operaciones de preparación del suelo. Del mismo modo, el contenido en N asimilable es moderado a bajo, por lo que no es frecuente observar un excesivo vigor de la planta. En este sentido va incrementándose la utilización de cubierta vegetal para controlar este vigor.

En cuanto a los contenidos en distintos elementos químicos (tabla 2) cabe resaltar que son variables en función de determinados factores como la naturaleza de los materiales de partida y los procesos edáficos acaecidos.

En CLM se encuentran formaciones geológicas muy antiguas, con largos períodos de exposición a la meteorización, que han dado lugar a la viticultura que se practica en algunos de los materiales y suelos más antiguos del mundo. En este sentido, existen suelos con un avanzado estado de meteorización, en los que abunda caolín y sesquióxidos (hierro y óxidos de aluminio) como minerales de arcilla dominante, con una baja capacidad de intercambio (3-5 cmol kg^{-1}, en comparación con 10-25 cmol kg^{-1} de suelos más jóvenes) y con una acidez creciente con profundidad (<4,0-5,0, medida en KCl 1N), ya que la mayoría de los cationes básicos se han perdido por lixiviación y posteriormente han sido reemplazados por iones de hidrógeno. Cómo afecta esto al carácter del vino en comparación con los suelos con pH más alto o los suelos que contienen cal libre, aún no se ha descifrado. Sin embargo, el efecto inhibidor de un pH demasiado bajo en

 https://doi.org/10.1051/e3sconf/20185001026

Tabla 2. Resultado de los análisis químicos elementales

		Media	Mínimo	Máximo			Media	Mínimo	Máximo
(g/kg)	Al	53,96	124,99	830,89	(mg/kg)	Rb	63,20	12,50	223,00
	Ca	108,7	290,18	8548,78		Sc	8,30	1,80	24,30
	Fe	22,05	68,98	207,51		Sn	2,36	0,03	10,40
	K	16,36	46,66	77,01		Sr	241,9	22,30	3602,7
	Mg	7,34	35,21	36,63		Ta	1,29	3,37	3,75
	Mn	0,38	5,68	0,39		Th	7,81	0,75	20,63
	Na	1,49	16,66	6,40		V	51,72	10,40	193,25
	P	0,60	2,78	0,14		W	2,89	0,12	88,20
	S	2,70	61,13	104,26		Y	13,83	5,20	27,65
	Si	233,8	459,35	10226,88		Zn	32,57	11,30	79,70
	Ti	2,92	21,43	7,14		Zr	158,1	12,30	495,40
(mg/kg)	As	7,25	0,06	35,50		Hf	5,67	1,00	16,50
	Ba	191,8	65,40	469,30		La	25,36	4,65	74,20
	Ce	45,54	11,90	143,13		Mo	0,61	0,52	2,45
	Co	6,22	1,85	25,20		Nd	21,25	5,50	56,40
	Cu	10,87	3,55	30,50		Nb	9,96	5,05	17,10
	Cs	4,89	0,11	20,67		Ni	18,59	1,65	80,75
	Cr	36,19	15,45	115,40		Pb	16,18	7,30	27,75
	Ga	8,87	1,30	20,90					

el desarrollo y funcionamiento de la raíz está bien documentado y las medidas para aumentar el pH del suelo a 5,5 incorporando cal a profundidades de 1,0-1,2 m mediante arado de zanja antes de la siembra, son prácticas estándar.

Partiendo del muestreo y caracterización más de 150 perfiles de suelos de las zonas de mayor densidad y representatividad, en CLM se identifican, según Soil Taxonomy [9], cuatro órdenes de suelos: Entisoles, Inceptisoles, Alfisoles y Ultisoles; el número de subgrupos

llega hasta 15. Muchos suelos derivan de procesos relacionados dinámica fluvial; se trata de Entisoles o Inceptisoles que muestran una equilibrada proporción de arcilla, limo, arena, pero sobre todo con determinados contenidos en grava, unas veces calizas y otras veces cuarcíticas. Son Entisoles en terrazas recientes e Inceptisoles en terrazas no tan recientes; en las terrazas antiguas tienden a formarse Alfisoles. Dada la importancia que le confiere la Soil Taxonomy al régimen de humedad, cabe señalar que prácticamente todo el viñedo de CLM está sometido a un régimen xérico, y solo excepcionalmente se ha admitido un régimen áquico.

Puestos a resaltar, los suelos más abundantes son los Typic Calcixerepts y Typic Rhodoxerafs. En ambos tipos de suelos, como sucede en otros tipos como Petrocalcic Calcixerept o Petrocalcic Rhodoxeralf, aparecen acumulaciones de carbonatos como nódulos con diferentes formas y grados de cementación, tales como micelios, pseudomicelios, calcanes, etc. También son relativamente frecuentes los Typic Xerorthents, que responden a las limitaciones provocadas por procesos erosivos, mientras que otras veces se deben a la presencia de roca dura próxima a la superficie.

Según FAO-ISRIC-ISSS [10], existen fundamentalmente: Cambisoles, Calcisoles, Luvisoles, Arenosoles, Gleysoles, Gypsisoles, Fluvisoles, Leptosoles, Regosoles, Acrisoles y Alisoles. Esta tipología de suelos la atribuimos fundamentalmente a procesos tales como alteración mineral parcial, calcificación o acumulación de carbonatos o yesos, argiluviación, rubefacción, etc. También cabe señalar, entre otros, dos procesos singulares y de moderada extensión territorial, como son la salificación y la gleización.

5 Conclusiones

Partiendo de la base de que los suelos juegan un papel trascendental en la fisiología de la vid y en la evolución bioquímica de la baya, definimos los suelos de CLM como muy edafodiversos. En efecto, se trata de coberturas edáficas, de edafotaxa y de una distribución espacial muy variada, a veces incluso a escala decamétrica. Dado que, si los suelos interaccionan con la vid, fundamentalmente a través de las propiedades físicas y químicas, de su nivel nutricional, por la humedad (disponibilidad de agua), por las propiedades térmicas y lumínicas y de la permisividad a la penetración radicular de la planta en el medio edáfico, no es pues extraño que esta notable edafodiversidad debe jugar un papel fundamental en la extraordinaria variedad de vinos que se producen en CLM.

Si el propósito de este trabajo era examinar el grado de edafodiversidad de CLM para comprender sus vinos, puede concluirse que, si los vinos traducen el papel del suelo, entonces el vino de CLM no debe ser único sino muy variado, lo que constituye un gran activo y un criterio interesante para los amantes del vino, vinicultores y viticultores.

Referencias

[1] Olego M.A, Garcia V.M., Garzón E. (2009). Terroir ¿una realidad telúrica? InfoWine 1/2, 1-6.

[2] Seguin G. (1986). Terroirs and Pedolgy of wine growing. Experientia, 42, 8: 861-873.

[3] Deloire A., Vaudour E., Carey V., Bonnardot V. and Van Leeuwen C. (2005). Grapevine responses to terroir: a global approach. J. International des Sciences de la Vigne et du Vin, 39, 4: 149-162.

[4] Tisseyre, B., Ojeda, H. and Taylor, J. (2007) New technologies and methodologies for site-specific viticulture. Journal international des Sciences de la Vigne et du Vin 41, 63-76

[5] Van Leeuwen, C., Friant, Ph., Choné, X., Trégoat, O., Koundouras, S. and Dubourdieu, D. 2004. "The influence of climate, soil and cultivar on terroir". In *American Journal of Enology and Viticulture* Vol. 55, 207–217.

[6] Cornelis Van Leeuwen & Gerard Seguin (2007) The concept of terroir in viticulture, Journal of Wine Research, 17:1, 1-10,DOI: 10.1080/09571260600633135

[7] Gibbons W, Moreno T (2002) The geology of Spain. The Geological Society, London. 649 pp

[8] Amorós, J.A; Bravo S.; García-Navarro, F.J.; Pérez-de-los-Reyes, C.; Chacón, J.L.; Martínez, J.; Jimenez-Ballesta, R. (2016). Atlas de suelos vitícolas de Castilla-La Mancha. Universidad de Castilla-La Mancha. Globalcaja

[9] Soil Survey Staff. 2014. Keys to Soil Taxonomy. 12th ed. Washington, D.C.: USDA-Natural Resources Conservation Service

[10] FAO-ISRIC-IUSS (2006) World reference base for soil resources. A framework for international correlation and communication. World soil resources report. FAO, Rome.

Caracterización geoedáfica del Valle de Secastilla, Huesca (España)

D. Badía[1], J.A. Cuchí[1], J. Casanova[1], C. Martí[1], J. Ferrer[2], J. Albar[2]

[1]Escuela Politécnica Superior de Huesca. Universidad de Zaragoza. Correo: jcasan@unizar.es

[2] Viñas del Vero S.A.

Introducción

Secastilla se sitúa en el NE de la Península Ibérica, entre los Pirineos y el Valle del Ebro. Es una población perteneciente a la Denominación de Origen Somontano, localizada en el interfluvio Cinca-Ésera.

Los suelos muestreados se disponen en una ladera (figura 1), entre los barrancos de Pallaruelo y Peralta (Secastilla, D.O. Somontano).

Material y métodos

Seleccionados los perfiles de suelos más representativos, se describen macromorfológicamente (FAO, 2009) y se muestrean para su análisis. De la fracción de tierra fina, se analizan sus principales propiedades físicas y químicas (pH, carbonatos, materia orgánica, CEe, iones en solución, porosidad, estabilidad estructural, CRAD, granulometría, CIC), se clasifican (IUSS, 2007) y se evalúa el potencial de vigor y de precocidad del terreno para la viña (Morlat, 2010).

Resultados

Dentro de un mismo contexto climático, las variaciones **litológicas** y del **relieve** conducen a la formación de diferentes tipos de suelos (figura 2).

1- Calcilutitas o margas en laderas. En estas condiciones se desarrollan Calcisoles háplicos.
2- Conglomerados en laderas convexas y cerros. Presentan Regosoles lépticos.
3- Conglomerados superpuestos a calcilutitas en laderas. Se clasifican como Regosoles háplicos.
4- Arcillas con yesos en laderas y rellenos. Se clasifican como Gipsisoles háplicos.
5- Depósitos coluviales. Se trata de Regosoles colúvicos.
6- Depósitos aluviales. Son Fluvisoles háplicos

https://doi.org/10.1051/e3sconf/20185001027

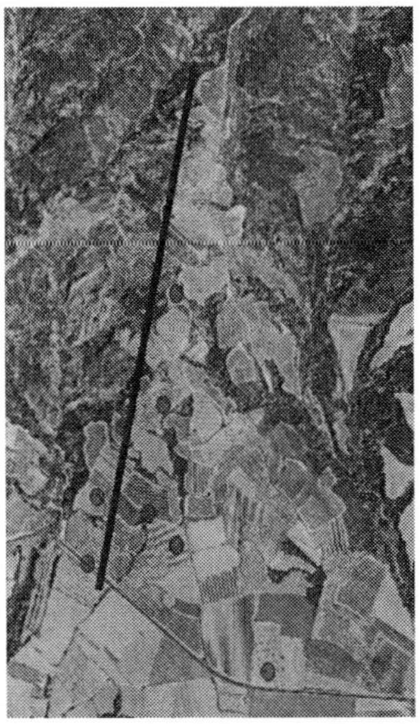

Figura 1Transecto de los perfiles caracterizados en Secastilla (DO Somontano)

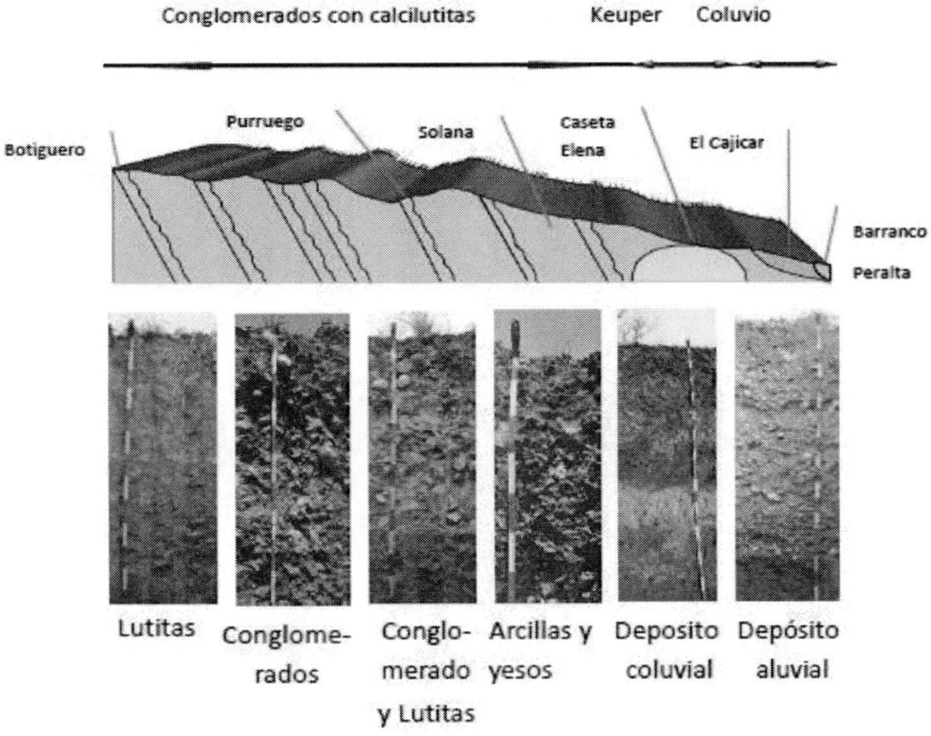

Figura 2: Tipos de suelos formados por la influencia del relieve y las variaciones litológicas en Secastilla (DO Somontano)

Conclusiones

Cada **suelo** posee unas propiedades y componentes (CRAD, caliza activa, pedregosidad, espesor...) con influencia sobre el potencial de vigor y precocidad del viñedo y, en definitiva, sobre **la calidad del vino**

Bibliografía

 https://doi.org/10.1051/e3sconf/20185001027

FAO (2009) Guía para la descripción de suelos. FAO, Roma

IUSS Grupo de Trabajo WRB. (2007). Base Referencial Mundial del Recurso Suelo. Primera actualización 2007. Informes sobre Recursos Mundiales de Suelos No. 103. FAO, Roma.

Morlat, R. (2010). Traité de viticulture de terroir. *Editions Tec&Doc*

https://doi.org/10.1051/e3sconf/20185001028

A climate index proposal for the wine sector: a descriptive statistical approach

Aníbal Galindro[1], Ana Alexandra Marta-Costa[2], Adelaide Cerveira[3]; João Matias[4 a]

[1] University of Trás-os-Montes e Alto Douro (UTAD, www.utad.pt) and Centre for Transdisciplinary Development Studies (CETRAD), Portugal, anibalg@utad.pt.

[2] University of Trás-os-Montes e Alto Douro (UTAD, www.utad.pt) and Centre for Transdisciplinary Development Studies (CETRAD), Portugal, amarta@utad.pt

[3] University of Trás-os-Montes e Alto Douro (UTAD, www.utad.pt) and INESC-TEC, Portugal, cerveira@utad.pt

[4] University of Trás-os-Montes e Alto Douro (UTAD, www.utad.pt), and CMAT-UTAD, Portugal, j_matias@utad.pt [a]

[a] Corresponding author

Abstract. Understanding the role of the climate on the wine production is one of the major concerns of this sector since the environment usually determines the output of this industry. There are only a few previous studies that attempted to compile these environmental effects as an index, usually considering the temperature and the precipitation as their core variables. The present study suggests a new climate index which is based on descriptive statistics. Our index tries to mimic the target region characteristics and avoid the past studies premise of imposing previously conceived restrictions such as a fixed optimal climate. We then used yearly production and daily temperature data (1950-2016) from the Portuguese Minho wine region to test our proposed index and compare it with Ribéreau-Gayon and Peynaud (RGP, Ribéreau-Gayon et al., 2003) and Growing Degree-Days (GDD, Winkler et al., 1974) indexes. Our results showed that the newly proposed index may outperform the explanatory power of the other indexes and, in addition, may output interesting and unknown characteristics such as the different ideal temperatures regarding the studied region.

1 Introduction

The climate plays an important role on the wine production since its quality and quantity is strongly dependant on the inherent weather conditions as stated on Malheiro et al. [1] and Salinger et al. [2]. That strong bond made the yearly grape quality become a feature included in the study of the effects of the climate change [3,4]. Even though many individual climate factors make their contribution in the grape development and the following grape quality (*e.g.* solar radiation, wind, humidity, among others), the temperature and the water supply are among the most important [5].

The vines are capable to grow in a wide range of climatic conditions, nonetheless the most remarkable wine-growing regions are located between the 35th and the 50th parallels in the Northern Hemisphere and between the 30th and the 45th parallels in the Southern Hemisphere [6]. During the growing season, the extreme high temperatures can affect the grape quality and the vineyard overall production in many ways, inducing high grape mortality or total failure of flavour ripening [7].

In the opposite way, low and frosting temperatures can also harm the grape flavour and colour due to the slower grape fermentation. In the worst-case scenario extreme cold can also cause the total wipe-out of the vineyard production. Summarily, Gladstones [8] suggests that the grape growing process usually requires mild and narrow ranged temperatures (avoiding the extreme heat or the extreme cold) and actually, according to Gladstones [9] and Jones [10] the most remarkable quality wines are associated with low frost damage and mild winters (January, February, March), early flowering, and warm springs (April, May, June) and optimal maturation associated with short range variability in the summer temperature (July, August, September).

The temperature has such a heavy impact on the grape growing process that the different vine stages can be predicted by simple single based temperature models [11]. The amount of the precipitation that a vineyard receives is also an important feature that determines the following grape quality and output. During the grape maturation period a vineyard should only be exposed to a certain amount of precipitation since its excess may cause major quality issues, including disease susceptibility and rotten grapes. The lack of precipitation in every non-irrigated farm may also cause production problems such as shrinking berries, and of course vineyard mortality and low production levels [12]. Nonetheless, a moderate water deficit may reduce the grape yield but improve the grape quality [13,14]. This work gathers information about the climate impact on the wine production and presents an overview of the current bibliography about the existent wine-oriented climate indexes. Afterwards a new climate index based on descriptive statistical parameters is created and evaluated using a simple case study of the Portuguese Minho wine production. In order to test the proposed index, we compare its explanatory power upon the Ribéreau-Gayon and Peynaud (RGP) [15] and Growing Degree-Days (GDD) [16] indexes.

The article is divided in four sections. Besides the overview of environmental effects on the vineyard production, the next section introduces the notion of climate index, with a few examples from the available bibliography, and presents in more detail two indexes (GDD and RGP indexes). The New Index is described in section three. Finally, section four summarizes the main conclusions and suggestions for further research.

2 Climate Index

The weather conditions are roughly random and regarding the vineyard farms there is not much to do to avoid extremal conditions, apart from irrigated farms that can overcome dry and low precipitation periods. The effects of the weather conditions, particularly the temperature and the precipitation have been the subject of many studies [8,17-19] due to their strong link with good grape quality and the following premium wines. With such randomness brought into the wine sector and overall agricultural production there's also a substantial research on risk protection using financial assets such as weather derivatives as stated by Zara [20] or Leggio [21]. Those studies also consider weather indexes to determine the asset price and the possible hedging strategies.

The great majority of the bioclimatic indexes that are typically used to evaluate the wine production calculate the accumulated heat over the growing season (often since October 1st until April 30th). This growing season index originally stated on Winkler et al. [16] is usually called Growing Degree Days (GDD) or Winkler Index (WI) and it is used to describe a general climate and determine if it is suitable for wine production. Another remark about the WI is that the author defines five possible temperature outcomes (1-very cold until 5-very hot).

The Huglin Index (HI), in Huglin [22], follows the same premise of WI, although, it differs in some features since it uses simultaneously the mean and maximum temperatures in the daytime temperature estimation. It is also calculated using a shorter 6-month period rather than seven. Furthermore, the HI incorporates a "length of the day" coefficient into their calculations which is self-explanatory.

The research by Stock et al. [23] showed that it is expected an overall northern hemisphere increase of 100 to 600 units in the HI until 2050. That fact may indicate a latitudinal shift of the grapevine cultivation in Europe, with new areas on the northern territories becoming more suitable to wine production opposed to the southern ones which may become inadequate due to the excessive heat.

Jones et al. [18] demonstrated that the simple average growing season temperatures on WI and HI are associated to quality vintages. The biologically effective degree-day index (BEDD) created by Gladstones [9] is also a heat sum index. One difference between HI and BEDD is the way that the daytime temperature is estimated. The BEDD index incorporates a new feature which is calculated based on the diurnal temperature range. This index is considered overly adjusted if the diurnal temperature range is greater than 13° C, and it is under adjusted when the temperature range is less than 10° C.

Other available indexes compile more than the cumulative temperature such as Branas et al. [24] which suggested an index that compiles the daily average temperature multiplied by the daily precipitation in order to estimate the risk of the downy mildew disease appearance. Ribéreau-Gayon et al. [15] created Ribéreau-Gayon and Peynaud bifactorial hydrothermal scale also known as the RGP index (RGP), even though it presents the similar cumulative temperature approach as its predecessors, the RGP also considers the daily precipitation and shifts the index cycle, starting at 1st April until 30th October. Fregoni [25] actually found a meaningful relationship between the RGP index and the Pinot Noir grape production. Similarly, Zara [20] also validated the RGP index utility for their hedging strategy development.

Other authors successfully connected different indexes such as Tonietto and Carbonneau [26]. They combined HI, the Dryness Index (DI) and the Cool Night Index (CI) to classify the climate stations in featured wine regions. Nemani et al. [27] also combined the WI and extreme event frequencies to highlight possible shifts and expansion/contraction of grapevine areas, while Fraga et al. [28] used the natural variables such as minimum/maximum temperature and precipitation as straight-forward model regressors in a logistic model. Even though many features can be considered and included when the climate suitability is evaluated, the temperature grabs the first place as the most relevant and unavoidable (at least on viticulture).

3 A new climate index

In this section we will purpose a new index (NI), but in order to furtherly compare it to the available previous studies, we describe the aforementioned indexes GDD (WI) and RGP. These two indexes were selected since they use the temperature value, which is the core variable of the model of this work.
The specific formulation of GDD and RGP lays on the expression (1) and (2) respectively.

$$GDD = \sum_{i=1}^{n} max\{T_i - 10{,}0\} \quad (1)$$

$$RGP = \sum_{i=1}^{m} max\{T_i - 10{,}0\} - P_i \quad (2)$$

where n is equal to 214, corresponding to the number of days between April 1st and October 31th, and m is equal to 212 or 213, corresponding to the days between 1 October and 30 April, for each year.
Each index i corresponds to a single day, and T_i is the average daily temperature recorded during i^{th} day; P_i is the amount of rainfall in the ith day.

The RGP and GDD have quite similar expressions, in fact the RGP only shifts the analysed timeline and adds up the precipitation variable. One feature that immediately comes upfront is their lethargic nature since they both have a fixed reference temperature (10° C). According to Nemani et al. [27] the ideal temperature for grape growing changes across the whole production process, therefore, settling an average temperature (RGP and GDD) may generalize the calculations too much.
The main goal of this work is to find a more suitable way to consider the weather effects on the wine production and overcome some flaws pinpointed on the featured two indexes. The NI is suggested based on the descriptive statistics of the data, such as Kurtosis and Skewness which measures the tailedness and asymmetry of the distribution [29].

$$Kurtosis = \frac{\sum_{i=1}^{N}(Y_i - \bar{y})^4/N}{s^4} \quad (3)$$

$$Skewness = \frac{\sum_{i=1}^{N}(Y_i - \bar{y})^3/N}{s^3} \quad (4)$$

For univariate data $Y_1, Y_2, \ldots, Y_N$ the Kurtosis and Skewness values are given by equations (3) and (4) respectively, where s represents the standard deviation of the data, while $\bar{y}$ represents its mean value. Apart from the newly introduced parameters we also adopt a different timeline suggested by Gladstones [9] and Jones [10] in order to better define the vineyard lifecycle. Therefore, we consider three different periods of three months each (equation 5).

$$NI = \sum_{j=1}^{3} \left| \frac{\sum_{i=1}^{N_j} \frac{(Y_{ij} - \bar{y}_j)^4}{N_j}}{s_j^4} - k_j \right| + \left| \frac{\sum_{i=1}^{N_j} \frac{(Y_{ij} - \bar{y}_j)^3}{N_j}}{s_j^3} - \gamma_j \right| + |\bar{y}_j - t_j| \quad (5)$$

Where $j = 1$ corresponds to the first time interval with N_1 days, the period between the 1st of January and 30th of March; $j = 2$ corresponds to the second time interval with N_2 days, the period between the 1st of April and 30th of June and, finally, $j = 3$ with N_3 days, corresponds to the period between the1st of July and 30th of September. The parameter $\bar{y}_j$ is the average temperature for each period j and the parameters k_j, γ_j, t_j are, respectively, the optimal value of kurtosis, skewness and mean temperature for the correspondent period, which will be explained in the following.

The suggested index tries to mimic the descriptive statistical values of a target distribution. Therefore, to calculate these "optimal" values we fitted a polynomial single regression (equation 6).

$$Z = \beta_0 + \beta_1 X + \beta_2 X^2 + e \quad (6)$$

The dependant variable Z corresponds to the production values (in litres) and the single explanatory variable X is the kurtosis, skewness, or the mean temperature values for each period j ($j = 1,2,3$).
Considering 67 values for variable Z of Minho's green wine production from 1950 until 2016, collected from CVRVV [30,31], nine different polynomial single regression were performed to obtain the three optimal parameters for each period and each variable.
We gathered the daily averaged temperature data and precipitation values from the E-OBS observational interpolated/gridded dataset, version 15 [32]. Despite a few limitations explained at Hofstra et al. [33], this dataset provides uninterrupted and homogeneous gridded fields of daily average temperatures and daily cumulative precipitation over Europe from 01-01-1950 until 31-12-2016, on an approximate 27 km spatial

resolution. The climatic data is extracted from a 16 square grid-box (41.12° to 42.12° latitude and 7.87° to 8.87° longitude) that covers the Minho territory. Considering those 16 values we calculate their total average which in the end samples a good approximation to the Minho's overall weather conditions.

First, we test our available data in the previously mentioned GDD (WI) and RGP indexes.

There are some interpretation difficulties associated with the polynomial regression. These problems can be surpassed performing algebraic manipulations, according to Stimson [34], and equation (6) can be reparametrized as equation (7):

$$Z = M + \beta_2(F - X)^2 \quad (7)$$

where M is the minimum/maximum value of Z (equation 8) while F is the minimizing (for convex, $\beta_2 > 0$) or the maximizing (for concave, $\beta_2 < 0$) value of Z (equation 9).

$$M = \beta_0 - \frac{\beta_1^2}{4\beta_2} + e \quad (8)$$

$$F = \frac{-\beta_1}{2\beta_2} \quad (9)$$

Table 1 displays the nine Polynomial regression results and the target F calculated as Stimson [34]. Each regression was performed individually settling the dependant variable as the yearly production upon of kurtosis k_j, skewness γ_j and mean temperature t_j for each period $j = 1,2,3$ corresponding to each previously mentioned trimester. The kurtosis values that maximize the output tend to be higher than three (heavier tails than the normal distribution) while skewness values should linger around zero (similar to the normal distribution).

Table 1. Polynomial regression results of Production (dependent-variable) upon kurtosis, skewness and mean temperature for each period

Variable X	β_1	β_2	F
k_1	76012408*	-5590239	6.7987
k_2	92450299*	-10763386*	4.2947
k_3	75220789*	-5819973*	6.4623
γ_1	-102728129*	432115254*	0.1189
γ_2	135225197*	165552672*	-0.4084
γ_3	-68206445	570990174*	0.0597
t_1	65840653*	-5500094*	5.9854
t_2	54500304*	-2988560*	9.1182
t_3	47406145*	-2033064*	11.6588

* Statistically significant results for a 5% level of significance.
Source: Own elaboration with CVRVV [30,31] data.

There is not much to say about t_1, t_2, t_3 since they represent the target mean temperature values, although the increasing temperature $(t_3 > t_2 > t_1)$ displays the typical differences between Winter, Spring and Summer respectively. That is actually another concern, since the negative regression coefficient (β_2) illustrate that F accounts the value where the production is maximized. On the other hand, that doesn't apply to the skewness values where $\beta_2 > 0$, since F retrieves a minimum for γ_1, γ_2 and γ_3. Given that, we need to accommodate equation (5) for each situation and also ensure that the three variables contribute equally to the final result. To overcome that problem, we normalized each variable within [0,1] settling 1 as the major penalty to the overall result (Table 3) while 0 represents the optimal target. Since the parameter F absorbs different meanings we need to settle a set of rules for each F representation (while x denotes a single observation), presented in Table 2.

Table 2. Different boundaries per each polynomial regression result

	$x_{i.} = F - \omega$	$x_{i.} = F$	$x_{i.} = F + \omega$
F is a Maximum (A)	1	0	1
F is a Minimum (B)	0	1	0

Where:

$$\omega = \max\{|F_i - a_i|, |F_i - b_i|\}, \quad (10)$$
$$X_i = \{x_{i1}, \ldots, x_{in}\}, a_i = \min\{X_i\}, b_i = \max\{X_i\}$$

After gathering information about the F representation, we settle the normalization process accordingly (Table 3).

Table 3. Normalization scenarios per each F formulation

	Normalization Formula
Scenario A	$x'_{in} = \frac{\lvert x_{in} - F \rvert}{\omega}$
Scenario B	$x'_{in} = 1 - \frac{\lvert x_{in} - F \rvert}{\omega}$

The equation (5) only gives us the general sum and dispersion guideline when F represents a maximum. But technically each F can be a maximum (A) or a minimum (B) suggested value to place on equation (5), therefore the input calculation varies accordingly. Substituting each optimal $k_j; \gamma_j; t_j$ and the temperature values Y_{j_i} on equation (5) we gather a set of raw values that don't acquaint their F as a maximum/minimum value. By labelling them properly according to Table 2 and applying the normalization on Table 3 the final index data series (67 observations, 1950-2016) is obtained.

To test our index, we performed an Ordinary Least Squares such as Fregoni [25] and Cossu et al. [35] opposed to other indexes already presented during this work (RGP, GDD). Equation (11) relates each index value (*D*) with the production (*Z*).

$$Z = \beta_0 + \beta_1 D + e \qquad (11)$$

The Table 4 displays the OLS regression output coefficients for the three different indexes considering the Minho region production values as the dependent variable.

Table 4. OLS regression output for three different indexes

Index	β_0	β_1	R^2
RGP	1.421^8*	-1.651^3*	0.07545
GDD	482586786*	-231452*	0.251
NI	569342373*	-66472298*	0.3036

* Statistically significant results for a 5% level of significance.

The first remark that we've notice is the substantially low *R*-Squared values, that fact may be explained by the omission of many other explanatory variables that actually also affect the production output, such as the harvested land, the economic policies and market environment. Nonetheless, our suggested model manages to achieve a higher *R*-Squared upon the other two indexes. All three indexes reveal a negative coefficient β_1 indicating that increments on each index value may result in a decrease on the output variable (*Z*) – production as expected (since higher index values logically represent worse weather conditions). The straightforward interpretation of our suggested index (NI) is that the more the distributional weather values drift away from the settled as optimal, the smaller the production values should be. Other advantage of the NI against the other common indexes is that it doesn't settle arbitrary bounds since it is constructed to evaluate a specific production environment. Therefore, the optimal values $k_j; y_j; t_j$ should vary upon different regions and adapt to different environment in order to become a versatile and suitable index. A well calibrated index may play an important role in production hedging calculations resembling Zara [20] or efficiency related works where the climate is quite responsible for apparent outliers and general data noise. Nonetheless, we present a quite complex formulation reinforced by the necessary set of rules upon the *F* values.

4 Conclusions

This empirical work tries to understand the role of the climate effects upon the wine industry production. After considering the available bibliography regarding the environment effect measures (climate indexes) we noticed that even though each existing index underpins and justifies its features in solid theoretical background, all of them used pre-formulated values that generalizes the index application apart from the production geographic position. Therefore, our main goal was to provide an index that can be adjusted individually per each wine production that we are actually studying. Our suggested index assembles descriptive statistics features such as the temperature mean, the kurtosis and skewness with optimal values calculated according to Stimson [34] polynomial regression approach. Regarding the Minho region and our model formulations, generally heavy-tailed and non-skewed (roughly normal) temperature distributions during the three period subsets are suggested in order to maximize target region output. In order to evaluate our index performance, we made an OLS regression such as Fregoni [25], Zara [20] and Cossu [35] to study if the suggested index actually does have better explanatory capabilities than the others. Using Minho's region wine production dataset our proposed index reveals himself statistically significant and also outperforms GDD and RGP, considering explanatory power parameters. For further development it is suggested to add more environmental parameters such as the precipitation to the model in order to increase his accuracy, the NI also computes the observation dispersion linearly on the objective function (OB), so, it might be interesting to test non-linear OB formulations. All inputs (temperature, kurtosis and skewness) were weighted equally, changing the formulation weights may also display important and relevant information since we might overvaluing/undervaluing the parameters. Such study may zoom even further geographic characteristics and find the core parameter for each specific farm. It would be also interesting to see a further NI application on vineyard hedging problems or efficiency-based studies.

References

1. A.C. Malheiro, J.A. Santos, H. Fraga, J.G. Pinto, Clim. Res. **43**,3, 163–177.(2010)
2. M.J. Salinger, M. Baldi, D. Grifoni, Int J Biometeorol. **59**,1799 (2015)
3. G.V. Jones, L.B. Webb, Journal of Wine Research **21**, 103-106 (2010)

4. B. Seguin, *Coup de chaud sur l'agriculture*. Editions Delachaux et Niestlé, Paris. (2010)
5. B.G. Coombe, Acta Horticulturae, **206**, 23-35 (1987)
6. C. Leeuwen, P. Darriet, Journal of Wine Economics, **11**,1,150-167 (2016)
7. M.G. Mullins, A. Bouquet, L.E. Williams, *Biology of the Grapevine*, Cambridge University Press, Great Britain, 239 (1992)
8. J. Gladstones, *Wine, Terroir and Climate Change* . Kent Town, South Australia: Wake field Press (2011)
9. J. Gladstones, *Viticulture and environment* Winetitles: Adelaide (1992)
10. G.V. Jones, Arbor Phaenol **42**, 3-7 (1999)
11. A. Parker, C.A.I. Garcia, C. van Leeuwen, I. Chuine, Australian Journal of Grape and Wine Research, **17**(2), 206-216 (2011)
12. T. Popovic, S. Mijovic, R. Pajovic, Agriculture & Forestry, **59**, 2, 137-145 (2013)
13 B. Basile, J. Marsal, M. Mata, X. Vallverdu, J. Bellvert, J. Girona, American Journal of Enology and Viticulture, **62**, 452-461 (2011)
14 D.S. Intrigliolo, D. Perez, D. Risco, A. Yeves, J.R. Castel, Irrigation Science 30, 339-349 (2012)
15 P. Ribéreau-Gayon, *Trattato di enologia*, Ed agricole, Milan (2003)
16. A.J. Winkler, W. Cook, W. Kliewer, L. Lider. *General Viticulture*, University of California Press, Berkeley (1974)
17 G.V. Jones, Acta Horticulturae (ISHS), **689**, 41-60 (2005a)
18 G.V. Jones, M.A. White, O.R. Cooper, K. Storchmann, **73**, 319-343 (2005b)
19 G.V. Jones, E. Duchene, D. Tomasi, J. Yuste, O. Braslavksa, H. Schultz, C. Martinez, S. Boso, F. Langellier, C. Perruchot, G. Guimberteau, GESCO 2005 (2005)
20 C. Zara, International Journal of Wine Business Research, **22**, 3, 222-237, (2010)
21. K.B. Leggio, Managerial Finance, **33**, 4, 246-252 (2007)
22 .P. Huglin, C R Acad. Agric. France, 1117-1126, (1978)
23. M. Stock, F.W. Gerstengarbe, T. Kartshall, P.C. Werner, Acta Hort, **689**, 29-39 (2005)
24. J. Branas, G. Bernon, L. Levadoux, *Éléments de viticulture générale*. Delmas, Montpellier, France (1946)
25. M. Fregoni, *Viticulture of Quality*, Editore Tecniche Nuove, Milan (2005)
26. J. Tonietto, A. Carbonneau, Agricultural and Forest Meteorology, **124**, 81-97 (2004)
27. R.R. Nemani, M.A. White, D.R. Cayan, G.V. Jones, S.W. Running, J.C. Coughlan, Climate Research, **19**(1), 25-34 (2001)
28. H. Fraga, A.C. Malheiro, J. Moutinho-Pereira, J.A. Santos, Agric. For. Meteorol. **185**, 26-36 (2014)
29. D.J. Sheskin, *Handbook of parametric and nonparametric statistical procedures*. Chapman & Hall /CRC (2011)
30. CVRVV, Produção Vinho Verde 1934-1998. Availa-ble at http://portal.vinhoverde.pt/pt/estatisticas#! (accessed at 15/08/2017) (2017)
31. CVRVV, Produção Vinho Verde e Vinho Regional Minho 1999-2016. Available at http://portal.vinhoverde.pt/pt/estatisticas#! (accessed 15/08/2017) (2017)
32. M.R. Haylock, N. Hofstra, A.M.G.T. Klein, E.J. Klok, P.D. Jones, M. New, J. Geophys. Res. **113** (2008)
33. N. Hofstra, M. Haylock, M. New, P.D. Jones, J. Geophys. Res. **114** (D21) (2009)
34. J.A. Stimson, E.G. Carmines, R.A. Zeller R.A. Sociological Methods and Research, **6**, 515-523, (1978)
35. A. Cossu, A. Bodini, E. Fronteddu, LInformatore Agrario, **21** (2004)

Acknowledgements
This work was supported by the R&D Project INNOVINE & WINE – Vineyard and Wine Innovation Platform - Operation NORTE -01-0145-FEDER-000038, co-funded by the European and Structural Investment Funds (FEDER) and by Norte 2020 (Programa Operacional Regional do Norte 2014/2020).

Impactos ambientales y medidas de mitigación en el sector vitivinícola español

Environmental impacts and mitigation measures in the Spanish wine sector

Pilar Gargallo[1] y Nieves García-Casarejos[1]

[1]Facultad de Economía y Empresa, 50008 Universidad de Zaragoza, España

Resumen. Existe evidencia de que gran parte del cambio climático es causado por las emisiones de gases de efecto invernadero (GHG), siendo la vitivinicultura una de las actividades responsables. El uso del combustible en la maquinaria agrícola, el empleo de fertilizantes y el consumo de energía en las bodegas, entre otros motivos, contribuyen a estas emisiones. A su vez, el sector del vino es uno de los más vulnerables por su estrecha interacción con las condiciones ambientales, y concretamente en España es especialmente sensible por su situación geográfica y sus características socioeconómicas. Por tanto, el cambio climático constituye uno de los retos ambientales más importantes para este sector, y la puesta en marcha de medidas de mitigación y adaptación se plantea como una posible solución para cumplir con los objetivos de 2020 y favorecer la transición hacia una economía baja en carbono. El objetivo de este trabajo es ofrecer un panorama de los principales impactos ambientales provocados por la vitivinicultura y presentar las medidas de mitigación más comunes que permiten gestionar dichos impactos. Este trabajo pretende servir de guía de buenas prácticas para todas aquellas bodegas que han tomado conciencia de la importancia de llevar a cabo acciones concretas para aliviar el cambio climático.

1 Introducción

En la actualidad existe un consenso científico, casi generalizado, en torno a la idea de que nuestro modo de producción y consumo energético está generando una alteración climática global, que provocará, a su vez, serios impactos tanto sobre la tierra como sobre los sistemas socioeconómicos. El impacto de este cambio climático es enorme, con predicciones de falta de agua potable, grandes cambios en las condiciones para la producción de alimentos y un aumento en los índices de mortalidad debido a inundaciones, tormentas, sequías y olas de calor. En definitiva, el cambio climático no es un fenómeno sólo ambiental sino de profundas consecuencias económicas y sociales [1-2].

El clima y su variabilidad inciden en muchos ámbitos, pero el sector agrario, en general, y el sector vitivinícola, en particular, presentan una mayor vulnerabilidad ante el cambio climático [3]. El vino necesita de unas determinadas condiciones ambientales, por lo tanto, no es difícil deducir que un aumento de la temperatura o disminución de las lluvias puede traer consecuencias para los viñedos. Temperaturas más frías que las normales dan lugar a una maduración incompleta con alto contenido en ácido, bajo contenido de azúcar y sabores inmaduros; mientras que temperaturas más cálidas de lo normal crean fruta demasiado madura con bajo contenido de ácido y alto contenido de azúcar y de alcohol [4]. La capacidad de adaptación de la vid a estos cambios dependerá de muchos factores, unos intrínsecos, como la variedad de la uva, y otros extrínsecos como las medidas preventivas aplicadas durante el proceso de cultivo, así como, los métodos de producción utilizados para la elaboración del vino [5].

Aunque en apariencia el cultivo de la vid parece un proceso inofensivo desde el punto de vista ambiental, la realidad es bien distinta. De ahí que el sector vitivinícola deba hacer cambios en sus procesos, tanto de cultivo como de elaboración del vino para reducir sus emisiones de gases de efecto invernadero y contribuir de esta forma a la mejora del medio ambiente. Ante la seria amenaza del cambio climático, los productores vitivinícolas y los gestores de las bodegas deben a corto plazo considerar una gran variedad de medidas de adaptación y mitigación del cambio climático para preservar la calidad, identidad y rentabilidad de los vinos [6].

Es necesario ensamblar la cultura del vino con el respeto por la naturaleza, combinando con acierto el binomio experiencia y modernidad. En la elaboración de un buen vino deben ir unidas tradición e innovación. Por una parte, la tradición conlleva utilizar métodos de cultivo que permitan que la naturaleza siga su curso y respete el ecosistema y, en este sentido, el cultivo orgánico, donde se trabaja con elementos naturales, puede ser una de las soluciones. Por otra parte, la innovación tecnológica de la mano de las TIC, ofrece la posibilidad de una vitivinicultura climatológicamente inteligente mediante la construcción de edificios sostenibles, arquitecturas bioclimáticas, maquinaria y vehículos más eficientes energéticamente, así como tecnologías de ahorro de agua y de gestión de los suelos.

El objetivo de este trabajo es ofrecer un panorama de los principales impactos ambientales provocados por la vitivinicultura y presentar las medidas de mitigación más comunes que permiten gestionar dichos impactos. Este trabajo pretende servir de guía de buenas prácticas para todas aquellas bodegas que están tomando conciencia de la importancia de llevar a cabo acciones concretas para aliviar el cambio climático, de modo que, en base a contribuciones individuales, el sector en su conjunto sea más sostenible y se consiga un efecto escalable. Esto no siempre es fácil, debido, por una parte, a la falta de concienciación y disposición para actuar y, por otra, a las barreras existentes, que van desde el desconocimiento de las técnicas a aplicar y de su fiabilidad, hasta los elevados costes económicos que suponen la implementación de algunas de ellas.

El esquema de presentación del trabajo se organiza del siguiente modo. En la sección 2 se describen los efectos ambientales causados por los dos principales procesos de la elaboración del vino (cultivo y producción), analizando que subprocesos dentro de ellos contribuyen en mayor medida a estos efectos. En la sección 3 se presentan medidas de mitigación específicas para los efectos ambientales identificados en la sección 2. La sección 4 finaliza resumiendo las conclusiones más relevantes y proporcionando algunas recomendaciones.

2 Principales impactos en el proceso de elaboración del vino

El proceso de elaboración del vino, es por muchos considerado como un proceso más cercano al arte que a una técnica, ya que a pesar de los siglos de conocimientos adquiridos por el hombre, son elementos tan sutiles como el tipo de tierra o la combinación de cosechas lo que define las características que lo convertirán en único.

Aunque la imagen de un viñedo sugiere paisajes verdes idílicos [7], la mayoría de las técnicas, herramientas y equipos que se utilizan en su cultivo, son fuentes de gases de efecto invernadero que aceleran el calentamiento global [8-9]. Del mismo modo, tras bodegas que se sitúan en parajes con encanto y edificios emblemáticos, se acometen procesos de elaboración del vino no siempre respetuosos con el medio ambiente.

En esta sección se van a detallar los principales impactos medioambientales en un proceso genérico de elaboración del vino, teniendo en cuenta las dos principales etapas: el cultivo y la producción.

2.1 Impactos en el cultivo de la vid

En España, el cultivo de la vid comienza en enero con la poda y el tratamiento de sus heridas. En febrero se aplican los abonados para obtener un terreno rico en nutrientes, y el mes de abril es muy importante para los tratamientos preventivos de control de plagas y enfermedades. A partir de junio se monitoriza la humedad del suelo y se organizan los riegos de acuerdo con la estrategia hídrica seleccionada; además, se reduce la producción a través de la eliminación de racimos, si es necesario. En julio y agosto se inicia el control de la maduración de las uvas, de su pH, de su acidez, etc. Se estima la producción y se comunican los resultados a la bodega para organizar la disponibilidad de personal, vendimiadoras, remolques y transporte. El mes de septiembre es el mes de la vendimia tradicional y los meses de octubre y noviembre se destinan a la vendimia de las variedades de ciclo medio tardío y de ciclo largo, respectivamente. Por último, el mes de diciembre se destina a la cloración de las tuberías de riego para evitar la formación de algas, al control de malas hierbas y de la humedad del suelo. Es el momento de tomar las decisiones de densidad de plantación, para lo cual se arrancan algunas plantaciones antiguas y se realizan nuevas plantaciones.

Aunque el cultivo de la vid parece un proceso inocuo medioambientalmente, esto no es así; los efectos negativos se producen por el contacto con el suelo, el agua y el aire.

El cambio climático está ocasionando una extrema variabilidad de las lluvias, así como sequías estacionales, que provocan la aridez y la erosión de los suelos y, más concretamente, la desertificación del terrero. El viñedo se ha cultivado tradicionalmente en zonas de secano en pendiente y mediante la técnica del laboreo, lo que acrecienta los problemas de erosión del suelo.

El uso de fertilizantes y pesticidas contribuye de forma importante a las emisiones de los gases efecto invernadero. Los fertilizantes químicos pueden causar la infertilidad del suelo a través de su saturación, anulando la eficacia de otros nutrientes vitales. Además, su aplicación excesiva puede hacer que las plantas tengan unas extremidades más grandes y más gruesas que pueden afectar el follaje, por tanto, provocan un considerable aumento de peso, que puede poner presión sobre las raíces de la planta. Por último, y no menos grave, los compuestos químicos de los fertilizantes pueden contaminar los suministros de agua potable y alterar los ecosistemas. Por su parte, los pesticidas contribuyen a la eco toxicidad y muchos de ellos son perdurables en el tiempo, acumulándose en los lugares donde se aplican.

Los residuos producidos por pesticidas y fertilizantes son finalmente arrastrados, drenados o depositados en afluentes de agua o napas subterráneas, causando el incremento anormal de algas y disminuyendo el oxígeno necesario para la supervivencia de las especies.

El agua se utiliza, principalmente, para regar los viñedos. El mayor problema dentro de este proceso no es el hecho de utilizar el agua, ya que las plantas deben ser regadas para su crecimiento y el realizarlo de forma artificial es la única alternativa en muchos casos, sino que el problema real es que el riego se realiza en una cantidad muy superior al requerimiento hídrico de la planta, lo que provoca un derroche de agua.

El uso de combustibles fósiles en la maquinaria necesaria para el cultivo, tales como podadoras, tractores y máquinas, contribuyen al agotamiento de combustibles fósiles y a la toxicidad humana, debido a las emisiones causadas por la quema de combustibles, que son nitrógeno incombustible, dióxido de carbono (CO_2), monóxido de carbono (CO), óxidos sulfúricos (SOx), óxidos nitrosos (NOx), hidrocarburos volátiles (HC) y partículas sólidas (C+).

Todo lo anterior pone en evidencia que en el proceso vitícola, se generan impactos negativos como consecuencia del uso de fertilizantes y pesticidas, del despilfarro de agua, así como de la quema de combustibles fósiles y, por tanto, sobre estos *hotspots* será necesario actuar si se desea mitigar el cambio climático que surge como consecuencia del proceso de cultivo de la vid.

2.2 Impactos en la producción del vino

Tras la vendimia o recogida de la uva, comienza el proceso de elaboración del vino en la bodega. En primer lugar, a través del proceso de despalillado se separan las uvas del resto del racimo. Mediante el estrujado se rompe el hollejo y se extrae el mosto del interior de la fruta. A continuación, el proceso de maceración propicia que el mosto adquiera su color. El sangrado y desfangado permite separar el mosto y dejarlo reposar para que los restos sólidos decanten hacia el fondo del recipiente. A través del descube se transfiere el líquido a otro depósito. Comienza el proceso de fermentación, en el que el azúcar de las uvas termina transformándose en alcohol etílico. Tras el descube, el producto sólido de la fermentación aún contiene grandes cantidades de vino, por lo que es sometido a un prensado para extraer todo el líquido, obteniéndose el vino de prensa, rico en aromas y taninos, que no se mezcla con el obtenido en el descube. Estos dos vinos se utilizarán para la elaboración de productos diferentes. Los restos sólidos que resultan del prensado se emplean como subproducto en la elaboración de orujos y otros productos. El vino obtenido durante los pasos anteriores es sometido a un nuevo proceso de fermentación. Este proceso rebaja el carácter ácido del vino y lo hace mucho más agradable para su consumo. Después de las dos fermentaciones, comienza el proceso de envejecimiento o de crianza, en barricas de roble. Durante la estancia en la barrica se producen una serie de procesos físico-químicos, al mismo tiempo que la porosidad de la madera permite además la microoxigenación del vino. En paralelo a la crianza del vino en barrica se realizan dos trabajos adicionales el trasiego y el clarificado. Durante este periodo, el vino se cambia varias veces de recipiente, con el fin de ir eliminando los sedimentos sólidos y de airear el vino. Una segunda parte del período de crianza del vino tendrá lugar una vez que este es embotellado. Dependiendo de los tiempos de crianza que se den tanto en barrica como en botella, obtendremos como producto final un vino crianza, reserva o gran reserva. De ser necesario, el vino deberá pasar a procesos opcionales como son la refrigeración, cuando las temperaturas de las bodegas son muy altas o nuevos filtrados cuando se han generado cristales por los azúcares naturales de la bebida.

Son muchos los instrumentos y maquinaria que demandan energía en la producción de vino, la cual es mayoritariamente eléctrica de origen fósil y gasoil. En estos procesos se generan emisiones contaminantes cuyos efectos negativos serían: el agotamiento de combustibles fósiles, ya que estos elementos son no renovables y su utilización en la intensidad actual predice su agotamiento dentro de los próximos cien años, y los problemas de salud ocasionados por la formación del *smog* fotoquímico que produce un microambiente irritante y nocivo ya que bloquea el paso de aire hacia los pulmones de los seres vivos.

Otro *hotspot* del proceso de producción de vino es la utilización intensiva de elementos de limpieza y desinfección. Durante la fermentación, trasiego y clarificado, el vino se cambia varias veces de recipiente, con su consecuente limpieza que se realiza con elementos químicos. Una conservación efectuada en recipientes limpios y desinfectados permite al vino desarrollarse de manera armoniosa a lo largo de su proceso de envejecimiento en botella sin necesidad de tratamientos previos. Sin embargo, este proceso de limpieza y desinfección puede perjudicar al medio ambiente puesto que contiene sustancias químicas contaminantes y potencialmente peligrosas que, cuando son depositadas en la basura o vertidas por los desagües, contribuyen a la degradación ambiental.

Las bodegas consumen agua en casi todas las fases del proceso productivo, fundamentalmente por las operaciones de limpieza requeridas tras cada etapa de elaboración. La naturaleza de la contaminación hídrica es principalmente orgánica, consecuencia de la materia seca del mosto o del vino o bien de microorganismos; aunque también contiene residuos minerales, restos de tierra, polvo o barro procedente de la vendimia; detergentes o agentes de limpieza como ácidos fosfóricos, clorhídrico y carbonato sódico entre otros; desinfectantes como cloro y sus compuestos; grasas y aceites procedentes de la maquinaria y aperos; y residuos de productos fitosanitarios que se encuentran en la uva, muchas veces por una inadecuada utilización de los plaguicidas.

La última etapa en el proceso productivo que se realiza en la bodega es el embotellado y embalaje del producto. El proceso de embalaje considera desde que el vino a granel llega a la planta de embotellado hasta que sale en dirección al destino, en este proceso el vino es embotellado y asignado a los diferentes formatos de venta, cajas o pallets, utilizando insumos que producen variados efectos ambientales.

El proceso de embotellado aporta efectos nocivos al medio ambiente como consecuencia del uso de los diferentes materiales de *packaging* como corchos, botellas, cartón, papel y otros, con el fin de que el producto quede listo para su distribución. A pesar de que la mayoría de los residuos que se utilizan en las bodegas son reciclados, existe un factor de impacto ambiental por el mero hecho de utilizar estos elementos, ya que contribuyen al agotamiento de minerales. Esta etapa impacta sobre el deterioro de la capa de ozono en términos de CO_2 debido principalmente a la contribución que realiza la utilización de botellas de vidrio.

En resumen, este proceso vinícola causa efectos considerables en el agotamiento de recursos y en el cambio climático debido al uso de combustibles fósiles, al uso de elementos de limpieza y desinfección, a la contaminación hídrica y al uso de minerales en la fase de embotellado y al reciclaje en la etapa de embalaje. Además, todos estos *hotspots* impactan negativamente sobre la salud humana y, por tanto, es necesario actuar sobre ellos.

3 Medidas de mitigación del cambio climático

Esta sección busca examinar las iniciativas actuales en mitigación del cambio climático emprendidas por el sector vitivinícola organizadas en función del proceso en el que tienen lugar, es decir, cultivo de la vid y elaboración del vino.

3.1 Medidas en el cultivo de la vid

Una primera medida de mitigación a aplicar tiene que ver con el control de la cantidad de agua utilizada en el riego. Así, algunos viticultores hacen uso de técnicas agrícolas de riego que minimizan los caudales a emplear, como el riego por goteo, monitorizado con sondas de humedad [10]. También se puede optar por la tecnificación, mediante sistemas remotos, que permite monitorizar de forma óptima el viñedo y mejorar la toma decisiones en base a datos e información objetiva.

Otras medidas para disminuir el uso de agua en el riego de los viñedos, tienen relación con la reutilización de agua depurada de la bodega [11] y con el aprovechamiento de aguas pluviales. La principal forma de reutilizar agua es mediante plantas de tratamiento y sistemas de decantación. Esta depuración es posible porque los desechos, en su mayoría, son biodegradables y los elementos tóxicos se encuentran en pequeñas concentraciones.

El suelo es la principal razón de ser de la viticultura, debe ser rico en material orgánico, capaz de retener los iones y los minerales inorgánicos que la planta necesita. El laboreo tradicional permite airear el suelo, movilizar minerales y regular su régimen hídrico, sin embargo, al mullir el suelo, se favorece la erosión y la extracción de horizontes infértiles. Por tanto, una medida alternativa para evitar estos efectos desfavorables del laboreo es la creación de una cubierta vegetal que aporte los nutrientes más complejos a la planta; mediante el sembrado selectivo de semillas de especies silvestres autóctonas con selección espontánea (trebolina, amapolas, leguminosas, especies de raíz corta etc...), es decir, fijadoras de nitrógeno. Este suelo "vivo" con los microbios y los insectos es más resistente a la compactación y está más protegido contra la erosión, además, mantiene su humedad y aumenta su permeabilidad.

Para mitigar los efectos del uso de fertilizantes en la viticultura, se han desarrollado una serie de medidas como, por ejemplo, la sustitución de herbicidas por laboreo mecánico, insecticidas por técnicas de confusión sexual, fungicidas por operaciones en la vegetación, el uso de cobre y azufre, así como abonos orgánicos y no de síntesis. En este sentido, una medida de mitigación más genérica, es la llamada viticultura biodinámica u orgánica que busca la reducción e, incluso, eliminación total del uso de fertilizantes, herbicidas y plaguicidas nocivos para el ambiente con el fin de cuidar el entorno donde se desarrollan y mantener intactas las condiciones naturales de los factores abióticos (agua, suelo, etc). En un viñedo orgánico se intenta modificar mínimamente las condiciones naturales del entorno dónde se encuentra y, para ello, se trabaja sólo con elementos naturales de la flora y fauna, con el fin de combatir plagas, fertilizar y alejar depredadores. Se trata de dejar que la naturaleza siga su curso y respete el ecosistema. No se matan las hierbas y plantas que crecen alrededor de las vides. Con esta técnica se favorece la biodiversidad de plantas y animales en el terreno y se deja que los viñedos crezcan más sanos y fuertes evitando la mayoría de las plagas comunes en los cultivos.

El aumento de la temperatura reduce la temporada del cultivo de la vid, provocando que la uva madure antes, lo que obligaría a vendimiar durante los meses más calurosos del verano. Esta subida de las temperaturas también modifica los niveles de azúcar y acidez de la uva, reduciendo la calidad, provocando que los vinos contengan un mayor grado alcohólico y pH muy elevados. Con el fin de actuar desde el viñedo y de forma natural, para contrarrestar los efectos del cambio climático y conseguir una uva de calidad suficiente con un menor contenido en azúcares, se proponen diferentes guías de actuación, en lo que se conoce como arquitectura del cultivo de la vid. Algunas de estas estrategias son el retraso de la poda para postergar la maduración y forzar la vendimia en otoño, la poda mínima del viñedo o el cambio a cepas más resistentes al calor.

3.2 Medidas en la elaboración del vino

Las bodegas utilizan energía para satisfacer sus necesidades de agua caliente, en su mayoría para lavados de barricas, cubas y líneas de transmisión así como para el consumo eléctrico requerido. Actualmente, la energía utilizada procede principalmente de fuentes fósiles, tales como, GLP, diésel o electricidad. Con el fin de mitigar el cambio se propone sustituir estas energías por otras renovables como los sistemas fotovoltaicos, eólica, biomasa, geotermia, etc.

Otra actuación que se puede llevar a cabo es la reducción del consumo de energía puesto que también impacta positivamente en el cambio climático. Para ello, entre otras, se introducen medidas del tipo: aislamiento de las cubas de acero; aislamiento de paredes; construcción de almacenes subterráneos para aprovechar al máximo la gravedad en la recepción y transporte de la uva; incorporación de sistemas de eficiencia energética en los edificios; racionalización del sistema de renovación de aire favoreciendo la ventilación y su circulación natural; así como el diseño de edificaciones con iluminación natural.

Otro proceso que conlleva un elevado consumo de combustible es el que procede de la maquinaria y de los vehículos utilizados en la elaboración del vino. La implementación de mejoras, tanto en la maquinaria como en los vehículos, podría disminuir las emisiones de CO_2. Así, por ejemplo, podría utilizarse hidrógeno, procedente de la hidrólisis de agua a partir de energías renovables, como combustible en la maquinaria o sustituir los vehículos diésel actuales por híbridos.

Los productos químicos utilizados para la limpieza de cubas, barricas, líneas de transmisión, etc. pueden ser bastante nocivos tanto para el medio ambiente como para los trabajadores. La alternativa supondría la realización de la limpieza de barricas y depósitos con vapor de agua o agua caliente a presión en trenes de limpieza automáticos y/o la utilización de ozono para su desinfección.

La última etapa del proceso productivo que se realiza en la bodega, el embotellado y embalaje del producto, también precisa de la implantación de medidas de mitigación. Entre ellas destaca la innovación en envases más sostenibles que cumplan las '3 R': reducir, reutilizar y reciclar. La creación de botellas con un peso más reducido conlleva una disminución de los recursos necesarios para su producción ya que requieren menos energía en su fabricación y también en el transporte, suponiendo una reducción de las emisiones de CO_2.

Otra forma de contribuir al cuidado del medio ambiente es utilizando nuevos envases como el bag-in-box (envase de cartón con un envase de plástico conteniendo el vino en su interior) que poseen un factor de emisión de CO_2 de un 50% menor que el de una botella de vidrio tradicional.

Además del diseño de la botella, es importante el del tapón. Conviene que sean tapones más cortos, de polímeros mejorados de base vegetal (como caña de azúcar) y que cumplan al 100% con el reciclado. Estudios pioneros en España han demostrado que el tapón de corcho es el único cierre sostenible del mercado que protege el entorno natural, influye en la vinificación de los vinos y mejora sus propiedades organolépticas y su sabor en boca.

4 Conclusiones

Ese trabajo ha mostrado en primer lugar, los principales impactos ambientales provocados por la vitivinicultura, teniendo en cuenta las dos principales etapas en la elaboración del vino: el cultivo y la producción, y en segundo lugar, se han presentado algunas de las medidas de mitigación más comunes que permiten gestionar dichos impactos.

Como conclusión y con la finalidad de poder configurar una bodega respetuosa con el medio ambiente, el esfuerzo debería comenzar educando al personal de la bodega en materia de sostenibilidad, para poder favorecer el reciclaje, el uso de energías renovables, controlar los consumos (agua, electricidad, gasóleo,...), reducir la necesidad y carga de iluminación artificial, mejorar los aislamientos térmicos, reducir los pesos de los envases y diseñar nuevos modelos, adquirir maquinaria, instalaciones y vehículo de bajo consumo. En cuanto a las medidas a adoptar en el proceso del cultivo de la vid deberían ponerse en marcha medidas de gestión de riegos, modificación de fechas de las vendimias, cultivo de nuevas variedades de uva o cambios en la arquitectura del cultivo.

Este trabajo pretende servir de guía de buenas prácticas para todas aquellas bodegas que están tomando conciencia de la importancia de llevar a cabo acciones concretas para aliviar el cambio climático. Esto no siempre es fácil, debido, por una parte, a la falta de concienciación y disposición para actuar y, por otra, a las barreras existentes, que van desde el desconocimiento de las técnicas a aplicar y de su fiabilidad, hasta los elevados costes económicos que suponen la implementación de algunas de ellas.

Agradecimientos

Este trabajo ha sido financiado por los proyectos ECO2016-77-P y ECO2016-79392-P (AEI/FEDER, UE).

Referencias

1. A.B. Tate. Global warming's impact on wine. J. Wine Res. 12, 2, 95–109 (2001)
2. M. Venkataramanan. Causes and effects of global warming. Ind. J. Sci. Technol. 4, 3, 226–229 (2011)
3. K.L. Blackmore, I.D. Goodwin, Analysis of Past Trends and Future Projections of Climate Change and their Impacts on the Hunter Valley Wine Industry, Hunter and Central Coast Regional Environmental Management Strategy, Newcastle upon Tyne (2009)
4. J. Santisi. Warming up the wine industry. E: Environ. Mag. 22, 6, 15–17 (2011)
5. H.R. Schultz. Climate change and viticulture: research needs for facing the future. J. Wine Res. 21, 2/3 ,113–116 (2010)
6. I. Bernetti, S. Menghini, N. Marinelli, S. Sacchelli, V. AlampiSottini. Assessment of climate change impact on viticulture: economic evaluations and adaptation strategies analysis for the Tuscan wine sector. Wine Econ. Policy 1, 73–86 (2012)
7. K.L. Christ, R.L. Burritt. Critical environmental concerns in wine production: An integrative review. J. Clean. Prod. 53, 232–242 (2013)
8. A. Chel, G. Kaushik. Renewable energy for sustainable agriculture. Agron. Sustain. Dev. 31, 91–118 (2011)
9. U.A. Schneider, P. Smith. Energy intensities and greenhouse gas emission mitigation in global agriculture. Energy Effic. 2, 195–206 (2009)
10. A. Lereboullet, G. Beltrando, D.K. Bardsley. Assessing Vulnerability and Framing Adaptive Options of two Mediterranean Wine Growing Regions Facing Climate Change: Roussillon (France) and McLaren Vale (Australia). EchoGéo [En ligne], 23, mis en ligne le 05 juin 2013, consulté le 26 mars 2018. http://journals.openedition.org/echogeo/13384 DOI : 10.4000/echogeo.13384 (2013)
11. M. R. Mozell, L. Thach. El impacto del cambio climático en la industria mundial del vino: desafíos y soluciones. Economía del vino y política, 3, 2, 81-89 (2014)

 https://doi.org/10.1051/e3sconf/20185001030

Climate change impacts on thermal growing conditions of Portuguese grapevine varieties

João A. Santos[1], Ricardo Costa[1] and Helder Fraga[1]

[1]Centre for the Research and Technology of Agro-Environmental and Biological Sciences, CITAB, Universidade de Trás-os-Montes e Alto Douro, UTAD, 5000-801 Vila Real, Portugal

Abstract. Heat accumulation conditions of a collection of 44 grapevine cultivars currently grown in Portugal are assessed at very high spatial resolution (~1 km) and for 1981–2015. A Growing Degree Hours – GDH (February–October) index is used for this purpose. Three clusters of grapevine cultivars are identified, assembling varieties with close heat accumulation requirements (early, intermediate and late ripening). These clusters provide more physiologically consistent information when compared to previous studies, as non-linear plant-temperature relationships are herein taken into account. For the future climates in Portugal, ensemble mean projections under two anthropogenic-driven scenarios (RCP4.5 and RCP8.5, 2041–2070), from four EURO-CORDEX simulations, reveal a widespread increase of GDH, but with spatial heterogeneities. The spatial variability throughout Portugal is projected to decrease in GDH, with strongest increases in the coolest regions of the northeast. The typical heat accumulation conditions of each cluster are projected to gradually shift north-eastwards and to higher-elevation areas. An unprecedented level of detail for a large collection of grapevine varieties in Portugal is provided, which may promote a better planning of climate change adaptation measures in Portuguese viticulture.

1 Introduction

Viticulture is strongly subject to environmental factors, as grapevine (*Vitis vinifera* L.) physiological development is significantly influenced by terroirs [1], comprising climate, soils, variety-clone-rootstock combinations, vineyard biodiversity and a wide range of management and cultural practices. Climate is a major terroir element [2, 3], since air temperature largely controls grapevine phenology [4, 5], also influencing yields and quality attributes [6]. Furthermore, precipitation determines soil water balance [7]. Thus, adequate climate conditions are critical for producing balanced table wines. Additionally, climate change drives alterations in vineyard microclimates and, as a result, in regional/local suitability to a specific variety. Therefore, assessing regional climatic suitability for a given grapevine variety, in present and future climates, is critical for a strategic planning of the winemaking sector.

Grapevine thermal requirements play a key role on grapevine varietal distribution [8], as well as on growth, development and yields [9]. Nonetheless, broad differences among grapevine cultivars have also been underlined [4, 10, 11], highlighting the importance of variety-dependent analyses.

Thermal growing conditions can be evaluated by several bioclimatic indices, such as Growing Degree Days, GDD [12] or Growing Degree Hours, GDH [13], though the later provides more accurate assessments by considering an optimum temperature level, as is explained below. Hence, in the present study, heat accumulation conditions of main grapevine cultivars in Portugal are assessed under present and future climates.

2 Data and methods

Gridded daily maximum and minimum 2m air temperatures from the E-OBS dataset [14] were used as input for GDH calculations. The period of 1981–2015 (35 years) was selected. GDH were computed on the original 0.25° latitude × 0.25° longitude grid (~25 km spacing) and were subsequently downscaled to 1 km grid resolution by multivariate linear regression analyses, using latitude, distance to coastline and elevation as exploratory variables. The GTOPO30 digital elevation dataset (https://lta.cr.usgs.gov/GTOPO30) was used for this purpose. Further details about the downscaling methodology can also be found in [15].

For future climates, simulations generated by four Global Climate Model-Regional Climate Model (ESM-RCM) chains were obtained from the EURO-CORDEX project (http://www.euro-cordex.net/, Table 1). These simulations were forced by the representative concentration pathways 4.5 and 8.5 (RCP4.5 and RCP8.5) over the period of 2005–2100 [16]. In RCP4.5, CO2 emissions increase until the mid-21st century and decrease afterwards, whereas in RCP8.5 emissions steadily increase throughout the 21st century [17].

For climate change assessments, an intermediate future period (2041–2070) is compared to a baseline period (1981–2005). For all models, the simulated daily maximum and minimum temperatures were bias-corrected using E-OBS as baseline [16]. All indices from model simulations were computed on bias-corrected daily maximum and minimum temperatures for each period, model and future scenario, separately. Ensemble means were then produced to identify the climate change signal of each index.

Table 1. List of the ensemble members used in the present study (ESM-RCM chains), along with the acronyms used herein and the developer institutions.

ESM / RCM	Acronym	Institutions
CNRM-CERFACS / SMHI-RCA4	CNRM	Centre National de Recherches Météorologiques
IPSL-CM5A-MR / SMHI-RCA4	IPSL	Institut Pierre Simon Laplace
MOHC-HadGEM2 / SMHI-RCA4	MOHC	Met Office Hadley Centre
MPI-M-MPI / SMHI-RCA4	MPI	Max Planck Institute

GDH measures thermal accumulation on an hourly timescale, using two cosine functions between a base temperature (4°C) and an upper limit temperature (36°C), with an intermediate optimum at 26°C [18]. Limited growth under excessively high temperatures is thereby taken into account, which is a critical property either for regions with hot summers or for climate change assessments. Although these thermal thresholds for GDH computations were originally developed for temperate climate fruit trees, previous studies suggest similar values for grapevines, namely for the intermediate optimum and upper limit temperatures [19] or for the base temperature of several Portuguese grapevine cultivars [20].

In Portugal, grapevine budburst typically occurs from the beginning to the end of March, though it may occur already by the end of February, namely in the warmest areas of southern Portugal. Under future warmer climates an anticipation of phenological timings is also expected throughout Portugal and budburst are indeed projected to frequently occur in February [21]. Although several previous studies considered the heat accumulation period for grapevines starting in March, the aforementioned future conditions in Portugal justify the anticipation of this period from March to February. GDH is thereby calculated from 1 February to 31 October.

For the vineyard spatial distribution in Portugal, the fifth level of the COS2007 was used. This dataset provides detailed land use and land cover information over Portugal (Source: Direção-Geral do Território, http://www.dgterritorio.pt/). In addition, a digital atlas of 44 grapevine varieties currently grown in Portugal (Table 2) was used following a previous study [11]. Their spatial distributions were overlapped on the GDH maps to extract the corresponding thermal conditions for each variety.

A Ward's linkage hierarchical clustering, based on squared Euclidean distances between the GDH medians for each of the 44 varieties, was applied to identify clusters of grapevine varieties concerning their current thermal growing conditions in Portugal (1981–2015).

Table 2. List of grape varieties used in this study, their synonyms and respective cluster numbers, for current (1981–2015) and future scenarios (RCP4.5 and RCP8.5, 2041–2070).

Grape varieties	Current	RCP4.5	RCP8.5
Alfrocheiro	3	3	3
Alicante-Branco	3	3	3
Alvarinho	3	3	3
Antão-Vaz	3	3	3
Aragonez	3	3	3
Arinto	3	3	3
Avesso	2	3	3
Azal	2	3	3
Baga	3	3	3
Bastardo	1	2	2
Bical	2	3	3
Borraçal	2	3	3
Castelão	3	3	3
Diagalves	2	3	3
Encruzado	2	3	3
Espadeiro	2	3	3
Fernão-Pires	3	3	3
Gouveio	1	2	2
Jaen	2	3	3
Loureiro	2	3	3
Malvasia-Fina	1	2	2
Malvasia-Preta	1	2	2
Malvasia-Rei	2	2	3
Marufo	1	2	2
Moreto	3	3	3
Moscatel-Galego-Branco	1	2	2
Moscatel-Graúdo	3	3	3
Rabigato	1	2	2
Rabo-de-Ovelha	3	3	3
Rufete	1	2	2
Sercial	2	3	3
Síria	2	2	3
Tália	3	3	3
Tinta-Barroca	1	2	2
Tinta-Carvalha	1	2	2
Tinta-Miúda	3	3	3
Tinto-Cão	2	2	3
Touriga-Franca	1	2	2
Touriga-Nacional	3	3	3
Trajadura	2	3	3
Trincadeira	3	3	3
Vinhão	2	3	3
Viosinho	1	2	2
Vital	3	3	3

3. Results

Vineyards are spread over most of Portugal, though particularly high densities can be found in the Douro and Lisboa Wine Regions. Mainland Portugal is characterized by very complex topography in its northern half, with several mountain ranges, deep valleys and steep slopes, contrasting with southern Portugal, where flat terrain widely prevails. As a result, northern Portugal has a strong spatial climate variability, while relatively homogeneous climatic conditions are found in the south

https://doi.org/10.1051/e3sconf/20185001030

[22]. Portuguese vineyards are thus currently growing under a wide range of thermal conditions.

The pattern of GDH (Fig. 1a) for current climate conditions over Portugal reveal important spatial variability, with a clear signature of both elevation and latitudinal thermal gradient. Highly contrasting conditions between the inner high-elevation areas of northern/central Portugal and the low-elevation areas in the south are particularly noteworthy (Fig. 1b).

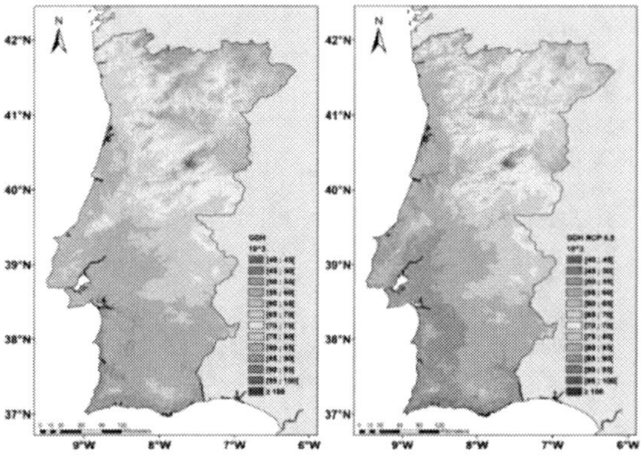

Figure 1. a) Growing degree hours (GDH) for present and b) future period (2041–2070) under RCP8.5.

The values of GDH for each grapevine variety were extracted for their current locations and their corresponding medians were then calculated. The varieties are subsequently ranked as a function of their respective GDH medians (Fig. 2). For current conditions, grapevine varieties are found between ~67,000 and ~83,000 GDH medians. Varieties such as Bastardo, Viosinho, Gouveio, Malvasia-Preta and Marufo are manifestly growing in areas with lower values of GDH (lower heat accumulation). Conversely, Tinta-Miúda, Castelão, Moscatel-Graúdo and Antão-Vaz are preferably located in areas with high values of GDH (higher heat accumulation).

In order to systematize the results, the 44 selected grapevine varieties were clustered according to their GDH medians (Fig. 2). Cluster 1 (C1) comprises 12 varieties (Bastardo, Viosinho, Gouveio, Malvasia-Preta, Marufo, Rabigato, Moscatel-Galego-Branco, Tinta-Barroca, Touriga-Franca, Tinta-Carvalha, Malvasia-Fina and Rufete), cluster 2 (C2) 15 varieties (Jean, Malvasia-Rei, Bical, Tinto-Cão, Encruzado, Avesso, Azal, Sercial, Síria, Vinhão, Diagalves, Espadeiro, Trajadura, Borraçal, Loureiro) and cluster 3 (C3) 17 varieties (Alvarinho, Tália, Alfrocheiro, Arinto, Baga, Touriga-Nacional, Rabo-de-Ovelha, Moreto, Aragonez, Vital, Fernão-Pires, Alicante-Branco, Trincadeira, Antão-Vaz, Moscatel-Graúdo, Castelão and Tinta-Miúda). The centroids of each cluster (averages over all cluster elements) are of approximately: 68,000 GDH, 74,000 GDH and 80,000 GDH for C1, C2 and C3, respectively (Fig. 2).

The climate change projections for GDH under RCP8.5 hint at an overall warming trend (Fig. 1b), with the previous range of 40,000–85,000 GDH changing to 60,000–95,000 GDH, apart from a few exceptions in high elevation areas. Nevertheless, there are remarkable spatial heterogeneities in the warming trend, with much stronger values over northern Portugal and even slightly negative changes in the inner-south, which is a remarkable outcome. In fact, the changes in the temperature distributions in future climates are expected to have a more effective effect on plants in northern than in southern Portugal, as temperature distributions will be closer to optimum in the north and clearly beyond optimum in the south (excessively warm conditions). As a result, the contrasts between GDH over northern and southern Portugal will be progressively lowered in future warmer climates (i.e. the lower GDH in the cooler areas of the north will gradually converge to the higher GDH in the warmer areas of the south).

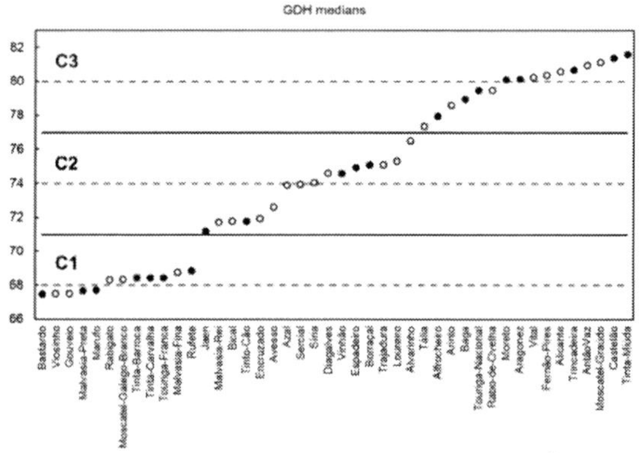

Figure 2. Medians of the growing degree hours (in 10^3 GDH) for the outlined 44 grapevine varieties over the period of 1981–2015. White (red) varieties are plotted as white (black) circles. The varietal clusters C1, C2 and C3 are also shown. Solid black lines correspond to the cluster limits, while dashed grey lines correspond to the centroid values of each cluster.

Lastly, the current and future projected thermal clusters for each variety are shown in Table 2, where shifts in clusters are frequent. For the 44 varieties, there are 24 class shifts (55%) under RCP4.5 and 27 shifts (61%) under RCP8.5. Of these shifts, 12 are from C1 to C2, while the remaining are from C2 to C3 (12 under RCP4.5 and 15 under RCP8.5). It is still worth noting that C1 is totally absent under both future scenarios, while C2 only occurs for 15 or 12 varieties under RCP4.5 and RCP8.5, respectively, thus highlighting the large prevalence of the C3 conditions.

4. Conclusions

The thermal growing conditions of a collection of 44 grapevine cultivars in Portugal are aggregated into three clusters, based on their current heat accumulation conditions (GDH): cluster 1 for varieties growing in the coolest conditions, cluster 2 for intermediate conditions and cluster 3 for the warmest conditions over Portugal. These clusters tend to isolate varieties with similar thermal requirements and provide more detailed information when compared to previous studies based on GDD or not considering a variety-dependent analysis.

Despite the identification of clusters and of their thermal niches, it cannot be strictly stated that grapevine varieties in C1 prefer colder climates, while those in C3 are more adapted to warmer climates. However, many of these varieties are either native or have been empirically selected in each region and are thus already growing in the most adequate climatic conditions to be economically viable and sustainable. Future alterations of the thermal conditions under two anthropogenic-driven scenarios (RCP4.5 and RCP8.5) clearly reflect a general warming over Portugal. As such, the typical conditions associated with these three clusters will undergo a general shift from the south to the northeast and higher-elevation areas.

This variety-dependent information reveals the changes that are expected to occur for each variety under climate change, but considering their current geographical distributions and not taking into account any possible adaptation measures. The successful adaptation of a given wine region to changing climates may be significantly controlled by the resilience of locally-relevant grapevine varieties. These changes may modify grapevine physiological processes, leading to an anticipation of the phenological timings, such as budburst, flowering, veraison and ripeness, or to heat, water and nitrogen stresses, implications on yields and modifications in berry composition, among others [21]. These impacts may affect the wine typicity of a given region and are an important challenge to the winemaking sector. Overall, climate change is a global problem, but adaptation should follow local solutions, as no general solution applies everywhere.

Acknowledgments

This work was supported by: the INNOVINE&WINE project (NORTE-01-0145-FEDER-000038), co-funded by the European Regional Development Fund through NORTE 2020 Programme; the ModelVitiDouro project (PA 53774), funded by the Agricultural and Rural Development Fund (EAFRD) and the Portuguese Government (Measure 4.1 – Cooperation for Innovation PRODER Programme – Rural Development Programme); European Investment Funds (FEDER/COMPETE/POCI), POCI-01-0145-FEDER-006958, and Portuguese Foundation for Science and Technology (FCT), UID/AGR/04033/2013. The postdoctoral fellowship of Helder Fraga, SFRH/BPD/119461/2016, is also acknowledged.

References

1. C. van Leeuwen, G. Seguin, J Wine Research **17**, 1-10 (2006)
2. G. V. Jones, R. E. Davis, Am J Enol Viticult **51**, 249-261(2000)
3. C. van Leeuwen, H. R. Schultz, I. Garcia de Cortazar Atauri, E. Duchêne, N. Ollat, P. Pieri, B. Bois, J. P. Goutouly, H. Quénol, J. M. Touzard, A. C. Malheiro, L. Bavaresco, S. Delrot, *Why climate change will not dramatically decrease viticultural suitability in main wine-producing areas by 2050* (Proc Natl Acad Sci U S A, 2013)
4. F. Orlandi, T. Bonofiglio, F. Aguilera, M. Fornaciari, Vitis **54**, 129-136 (2015)
5. M. C. Ramos, G. V. Jones, J. Yuste, Eur J Agron **70**, 57-70 (2015)
6. E. A. C. Costantini, R. Lorenzetti, G. Malorgio, Land Use Policy **57**, 53-63 (2016)
7. M. Permanhani, J. M. Costa, M. A. F. Conceicao, R. T. de Souza, M. A. S. Vasconcellos, M. M. Chaves, Theor Exp Plant Phys **28**, 85-108 (2016)
8. J. San-Miguel-Ayanz, D. de Rigo, G. Caudullo, T. Houston Durrant, A. Mauri, *European Atlas of Forest Tree Species* (Publication Office of the European Union 2016)
9. H. Fraga, J. A. Santos, Aust J Grape Wine R **23**, 296-304 (2017)
10. G. Fila, M. Gardiman, P. Belvini, F. Meggio, A. Pitacco, Agr Forest Meteorol **195**, 192-205 (2014)
11. H. Fraga, J. A. Santos, A. C. Malheiro, A. A. Oliveira, J. Moutinho-Pereira, G. V. Jones, Int J Clim **36**, 1-12 (2016b)
12. A. J. Winkler, *General viticulture* (University of California Press, USA, 1974)
13. S. Gu, Int J Biometeorol **60**, 1123-1134 (2016)
14. M. R. Haylock, N. Hofstra, A. M. G. K. Tank, E. J. Klok, P. D. Jones, M. New, J Geophys Res-Atmos **113** (2008)
15. J. A. Santos, R. Costa, H. Fraga, Clim Change **140**, 273-286 (2017)
16. S. Kotlarski, K. Keuler, O. B. Christensen, A. Colette, M. Déqué, A. Gobiet, K. Goergen, D. Jacob, D. Lüthi, E. van Meijgaard, G. Nikulin, C. Schär C. Teichmann, R. Vautard, K. Warrach-Sagi, V. Wulfmeyer, Geosci Model Dev **7**, 1297-1333 (2014)
17. IPCC, *Climate Change 2013: The Physical Science Basis. Contribution of Working Group I to the Fifth Assessment Report of the Intergovernmental Panel on Climate Change* (Cambridge University Press, United Kingdom 2013)
18. J. L. Anderson, E. A. Richardson, C. D. Kesner, International Society for Horticultural Science, **71-78** (1986)
19. N. Brisson, M. Launay, B. Mary, N. Beaudoin, *Conceptual Basis, Formalisations and Parameterization of the STICS Crop Model* (Editions Quae, France, 2008)
20. J. Lopes, J. E. Eiras-Dias, F. Abreu, P. Climaco, J. P. Cunha, J. Silvestre, Ciencia Tec Vitiv **23**, 61-71 (2008)
21. H. Fraga, I. Garcia de Cortazar Atauri, A. C. Malheiro, J. A. Santos, Glob Chang Biol **22**, 3774-3788 (2016a)
22. R. Costa, H. Fraga, P.M. Fernandes, J. A. Santos, Reg Environ Change **17**, 117-127 (2017)

Linking viticultural climatic indices to grape phenology in the South Tyrolean Alps

Andrea Balotti[1,a], Simon Tscholl[1] and Lukas Egarter Vigl[1]

[1]Eurac Research, Institute for Alpine Environment, 39100 Bolzano, Italy

Abstract. Climate indices based on heat accumulation, e.g. the Winkler index, are widely used to define the climatic niches for vines. In this study, we investigate how a combined use of high-resolution (25 m) climate index maps and phenology records from 30 vineyards in South Tyrol (in the Italian Alps) can help to *(i)* assess viticultural suitability across a mountainous landscape and *(ii)* estimate the timing of physiological processes of Pinot Noir development (ripening and must weight) at various sites across the region. First, the best interpolation method is chosen (from multiple linear regression (MLR), regression kriging or support vector regression) to create maps of climate indices averaged over the time period 1991-2010. Second, correlation is calculated between the timing of phenological stages of Pinot Noir for the year 2017 and various climate indicators, such as temperature-based indices (Winkler, Huglin, Biologically Active Degree Days, Cool Night, Fregoni) and average (GST), minimum and maximum temperature over the growing season. The MLR method is shown to yield the best interpolations of the climate indices across the complex terrain of the study area. The Winkler and GST indices correlate most precisely with the late-season phenological events of the study sites, and are thus the most predictive. These findings demonstrate the potential of climatic maps to effectively define suitable areas for grape growing and estimate ripening dates in South Tyrol.

1 Introduction

Unlike other famous Italian wine regions, such as Tuscany or Veneto, which are characterized mainly by hilly topography, South Tyrol is characterized by complex topography over a large elevation range. On the one hand, this is an advantage because many different climatic niches are available for grape growing. On the other hand, the amount of available cultivation area in each niche is limited. The latter is especially true in South Tyrol, where lower elevations are widely used for other cultures, like orchards, and higher elevations are often characterized by conditions that are too harsh for viticulture.

Viticultural suitability is strongly influenced by temperature [1], and thus having accurate maps of temperature-related indices is the first step in obtaining realistic climatic zoning of an area.

The complex topography of mountainous landscapes, however, represents a large challenge when spatializing environmental variables. In recent years, different efforts have been made to develop gridded temperature datasets specifically designed for complex topography. However, these maps are usually in spatiotemporal resolutions that are too coarse for effective use in small-scale viticulture studies. Thus, in the present study we decided to perform an in-house interpolation of temperature-based indices to create a set of maps that are suitable for defining climatic zones for this purpose.

High-resolution maps (25 m) of climatic indices are interpolated using different geostatistical approaches (multiple linear regression, regression kriging, support vector regression) and data from the official national meteorological service from 1991-2010. The best interpolation method is selected to create maps of climate indices that define average climate zones in South Tyrol for this period.

Using data from meteorological stations installed in 30 different vineyard plots, correlation is calculated between various temperature-based indices and the timing of phenological stages of Pinot Noir for the year 2017. The goal is to find the best explanatory variables to estimate ripening date at any specific location.

The 30 vineyard stations were installed as part of the transnational EU-funded "REBECKA" project (Interreg V-A IT-AT:ITAT1002, duration: 2017-2019) between South Tyrol and Carinthia (Austria). The goal of the project is to assess the parameters that influence the vines of Pinot Noir in these Alpine regions and how climate change can impact them. One of the aims of the project is to monitor the phenology of grapevines and the chemical characteristics of grapes, and to correlate them with temperature-based indices.

This work presents the first results of the project, focusing on South Tyrol and on the parameters that have the largest influence on viticulture.

 https://doi.org/10.1051/e3sconf/20185001031

2 Material and methods

2.1 Study area

South Tyrol is located in Northern Italy, in the central-eastern Alps. From a topographic point of view, South Tyrol has a large elevation gradient, with a range from 200 to 3900 m, but with most of the territory situated between 1000 and 2500 m. This mountainous area is crossed by three interconnected main valleys that are relatively wide and with different orientation. The vineyards are distributed along these three main valleys from 200 to 1300 m. Due to its position in the middle of the Alps, the climate is influenced by both continental and Mediterranean factors, but in general conditions typical of subcontinental and continental areas can be found.

If we look at the climate in South Tyrol, and more generally in the central-eastern Alps, we find a warming trend in the annual mean temperature from 1980 to 2000, which subsequently flattens out.

Seasonal temperature trends are less consistent: spring and summer have seen increased mean temperatures, while winter and autumn have been much more variable. Additionally, hot days (maximum temperature above 25 °C) are occurring more frequently, while frost days (maximum temperature below 0 °C), on the other hand, show a decreasing trend [3].

Wind fields are complex and driven by general circulation, pressure gradient and slope. Pressure driven winds from the north, called Föhn, have an important role in northwestern areas of South Tyrol, where the driest part of the study area is located.

Precipitation in South Tyrol is driven mainly by mesoscale systems and, during summer, also by convective activity [3]. There are no discernable trends in climatic analysis, yet there exist differences in the spatial distribution of precipitation in South Tyrol.

2.2 Datasets

The values of dependent variables (i.e. climatic parameters) were obtained by processing the data from existing meteorological stations in South Tyrol. We have collected the daily value of minimum and maximum temperatures from 1990 to 2017 from a pool of 82 stations, all below 2000 m. Using the package *ClimClass* [4] in *R environment*, we have calculated various climate indices, for each year and station. The calculated climate indices and their abbreviations are reported in Table 1.

The years between 1991 and 2010 are the reference period over which climate averages are computed. This timespan was selected because it aligns with the majority of the WMO climatological "standard normal" period (1981-2010) [5]. The climate indices are also calculated for 2017 to find a correlation with phenological stages.

Of the pool of 82 stations, three were used as final validation points, whereas the remaining stations were used as training points for the models. The number of training stations with meaningful data each year varies from 44 to 58 (if we consider the period 1991-2010) and gives a sample density of 0.23-0.30 stations per square kilometer.

Table 1. Acronyms of climate indices used.

Climate index name	Abbr.
Winkler index	WI
Huglin index	HI
Biologically Active Degree Days	BEDD
Cool Night Index	CNI
Fregoni's Simplified index	FSI
Growing Season Temperature	GST
Maximum Growing Season Temperature	GSTmax
Minimum Growing Season Temperature	GSTmin

2.3 Interpolation methods

In order to map climate indices across the landscape, data from meteorological stations must be interpolated using the most appropriate method, either multiple linear regression (MLR), regression kriging (RK), or support vector regression (SVR).

The MLR assumes the existence of a relationship between a dependent variable and its (independent) predictors that can be explained using a linear function [6]. Using independent predictors doesn't mean that all of them are statistically significant to describe the dependent variable. For each dependent variable, the statistically significant predictors were assessed using the R package *leaps* and the method for cross-validating all the possible models described by James et al. [7]. The statistic used to select the best predictors is the BIC value. This method retains only the predictors that are statistically significant for the model.

The RK method combines a linear regression and a variogram analysis of the linear regression residuals [8], [9]. The variogram is used to construct a distance-based model for the residuals that are later interpolated with ordinary kriging, and then added to the linear regression interpolation. If the variogram of residuals has no spatial correlation, theoretically, the RK is composed only by the linear part as a normal linear regression. But in practice, it is still possible to model the residuals as a constant value added to the linear regression (nugget model). Our interpolation script is constructed to first try a variogram fit with an exponential model and, in case of failure, to adopt a nugget model. In this way, it is always possible to construct a model for the residual.

The SVR method uses a non-linear regression formula [10]. Since the predictor selection method used in MLR (and consequently in RK) is developed for linear regression, the same method is not applicable for SVR and a specific predictor selection algorithm should be developed. In literature, this topic is still debated (see [11], [12]), so we decided to retain all of the predictors for SVR.

2.4 Predictors

The initial predictors we had selected for the interpolation process were the same used in Le Roux et al [13]: elevation, slope, aspect (North-South and West-East components), longitude, latitude, and potential solar radiation and direct solar radiation duration, both summed over the vegetation period (April 1st – October 31st). All these predictors are computed directly in R except for potential solar radiation and solar radiation duration, which are created within ArcGIS using the *Area Solar Radiation* tool and its default parameters.

Because vineyards in South Tyrol cover a relatively small area (only 0.2 hectares on average), we have decided to use a resolution of 25 m as a good compromise between computational time and accurately assessing spatial variability. Aspect was processed in order to remove circular geometry and avoid problems in computation [13]. In particular, we have split North-South and West-East components, computing sine and cosine of the aspect angle respectively. Longitude and latitude are UTM coordinates expressed in meters. Potential solar radiation and direct solar radiation duration over vegetation period, were computed first at 50 m of resolution and then resampled to 25 m using bilinear interpolation in order to speed up computation.

Before using these predictors in the interpolation process, a correlation analysis is necessary to detect signs of multicollinearity. In general, the predictors don't show signs of strong correlation that could influence the regression, but potential solar radiation shows a moderate correlation with elevation, irradiance duration and slope with values of 0.63, 0.44 and 0.43 respectively. Moreover, slope has a correlation of 0.44 with elevation. We have therefore decided to apply a principal components analysis on predictors in order to reduce the multicollinearity effects as suggested in Hengl et al [8]. As expected, the new predictors show a negligible correlation and are thus used in the interpolation process.

Climate conditions above a certain elevation become too cold for effective viticulture. We therefore decided to restrict all subsequent computations to elevations below 2000 meters.

2.5 Phenology data

In 2017, in the scope of the "REBECKA" project, we installed 30 meteorological stations: 27 in vineyards with Pinot Noir vines and 3 near meteorological stations of the South Tyrolean provincial network to assess the measurement quality of our stations. Our meteorological stations are equipped with two temperature sensors: one for the standard 2-m air temperature measurement, and one positioned at 0.9 m, at the base of the canopy. In addition, there are solar radiation sensors, and soil temperature and soil water potential sensors installed at 40 cm below the soil surface.

Due to time constraints, it was not possible to start the measurements at the beginning of the vegetation period (April 1st), so a gap filling procedure has been adopted to reconstruct minimum and maximum 2-m air temperature until that date. To do this, a linear regression model is trained between the data of each "REBECKA" station and the data of the nearest provincial station. The training process uses the data from the whole period where the "REBECKA" station was recording. The model is then used to estimate the missing data of our stations from the measured temperature of the respective nearest provincial station. The performance of the model is validated via a 5-fold cross-validation, which is repeated 3 times. This process proves quite effective for most stations. The highest error rate is around 1.7 °C, average mean absolute error is 0.86, average r^2 is 0.97 and 0.98 is the average Pearson's correlation. The result of this gap-filling procedure is a table with daily measurements of maximum, minimum and average temperature starting from April 1st for every "REBECKA" station.

For each site with a meteorological station, surveys to detect the day-of-year (DOY) of phenological stages are performed. In particular, DOY of budburst, flowering and ripening (respectively the level 8, 65, 81 on the BBCH scale [14]), as well as DOY of reaching 16, 17 and 18 °KMW for must weight are assessed.

3 Data analysis

3.1 Interpolation results

To compare the performance of interpolation methods, we have computed RMSE and MAE statistics, as well their difference, averaged over the years 1991-2010 (Figure 1). Since the indices have different units of measurement, statistics are normalized using the range of observed values used to compute the statistics. The results show that, looking at the model training, the SVR is the best method for temperature-sum indices (WI, HI, BEDD) and RK is best for pure temperature indices (CNI, GST). But if we look at the statistics of validation points, MLR and RK show the best performance, as reported in Table 2.

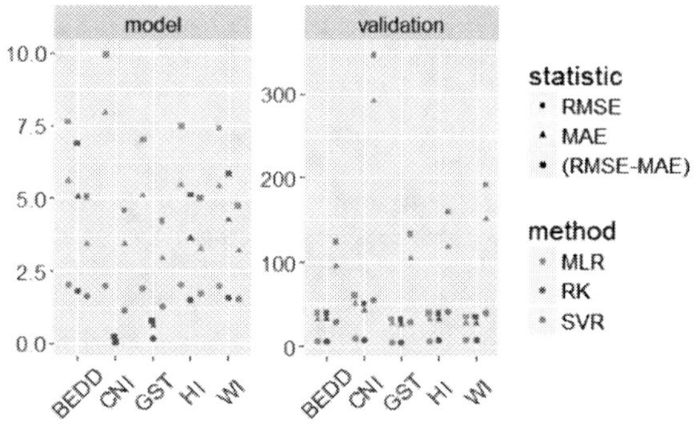

Figure 1. Average performance of models over training and validation points. Statistics are normalized to permit comparison between indices.

Pearson's correlation is computed and values are reported in Table 2. However, three of the twenty years analyzed have observed values for only two validation points instead of the three we have chosen. In these cases, Pearson's correlation is equal to one and we discarded these points from the average correlation value computation. Using only three points for validation may underestimate the true quality of the interpolations, but this choice was required due to the low density of available station data.

Table 2. Normalized RMSE and Pearson's correlation average over 1991-2010. Correlation average is computed only over years where all three validation points are available.

Index	Method	Normalized RMSE	Pearson's correlation
BEDD	MLR	38.81	0.98
	RK	39.47	0.98
	SVR	124	-0.96
CNI	MLR	60.1	0.82
	RK	50.78	0.84
	SVR	345.79	-0.76
GST	MLR	32.07	0.98
	RK	31.68	0.98
	SVR	133.3	-0.95
HI	MLR	39.26	0.94
	RK	39.25	0.94
	SVR	159.84	-0.93
WI	MLR	35.36	0.98
	RK	35.57	0.98
	SVR	191.7	-0.94

3.2 Phenology analysis

To be able to predict the suitability of an area for viticulture, it is necessary to know the relationship between climate and the phenological and chemical characteristics of the vines and grapes. This relationship is quantified via a linear regression between climate indices, calculated from the gap-filled time series of the "REBECKA" stations, and measured phenological and chemical properties of the grapes. The sampling of grapes for the assessment of chemical characteristics is done in the same plots where the "REBECKA" stations are located. These values are used to evaluate different climate indices in predicting the day of ripening for vines. Once the relation for the most accurate index is known, it is possible to spatialize the estimated ripening date using only an interpolated index map. Knowledge about the estimated timing of ripening can then be used in assessing the suitability of an area for viticulture.

The climate indices evaluated here are temperature-sum indices (BEDD, HI, WI) and average temperature (GST, GSTmin, GSTmax). Since they are computed from April 1st, we decided to focus our attention on end-of-season phenological stages, in particular ripening (defined as level 81 in BBCH scale), and when must weight reaches 16-17 degrees on the KMW scale.

From Figure 2 it is possible to recognize trends: higher temperatures, or temperature-sums, are linked to earlier phenological developments. We have also tried to fit linear models to the distributions. As indicated in Table 3, a statistically significant linear relationship (p-value < 0.001) exists for almost all of the indices. Looking at r^2 values, models with WI and GST are equally effective in explaining most of the variability for

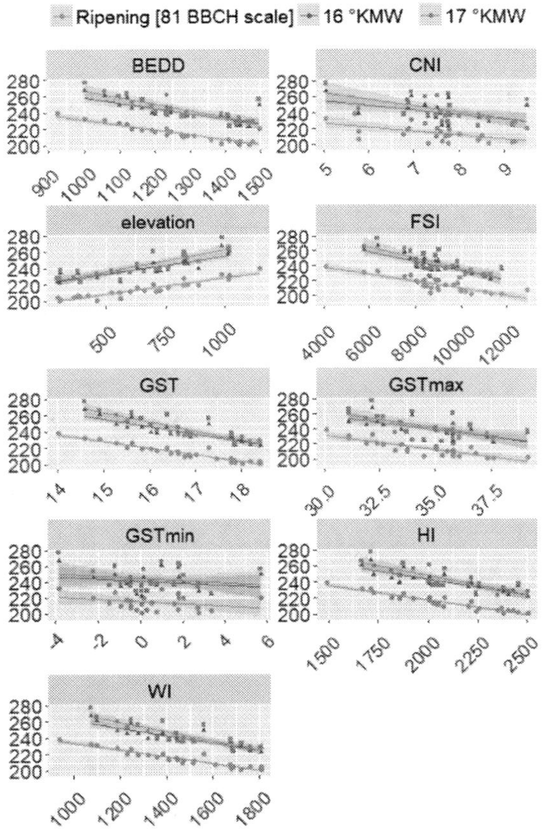

Figure 2. Correlation between day-of-year of phenological stages versus climatic indices and elevation. The lines represent a linear relationship between points and grey area is the 0.95 confidence interval.

all phenology stages, but other indices perform worse. It must also be noted that linear regression explains ripening far better than the KMW stages.

4 Conclusions

From the interpolation statistics we can state that MLR can be used to create high-resolution maps for climate indices with a significant level of accuracy. Despite RK and MLR performing similarly, the decision to select the MLR method is supported by two facts. First, computational time for a single map using MLR is four times faster than RK. Second, RK requires the model to fit over a variogram and automating this process does not always leads to satisfying results because our script can't detect all the possible variogram patterns and fits the model accordingly. The average map with interpolated Winkler zoning over 1991-2010 is shown in Figure 3.

From the phenology analysis, we have found that WI and GST are good indicators for estimating later phenological stages. It is now possible to apply the linear model of phenological stages to the interpolated average maps in order to obtain interpolated phenology maps. For example, Figure 4 represents the map of estimated DOY for ripening, obtained from 1991-2010 Winkler index average.

 https://doi.org/10.1051/e3sconf/20185001031

Table 3. Linear model estimation between phenology stages and climatic indices. Only statistically significant models (p-value < 0.001) are reported.

Phenology stages	Index	Adj. r^2	Residual std. dev.	p-value
Ripening	WI	0.91	3.09	9.39E-15
	GST	0.91	3.15	1.50E-14
	HI	0.89	3.37	8.19E-14
	elevation	0.88	3.55	3.05E-13
	BEDD	0.8	4.6	2.09E-10
	FSI	0.71	5.49	1.79E-08
	GSTmax	0.66	5.99	1.65E-07
16 °KMW	WI	0.76	5.98	1.64E-07
	GST	0.76	5.98	1.64E-07
	elevation	0.73	6.34	5.13E-07
	HI	0.71	6.56	9.97E-07
	BEDD	0.62	7.47	1.22E-05
	FSI	0.61	7.57	1.58E-05
	GSTmax	0.47	8.85	3.47E-04
17 °KMW	WI	0.74	7.31	2.15E-08
	GST	0.74	7.35	2.48E-08
	elevation	0.71	7.72	7.75E-08
	HI	0.67	8.26	3.75E-07
	BEDD	0.61	8.92	2.27E-06
	FSI	0.51	10.02	3.54E-05

The "REBECKA" project will collect data until the end of 2019, providing two full years worth of data beginning January 1st. Further efforts have already begun to move the interpolation from a yearly basis (vegetation-season) to a daily basis. This will allow us to better evaluate the earlier phenological stages using specific phenology models: models that are usually based on daily temperature sums.

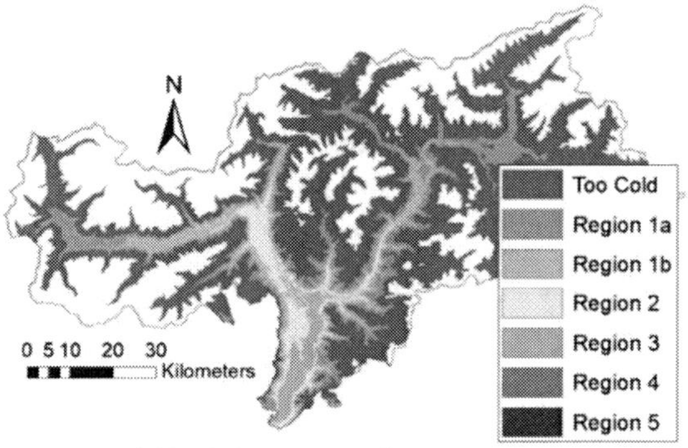

Figure 3. Winkler index average of years 1991-2010

In addition, the effects of climate change may play an important role in the upwards or downwards shift of the climatic niches for specific varietals [1]. This includes the risk, in the long-term, that suitable areas currently inside DOC zones in South Tyrol, will shift outside of them. Comparing the climate zoning computed over 1991-2010 with the average over a suitable climatic period (at least 10 years, e.g. 2011-2020) will provide information about a possible shift of the climatic niches suitable for viticulture.

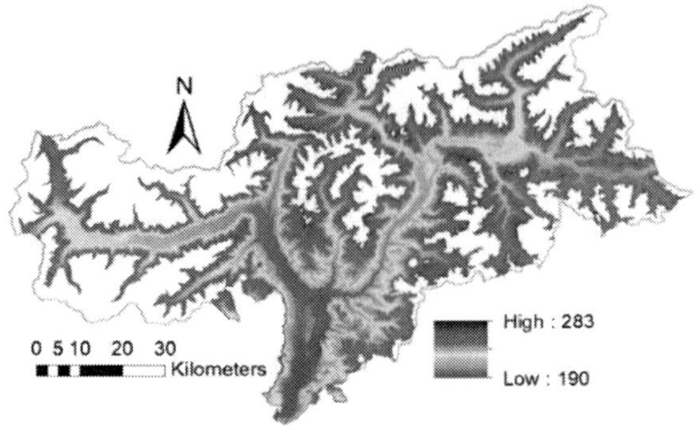

Figure 4. Estimated ripening day-of-year using a linear model with Winkler index as predictor.

References

1. G. V. Jones, "Climate, Grapes, And Wine: Structure And Suitability In A Changing Climate," *Acta Horticulturae* (931) (2012): 19–28, doi:10.17660/ActaHortic.2012.931.1.
2. C. Frei, "Interpolation of temperature in a mountainous region using nonlinear profiles and non-Euclidean distances," *International Journal of Climatology* 34(5) (2014): 1585–1605, doi:10.1002/joc.3786.
3. Zentralanstalt für Meteorologie und Geodynamik, Agenzia Regionale per la Prevenzione e Protezione Ambientale del Veneto, and Institut für Interdisziplinäre Gebirgsforschung, *Il Clima del Tirolo - AltoAdige - Bellunese* (2015).
4. E. Eccel, E. Cordano, and G. Toller, *ClimClass: Climate Classification according to Several Indices* (2016).
5. World Meteorological Organization, *Guidelines on the Calculation of Climate Normals* (Geneva, Switzerland: World Meteorological Organization, 2017).
6. M. H. Kutner, ed., *Applied linear statistical models*, 5th ed. (Boston: McGraw-Hill Irwin, 2005).
7. G. James et al., eds., *An introduction to statistical learning: with applications in R* (New York: Springer, 2013).
8. T. Hengl et al., *A practical guide to geostatistical mapping of environmental variables* (Luxembourg: Publications Office, 2007).
9. T. Hengl, G. B. M. Heuvelink, and D. G. Rossiter, "About regression-kriging: From equations to case studies," *Computers & Geosciences* 33(10) (2007): 1301–1315, doi:10.1016/j.cageo.2007.05.001.
10. S. R Gunn, *Support Vector Machines for Classification and Regression*, 1998.
11. A. J. Miller, *Subset selection in regression*, 2nd ed. (Boca Raton: Chapman & Hall/CRC, 2002).
12. M. H. Nguyen and F. de la Torre, "Optimal feature selection for support vector machines," *Pattern Recognition* 43(3) (2010): 584–591, doi:10.1016/j.patcog.2009.09.003.
13. R. Le Roux et al., "Comparison of two fine scale spatial models for mapping temperatures inside winegrowing areas," *Agricultural and Forest Meteorology* 247 (2017): 159–169, doi:10.1016/j.agrformet.2017.07.020.
14. D. H. Lorenz et al., "Growth Stages of the Grapevine: Phenological growth stages of the grapevine (Vitis vinifera L. ssp. vinifera)—Codes and descriptions according to the extended BBCH scale†," *Australian Journal of Grape and Wine Research* 1(2) (1995): 100–103, doi:10.1111/j.1755-0238.1995.tb00085.x.

Classification of Croatian winegrowing regions based on bioclimatic indices

Marko Karoglan[1], Maja Telišman Prtenjak[2], Silvio Šimon[3], Mirela Osrečak[1], Marina Anić[1], Jasminka Karoglan Kontić[1], Željko Andabaka[1], Ivana Tomaz[1], Branko Grisogono[2], Andreina Belušić[2], Antun Marki[2], Željka Prša[2], Branimir Omazić[2], Damjan Jelić[2], Željko Večenaj[2], Višnjica Vučetić[4], Damir Počakal[4], Ivana Vladimira Petric[3] , Renata Leder[3], Ivan Prša[3].

[1] Faculty of Agriculture, Department of Viticulture and Enology, University of Zagreb, Zagreb, Croatia

[2] Department of Geophysics, Faculty of Science, University of Zagreb, Zagreb, Croatia

[3] Centre for Agriculture, Food and Rural Affairs of Croatia, Zagreb, Zagreb, Croatia

[4] Meteorological and Hydrological Service of Croatia, Zagreb, Croatia

Abstract. The aim of this work was to classifie vine growing regions of Croatia using bioclimatic indices. For the analysis of climatic conditions, linear trends of bioclimatic indices were determined using meteorological observations for all avaliable climatological stations located in vine growing regions of Croatia. Analysis were performed for two different climatological periods: 1961-1990 and 1988-2017. Four commonly used bioclimatic indices were determined: the Winkler index, the Huglin index, Cool night index and Growing season average temperature.

Key words: grapevine, bioclimatic indices, vineyard zoning, Croatia

1 Introduction

Viticulture and enology are important agricultural practices and one of the key economic activities in Croatia. Climate is the critical component of *terroir* limiting grape and wine production, which also determine suitability of particular grapevine (*Vitis vinifera* L.) varieties for wine production in the particular winegrowing region (Gladstones, 1992). Several bioclimatic indices are commonly used in vineyard zoning and in aim to describe suitability of climate of different winegrowing regions (Jones et al. 2010). Bioclimatic indices are also useful metrics to provide the information about climate changes impact on viticulture (Malheiro et al., 2010). One of the most widely used indices was temperature based Winkler index (WI), using a growing degree base of 10°C (growing degree-days; GDD), to place viticulture in the context of climate suitability (Amerine and Winkler, 1944). Correlating the WI with quality of grapes and wines production, they came up with five viticulture regions (Winkler et al, 1974). Despite the fact that their system has been changed among the years, WI still represent one of the most relevant bioclimatic indices. Jones (2006) defined five climate maturity groupings, ranging from very cool to very hot, based upon the average growing season temperature (TGS) and ability to ripen a given variety. The Cool Night Index (CI) which acounts for minimum temperatures during grapevine maturation period is also one of the strictly thermal indices (Tonietto 1999, Tonietto & Carbonneau 2004). Using a degree-day approach, with the inclusion of a day-lenght factor as a proxy for radiation, the Huglin Heliothermal Index (HI; Huglin, 1978) allows assessing the thermal potential of a given region. HI can help in determination of the thermal demands for the ripening of each grape variety, also reflecting the potential grape sugar content.

The viticultural zoning in Croatia has been established in 1978, and was based on the several factors including climate, soils and topography of Croatian vineyards. In the mean time, we have witnessed that the Earth's average surface temperature has increased by approximately 0,7°C since the beggining of the century. Along with that, climate models forecast that the average global temperature will continue to rise (Branković et al., 2010).

The main goal of this study was to examine the curent spatial climate structure of the Croatian viticultural regions, using four standard grape-based bioclimatic indices, such as WI, TGS, CI and HI which were calculated for two different climatological periods.

2 Material and methods

Temperatures observation data for 29 meteorological stations located in Croatian viticultural regions were collected for the reference time period (1961-1990) for the present climate according to the World Meteorological Organisation. Beside that, the same observations were collected for the period of last 30 years (1988-2017), in aim to compare the data with the reference period. Data were provided by the Meteorological and Hydrological Service of Croatia.
Bioclimatic indices were calculated as folows:

Growing degree-days (GDD or WI) according to Winkler et al. (1974):

$$WI = \Sigma((Tmax + Tmin)/2) - 10°C \qquad (1)$$

Tmax – maximum daily temperature

Tmin – minimum daily temperature

Huglin Index (HI) according to Huglin (1978):

$$\sum = ((Tavg - 10°C) + (Tmax - 10°C))/2*k \quad (2)$$

Tavg – average daily temperature

k – latitude, daylenght adjustment factor

Cool night index (CI) according to Tonietto(1999) and Tonietto and Carbonneau (2004):

$$CI = 1/N \sum T\,min \quad (3)$$

N – number of days in the period

Average growing season temperature according to Jones (2006):

$$1/N\, \Sigma((Tmax + Tmin)/2) \quad (4)$$

In order to calculate agro-climatic indices, the interpolation method (using neighboring stations data) will be compiled by strings at those times and for those data where data is missing.

The bioclimatic indices were evaluated on the basis of the class levels listed in Table 1.

Table 1. Bioclimatic indices, formulas and classes definition

INDEX	PERIOD	CLASSES DEFINATION	
Average growing season temperature (TGS) Jones (2006)	April - October	Too cool	< 13 °C
		Cool	13 - 15 °C
		Intermediate	15 -17 °C
		Warm	17 - 19 °C
		Hot	19 - 21 °C
		Very hot	21 - 24 °C
		Too hot	> 24°C
Growing degree - days (GDD or WI) Winkler et al. (1974)	April - October	Region I	< 1390
		Region II	1391 - 1670
		Region III	1671 - 1940
		Region IV	1941 - 2220
		Region V	> 2220
Huglin index (HI) Huglin (1978)	April - September	Very cool (HI-3)	< 1500
		Cool (HI-2)	1500 - 1800
		Temperate (HI-1)	1800 - 2100
		Temperate warm (HI+1)	2100 - 2400
		Warm (HI+2)	2400 - 2700
		Very warm (HI+3)	> 2700
Cool night index (CI) Tonietto (1999)	September	Very cool nights (CI+2)	< 12 °C
		Cool nights (CI+1)	12 - 14 °C
		Temperate nights (CI-1)	14 - 18 °C
		Warm nights (CI-2)	> 18 °C

3 Results and discussion

Bioclimatic indices were analyzed for two climatic periods and for each of 29 meteorological stations individually. Results obtained were grouped according to main Croatian viticulture regions and subregions. There were 9 meteorological stations in Western continental Croatia region, 4 were placed in Eastern continental Croatia region and 16 in the Coastal Croatia region.

The areas of Croatian vineyards, according to these four bioclimate indices, belong to different classifications, depending on the region and subregion (Table 2).

Regarding reference time period, both Western and Eastern continental Croatia regions, ranging from 1086,7 to 1480,9 GDD belong to Winkler Region I and II. The same is for the recent climatological period (1988-2017), but with obvious higher temperature values ranging from 1323,9 to 1652,5 GDD. Coastal Croatia region ranged from 1325,1 to 2289,1 GDD for the reference period, thus belonging to Winkler regions I to V. Regarding recent climatological period, values ranged from 1496,5 to 2483,5 GDD, which means that Coastal Croatia region belongs to Winkler region II to V.

 https://doi.org/10.1051/e3sconf/20185001032

Both Western and Eastern continental Croatia regions belonged to intermediate climate based on TGS regarding reference period, ranging from 15,1 to 16,9°C. Regarding the period from 1988-2017, values ranged from 16,2 to 17,7 meaning that some parts of Continental Croatia region fall into warm climate according to TGS. The Coastal Croatia region for the reference period belonged to warm and hot category according to TGS ranging from 17,5 to 20,7°C. Only one (Pazin) meteorological station measured lower TGS values (16,2°C). Regarding recent climatological period, TGS values ranged from 17,0 to 21,6°C, thus placing Coastal Croatian region in warm, hot and very hot climate.

According to the Huglin index and for the reference period, both Western and Eastern continental Croatia regions belonged to cool (HI-2) and temperate (HI-1) climate, ranging from 1611,3to 2050,6. In the period from 1988 – 2017, Huglin index ranged from 1855,4 to 2230,0, placing both Continental Croatia regions in temperate (HI-1) and temperate warm (HI+1) climate. Huglin index in the reference period for the Coastal Croatian region ranged from 1919,9 to 2453,1 thus placing the region in the range from temperate (HI-1) to warm (HI-2) climate. When compared to the recent climatological period, the temperate climate class is missing, since the Huglin index ranged from 2100,5 to 2660,5 placing Coastal Croatian region in temperate warm (HI-1) and warm (HI-2) climate.

Cool night index for both Western and Eastern Continental Croatia did not change in either climatological periods and the whole region retained very cool nights (CI+2) with the measured maximum of 11,6°C. Regarding reference period Coastal Croatian region CI ranged from 9,8°C to 18,4°C. In the recent period CI ranged from 10,2 to 19,0°C. Only one meteorological station (Pazin) measured very cool nights (CI+2) in both climatological periods.

The more precision results are obtained regarding subregions within three main regions., regarding recent climatological period 1988-2017.

According to Winkler index, the coolest winegrowing subregion in Croatia is Pokuplje, ranging between Region I and II. Subregions Podunavlje, Slavonija, Zagorje-Međimurje, Prigorje-Bilogora and heartland of Istra belong to Region II. Dalmatinska zagora fall into Region III. The coastal part of Istra belong to Region IV, while subregions Hrvatsko-Primorje and Dalmacija range between Region IV and V.

According TGS index, the coolest winegrowing subregion in Croatia is Zagorje-Međimurje, followed by Pokuplje and Prigorje-Bilogora. Subregions Podunavlje, Slavonija and Dalmatinska zagora belong to warm climate, followed by Istra subregion ranging from warm to hot climate. Subregion Hrvatsko Primorje belong to hot climate, while the subregion Dalmacija ranges from hot to very hot climate.

According to Huglin index, winegrowing subregion Zagorje-Međimurje belongs to temperate climate, followed by Prigorje-Bilogora and Pokuplje subregions, which range between temperate and temperate warm climate. Subregions Podunavlje and Slavonija clearly fall into temperate warm climate. Subregions Istra, Hrvatsko Primorje, Dalmatinska zagora and Dalmacija vary between temperate warm and warm climate, depending on the location of meteorological station.

According to Cool night index, winegrowing subregions Podunavlje, Slavonija, Zagorje-Međimurje, Prigorje-Bilogora, Pokuplje as well as the heartland of Istra, all have very cool nights. In the subregion Dalmatinska zagora, night temperatures vary from very cool to cool. The coastal part of Istra and subregion Hrvatsko primorje have temperate hights, while night temperaturs in Dalmacija vary from temperate to even warm.

Calculated indices for Croatia winegrowing regions was compared to those of the other worldwide known winegrowing regions, according to Jones (2009). Considering TGS, WI and HI, all Croatian subregions are much warmer than all German regions, as well as than Bourgogne and Champagne regions. Winegrowing subregions from the continental part of Croatia have very similar climatic characteristics to Bordeaux, Côte du Rhône Méridionales, Barolo, Vatellina Superiore, Vino Nobile di Montepulciano and Rioja. On the other hand, winegrowing subregions from the coastal part of Croatia have very comparable climatic characteristics to the European regions Chianti Classico, Jerez, La Mancha, Porto and Vinho Verde, as well as to the Californian regions Napa Valley and Paso Robles.

Considering HI, there are certain limits regarding growing of different grape varieties. According to Huglin (1978), varieties such as Cinsault, Syrah, Grenache, Carignan, Mourverdre needs to be grown in temperate warm (HI+1) climate to reach maturity. Regarding this, such varieties are still not suitable for growing in Zagorje-Međimurje subregion as welll as in some parts of Pokuplje and Prigorje-Bilogora subregions.

There is a certain concerne regarding CI values for Dalmacija subregion. Warm nights very often have a negative impact on grape aroma, colour and acidity content. Night temperatures in other winegrowing regions of Croatia are favorable in the term of retaining satisfactory levels of organic acids, as well as aromatic and phenolic compounds.

4 Consclusion

This study provides detailed analysis of bioclimatic indices in Croatian winegrowing regions. Significant changes were noticed when comparing recent climatologic period (1988-2017) to the reference climatological period (1961-1990), thus pointing out the presence of climate changes in Croatia.

Bioclimatic indices values were compared to those from the other worldwide known winegrowing regions in order to evaluate viticulural potential in Croatia. But, for more precise comparison, more viticultural factors must be included (e.g. soils, precipitation etc.) since the same climatic group does not necessarily mean the same climate.

Further investigations are going toward phenology observations and grape quality measurements within single productive year.

However, this study provides accurate information needed for the questioning of existing and establishing a new viticulture zoning system in Croatia.

Table 2. Range values of bioclimatic indices of major Croatian winegrowing regions

		TGS		GDD		HI		CI	
REGION	**SUBREGION**	**1961-1990**	**1988-2017**	**1961-1990**	**1988-2017**	**1961-1990**	**1988-2017**	**1961-1990**	**1988-2017**
EASTERN CONTINENTAL CROATIA	**Podunavlje**	16,2 - 16,8	17,3 - 17,7	1329,6 - 1461,0	1567,4 - 1652,5	1964,0 - 2039,7	2200,3 - 2226,6	9,8 - 10,9	10,5 - 11,3
	Slavonija	16,9	17,6	1480,9	1625,8	2050,6	2204,9	10,9	11,2
WESTERN CONTINENTAL CROATIA	**Zagorje - Međimurje**	15,5	16,5	1184,8	1395,8	1776,3	1997,1	9,9	10,5
	Prigorje - Bilogora	15,4 - 16,2	16,6 - 17,4	1149,4 - 1317,9	1405,2 - 1590,6	1738,0 - 1884,5	1990,6 - 2142,9	9,3 - 10,7	10,1 - 11,5
	Pokuplje	15,1 - 16,8	16,2 - 17,4	1086,7 - 1449,2	1323,9 - 1574,7	1611,3 - 1979,9	1855,4 - 2142,7	10,1 - 11,4	10,5 - 11,2
COASTAL CROATIA	**Istra**	16,2 - 18,5	17,0 - 19,8	1325,1 - 1813,9	1496,5 - 2095,8	1919,9 - 2149,8	2100,5 - 2414,5	9,8 - 14,9	10,2 - 15,7
	Hrvatsko Primorje	18,4 - 19,7	19,5 - 20,8	1805,5 - 2072,6	2042,7 - 2306,3	2104,4 - 2296,6	2371,3 - 2529,6	15,0 - 17,6	15,0 - 17,7
	Dalmatinska Zagora	17,5 - 18,2	18,2 - 18,9	1601,1 - 1759,3	1752,6 - 1867,2	2129,7 - 2266,8	2311,0 - 2438,6	11,4 - 12,4	11,4 - 12,7
	Dalmacija	19,1 - 20,7	20,3 - 21,6	1948,4 - 2289,1	2193,9 - 2483,5	2221,3 - 2476,0	2433,5 - 2660,5	16,0 - 18,0	16,6 - 18,2

5 Literature

1. Amerine M.A., Winkler A.J. Composition and quality of musts and wines of California grapes. Hilgardia **15**, 493-675 (1944)
2. Branković Č., Srnec L., Patarčić M. An assessment of global and regional climate change based on the EH5OM climate model ensemble. Clim. Change. **98**, 21–49. (2010).
3. Gladstones J. Viticulture and environment. WineTitles, Adelaide. 310 pp. (1992)
4. Huglin P. Nouveau Mode d'Evaluation des Possibilites Heliothermiques d'un Milieu Viticole. C. R. Acad. Agr. France, 1117-1126. (1978)
5. Jones G.V. Climate and Terroir: Impacts of Climate Variability and Change on Wine. In Fine Wine and Terroir- The Geoscience Perspective. Macqueen R.W., and Meinert L.D. (eds.), Geoscience Canada Reprint Series Number 9, Geological Association of Canada, St. John's, Newfoundland, 247 p (2006)
6. Jones G.V., Moriondo M., Bois B., Hall A., Duff A. Analysis of the Spatial Climate Structure in Viticulture Regions Worldwide. *Proceedings of the 32nd World Congress of the Vine and Wine and 7th General Assembly of the International Organisation of Vine and Wine, June 28- July 3, 2009. Zagreb, Croatia* (2009)
7. Malheiro A.C., Santos J.A., Fraga H., Pinto J.G. Climate change scenarios for viticultural zoning in Europe. Clim. Res. **43**, 163–177. (2010).
8. Tonietto J. Les macroclimats viticoles mondiaux et l'influence du mesoclimat sur la typicite de la Syrah et du Muscat de Hambourg dans le sur de la France:

methodologie de caracterisation. These Doctorat. Ecole Nationale Supericure Agronomique, Montpellier (1999)
9. Tonietto J., Carbonneau A. A multicriteria climatic classification system for grape-growing regions worldwide. Agric. Forest Meteorol. **124**, 81–97. (2004).
10. Winkler A. J., Cook A., Kliewere W. M., Lider L. A. General Viticulture, (4th ed.), University of California Press, Berkeley, 740 p. (1974).

Efectos de cambio climático en el comportamiento de variedades tintas de Mallorca

José M. **Escalona**, María Lluisa **Oliver**, Hipólito **Medrano**

Departamento de Biología. Universidad de las Islas Baleares. INAGEA. Ctra Valldemossa Km 7,5. 07122 Palma. Baleares. España

Resumen

En los últimos cincuenta años se ha producido un incremento de las temperaturas medias y una modificación del régimen de precipitación como consecuencia del Cambio climático. Además, las previsiones climáticas a futuro indican que los períodos de sequía extrema y aridez serán cada vez más frecuentes asociadas a incrementos progresivos de la temperatura. La cuenca mediterránea es un área de especial vulnerabilidad frente al cambio climático. En este escenario, se ha comprobado que la vid modifica su comportamiento. Numerosos estudios indican un efecto en la fenología, con brotaciones más tempranas, acortamiento de los ciclos de maduración y un retraso en la finalización del ciclo vegetativo. Además, se está produciendo un cambio en la composición de la uva y de los vinos producidos, que comprometen la identidad de los productos. Todo ello, compromete la definición de Terruño y la tipicidad de los productos de las regiones vitícolas. En Mallorca, la actividad vitícola y la producción de vinos tiene una extensa historia. En la actualidad conviven variedades locales con variedades foráneas. Hasta ahora no se han realizado estudios de los efectos del cambio climático en el comportamiento de estas variedades en las condiciones particulares de la Isla. Por ello, este trabajo pretende aportar nuevos conocimientos sobre dichos efectos, Los resultados muestran que en los últimos 50 años se ha producido un incremento de 1,5 ºC en la temperatura media anual y se han repetido períodos de sequía cada vez más prolongados en el tiempo. Como consecuencia, se ha producido un incremento significativo de la concentración de azúcares del mosto y, por ende, de la graduación de los vinos. Por otra parte, se han adelantado las vendimias sobre todo en variedades de ciclo medio y largo de maduración. De todas las variedades estudiadas, la variedad local Manto Negro es la que ha mostrado una mayor capacidad de adaptación a las nuevas condiciones climáticas.

Estudio de las propiedades de retención de humedad de suelos vitícolas en Castilla- La Mancha (España)

Study of the moisture retention properties of vineyard soils in Castilla- La Mancha region (Spain)

Caridad Pérez-de-los-Reyes[1,a], María Luisa Pérez-de-los-Reyes[1], David Chocano[1], Mónica Sánchez-Ormeño[1], Sandra Bravo[1], José Ángel Amorós[1] y Francisco Jesús García Navarro[1]

[1]Universidad de Castilla- La Mancha, Escuela Técnica Superior de Ingenieros Agrónomos, Ronda de Calatrava, 7, Ciudad Real, España

Resumen. En el presente trabajo se estudió la relación entre las principales características del suelo que influyen en la retención de humedad (densidad aparente y porcentajes de arena, limo, arcilla, materia orgánica y carbonato cálcico) y la humedad gravimétrica a capacidad de campo (CC), punto de marchitez permanente (PMP) y capacidad de retención de agua disponible (CRAD) o agua útil de los horizontes superficiales de 101 suelos vitícolas de Castilla-La Mancha (España). Los suelos se describieron morfológicamente y se realizaron los análisis pertinentes de establecimiento de las principales propiedades de los mismos. La CC y PMP se determinaron con el método de placas Richards (aplicando -33 KPa y -1500 KPa, respectivamente) y la CRAD fue deducida por diferencia (CC-PMP). Del Análisis de Componentes Principales se deduce que dos componentes explican el 61,76% de la variabilidad de los datos (porcentaje de contenido en arena y densidad aparente). Las propiedades que mostraron una mayor relación lineal, estadísticamente significativa, con CC, PMP y CRAD fueron el porcentaje de arena (R= -0,68, -0,68 y -0,47) y de limo (0,5, 0,49 y 0,37) respectivamente. Se utilizó el programa Surfer para la elaboración de mapas de distribución espacial de propiedades del suelo.

Abstract. In the work described here, the relationship between soil characteristics that influence moisture retention properties (bulk density and percentages of sand, silt, clay, organic matter and calcium carbonate) and gravimetric moisture at Field Capacity (FC), gravimetric moisture at the Permanent Wilting Point (PWP) and Available Water Capacity (AWC) was studied for 101 surface horizons of vineyard soils in the Castilla-La Mancha region (Spain). The soils were described morphologically and the necessary analyzes were carried out to establish the main characteristics of the soils. FC and PWP were determined according to Richards' method (applying –33 kPa and –1500 kPa, respectively). Available Water Capacity was calculated by subtraction (AWC = FC – PWP). From the Principal Component Analysis two components were found that can explain 61.76% of the variability of the original data (sand content and bulk density). The properties that showed a higher significant linear relationship with FC, PWP and AWC were sand content (R= -0.68, -0.68 y -0.47) and silt content (R= 0.5, 0.49 y 0.37) respectively. A GIS tool (Surfer) was used to prepare spatial distribution maps of soil properties.

1 Introducción

El conocimiento de las propiedades de retención de humedad de los suelos vitícolas permite la definición de parámetros necesarios para la determinación de las dosis de riego de los mismos. La retención de agua es la propiedad hidrofísica del suelo que puede ser descrita por la dependencia entre el contenido de agua del suelo y el potencial del agua del suelo [1]. El hecho de que dos suelos tengan el mismo potencial mátrico no significa que posean la misma cantidad de agua y puede ocurrir que no toda el agua retenida en el suelo esté a disposición de las plantas de vid. Lo que interesa desde el punto de vista del viticultor es conocer qué cantidad de agua (de la que haya acumulada en el suelo) podrá ser utilizada por las plantas y para ello se utiliza el concepto de capacidad de retención de agua disponible (CRAD, también conocida como agua útil). Se calcula como la diferencia entre la humedad gravimétrica a Capacidad de Campo (CC), porcentaje de agua que hay en el suelo a un potencial de -33 kPa y la humedad gravimétrica en el Punto de Marchitez Permanente (PMP), porcentaje de agua que hay en el suelo a -1500 kPa.

Las principales propiedades del suelo que influyen en el valor de estos parámetros según los distintos estudios realizados por numerosos investigadores son, entre otros, la textura (porcentaje de fracciones texturales arena, limo y arcilla), la materia orgánica, la densidad aparente y el contenido en carbonato cálcico.

El uso de funciones de edafotransferencia también ha contribuido al conocimiento de las propiedades del suelo que influyen en las propiedades de retención de humedad. Wösten et al. [2] realizaron en 2001 una profunda revisión bibliográfica sobre las principales funciones de edafotransferencia que se habían desarrollado hasta ese momento y las principales propiedades usadas en las mismas: el porcentaje de agua a potenciales a -33 kPa y -1500 kPa eran las más usadas como variables dependientes y el contenido en arena, limo, arcilla, materia orgánica y densidad aparente eran las más utilizadas como variables independientes. Walczak et al. (2006) [1] establecieron una función de edafotransferencia que relacionaba el contenido de agua del suelo a determinados valores de potencial mátrico, con arena, limo, densidad aparente y porosidad total, sin considerar el contenido en materia orgánica ni en $CaCO_3$.

Del estudio de la bibliografía especializada se puede deducir que no hay un inventario definitivo de propiedades del suelo que influyen sobre las propiedades de retención de humedad. Ciertas propiedades del suelo parecen tener influencia sobre los parámetros de CC, PMP y CRAD, pero esta influencia puede variar según las condiciones de estudio y las posibles interacciones que se produzcan sobre ellas. El objetivo de este trabajo es contribuir al conocimiento de las propiedades de los suelos vitícolas de la región de Castilla La Mancha (España) que influyen sobre las propiedades de retención de humedad. Se permite, de ese modo, mejorar la caracterización de dichos suelos, lo que facilitará un manejo agronómico adecuado de los mismos en una región con gran variedad y cantidad de suelos vitícolas. Para ello, se estudiaron 101 perfiles de suelos vitícolas y se determinaron, en el horizonte superficial, los valores de CC y PMP con placas Richards, obteniendo el valor del CRAD por diferencia entre los anteriores.

Finalmente, se utilizó una herramienta de SIG (Surfer, © 1993-2009 Golden Surfer, Inc.) para la elaboración de mapas de distribución espacial de CRAD y otras propiedades del suelo, con el objeto facilitar la consulta de los mismos por parte de agricultores y técnicos que deban tomar decisiones respecto a las dosis de riego del cultivo.

2 Materiales y métodos

El trabajo se desarrolló sobre 101 perfiles de suelos vitícolas de la región de Castilla- La Mancha. Todos los perfiles estudiados se dedicaban al cultivo de vid, con diferentes variedades y con un manejo del cultivo también distinto. Los puntos de muestreo se eligieron en zonas representativas de las diferentes denominaciones de origen, pagos y bodegas con indicación geográfica Vinos de la Tierra de Castilla, buscando la mayor diversidad de suelos posible.

En las zonas seleccionadas, los perfiles se abrieron con una pala retroexcavadora y fueron descritos de acuerdo con los criterios de la FAO [3]. Posteriormente se tomaron muestras de cada horizonte que, en el laboratorio, se secaron al aire y se tamizaron (< 2mm) para realizar los análisis. Se tomó así mismo una muestra del horizonte superficial en cilindros de 100 cm^3 para la determinación de la densidad aparente [4]. Las determinaciones analíticas se realizaron de acuerdo con SCS-USDA, por triplicado. Para este estudio se determinaron: porcentaje de fracciones texturales arena, limo y arcilla usando el método del hidrómetro [5], materia orgánica por oxidación de dicromato potásico y valoración del dicromato sobrante con sulfato ferroso amónico [6] y carbonato cálcico con un calcímetro de Bernard [7].

Para cada uno de los horizontes superficiales, se determinaron por triplicado los valores de la humedad gravimétrica (en porcentaje) de las muestras en función de los potenciales mátricos según el método de Richards [8]: humedad gravimétrica a -33 kPa para determinación para determinación de la Capacidad de Campo (CC) y humedad gravimétrica a -1500 kPa para determinación del Punto de Marchitez Permanente (PMP). De los valores obtenidos se calculó la capacidad de retención de agua disponible como diferencia de las anteriores (CRAD=CC-PMP).

El análisis estadístico de los datos se realizó usando los programas STATGRAPHICS Plus 5.1 (Copyright© 1994-2000 Statistical Graphics Corp) y Excel 2013 (Copyright© 2014 Microsoft). Se realizó un Análisis de Componentes Principales (ACP) con los diferentes valores obtenidos de todas las propiedades estudiadas (excepto CRAD) y un análisis de regresión lineal para la determinación del coeficiente de correlación entre características del suelo y propiedades de retención de humedad. El p-valor del test ANOVA indicaba si había relación estadísticamente significativa entre las propiedades para un nivel de significación de $P<0,05$.

3 Resultados y discusión

En una aproximación inicial al estudio de las propiedades de los suelos vitícolas estudiados, se realizó, como se ha indicado anteriormente, un Análisis de Componentes Principales incluyendo todas las propiedades excepto

CRAD. Se dedujo que dos componentes explican el 61,76% de la variabilidad de los datos: porcentaje de contenido en arena, con un peso del 45,50% y densidad aparente con un peso de 71,62% como segundo componente. Se calcularon los coeficientes de correlación lineal entre las propiedades estudiadas para determinar cuáles de ellos tenían una relación estadística significativa (Tabla 1)

Tabla 1. Matriz de coeficientes de correlación lineal entre diferentes propiedades del suelo y propiedades de retención de humedad (en color rojo cuando existe relación estadísticamente significativa)

	Densidad Aparente	Arena	Limo	Arcilla	Materia Orgánica	$CaCO_3$	CC	PMP	CRAD
Densidad Aparente	1,00	0,11	-0,04	-0,08	-0,13	-0,36	-0,34	-0,27	-0,33
Arena		1,00	-0,79	-0,56	-0,30	-0,34	-0,68	-0,68	-0,47
Limo			1,00	0,25	0,17	0,37	0,50	0,49	0,37
Arcilla				1,00	0,14	0,11	0,36	0,39	0,22
Materia Orgánica					1,00	0,10	0,38	0,35	0,31
$CaCO_3$						1,00	0,45	0,31	0,49
CC (%)							1,00	0,89	0,84
PMP (%)								1,00	0,50
CRAD (%)									1,00

La densidad aparente, propiedad de suelo ampliamente utilizada para determinar propiedades de retención de humedad a través de funciones de edafotransferencia [1][2] parece tener relación débil con las propiedades de retención de humedad en los suelos estudiados (R= -0,34; -0,27 y -0,33 para CC, PMP y CRAD respectivamente). En este sentido, Walczak et al. (2006) [1] propusieron una función de edafotransferencia en la que el contenido de humedad gravimétrica a CC y en el PMP estaba inversamente relacionado con la densidad aparente (menor humedad a mayor densidad aparente) y existía una relación estadísticamente significativa, con coeficientes de correlación de -0,93 y -0,54 para CC y PMP respectivamente. En el presente estudio de suelos vitícolas los Coeficientes de Correlación son inferiores a los referenciados anteriormente.

La presencia de arena en el suelo está inversamente relacionada con la CC, el PMP y la CRAD con coeficientes de correlación de -0,69, -0,68 y -0,46 respectivamente. La relación lineal es estadísticamente significativa en los tres casos y el valor de R supone una relación entre las variables moderadamente fuerte para el caso de CC y PMP y más débil para el caso de CRAD. La relación negativa entre el contenido de arena del suelo y las propiedades de retención de humedad del suelo ya ha sido demostrada por otros investigadores al estudiar suelos de regiones semiáridas o de regiones de clima templado [1] y puede observarse comparando las figuras 1 y 2 donde las zonas con mayor CRAD coinciden con de menor contenido en arena. Los suelos vitícolas estudiados presentan coeficientes de correlación mayores entre el contenido de limo y las propiedades de retención de humedad (0,50; 0,49 y 0,37 respectivamente para CC, PMP y CRAD) que entre el contenido en arcilla y las mismas propiedades, en las que la relación es estadísticamente significativa, pero débil.

En estudios previos, la materia orgánica se había destacado como la propiedad más importante para definir las propiedades de retención de humedad de un suelo, aunque estos trabajos presentaban elevados rangos en el contenido de materia orgánica entre las muestras que se comparaban: de 0,1 a 20% en Rawls et al. [9] y de 1,7 a 9,3% en Pérez-de-los-Reyes et al. [10]. En nuestro caso, el rango en el contenido de materia orgánica en los 101 suelos es de 0,1 a 6,2%, con una media de 2 % (y una desviación típica de 1) y 92 muestras con un contenido en materia orgánica menor del 3%. Este hecho puede explicar coeficientes de correlación relativamente débiles en las muestras de este estudio.

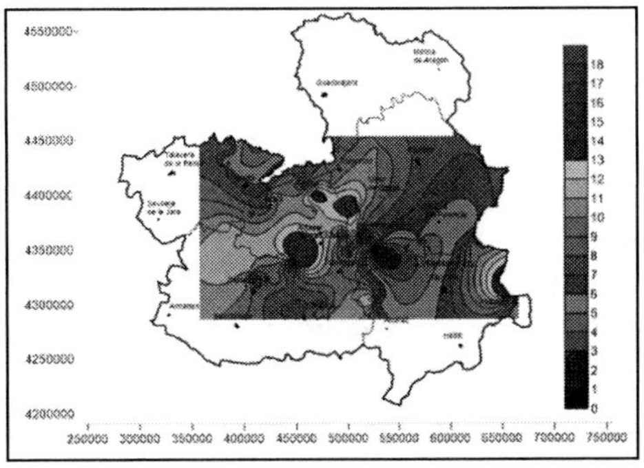

Figura 1. Mapa de distribución espacial CRAD (%) en 101 suelos vitícolas de Castilla La Mancha (España)

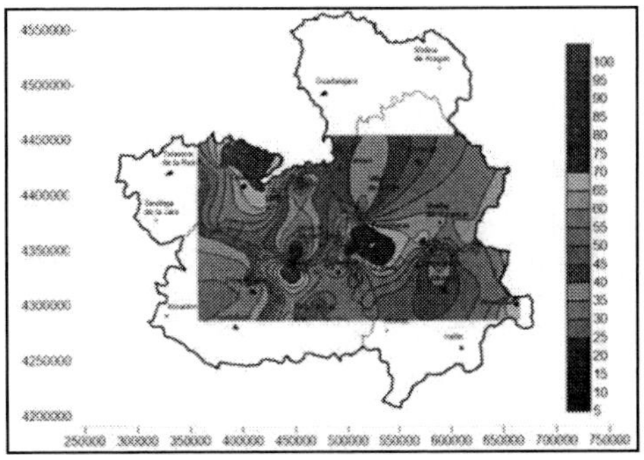

Figura 2. Mapa de distribución espacial de porcentaje de arena en 101 suelos vitícolas de Castilla La Mancha (España)

Finalmente, el contenido de carbonato cálcico puede tener influencia en las propiedades de retención de humedad para suelos de zonas semi áridas donde abundan los suelos calizos, como es el caso de los suelos vitícolas de Castilla La Mancha (donde alrededor de un 40 % de los suelos vitícolas de la región son calizos). Walczak et al. [1] determinaron coeficientes significativos de correlación entre el contenido de carbonato cálcico y la CC (R=0,59), así como con PMP (R=0,40) en 74 muestras de suelos de Sevilla (España). Por otro lado, Pérez-de-los-Reyes et al. [11] detallaron el aumento del CRAD en un suelo tratado con espumas de azucarera (de 4% en un suelo sin tratar a 34,5% en un suelo con

https://doi.org/10.1051/e3sconf/20185001034

espumas de azucarera) como consecuencia del incremento en carbonato cálcico y materia orgánica en el mismo. En el conjunto de los suelos vitícolas estudiados, el carbonato cálcico se presenta como una propiedad relacionada significativamente con la CC, PMP y CRAD (R= 0,45; 0,31 y 0,49 respectivamente) aunque dicha relación es relativamente débil. El cotejo de los mapas de distribución espacial (Figura 1 y Figura 3) refuerza esta conclusión.

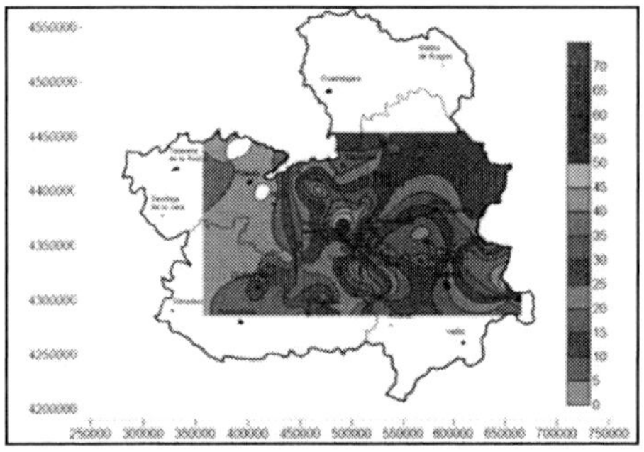

Figura 3. Mapa de distribución espacial de contenido en $CaCO_3$ (%) en 101 suelos vitícolas de Castilla La Mancha (España)

4 Conclusiones

La variabilidad del recurso natural suelo se manifiesta en este estudio en el que, como ha ocurrido con otros realizados anteriormente, las propiedades que se relacionan con la retención de la humedad en el suelo no son siempre las mismas. En este trabajo, las propiedades que han mostrado una relación lineal significativa con CC, PMP y CRAD para los 101 perfiles estudiados conjuntamente, son la densidad aparente, el porcentaje de arena y limo, el contenido en materia orgánica y el contenido en carbonato cálcico, siendo las que mostraron una mayor relación lineal el porcentaje de arena (R= -0,68; -0,68 y -0,47) y de limo (R= 0,5; 0,49 y 0,37), respectivamente.

Los mapas de distribución espacial permiten la comparación de estas propiedades visualmente y, en el futuro apoyarán y facilitarán investigaciones más amplias sobre las propiedades de retención de humedad de suelos vitícolas.

Referencias

1. R.T. Walczak, F. Moreno, C. Sławiński, E. Fernandez, J.L. Arrue. Journal of Hidrology **329,** (3-4) (2006).
2. J.H.M.Wösten, Y. A. Pachepsky, W.J. Rawls: Journal of Hidrology **251** (2001)
3. FAO ISRIC ISSS. 2006. World soil resources report 103, FAO, Rome, 132 pp (2006)
4. G.R. Blake, K.H. Hartge, In A. Klute (Ed) *Methods of Soil Analysis*. Agronomy Monograph No.9, ASA-SSSA, Madison (1986)
5. G.W. Gee, J.W. Bauder, In: Klute A. (Ed) *Methods of Soil Analysis*. Agronomy Monograph No.9, ASA-SSSA, Madison (1986)
6. A. Anne, Annals of Agronomy **2,** (1945)
7. J. Porta, *Técnicas y experimentos en Edafología*. (1986)
8. L.A. Richards. Soil Science **66**, (1948)
9. W.J. Rawls, Y.A. Pachepsky, J.C. Ritchie, T.M. Sobecki, H. Bloodworth, Geoderma **116**: (2003)
10. C. Pérez-de-los-Reyes, J.A. Amorós, F.J. García-Navarro, S. Bravo, R..Jiménez-Ballesta. Commun. Soil Sci. Plant Anal. **46** (10), (2015)
11. C. Pérez-de-los-Reyes, J.A. Amorós, F.J. García Navarro, S. Bravo, C. Sánchez, D. Chocano, R. Jiménez-Ballesta. Agricultural Water Management **98**, (2011)

Estado hídrico y actividad fisiológica en cv. Cabernet Sauvignon: respuesta a la aplicación de distintas dosis de riego

Water status and physiological activity in cv. Cabernet Sauvignon: response to application of different irrigation doses

Jesús Yuste[1,a], Alejandro Vicente[2] y Daniel Martínez-Porro[1]

[1]Instituto Tecnológico Agrario de Castilla y León, 47071 Valladolid, España
[2]Actualmente: actividad vitivinícola externa, España

Resumen. La producción y la calidad de la uva en las zonas de clara escasez hídrica dependen del aporte de agua de riego. A lo largo del periodo 2012-2014 se estudió la respuesta hídrica y fisiológica a 3 tratamientos de régimen hídrico: secano (P0), riego del 25% ETo (P1) y riego del 50% ETo (P2), aplicado desde tamaño guisante hasta vendimia. El ensayo se llevó a cabo con cv. Cabernet Sauvignon sobre portainjerto 110R, plantado en 2004 y conducido en espaldera en cordón Royat bilateral, con marco de plantación de 2,2 m x 1,2 m, en Valladolid. La variación de dosis de riego generó valores notablemente menos negativos de potencial hídrico, así como valores de los parámetros básicos de intercambio gaseoso significativamente más altos, en el tratamiento regado con mayor dosis que en el de menor dosis, y a su vez en éste que en el no regado, tanto a las 9 como a las 12 hs. Las diferencias entre tratamientos fueron estadísticamente significativas en general, aunque se observaron algunas excepciones, más frecuentes entre el tratamiento regado con menor dosis y el no regado, que variaron en fecha, hora y parámetro fisiológico. La capacidad de discriminación entre tratamientos de riego a través del potencial hídrico se ve bien reflejada en la actividad fisiológica del viñedo, con una tendencia creciente según aumenta el aporte de agua, lo cual debe servir para orientar el riego del viñedo.

Abstract. *Grapevine yield and quality in areas of remarkable hydric scarcity depend on irrigation. Throughout the period 2012-2014, the water status and physiological vine response to 3 treatments of water regime was studied: rainfed (P0), irrigation of 25% ETo (P1) and irrigation of 50% ETo (P2), from pea size state until harvest. The trial was developed with vines of cv. Cabernet Sauvignon onto 110R rootstock, planted in 2004 and vertically trellised as a bilateral Royat cordon. The vine distances were 2.2 m x 1.2 m. It was located in Valladolid (Spain). The variation of irrigation doses generated significantly less negative values of water potential, as well as values of the basic parameters of gas exchange significantly higher, in the irrigated treatment with a higher dose than in the lower dose treatment, and in the latter than in that not watered, both at 9 and at 12 solar hours. The differences between treatments were statistically significant in general, although some exceptions were observed, more frequent between the irrigated treatment with lower dose and the non-irrigated treatment, which varied in date, time and physiological parameter. The discrimination capacity between irrigation treatments through the water potential is well reflected in the physiological activity of the vineyard, with an increasing trend as the water supply increases, which should be used to focus the irrigation of the vineyard.*

1 Introducción

La producción y la calidad de la uva en las zonas en que el balance hídrico es claramente deficitario dependen de los aportes de agua de riego, pero la cantidad de agua que debe ser aportada dependerá tanto del nivel productivo como de la calidad de uva perseguidos [1], habiendo diversos factores que pueden influir en la respuesta, tales como la reserva hídrica del suelo, el estado del viñedo, la variedad y las características edafoclimáticas del lugar de cultivo [3, 9]. En general, el riego debe ser aplicado con moderación tras la parada de crecimiento del pámpano principal, con el objetivo de que la disponibilidad hídrica ayude a mantener una actividad fisiológica adecuada que facilite la síntesis de azúcar y la translocación de sustancias elaboradas a las bayas [4].

El estado hídrico de la cepa es el factor más limitante de su actividad fisiológica, y fotosintética en particular [2], por lo que su estimación resulta fundamental para aplicar una estrategia de riego adecuada, a través de la toma rápida de decisiones objetivas y eficaces. La estimación del estado hídrico a través de la medida del potencial hídrico de la planta mediante la cámara de presión se ha difundido ampliamente por ser un método objetivo sencillo, pero depende parcialmente no sólo de la conductancia estomática sino también de la tasa de transpiración foliar en la hora de medida [6], aunque más en plantas poco estresadas, pues según observaron Williams y Baeza [7], en plantas estresadas, con potencial a las 12 hs inferior a -1,2 MPa, la humedad disponible en el suelo fue el factor más influyente de su estado hídrico, atenuándose la influencia de las condiciones atmosféricas. Por tanto, dada la influencia que las condiciones atmosféricas pueden ejercer en los valores, debe tenerse en cuenta la hora de medida tanto del potencial hídrico como del intercambio gaseoso para valorar su viabilidad en cada situación de cultivo [10].

Los valores de potencial hídrico soportados por la planta a lo largo del ciclo vegetativo, así como sus variaciones, repercuten en el comportamiento fisiológico, que a su vez influye a nivel global en la planta, modificando su crecimiento y productividad, así como la composición de la uva [5]. Por tanto, es conveniente conocer la respuesta del potencial hídrico y de los parámetros de intercambio gaseoso a lo largo del ciclo en cada situación de cultivo para mejorar la eficiencia fisiológica en la utilización del agua de riego [8].

Ante la escasez de trabajos relativos al riego desarrollados en el ámbito del valle del río Duero con la variedad Cabernet Sauvignon, el objetivo del trabajo es estudiar las consecuencias hídricas y fisiológicas de la aplicación de 3 regímenes hídricos durante el verano, mediante riego por goteo, en Valladolid (España), para mejorar, consecuentemente, la eficiencia hídrica global en el manejo del riego del viñedo de dicha variedad en el valle del río Duero, en Castilla y León.

2 Material y Métodos

El trabajo se ha llevado a cabo en los años 2012, 2013 y 2014 en un viñedo localizado en Valladolid, Castilla y León (España). El material empleado es *Vitis vinifera* L, cv. Cabernet Sauvignon, sobre portainjerto 110 Richter, plantado en 2004, con un marco de 2,2 m x 1,2 m (3.788 plantas.ha^{-1}). El sistema de conducción es espaldera vertical, mediante cordón Royat bilateral y poda con 3 pulgares de 2 yemas en cada brazo (12 yemas por cepa).

El diseño experimental es en bloques al azar, con 4 repeticiones por tratamiento y una parcela elemental de 27 cepas, 7 de control, con líneas contiguas a cada lado destinadas al efecto borde. El ensayo experimental consiste en el establecimiento de 3 regímenes hídricos basados en la dosis de riego, que son: secano, 0% ETo (P0), riego de dosis baja, 25% ETo (P1) y riego de dosis media, 50% ETo (P2). El riego fue aplicado con frecuencia semanal, desde de la parada de crecimiento de pámpanos principales hasta vendimia, en la línea de cepas, con goteros autocompensantes de caudal 4,0 L.h^{-1}, colocados cada 60 cm, a 30 cm del tronco de la cepa.

En la tabla 1 aparecen reflejados los datos de riego y precipitación del periodo de estudio.

Tabla 1. Agua de riego aplicada y Fechas de inicio y fin en los tratamientos P0, P1 y P2; Precipitación anual de 1-octubre a 30-septiembre y Precipitación de 1-abril a 30-septiembre en 2012-2014.

Trat.	Riego (mm)				Año	Fechas de Riego		Precipitación (mm)	
	2012	2013	2014	Media		Inicio	Fin	Anual	1 abr - 30 sep
P0	-	-	-	-	2012	18 jul	1 oct	263	134
P1	89	97	93	93	2013	16 jul	23 sep	469	129
P2	177	193	186	185	2014	17 jul	26 sep	259	148
					Media	17 jul	27-sep	330	137

El suelo del ensayo presenta alta pedregosidad (más de 65% de elementos gruesos), lo que le confiere un buen drenaje, con una capacidad potencial de retención de agua estimada en 70 mm / m de profundidad. Es de tipo arcillo-arenoso en el 1er horizonte (20 cm) y franco-arcillo-arenoso en los dos horizontes siguientes (20-100 cm), llano, sin limitaciones físicas ni químicas para el cultivo de viñedo. En la tabla 2 aparecen reflejados los datos termopluviométricos del periodo 2012-2014.

Tabla 2. Datos medios mensuales y anuales (1-octubre a 30-septiembre) de Temperatura (ºC) máxima (Tmax), mínima (Tmin) y media (Tm), y Precipitación (mm) del periodo 2012-2014.

	T_{max}			T_{min}			T_m			P		
	2012	2013	2014	2012	2013	2014	2012	2013	2014	2012	2013	2014
Oct	22,1	18,5	19,2	5,5	6,2	8,3	13,1	12,0	13,2	17,2	72,6	23,2
Nov	12,8	12,3	11,8	5,0	3,0	2,2	8,5	7,4	6,8	64,4	60,4	3,2
Dic	9,2	9,4	8,4	0,0	1,3	-1,6	4,0	5,1	2,7	11,0	21,4	1,8
Ene	7,0	9,0	10,1	-1,4	0,7	2,3	2,2	4,6	5,9	28,4	36,8	22,8
Feb	10,1	9,8	10,0	-3,9	-0,9	1,3	2,6	4,0	5,4	1,0	31,2	48,1
Mar	17,1	12,2	15,8	0,6	2,9	2,2	8,6	7,1	8,7	7,6	117,9	11,0
Abr	13,7	15,4	20,0	3,9	3,8	6,4	8,6	9,5	13,0	66,8	28,6	22,4
May	23,4	17,5	21,5	8,3	4,8	6,6	16,0	11,0	14,1	20,2	27,9	18,8
Jun	26,7	23,9	26,4	11,3	8,7	9,9	19,2	16,3	18,5	12,6	33,6	9,6
Jul	28,8	31,9	28,0	11,7	14,0	12,8	20,4	23,0	20,3	12,2	9,4	36,2
Ago	30,2	29,9	28,6	12,7	12,4	12,4	21,5	20,9	20,6	0,0	29,0	0,2
Sep	25,2	26,2	26,0	10,1	10,3	12,3	17,4	17,6	18,5	21,8	0,0	61,2
AÑO	18,9	18,0	18,8	5,3	5,6	6,3	11,8	11,5	12,3	263,2	468,9	258,5

Se realizaron medidas de potencial hídrico foliar, con una cámara de presión de Soilmoisture Corp. (USA), y de intercambio gaseoso, con un IRGA Li-COR 6400, en 2 cepas de cada repetición, tomando 1 hoja por cepa. Las horas (solares, hs) de medida fueron: 9:00 y 12:00 hs. Las medidas se hicieron aproximadamente con periodicidad quincenal, durante la fase de maduración de la uva. El análisis estadístico de los resultados se ha realizado mediante análisis de varianza con el programa SPSS 16.

3 Resultados y Discusión

3.1 Potencial hídrico y fisiología en 2012

Las medidas de potencial hídrico realizadas en 2012, tanto a las 9 como a las 12 hs, mostraron valores significativamente menos negativos en el tratamiento regado con mayor dosis, P2, que en el de menor dosis, P1, y a su vez en éste que en el no regado, P0, excepto a las 12 hs el día 8, en que las diferencias entre tratamientos no fueron estadísticamente significativas, y el 22 de agosto, en que la diferencia entre P1 y P0 tampoco fue estadísticamente significativa (tabla 3).

Las medidas de fisiología realizadas, tanto a las 9 hs como a las 12 hs, mostraron valores de los parámetros analizados significativamente más altos en el tratamiento regado con mayor dosis, P2, que en el de menor dosis, P1, y a su vez en éste que en el no regado, P0, excepto a las 12 hs los días 5 y 20 septiembre, en cuanto a conductancia estomática y transpiración (tabla 4). En dichas medidas, la tendencia observada entre tratamientos fue la misma, creciente con el aporte de riego, pero las diferencias entre el tratamiento P1 y el P0 no fueron estadísticamente significativas en conductancia estomática ni transpiración.

Tabla 3. Potencial hídrico foliar (**Ψ**, MPa) a las 9 y a las 12 hs en **2012**, de los tratamientos P0, P1 y P2. Significación estadística (Sig.): * = $p<5\%$ (para todas las tablas en adelante).

Trat.	**Ψ 9 hs**			**Ψ 12 hs**				
	22-ago	*5-sep*	*20-sep*	*8-ago*	*22-ago*	*5-sep*	*20-sep*	*3-oct*
P0	-1,67 *c*	-1,64 *c*	-1,74 *c*	-1,51	-1,67 *b*	-1,79 *c*	-1,91 *c*	-1,88 *c*
P1	-1,37 *b*	-1,23 *b*	-1,50 *b*	-1,41	-1,55 *b*	-1,47 *b*	-1,69 *b*	-1,44 *b*
P2	-0,88 *a*	-0,87 *a*	-1,23 *a*	-1,27	-1,27 *a*	-1,11 *a*	-1,50 *a*	-1,01 *a*
Sig.	*	*	*	-	*	*	*	*

Tabla 4. Fotosíntesis (A_n, µmol $CO_2.m^{-2}.s^{-1}$), Conductancia estomática (g_s, mmol $H_2O.\ m^{-2}.s^{-1}$) y Transpiración (E, mmol $H_2O.\ m^{-2}.s^{-1}$) a las 9 y a las 12 hs en **2012**, de los tratamientos P0, P1 y P2.

Trat.	**A_n 9 hs**			**g_s 9 hs**			**E 9 hs**		
	22-ago	*5-sep*	*20-sep*	*22-ago*	*5-sep*	*20-sep*	*22-ago*	*5-sep*	*20-sep*
P0	2,49 *c*	4,62 *c*	4,50 *c*	27,8 *c*	23,8 *c*	35,3 *c*	1,17 *c*	0,61 *c*	0,59 *c*
P1	10,79 *b*	10,42 *b*	12,90 *b*	156,6 *b*	80,2 *b*	133,3 *b*	4,80 *b*	1,88 *b*	1,84 *b*
P2	14,41 *a*	14,25 *a*	17,25 *a*	284,1 *a*	134,9 *a*	206,9 *a*	6,92 *a*	2,88 *a*	2,56 *a*
Sig.	*	*	*	*	*	*	*	*	*
	A_n 12 hs			**g_s 12 hs**			**E 12 hs**		
P0	2,21 *c*	3,12 *c*	2,96 *c*	16,6 *c*	12,4 *b*	7,8 *b*	1,13 *c*	0,47 *b*	0,24 *b*
P1	6,60 *b*	9,06 *b*	8,96 *b*	74,5 *b*	31,1 *b*	32,5 *b*	4,23 *b*	1,08 *b*	0,95 *b*
P2	11,86 *a*	18,39 *a*	15,39 *a*	184,0 *a*	114,2 *a*	94,8 *a*	8,19 *a*	3,07 *a*	2,24 *a*
Sig.	*	*	*	*	*	*	*	*	*

3.2 Potencial hídrico y fisiología en 2013

Las medidas de potencial hídrico realizadas en 2013, tanto a las 9 como a las 12 hs, mostraron en general valores progresivamente menos negativos con el aumento de aporte de agua de riego.

Tabla 5. Potencial hídrico foliar (**Ψ**, MPa), Fotosíntesis (A_n, µmol $CO_2.m^{-2}.s^{-1}$), Conductancia estomática (g_s, mmol $H_2O.\ m^{-2}.s^{-1}$) y Transpiración (E, mmol $H_2O.\ m^{-2}.s^{-1}$) a las 9 y a las 12 hs en **2013**, de los tratamientos T0, T2, P1 y P2.

Trat.	**Ψ 9 hs**			**A_n 9 hs**			**g_s 9 hs**			**E 9 hs**		
	22-ago	*4-sep*	*19-sep*	*22-ago*	*4-sep*	*19-sep*	*22-ago*	*4-sep*	*19-sep*	*22-ago*	*4-sep*	*19-sep*
P0	-1,40 *b*	-1,52 *c*	-1,50 *b*	6,35 *c*	6,68 *b*	8,03 *c*	29,5 *c*	32,2 *c*	51,8 *c*	1,30 *c*	1,35 *c*	1,63 *c*
P1	-1,24 *b*	-1,34 *b*	-1,36 *b*	13,73 *b*	17,23 *a*	15,16 *b*	73,9 *b*	99,6 *b*	111,1 *b*	2,81 *b*	3,70 *b*	2,99 *b*
P2	-0,99 *a*	-0,99 *a*	-0,95 *a*	19,65 *a*	20,74 *a*	19,04 *a*	137,5 *a*	146,9 *a*	167,1 *a*	4,53 *a*	5,14 *a*	4,26 *a*
Sig.	*	*	*	*	*	*	*	*	*	*	*	*
	Ψ 12 hs			**A_n 12 hs**			**g_s 12 hs**			**E 12 hs**		
P0	-1,40 *b*	-1,57 *b*	-1,72 *c*	5,94 *c*	4,31 *b*	6,86 *c*	15,6 *c*	16,1 *c*	34,9 *c*	1,04 *c*	1,20 *c*	1,40 *c*
P1	-1,42 *b*	-1,53 *b*	-1,50 *b*	12,83 *b*	9,51 *b*	14,39 *b*	51,4 *b*	63,0 *b*	92,9 *b*	3,01 *b*	4,00 *b*	3,14 *b*
P2	-1,12 *a*	-1,34 a	-1,24 *a*	17,96 *a*	18,31 *a*	20,34 *a*	109,4 *a*	170,0 *a*	158,0 *a*	5,28 *a*	8,22 *a*	5,08 *a*
Sig.	*	*	*	*	*	*	*	*	*	*	*	*

Las diferencias entre el tratamiento regado con mayor dosis, P2, y el tratamiento no regado, P0, siempre fueron estadísticamente significativas, pero las diferencias entre el tratamiento regado con menor dosis, P1, y el no regado, P0, sólo fueron estadísticamente significativas el 4 de septiembre a las 9 hs y el 19 de septiembre a las 12 hs (tabla 5).

Las medidas de fisiología realizadas, tanto a las 9 hs como a las 12 hs, mostraron valores de los parámetros analizados significativamente más altos en el tratamiento regado con mayor dosis, P2, que en el de menor dosis, P1, y a su vez en éste que en el no regado, P0, excepto el 4 de septiembre, tanto a las 9 hs como a las 12 hs en cuanto a fotosíntesis neta (tabla 5). En dichas medidas, la tendencia observada entre tratamientos fue la misma, creciente con el aporte de riego, pero la diferencias no fueron estadísticamente significativas en fotosíntesis, entre los tratamiento P2 y P1 a las 9 hs y entre los tratamientos P1 y P0 a las 12 hs.

3.3 Potencial hídrico y fisiología en 2014

Las medidas de potencial hídrico realizadas a las 9 hs en 2014 mostraron valores significativamente menos negativos en el tratamiento regado con mayor dosis, P2, que en el de menor dosis, P1, y a su vez en éste que en el no regado, P0 (tabla 6). Las medidas realizadas a las 12 hs mostraron en general una tendencia similar, pero las diferencias entre los tratamientos P1 y P0 no fueron estadísticamente significativas ni el 7 ni el 28 de agosto. Tampoco fue estadísticamente significativa la diferencia favorable al tratamiento P2 con respecto al P1 el 18 de septiembre.

Las medidas de fisiología realizadas, tanto a las 9 hs como a las 12 hs, mostraron valores de los parámetros analizados notablemente más altos en el tratamiento regado con mayor dosis, P2, que en el de menor dosis, P1, y a su vez en éste que en el no regado, P0 (tabla 6). No obstante, las medidas realizadas a las 9 hs reflejaron que las diferencias entre los tratamientos P2 y P1, favorables al primero, sólo fueron estadísticamente significativas el 28 de agosto en cuanto a conductancia estomática y transpiración. Las medidas realizadas a las 12 hs mostraron diferencias estadísticamente significativas crecientes en función del aporte de agua, excepto en fotosíntesis entre P2 y P1 el 18 de septiembre y en conductancia entre P1 y P0 el 7 de agosto.

Tabla 6. Potencial hídrico foliar (Ψ, MPa), Fotosíntesis (A_n, µmol $CO_2.m^{-2}.s^{-1}$), Conductancia estomática (g_s, mmol $H_2O.\ m^{-2}.s^{-1}$) y Transpiración (E, mmol $H_2O.\ m^{-2}.s^{-1}$) a las 9 y a las 12 hs en **2014**, de los tratamientos T0, T2, P1 y P2.

Trat.	**Ψ 9 hs**			**A_n 9 hs**			**g_s 9 hs**			**E 9 hs**		
	7-ago	*28-ago*	*18-sep*	*7-ago*	*28-ago*	*18-sep*	*7-ago*	*28-ago*	*18-sep*	*7-ago*	*28-ago*	*18-sep*
P0	-1,31 c	-1,58 c	-1.54 *c*	12,92 *b*	7,62 *b*	8,49 *b*	47,2 *b*	31,1 *c*	28,4 *b*	2,24 *b*	1,43 *c*	0,92 *b*
P1	-1,14 *b*	-1,44 *b*	-1.05 *b*	18,55 *a*	16,83 *a*	18,31 *a*	85,1 *a*	82,8 *b*	69,1 *a*	3,58 *a*	3,35 *b*	2,13 *a*
P2	-0,96 *a*	-1,02 *a*	-0.88 *a*	21,91 *a*	20,06 *a*	19,92 *a*	111,8 *a*	123,2 *a*	78,5 *a*	4,48 *a*	4,53 *a*	2,31 *a*
Sig.	*	*	*	*	*	*	*	*	*	*	*	*
	Ψ 12 hs			**A_n 12 hs**			**g_s 12 hs**			**E 12 hs**		
P0	-1,34 *b*	-1,50 *b*	-1,61 *b*	7,86 *c*	4,62 *c*	7,03 *b*	29,8 *b*	16,9 *c*	15,6 *c*	1,98 *c*	1,07 *c*	0,61 *c*
P1	-1,32 *b*	-1,52 *b*	-1,31 *a*	13,93 *b*	13,00 *b*	17,04 *a*	60,5 *b*	59,4 *b*	59,1 *b*	3,68 *b*	3,34 *b*	2,16 *b*
P2	-1,12 *a*	-1,27 *a*	-1,13 *a*	21,11 *a*	19,61 *a*	19,76 *a*	127,5 *a*	122,2 *a*	79,8 *a*	6,46 *a*	5,68 *a*	2,84 *a*
Sig.	*	*	*	*	*	*	*	*	*	*	*	*

4 Conclusiones

Las medidas de potencial hídrico foliar realizadas tanto a las 9 como a las 12 hs mostraron valores notablemente menos negativos en el tratamiento regado con mayor dosis, P2, que en el de menor dosis, P1, y a su vez en éste que en el no regado, P0. Las diferencias entre tratamientos fueron estadísticamente significativas en general, aunque se observó alguna excepción en este sentido entre el tratamiento regado con menor dosis y el no regado. Las medidas de intercambio gaseoso realizadas tanto a las 9 hs como a las 12 hs reflejaron en general valores de los parámetros analizados significativamente más altos en el tratamiento regado con mayor dosis, P2, que en el de menor dosis, P1, y a su vez en éste que en el no regado, P0. Hubo algunas excepciones en el registro de diferencias estadísticamente significativas entre los tratamientos comparados, que variaron en fecha, hora y parámetro fisiológico, siendo más frecuente la ausencia de significación estadística entre P1 y P0 en 2012 y entre P2 y P1 en 2014.

La capacidad de discriminación entre tratamientos de riego observada en las medidas de potencial hídrico se ve bien reflejada en general en la actividad fisiológica del viñedo, con una tendencia creciente en la medida que el aporte de agua aumenta. Esta respuesta de la planta debe servir para enfocar el riego del viñedo hacia la optimización fisiológica y productiva del uso del agua.

Referencias

1. M.V. Alburquerque, J. Yuste. Efectos de la dosis de riego en el estado hídrico, el desarrollo y la calidad de la uva de la variedad Tempranillo en el valle del río Duero. Enoviticultura **10**: 24-31 (2011)
2. M.V. Alburquerque, R. Yuste, J. Yuste. Influencia de la dosis de riego en Tempranillo en el Valle del Duero: fisiología, superficie foliar, productividad y calidad de la uva. La Semana Vitivinícola **3476**: 1334-1340 (2016)
3. J.R. Castel, M.E. Valdés, M.H. Prieto, D. Uriarte, L. Mancha, A. Montoro, F. Mañas, R. López-Urrea, P. López-Fuster, J. Yuste, M.V. Alburquerque, J.R. Yuste, E. Barajas, A. Yeves, D. Pérez, D.S. Intrigliolo. Efectos de clima y suelo sobre la respuesta al riego de Tempranillo. I. comportamiento agronómico y relaciones hídricas en cuatro localidades de España. La Semana Vitivinícola **3393**: 198-206 (2013)
4. D.S. Intrigliolo, J.R. Castel. Response of grapevine cv. Tempranillo to timing and amount of irrigation: water relations, vine growth, yield and berry and wine composition. Irri. Sci. **28**: 113-125 (2010)
5. J.L. Salón, C. Chirivella, J.R. Castel. Response of cv. Bobal to timing of deficit irrigation in Requena, Spain: water relations, yield, and wine quality. Am. J. Enol. Vitic. **56**: 1-8 (2005)
6. L.G. Santesteban, C. Miranda, J.B. Royo. Suitability of pre-dawn and stem water potential as indicators of vineyard water status in cv. Tempranillo. Aust. J. Grape Wine Res. **17**: 43-51 (2011)
7. L.E. Williams, P. Baeza. Relationships among ambient temperature and vapor pressure deficit and leaf and stem water potentials of fully irrigated, field-grown grapevines. Am. J. Enol. Vitic. **58**: 173-181 (2007)
8. J. Yuste, A. Vicente. Manejo del riego y control del rendimiento del viñedo en un marco de limitación hídrica y cambio climático. Aplicación en la variedad Verdejo (*Vitis vinifera* L.). Phytoma España **274**: 116-120 (2015)
9. J. Yuste, A. Vicente, E. Barajas. Efectos del tipo de riego por goteo (tradicional y deshidratación parcial de raíces) en la calidad y la producción de uva del cv. Cabernet Sauvignon (*Vitis vinifera* L.) en el valle del río Duero. Enoviticultura **40**: 28-35 (2016)
10. J. Yuste, A. Vicente, E. Barajas. Efectos anuales de la dosis de riego en Verdejo (D.O. Rueda) en situación de elevada sequía. Aspectos fisiológicos y agronómicos. La Semana Vitivinícola **3498**: 1094-1099 (2017)

Agradecimientos

La elaboración de este trabajo ha sido posible a través de la base de proyectos INIA previos y del RTA2014-00049-C05-01, fondos FEDER y de la Junta de Castilla y León, así como la colaboración del Grupo de Viticultura del ITACYL.

Herramientas para la transferencia del conocimiento geo-edáfico: aplicación a los viñedos de Secastilla (DO Somontano, NE España)

D. Badía[1], J.A. Cuchí[1], J. Casanova[1], C. Martí[1], J.M. Rodríguez[1], J.M. Ayuso[2], J. Albar[2]

[1]Escuela Politécnica Superior de Huesca. Universidad de Zaragoza; correo: jcasan@unizar.es

[2] Viñas del Vero S.A.

Introducción

El suelo, dentro del contexto vitícola, puede considerarse como el nexo de unión entre las viñas y el material parental. El suelo, junto al clima, el relieve y el viñedo, permiten diferenciar un determinado "terroir", un espacio geográfico singular debido a las interacciones entre el medio físico y biológico y las prácticas vitivinícolas que en él se aplican (Morlat, 2010). En este trabajo se proponen diversas acciones para divulgar el conocimiento de los suelos de viñedos en el valle de Secastilla (D.O. Somontano), una zona de interés geomorfológico, geológico, edafológico y también histórico.

En este trabajo se muestran diversas herramientas con las que dar a conocer los suelos y su formación en el valle de Secastilla (D.O. Somontano).

Material y métodos

Seleccionados los perfiles de suelos más representativos, se describen macromorfológicamente (FAO, 2009) y micromorfológicamente (Benyarku & Stoops, 2005), y se muestrean para su análisis. De la fracción de tierra fina, se analizan sus principales propiedades físicas y químicas (pH, carbonatos, materia orgánica, CEe, iones en solución, porosidad, estabilidad estructural, CRAD, granulometría, CIC), se clasifican (IUSS, 2007) y se evalúa el potencial de vigor y de precocidad del terreno para la viña (Morlat, 2010). Además, se toman varios kilos de cada horizonte edáfico para disponerlos en cajas de metacrilato y elaborar los monolitos de perfiles. Se identifican los cantos y gravas presentes en los suelos, organizándose en cajas de madera, con una leyenda que indica su naturaleza, edad, origen y localización en la zona de estudio.

Resultados

La evolución geológica en el valle de Secastilla se sintetiza mediante bloques diagrama con el fin de explicar la presencia y distribución de los materiales parentales de los suelos de la zona. Por otro lado, la diversidad de litologías en el valle de Secastilla, y especialmente la alteración de los conglomerados que aportan multitud de cantos rodados, ha permitido elaborar cajas de rocas.

La posición topográfica, el sustrato geológico conjuntamente con la acción del clima y la vegetación a lo largo del tiempo dan origen a las diferentes clases de suelos. Por otro lado, las imágenes obtenidas con polarizadores paralelos permitieron observar los procesos de calcificación, humificación y gipsificación, Con la información generada se publicaron diferentes paneles y estructuras informativas (figuras 1, 2, 3)

https://doi.org/10.1051/e3sconf/20185001036

Figura 1: Caja de rocas representando los diferentes orígenes geológicos de los suelos de Secastilla

Figura 2: Monolitos de suelos mostrando los diferentes horizontes en cajas de metacrilato

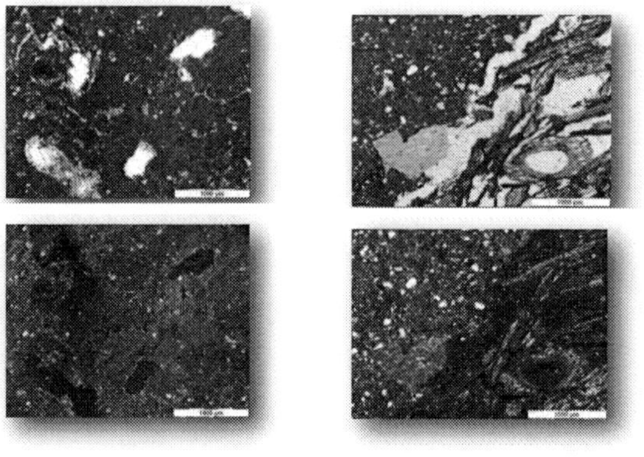

Figura 3: Paneles de microfotografías de los materiales edáficos en la actualidad. Calcificación (Izquierda) y Humificación (Derecha)

Conclusiones

A la tradicional caracterización de las propiedades de suelos y su distribución, le acompaña la elaboración de cajas de piedras, monolitos de perfiles de suelos (para su observación en sala y en campo) y paneles que muestran desde la evolución geológica hasta el potencial de vigor del terreno. Suelos y geología se incorporan, así como elementos con los que potenciar el turismo cultural y el eno-turismo en el mundo rural.

Bibliografía

Benyarku, C. A., & Stoops, G. (2005). *Guidelines for preparation of rock and soil thin sections and polished sections*. Departament de Medi Ambient i Ciències del Sòl, Universitat de Lleida.

FAO (2009) Guía para la descripción de suelos. FAO, Roma

IUSS Grupo de Trabajo WRB. (2007). Base Referencial Mundial del Recurso Suelo. Primera actualización 2007. Informes sobre Recursos Mundiales de Suelos No. 103. FAO, Roma.

Morlat, R. (2010). Traité de viticulture de terroir. *Editions Tec&Doc*

Impact of sown fallows on the *Xiphinema index* populations in different soil types

Coralie Laveau[1], Séverine Mary[1], Guillaume Darrieutort[1], Cornelis van Leeuwen[2] and Maarten van Helden[3]

[1] : *Univ. Bordeaux, Vitinnov, Bordeaux Sciences Agro, ISVV, 1 cours du Général de Gaulle, 33170 Gradignan, France*
[2] : *Bordeaux Sciences Agro, Univ. Bordeaux, ISVV, Ecophysiology and fonctional genomics of grapevines, GFV, INRA, UMR 1287, F-33140 Villenave d'Ornon, France*
[3] : *Bordeaux Sciences Agro, Univ. Bordeaux, 1 cours du Général de Gaulle, 33170 Gradignan, France / University of Adelaide, SARDI Sustainable Systems GPO Box 397 Adelaide SA 5001*

**Corresponding author: Coralie Laveau, coralie.laveau@agro-bordeaux.fr*

The nematode *Xiphinema index* is, economically, the major virus vector in viticulture, transmitting specifically the Grapevine Fanleaf Virus (GFLV), the most severe grapevine virus disease worldwide. The management of this disease has long been to use soil fumigation, harmful for both the applicator and the environment. The objective of this study was to evaluate an alternative approach using plants to reduce vector nematode populations between uprooting and replanting. Of thirty botanical species, tested in previous greenhouse trials, the seven best performing plants were evaluated for their capacity to reduce *X. index* populations in soil compared to bare soil in 5 field trials on different soil types in Bordeaux and Burgundy. In most trials, sown fallows reduced the number of *X. index* nematodes more efficiently than bare soil. All plants tested in field, except *Trifolium pratense*, showed their efficacy in field on survival of nematodes *X. index* but this efficiency varied according to species and site. The best results were obtained with *Medicago hybride, Tagetes minuta, Avena sativa* and *Vicia villosa*. Over the following years we will be evaluating if a decrease of the populations of the nematode vector does lead to a significant drop or delay of GFLV contamination for the newly planted vines.
Keywords : Fallow, *Xiphinema*, Virus, Grapevine fanleaf, soil type

Introduction

Fanleaf degeneration is a viral disease affecting the vines longevity, the production potential and the quality of wine grapes. Two-third of French Vineyards is affected by this viral disease [1].
The two main viruses responsible for the fanleaf disease are the grapevine fanleaf virus (GFLV) and the arabic mosaic virus (ArMV). They are respectively transmitted by *Xiphinema index* and *Xiphinema diversicaudatum* which often living deep in the soil [2]. Among the virus vector's nematodes, *X. index* has the most significant economic impact. The worldwide distribution of *X. index* is closely related to that of grapevine.
The propagation of GFLV is possible either by plant material or by nematodes from grapevine to grapevine. As the dissemination by propagation material has been reduced by clonal selection and certification for plants, it is now necessary to find strategies to control the propagation of GFLV in vineyard conditions. An effective strategy is needed to reduce nematode populations in the soil. For a long time, winegrowers used chemical products to fumigate the soil between two successive grapevine crops. One of the alternatives to the use of these toxic synthetic products is the use of antagonistic plants during fallow period for their effect on the reduction of nematodes vectors.
Several plants are known to their nematicidal effect on different crops. For instance, *Tagetes spp.* have been identified to their effective control of some species of root-infesting nematodes as *Pratylenchus penetrans* or *Meloidogyne hapla* [3, 4]. *Sorghum vulgare* incorporated as green manure before planting lettuce resulted in lower reproduction of M*eloidogyne hapla* [5].
In *vitro* biotests, *Thymus vulgaris* or *Ruta graveolens* extracts showed a nematicidal activity against *X. index* and *X. americanum* [6] but this effect was not confirmed in field tests. *Chenopodium ambrosioides* extracts have an effect on *X. index* and *X. americanum* in vitro but also against *Meloidogyne spp.* or *Ditylenchus dipsaci* [7, 8]. Aballay showed that all the plants tested in greenhouse trial (*Chenopodium ambrosioides, Ruta graveolens, Thymus vulgaris and Brassica juncea)* were able to significatively reduce *X. index* populations but the final densities were significantly higher than chemical control [9]. Field tests with *Brassicus napus, Cosmos bippinatus, Lupinus albus, Thymus vulgaris, Zinnia elegans, Hordeum vulgare or Calendula officinalis*, showed none of these plants were able to control the population of nematodes during their development or incorporation. Only Brassica significantly decreased populations after incorporation [10].
This study evaluates the use of a few plant species, previously selected through greenhouse screening, for their capacity to reduce *X. index* populations [11, 12] during fallow in the field.

Material and method

Seven plant species able to decrease the nematodes populations by about 50% compared to bare soil (negative control) in controlled conditions [11, 12], were tested in field conditions. Several trials were carried out between 2007 and 2012 with the same plants

https://doi.org/10.1051/e3sconf/20185001037

to take into account the effect of sowing date and pedoclimatic conditions.

- **Experimental field design**

The trials were carried out in the Bordeaux and Burgundy vineyard regions. For each trial, the protocol design encompassed 10 to 20 replicates of each tested plant. In the chessboard-like arrangement, each replicate (3 x 3m plot) was surrounded by several control plots (negative control = bare soil). A total of 98 to 142 plots was required to evaluate 3 to 4 plants per trial. Mechanical weeding was performed on the bare soil plots.

The sowing of the plant plots was carried out at the end of April for the first trial and in September/October for the four others trials with a SEMBONER manual single row seeder.

The plants were grown from 4 to 9 month depending on sowing date and vine planting date. Table 1 shows the characteristics of each trial.

Table 1 : Characteristics of the five trials

Trial	A	B	C	D	E
Site	Margaux (33)	Saint Emilion (33)	Gevrey Chambertin (21)	Saint Emilion (33)	Arsac (33)
Climate	Oceanic	Oceanic	Continental	Oceanic	Oceanic
Soil	Sandy gravel soil	Sandy gravel soil	Clay limestone soil	Clay limestone soil	Sandy gravel soil
Uprooting date	March 2007	March 2009	May 2012	July 2012	March 2012
Plants	*Vicia villosa, Lupinus albus, Tagete minuta, Medicago hybride*	*Avena sativa, Vicia villosa, Medicago hybride, Trifolium pratense*	*Avena sativa, Vicia villosa, Medicago hybride*	*Onobrychis viciifolia, Avena sativa, Medicago hybride*	*Avena sativa, Vicia villosa, Medicago hybride*
Control	Bare soil	Bare soil	Bare soil	Bare soil	Bare soil
Number of replicates	10	12	20	20	20
Samples number	98	142	110	110	114
Sowing	April 2007	September 2009	October 2012	September 2012	October 2012
Sampling date	August 2007	July 2010	June 2013	June 2013	April 2013
Period	4	9	8	9	6

- **Nematodes sampling and extraction**

At each sampling point, a trench (1.5 m long, 0.4 m wide, 1 m deep) was dug with an excavating machine. A 4 liter sample of soil was collected manually from 5-6 points from the undisturbed sides of the trench at depths of 0.3 to 0.8 depending on soil profile.

After mixing, nematodes were extracted in the laboratory from a 2 liter subsample, using the method described by Villate [12]. Adults and juveniles were counted from the final suspension under a dissecting microscope. Maximum observed rooting depth was observed for each plot

- **Data analysis**

For each site, the effect of sown fallows on the rooting depth was analyzed using ANOVA. When a significant effect was found, multiple comparisons were conducted to test differences using Tukey's HSD test. For each site, number of *X. index* on soil and number of *X. index* on bare soil were defined as a matched pair of counts. Effect of plants on *X. index* were analyzed as proportion data using a GLM with binomial errors and logit link and to test the difference to zero. Soil effect was tested with mixed-model. We used fixed effects for plants and soils and a random effect to take in account the site.

Results

- **Vine root system**

To control *X. index*, the first challenge for the plants tested is to reach the deeper soil layers where the nematodes live. So rooting depth was measured for each plant on each plot. The results are presented in figure 1.

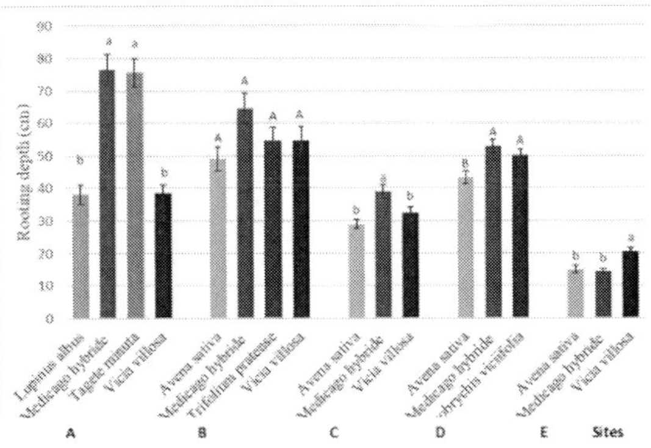

Figure 1: Depth of rooting of each plant on the 5 test sites.

In trial A, *Medicago hybride* and *Tagetes minuta* roots reached 75 cm mean depth whereas those of *Lupinus albus* and *Vicia villosa* reached only 38 cm. In trial B, there was no difference between rooting depths, *Medicago hybride* obtained 65 cm mean depth which was less than in trial A. For the same growing time, roots of *Medicago hybride* were even less developed in trial C and D with respectively 39 cm and 53 cm mean depth compared to the two first trials but *Medicago* was still the deepest rooting plant. In trial C, the difference was significant with *Vicia villosa* (32 cm) and *Avena sativa* (29 cm). In trial D, *Onobrychis viciifolia* obtained results similar to *Medicago hybride* with 50 cm mean depth while roots of *Avena sativa* reached 43 cm . In the last trial E, the plants were poorly developed the day of the observations in April, their roots had still reached 20 cm mean depth for Vicia villosa and around 15 cm mean depth for *Avena sativa* and *Medicago hybride*. Based on these results, it seems that the rooting depth is more dependent on the soil type than the length of the crop cycle (except for test E). For the same species, rooting will be deeper in sandy gravel soils than in clay limestone soils.

- **Antagonistic effect of plant species on *Xiphinema index* numbers**

All tested plant were able to reduce the populations of *X. index* compared to bare soil during at least one trial (one vegetative cycle in field) except *Trifolium pratense* (Figure 2). However, there are differences depending on the test sites.

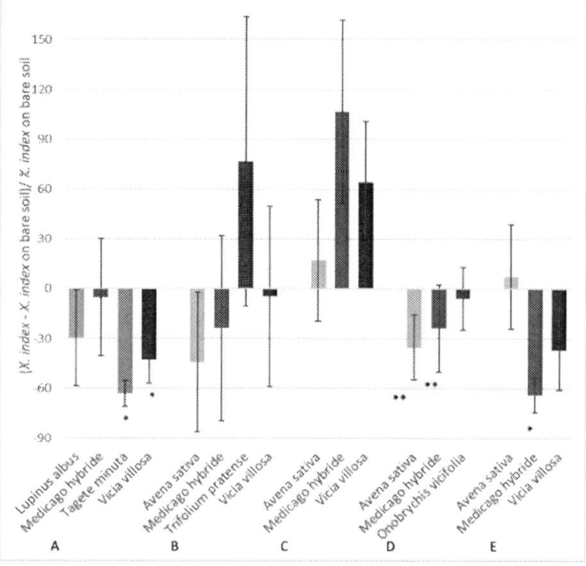

Figure 2: Efficiency of fallow plants on the reduction of *X. index* populations compared to bare soil

In the first trial (Site A), sown in spring and harvested in August, revealed a good efficiency of all 4 plants. The strongest reduction of *X. index* populations was obtained with *Tagetes minuta* that reached more than 60% decrease in *X. index*. A decrease in *X. index* populations of 40% and 30% was observed on plots sown with *Vicia villosa* and *Lupinus albus*, respectively, whereas only 5% reduction on *X. index* populations was obtained for *Medicago hybride* compared to bare soil. But only *Tagetes minuta* and *Vicia villosa* show a significant difference compared to the bare soil. In the second trial (Site B) was sown in autumn and harvested the next spring, which prevented the sowing of (frost sensitive) summer crops *Tagetes* and *Lupinus albus*. Only *Medicago hybride* and *Vicia villosa* species could be evaluated again. *Trifolium pratense* and *Avena sativa* completed the design trial. The best result in this trial was obtained with *Avena sativa* which reached more than 40% decrease of *X. index* compared to bare soil but the difference was not significant. *Vicia villosa* and *Medicago hybride* were less successful with only 24 % decrease in *X. index* numbers for *Medicago hybride* and no significant difference with bare soil. *Trifolium pratense*, that gave very good results in the greenhouse, [11, 12] did not confirm this in the field.

The other three trials were sown in October 2012 and harvested in late spring 2013. The trial on the site E was harvested earlier in April. On site C, the plants fallow did not reduce the populations of nematodes and on the contrary, there was more *X. index* than the bare soil control at the end of the trial (no statistical difference). On site D, all plants showed a decrease of *X. index* populations after 9 month in the same proportion with *Avena sativa* and *Medicago hybride* as in trial site B but the differences were significantes compared to the bare soil on this trial. *Onobrychis viciifolia* did not confirm in the field the results obtained in greenhouse. In site E the plants didn't have time to develop but *Medicago hybride* still allowed a very strong reduction in populations compared to the control with a statistical difference.

The period of plant development does not seem to have a strong impact on the percentage reduction of *X. index* populations. On the other hand, soil type seems to play a larger role in the nematodes population decrease with a stronger decrease on sandy gravel soils than on clay limestone soils.

Discussion

All the plants tested in field conditions had shown reduced nematode populations in greenhouse tests but they did not confirm these results in all field trials. In the field, the highest reduction was obtained with *Medicago sativa* (trials B, D and E) but in trial A this plant showed no effect and in trial C there was more *X. index* on average on the plots of this plant compared to the bare soil plot. Conversely, *Avena sativa* gave the most regular results in the different trials with reduction between 30% and 45% on trials B and D but no effect of this plant compared to bare soil in trial C and E. *Vicia villosa* did not confirm the strong diminution of *X. index* populations compared to the bare soil, obtained in the first trial, in the next trials. The decrease of *X. index* populations were better in greenhouse than those in field for *Tagetes minuta* and a little less for *Lupinus albus* but these two plants have been set up on only one trial [11]. However *Tagetes spp.* is known for its strong nematicidal effect on several species of nematodes and the results obtained in this test confirms once again this property. *Lupinus albus* reduced *X. index* population in our conditions while Aballay and Insunza [10] had not shown efficacy. Likewise, *Trifolium pratense* did not reduce populations of nematodes in the field, but was used in only one trial.

We can not explain why *Vicia villosa, Medicago hybride*, and to a lesser extent *Avena sativa,* have led to an *increased* survival of the populations of nematodes on the site C in Burgundy. The implantation of more trials in this geographic area would have been necessary to conclude. It could be that anaerobic conditions due to waterlogging in the bare soil reduced nematode numbers, while waterlogging was prevented by the plants. The success of one of a plant should be connected with its development and particularly with its rooting depth. However, the best results have not always been obtained with the plants which had the deepest root system.

Many unexplained points persist and the implantation of new trials is necessary to better qualify the link between the pedoclimatic conditions, the implantation of the species and their efficiency to reduce the populations of *X. index*. New experiments will be conducted to better understand the mechanisms of action of the best plants to reduce populations of *X. index* in the field. New trials will be set up to follow the rate of recontaminations of a new vine by the GFLV after two years of fallow with these nematicidal plants.

Acknowledgements

We thank the estates and their vineyard managers that allowed the realization of the trials on their plots and have funded the first one. We also thank France AgriMer that financially supported the three last trials.

References

1- G. Demangeat, D. Esmenjaud, R. Voisin, J.-M. Bidault, S. Grenan, M. Claverie. Le court-noué de la vigne: l'état des connaissances sur cette maladie lors de la journée "Alternative" du 28 avril 2005. Phytoma-La Défense des végétaux: 38-42. (2005)
2- P. A. Andret-Link, C. Laporte, L. Valat, C. Ritzenthaler, G. Demangeat, E. Vigne, V. Laval, P. Pfeiffer, C. Stussi-Garaud, M. Fuchs. Grapevine fanleaf virus : still a major threat to the grapevine industry. J. Plant Pathol. **86** : 183-195. (2004)
3- M. Oostenbrink. *Tagetes patula* L. als voorvrucht van enkele land- en tuinbouwgewassen op zand- en dalgrond. Meded. Landb.Hoogesch. Gent, **25** : 1065-1075 (1960)
4- M.J. Hijink and R. Winoto Suatmadji. Influence of different compositae on population density of *Pratylenchus penetrans* and some other root-infesting nematodes. Neth. J. Pl. Path. **73**:71-82 (1967)
5- N. M. Viaene and G. S. Abawi. Management of *Meloidogyne hapla* on Lettuce in organic soil with sudangrass as a cover crop. Plant disease **82** : 945-952 (1998)
6- B.V Insunza, E. Aballay and J. Macaya. Nematicidal activity of aqueus plants extracts on *Xiphinema index*. Nematologia mediterranea **29** (1) : 35-40 (2001)
7- R. Garcia Espinoza. Chenopodium ambrosioides, planta con uso potencial en el combate de nematodos fitoparasitos. Agricultura tropical **2**: 92-97. (1980)
8- B.V. Insunza and A. Valenzuela. Control of *Ditylenchus dipsaci* on garlic (Allium sativum) with extracts of medicinal plants from Chile. Nematropica **25**: 35-41(1995)
9- E. Aballay, R. Sepúlveda and V. Insunza. Evaluation of five nematode-antagonistic plants used as green manure to control *Xiphinema index* Thorne et Allen on Vitis vinifera. Nematropica **34** (1) : 45-51(2004).
10- E. Aballay and B. Insunza. Evaluation of plants with nematicidal properties in the control of *Xiphinema index* on table grapes cv. Thompson Seedless in the central zone of Chile. Agricultura Técnica **62** (3) (2002)
11- C. Laveau, C. van Leeuwen and M. van Helden. Management of grapevine fanleaf virus vectors. Proceedings of the 18th GiESCO conference 7-11 July 2013 PORTO Portugal (2013).
12- L. Villate. Origine, variabilité et gestion des populations de *Xiphinema index*, le nématode vecteur du Grapevine fanleaf virus (GFLV). Thèse de l'Ecole Doctorale "Vie-Agro-Santé", ENSA et Université de Rennes, 160 p. (2008).

Leaf morphoanatomy of four red grapevine cultivars grown under the same terroir

Ana Monteiro[1a], Generosa Teixeira[2], Cristina Santos[3] and Carlos M. Lopes[2]

[1a] Linking Landscape, Environment, Agriculture and Food (LEAF), Instituto Superior de Agronomia, Universidade de Lisboa, Tapada da Ajuda, 1349-017 Lisboa, Portugal. anamonteiro@isa.ulisboa.pt
[2] Centre for Ecology, Evolution and Environmental Changes (CE3C), Faculdade de Farmácia, Universidade de Lisboa, Av. Prof. Gama Pinto, 1649-003, Lisboa, Portugal, gteixeira@ff.ulisboa.pt
[3] IQV Agro Portugal SA, R. Bernardo Sequeira 16, Sala 5, 4715-671 Braga, Portugal, csantos@iqvagro.com

Abstract. This study compare leaf morphoanatomical characteristics of four red cultivars - 'Touriga Nacional', 'Trindadeira', 'Cabernet Sauvignon' and 'Syrah' -, grown side by side at the same terroir. The analyzed leaf traits, under light and scanning electron microscopy, showed large variability among genotypes. 'Trincadeira' has the biggest single leaf area and 'Cabernet Sauvignon' the smallest one. 'Touriga Nacional' showed the lowest leaf dry weight and 'Trincadeira' the highest one, nonetheless there was no significantly differences in leaf dry mass per area and in leaf density. Leaf dry mass per area was positively correlated with leaf density but showed no correlation with leaf thickness. The French genotypes presented higher thickness of the leaf anatomical traits than the two Portuguese ones. 'Trincadeira' showed significantly highest stomata density while the other cultivars showed no significant differences among them. The analyses of the three types of stomata revealed that 'Trincadeira' has the lower percentage of raised above and the highest percentage of sunken stomata while 'Cabernet Sauvignon' showed the opposite behaviour. The hairs on the lower surface presented a similar woolly aspect in all cultivars. The possible role of leaf morphoanatomical characteristics in determining the cultivars adaptation to abiotic stresses is suggested and discussed.

1 Introduction

Leaf morphoanatomical characteristics are functional and adaptive traits, inherent to genetic factors but also influenced by abiotic ones, which can regulate plant water use, heat dissipation and photosynthesis [1, 2, 3, 4]. The capacity of a plant to adjust leaf area to the water availability is also an important trait for drought avoidance. Some authors [5, 6, 7] stated that mesophyll architecture can influence leaf water movement by affecting water flux in the mesophyll and water evaporation at the cell wall surface. Tomás *et al.* [8] also underline the role of anatomy in mesophyll diffusion conductance and, hence, in the variability of photosynthetic capacity among species. Leaf hairs are also implicated in plant protection from biotic stresses as they can decrease pathogen germination rates by decreasing leaf wet ability [9]. The stomata type and size, width and length, are important traits for plant growth, with small, sunken or sparsely distributed stomata being considered adaptations to minimize transpiration [10].

Regarding the importance of leaf morphoanatomical characteristics for plant resistance to biotic stresses, some authors [11, 12, 13, 14, 15] have showed that different *Vitis* cultivars and even clones presented different susceptibility to downy mildew (*Plasmopara viticola* (Berck. & M.A. Curtis) Berl. & De Toni in Sacc.) and powdery mildew (*Erysiphe necator* Schwein.). The epidermis is the first barrier for invading pathogens and trichomes can act as mechanical barriers to insects. More studies concerning less susceptible cultivars to biotic stresses are needed to further understand the impact of anatomical traits for example on plant resistance to diseases and pests in order to reduce the enormous pesticide consumption in Viticulture [16, 17].

Since microcharacters are less plastic than the macro ones [18], microcharacters could be more important for taxonomical and physiological studies, in particular for

 https://doi.org/10.1051/e3sconf/20185001038

understanding grapevine cultivars adaptation to different abiotic and biotic stress and, hence, to allow the selection of genotypes better adapted to face the challenges of the climate change expected scenarios for the near future [19].

This study aims to compare the leaf morphoanatomical characteristics of four grapevines red cultivars (*Vitis vinifera* subsp. *vinifera*) - the Portuguese 'Touriga Nacional' (TN) and 'Trindadeira' (Tr) *vs* the French 'Cabernet Sauvignon' (CS) and 'Syrah' (Sy) grown at the same terroir. Despite being major cultivars, little is known about their tolerance to biotic and abiotic stresses. The cultivar TN is known as resistant to diseases and some pests but also as susceptible to water stress and excoriose; Tr is considered susceptible to downy and powdery mildew and *Botrytis* bunch rot and also as tolerant to drought stress [20, 21, 22]; CS is tolerant to diseases (powdery and downy mildews, *Botrytis* bunch rot) and to water stress but is susceptible to wooden diseases; Sy is susceptible to diseases (powdery and downy mildews, excoriose) and mites and tolerant to water stress (23).

2 Material and Methods

2.1 Plant material and growth conditions

Leaves of eight-year-old field-grown grapevines of the red cultivars from different geographical origins, the Portuguese 'Touriga Nacional' (TN) and 'Trincadeira' (Tr) and the French cultivars 'Cabernet Sauvignon' (CS) and 'Syrah (Sy), were studied. The 15 years old grapevines were grafted onto the 140 RU rootstock and grown in an rainfed vineyard located at Tapada da Ajuda, Lisbon, Portugal, within the Lisbon Winegrowing Region (38° 42' 27.5'' N, 9° 10' 56.3'' W and 62 m above sea level). The soil is a clay loam with 1.6% organic matter and a pH of 7.8. The vines are spaced 1.2 m within and 2.5 m between north-south oriented rows and trained on a vertical shoot positioning trellis with two pairs of movable wires, and spur-pruned (18-20 nodes per vine) on a bilateral Royat Cordon system. . Shoots were trimmed once, at berry pea-size. The 1.0 ha vineyard is a varietal collection where the four studied cultivars are planted side by side, (6 adjacent rows of ~30 plants per cultivar) under the same trellis and standard cultivation management. For data collection, four blocks of 10 plants per cultivar were established along a central row. For the macro and microscopic observations twelve fully expanded leaves per block from each cultivar were collected at veraison (end of July), from the 8-9th shoot node.

2.2 Leaf area, dry mass and leaf density

For the determination of the single primary leaf area, 12 leaves were scanned at 300 dpi and leaf area was determined from the images using the image processing program ImageJ®. Those 12 leaves were then oven dried at 70 °C for 48-h and weighed. From these measurements dry mass per unit area and specific leaf area were calculated. Using leaf thickness data from anatomical measurements, leaf density was calculated as dry mass per unit leaf mesophyll thickness [1].

2.3 Anatomical measurements

Five small leaf blade sections were cut from the central part of each fresh leaf, between the primary and secondary veins, and fixed following the usual processes [24, 25]. The cross-sections (8-10 μm) were done using the paraffin histological technique (26; 25) at a Minot Leitz 1512 microtome and examined with a Nikon Labophot 2 photomicroscope. Images were captured with a Nikon FX-35W camera equipped with a semiautomatic Nikon PFX adapter (Nikon®). The light microscopy (LM) observations focused on 22 leaf transverse sections were addressed to thickness of total lamina, mesophyll, palisade tissue, spongy tissues, upper cuticle, upper and lower epidermal cells and stomata dimensions (length and width). For Scanning electron microscopy (SEM) analysis plant material was fixed as above, critical-point dried in a Critical Point Polaron BioRad E3500, and coated with gold in a Jeol JFC-1200. Observations were carried out at 15 kV on a Jeol JSM-5220 LV scanning electron microscope with computer-assisted image analysis. SEM observations focused on the upper and lower epidermis surface details - type of indumentum, epicuticular waxes, stomata density (stomata mm^{-2}) and type [27]. All measurements and counts were done on random fields under LM and SEM images, always at comparable leaf situations and magnifications. Quantitative characters are the average of 12 observations for macroscopic parameters.

Statistical analysis were carried out in accordance with the SAS® program package (SAS Institute, Cary, NC, USA). Before ANOVA, the percent stomata type was transformed using arcsine of the square root of the percentage. Differences between means were assessed by LSD test ($p < 0.05$).

3 Results

The average values of the single primary leaf area, dry weight, specific leaf area and leaf density, inserted at the 8-9th node, are presented in Table 1. 'Trincadeira' presented the significantly largest single leaf area and CS showed the smallest leaves while the remaining cultivars returned intermediate values with no significant differences among them. 'Trincadeira' also presented the significantly highest leaf dry weight and TN the lowest one while the French genotypes showed intermediate values. Regarding the specific dry weight and leaf density, no significant differences were observed among genotypes. Specific leaf dry weight showed a significant positive correlation with leaf density ($r=0.95$, $p<0.001$, data not shown), but was not correlated with mesophyll thickness ($r=0.004$, $p>0.05$, data not shown). In all the studied cultivars the upper epidermis showed polygonal cells with straight walls, epicuticular waxes, some striate as well and pearl glands (Fig. 1A). In the lower epidermis the cell organization of all cultivars was similar, with several stomata types (hypostomatous leaves) (Fig. 1B) and one or two types of leaf hairs: i) small, multicellular, erect or slightly curved and ii) very long, probably unicellular, usually flat with a helical rolling, producing a fluffy effect (Fig. 1C, 1D). Both kinds of hairs presented an irregular distribution, no orientation and a variable density over the abaxial epidermal surface within each cultivar. These features and the fluffy long hairs unable the assessment of trichome density.

Considering the leaf cross sections, the studied genotypes showed dorsiventral leaves with an asymmetric mesophyll in structure limited in both upper and lower epidermis by a single layer of flattened cells (Fig. 1E and 1F). The thickness of upper epidermal cells presented the significantly higher values in the cultivars CS, Sy and Tr, and the lowest one in the cultivar TN. Regarding the thickness of the lower epidermal cells, Tr presented the significantly highest values and CS, Sy and TN the smallest ones (Table 2). The analysis of the upper cuticle thickness showed that the French cultivars presented significantly higher values than TN. Considering the lower cuticle thickness CS showed the significant highest value and Tr the lowest one (Table 2).

The asymmetric mesophyll was composed of two distinct parts, one close to the upper epidermis and the other one close to the lower epidermis. The palisade parenchyma, consisted of 1 to 3 layers of relatively elongated and close cells. The spongy parenchyma showed small and loosely arranged oval to round cells and irregular intercellular air-spaces (Fig. 1E and 1F). The thickness values of the anatomical traits of total lamina, mesophyll, palisade tissue and spongy tissue showed highly significant differences between cultivars (Table 2). In general the Portuguese cultivars presented the lowest values for all the above mentioned characters and Sy the highest ones. The palisade tissue varied from 40-45% and the spongy tissue from 55-63% of total mesophyll.

The stomata density (n° mm^{-2}) of Tr was significantly higher than the other three cultivars, which return similar values among them (Fig. 2A). Considering the insertion of stomata in the epidermis, three different types coexisted in all cultivars (Fig. 1B): i) raised above - the guard cells are above and each stomata is flanked by curved subsidiary cells; ii) same level - the guard cells are flattened with the subsidiary cells; and iii) sunken - the guard cells are buried between the subsidiary cells. All the three types of stomata presented a random distribution and an irregular orientation. Significant differences between cultivars were only observed in the raised above and sunken types of stomata (Fig. 2B): Tr presented the significantly highest percentage of sunken stomata (38%) and CS the lowest one (18%). Regarding the raised above stomata the higher percentages were presented by CS and TN (36%). In all the genotypes the dominant type of stomata was the same level stomata which varied from 41% in TN and Tr to 48% and 52% of the total stomata in Sy and CS, respectively, although no significant differences were detected (Fig. 2B). When analyzing the dimensions of the three types of stomata no significant differences were observed between cultivars, with the exception of the length of the sunken stomata where Tr showed the lowest value. In all cultivars the raised above stomata was the type that featured larger dimensions (length and width) whereas the same level stomata were in-between and the sunken stomata presented always the smallest dimensions (Table 3). Idioblasts with calcium oxalate crystals were also seen in leaf cross sections. They were widely distributed in both parenchyma tissues and two different types of calcium oxalate crystals were found inside: raphids and druses.

3 Discussion

Leaf size is an important feature regarding water use and heat dissipation with smaller leaves being known to dissipate heat to the ambient more quickly than larger leaves because of a smaller boundary layer resistance (10). As compared with data from similar studies carried out in other Portuguese terroirs ([20, 26]), our results displayed lower single leaf area, showing that the terroir could have a strong influence on leaf size. The observed significant correlation between specific dry weight and leaf density indicate that leaf density was the main responsible for the variation in specific dry weight as reported by Tomás *et al.* [8]. It is also reported that mesophyll architecture namely leaf thickness and density, which are the two components of leaf dry mass per area, might be related with mesophyll conductance to CO_2 [28] and with leaf water movement by affecting water flux in the mesophyll and water evaporation at the cell wall surface [6, 7]. This pointed out for further studies where ecophysiologic responses need to be correlated with leaf anatomy.

All cultivars displayed upper and lower unistratified epidermal cells with thin walls and a thin cuticle. They had almost the same rectangular or slightly polygonal cell shape. Cuticular striations were seen on both upper and lower surfaces, especially around the stomata and wax deposits were very few or scarce – for example in CS, similar observations were reported by other authors in different terroirs [11, 25]. Martin and Juniper [29] stated that cuticle can provide important information about plant habitat and its response to abiotic stresses. Average cuticular thickness varied between 1.9 μm in Tr to 2.5 μm in CS, values that are lower than the ones reported by Monteiro *et al.* [26]. Since three of the four genotypes used by Monteiro *et al.* [26], were the same, (TN, Sy and CS), these differences might have been determined by the growth conditions, namely the higher drought and heat stress observed in our study. Indeed, under heat stress conditions Ben Salem-Fnayou *et al.* [30] reported a significant decrease in cuticle thickness in their cultivars, followed by an increase in epidermal cell wall thickness. In our study only the thickness of the upper and lower epidermal cells was measured and for the same genotypes, the values, were higher in Lisbon terroir than in Torres Vedras [26].

In *Vitis* leaves the stomata are hypostomatous with a stomata density varying with leaf position and growth conditions (50 to 400 stomata mm^{-2}; [31]). In our study the obtained stomata density values are within that range but showed a smaller amplitude (from 179 stomata mm^{-2} in CS to 256 stomata mm^{-2} in Tr). A higher stomatal density is a feature of xerophytic plants as the presence of a huge number of stomata allows more efficient control of leaf gas exchanges [32]. As a rule, the size of stomata is negatively correlated with the number and the size of epidermal cells [33] which is also confirmed in our experiment for all the studied cultivars. The differences observed between the four genotypes on stomata features (density, type and dimensions) might have been an adaptation to the environmental conditions namely for the tolerance to water stress [10, 34].

Table 1. Average ± standard error of single leaf area, specific leaf area and leaf density for the four red cultivars of *Vitis vinifera* L. subsp. *vinifera*, cultivated at the Lisbon winegrowing region, Portugal.

Grapevine cultivars	Single Leaf Area (cm^2)	Dry Weight (g)	Specific leaf dry mass (mg cm^{-2})	Leaf Density (g cm^{-3})
'Cabernet Sauvignon'	123.2 (9.67) c	1.24 (0.13) bc	11.2 (1.61) a	0.61 (0.10) a
'Syrah'	163.9 (11.35) ab	1.60 (0.16) ab	10.6 (1.65) a	0.54 (0.09) a
'Touriga Nacional'	137.8 (9.99) bc	1.21 (0.10) c	9.4 (1.10) a	0.60 (0.07) a
'Trincadeira'	174.9 (12.87) a	1.72 (0.15) a	10.6 (1.40) a	0.63 (0.08) a

In each column different letters indicate significant differences by LSD test at $p<0.05$.

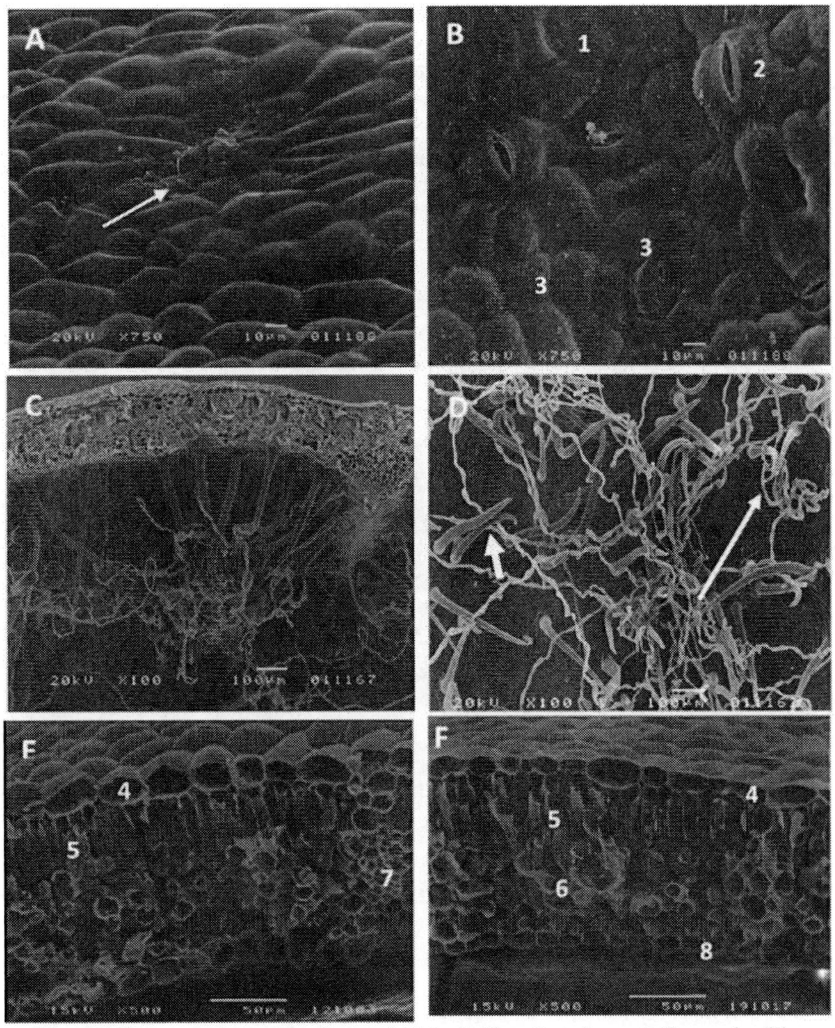

Figure 1. Leaves epidermal tissues and cross sections of *Vitis vinifera* L. subsp. *vinifera* red cultivars, upper and lower surface (SEM) – A – 'Cabernet Sauvignon' (x750) upper surface with a pearl gland (arrow); B – 'Trincadeira' (x750) lower surface with the three types of stomata:1 – sunken; 2 – raised above; 3 – at the same level; C – Leaf cross section of 'Touriga Nacional' (SEM X100). D – 'Trincadeira' (SEM X100) lower surface showing plant hairs of two types – slightly curved (small arrow) and long helicoidal (long arrow); E and F - Leaf cross section of 'Cabernet Sauvignon' and

 https://doi.org/10.1051/e3sconf/20185001038

'Touriga Nacional', respectively (SEM X500): 4 – Upper epidermal cells; 5 – Palisade tissue; 6 – Spongy tissue; 7 – Vascular bundle; 8 - Lower epidermal cells.

Table 2. Thickness of leaf anatomical traits (average ± standard error) of the four red cultivars of *Vitis vinifera* L. subsp. *vinifera,* cultivated at the Lisbon winegrowing region.

Thickness of the anatomical character (µm)	Grapevine cultivars			
	'Cabernet Sauvignon	'Syrah'	'Touriga Nacional'	'Trincadeira'
Total lamina	188.8 (5.62) b	205.9 (7.52) a	162.1 (4.77) c	166.7 (1.96) c
Mesophyll	153.2 (5.80) b	171.4 (7.17) a	128.9 (3.70) c	125.2 (2.59) c
Palisade tissue	61.1 (1.85) b	71.8 (3.38) a	48.5 (1.29) c	56.5 (1.79) b
Spongy tissue	91.9 (4.65) ab	98.7 (5.27) a	81.3 (3.23) b	68.8 (2.47) c
Upper cuticle	2.5 (0.04) a	2.4 (0.03) a	2.2 (0.09) b	2.3 (0.04) ab
Lower cuticle	2.6 (0.16) a	2.2 (0.09) bc	2.5 (0.13) ab	1.9 (0.09) c
Upper epidermal cells	18.3 (0.77) a	18.6 (0.60) a	15.4 (0.60) b	19.5 (0.70) a
Lower epidermal cells	18.3 (1.34) b	17.0 (0.62) b	18.3 (0.75) b	21.8 (0.78) a

In each line different letters indicate significant differences by LSD test at $p < 0.05$.

Table 3. Length and width of each stomata type (average ± standard error) for four red cultivars of *Vitis vinifera* L. subsp. *vinifera* cultivated at Lisbon winegrowing region.

Stomata dimensions (µm)	Grapevine cultivars			
	'Cabernet Sauvignon'	'Syrah'	'Touriga Nacional'	'Trincadeira'
Raised above				
Length (µm)	37.4 (1.11) a	39.9 (0.97) a	45.4 (10.41) a	37.3 (1.02) a
Width (µm)	25.4 (0.78) a	28.5 (2.19) a	31.9 (8.92) a	28.9 (2.24) a
Same level				
Length (µm)	28.8 (0.65) a	29.9 (0.87) a	28.5 (0.70) a	29.2 (1.03) a
Width (µm)	18.5 (0.70) a	18.9 (0.89) a	18.7 (0.87) a	20.5 (0.88) a
Sunken				
Length (µm)	21.6 (0.87) ab	25.3 (1.85) a	24.6 (1.04) ab	21.4 (1.01) b
Width (µm)	14.1 (0.92) a	15.7 (1.40) a	12.6 (0.79) a	12.8 (0.97) a

In each line different letters indicate significant differences by LSD test at $p < 0.05$

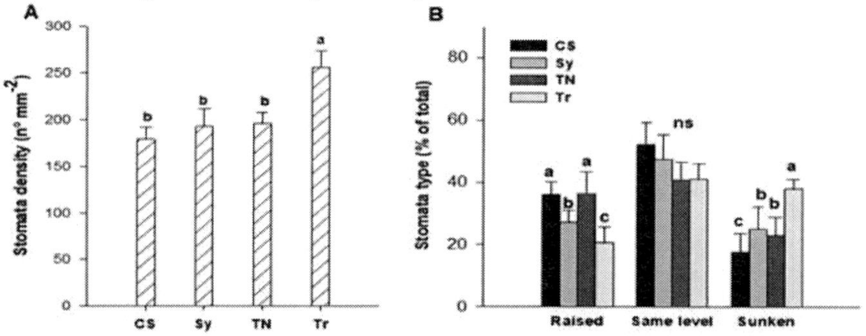

Figure 2. A - Stomata density and B – Percentage of stomata type – raised above, same level and sunken - for the four red cultivars of *Vitis vinifera* L. subsp. *vinifera*, cultivated at Lisbon winegrowing region. CS – 'Cabernet Saugivnon'; Sy – 'Syrah'; TN – 'Touriga Nacional'; Tr – 'Trincadeira'. Bars indicate standard error of the mean. Means with same letters are not significantly different ($p < 0.05$) for each cultivar (A) or for each type of stomata (B).

In our study, stomata density varied significantly with genotype, with values of the same order of magnitude as those presented for the same grapevine red cultivars studied by Monteiro *et al.* [26] at Torres Vedras WR (coastal area 60 km north from Lisbon) terroir but, with higher values than the ones observed by Costa *et al.* [20] in Alentejo WR inland south Portugal), where Tr was considered more drought resistant [20, 21]. The trichomes density and length varied with grapevine cultivars. A thick layer of trichomes will influence leaf radiation balance and probably will help to reduce water loss by reducing leaf surface temperature and increasing the boundary layer resistance [31]. The leaf hairs observed in the lower epidermis might have implications in epidermal temperature [36b] but not in the tolerance to biotic stresses [14].

The presence of calcium oxalate crystals in the foliar parenchyma can promote a more uniform distribution of light throughout the mesophyll, improving the photosynthetic rate. However, this feature might be related to the plant needs for large amounts of light, since the crystals prisms can increase the light interception and, therefore the photosynthetic rate as light will be more evenly distributed around the mesophyll [18]. The crystals can also provide mechanical support and physical and chemical protection against herbivores [18, 35].

4 Conclusions

The observed significant differences among genotypes in leaf traits suggest a major role of leaf anatomy in determining grapevine capacity for coping to specific environments. In particular, the genotypes which showed small leaves, high leaf density, high stomata density and small and/or sunken stomata may have a comparative advantage in dry and hot regions to support abiotic stresses. The French cultivars revealed most of those xeromorphic leaf anatomical features while the Portuguese ones, in general, tend to present more mesophytic leaf features. Further studies are needed to confirm the suggested association between some of these leaf traits – specific dry weight, leaf density, mesophyll structure and stomata type, for example – and the physiological behavior observed under field conditions.

Acknowledgements

We would like to thank Telmo Nunes, Faculdade de Ciências da Universidade de Lisboa, for his technical assistance with SEM.

References

1. Ü Niinemets, New Phytologist **144**(1), 35–47 (1999)
2. I. Terashima, YT Hanba, D Tholen, U Niinemets, Pl Physiol, **155**, 108-116 (2011)
3. A Palliotti, S Tombesi, T Frioni, F Famiani, O Silvestroni, M Zamboni, S Poni, Funct Plant Biol, **41**(6), 634-647 (2014)
4. A Montoro, R López-Urrea, E Fereres, Acta Hortic, **1115**, 41-47 (2016).
5. M Gómez-del-Campo, C Ruiz, P Baeza, JR Lissarrague, J Int Sci Vigne Vin, **37**, 131-143 (2003)
6. M Tomás, H Medrano, A Pou, JM Escalona, S Martorell, M Ribas-Carbó, J Flexas, Aust. J Grape. Wine. Res., 18 (2), 164–172 (2012)
7. J Flexas, C Scoffoni, J Gago, L Sack, J. Exp. Bot., **64** (13), 3965–3981 (2013)
8. M Tomás, J Flexas, L Copolovici, J Galmés, L Hallik, H Medrano, M Ribas-Carbó, T Tosens, V Vislap, Ü Niinemets, J Exp Bot **64** (8), 2269–2281 (2013)
9. CA Brewer, WK Smith, Plant Cell Environ. *20*, 1-11 (1997)
10. HG Jones, *Plants and Microclimate: A Quantitative Approach to Environmental Plant Physiology.* Cap. 10, Drought and other abiotic stresses, p. 255-289. (Cambridge University Press, Cambridge, 2014)
11. S Boso, V Allonso-Villaverde, JL Santiago, P Gago, M Durenberber, M Duggelin, Vitis, **49**, 43-50 (2010)
12. S Boso, V Allonso-Villaverde, P Gago, BJL Santiago, MC Martínez, Aust J Grape Wine Res, **17**, 304-400 (2011a)
13. S Boso, P Gago, V Allonso-Villaverde, JL Santiago, J Mendez, I Pazos, MC Martínez, Sci Hort, **128**, 228-238 (2011b)
14. S Boso, V Allonso-Villaverde, P Gago, BJL Santiago, MC Martínez, Crop Prot, **63**, 26-35 (2014)
15. A Atak, M Akkurt, Z Polat, H Celik, KA Kahraman, DS Akgül, N Özer, G Söylemezoğlu, G Sire, R Eibach, Ciência Téc Vitiv, **32**(1), 23-32 (2017)
16. Eurostat, *Eurostat statistics explains* http://ec.europa.eu/eurostat/statistics-explained/index.php/Agri-environmental_indicator_consumption_of_pesticides (2016)
17. I Pertot, T Caffi, V Rossi, L Mugnai, C Hoffmann, MS Grando, C Gary, D Lafond,C Duso, D Thiery, V Mazzoni, G Anfora, Crop Prot, 1-15 (2016)
18. C Metcalfe, L Chalk, *Anatomy of Dicotyledons*, Vol I. 2nd ed. (Oxford Univ. Press, 1979)
19. IPCC, *Climate Change 2013: The Physical Science Basis. Contribution of Working Group I to the Fifth Assessment Report of the Intergovernmental Panel on Climate Change* (Cambridge University Press, Cambridge, 2013)
20. JM Costa, MF Ortuño, CM Lopes, MM Chaves, Funct Pl Biol, **39**, 179-189 (2012)
21. M Vaz, R Coelho, A Rato, R Samara-Lima, LL Silva, E Campostrini, JB Mota,) Theor Exp Plant Physiol, **28** (2), 205-220 (2016
22. IVV, *Lista de castas.* (http://www.ivv.gov.pt/np4/111/, 2017)
23. IFV, *Catalogue des variétés et clones de vigne cultivés en France.* (Institute Français de la Vigne et du Vin, Ministère de l'Agriculture et da la Pêche – CTPS, 2ème edition. Montpellier, France, 2007)
24. M Hayat, *Principles and techniques of electron microscopy. Biological Applications.* 2nd ed. (Ed. Arnold Publ. London. 1981)
25. SE Ruzin, *Plant microtechnique and microscopy.* (Oxford University Press. 1999)
26. A Monteiro, G Teixeira, CM Lopes, Ciência Tec. Vitiv., **28**, 19-28 (2013)
27. C Pratt, Am. J. Enol. Vitic, **25**, 131-150 (1974)
28. Ü Niinemets, IJ Wright, JR Evans, J Exp Bot, **60**, 2433-2449 (2009)
29. TJ Martin, BE Juniper, *The Cuticles of Plants.* (Edward Arnold, London, 1970)
30. A Ben Salem-Fnayou, B Bouamama, A Ghorbel, A Mliki, *Microsc Res Tech,* **74**(8), 756-62 (2011)
31. M Keller, *The Science of Grapevines. Anatomy and Physiology.* (Academic Press, Burlington, MA. 2010)
32. HP Wilkinson, The Plant Surface (Mainly Leaf). Part I: Stomata. In: *Anatomy of Dicotyledons.* Vol I. 98-117. 2nd ed. (Oxford Univ. Press, 1979)
33. T Doheny-Adams, L Hunt, PJ Franks, DJ Beerling, JE Gray, Philos Trans R Soc Lond B Biol Sci, **367** (1588), 547–555 (2012)
34. I Serra, A Strever, P Myburgh, M Schmeisser, PA Deloire, Acta Hort, **1157**, 177-181 (2017)
35. P Nakata, Plant Sci, **164**, 901-909 (2003)

 https://doi.org/10.1051/e3sconf/20185001038

36 G Karabourniotis, D Kotsabassidis, Y Manetas, Can J Bot, **73**, 376-383 (1995)

Row Orientation and Defoliation Effects on Grape Composition of *Vitis vinifera* L. Agiorgitiko in Nemea (Greece)

Evangelia Chorti[1], Serafeim Theocharis[2], Konstantinos Boulokostas[2], Stamatina Kallithraka[3], Yorgos Kotseridis[3] and Stefanos Koundouras[2]
[1] Gaia Wines, 20500 Koutsi - Nemea, Greece, E-mail: echorti@hotmail.com
[2] Laboratory of Viticulture, School of Agriculture, Aristotle University of Thessaloniki, 54124 Thessaloniki, Greece, E-mail: mpoulokk@agro.auth.gr, sertheo@agro.auth.gr, skoundou@agro.auth.gr
[3] Laboratory of Enology, Department of Food Science & Technology, Agricultural University of Athens, 75 Iera Odos, 11855 Athens, Greece, stamatina@aua.gr; ykotseridis@aua.gr
*Corresponding author: S. Koundouras

Abstract. Vineyard row orientation and canopy side exert a significant role in determining grape microclimate. The latter can be further manipulated by selective defoliation in the bunch zone. The aim of this study was to investigate the combined effects of row orientation and basal leaf removal on grape ripening of *Vitis vinifera* L. cv. Agiorgitiko in Nemea, Southern Greece. The experiment was conducted in 2017, on two adjacent vineyard blocks planted in 1990 on a flat site with two row orientations, North-South and East-West. Both blocks were grafted onto 41B rootstock and trained on a double cordon vertical trellis with uniform vineyard operations. Defoliation treatments included full leaf removal in the bunch zone performed at berry set and a non defoliated control, replicated three times in both blocks. Grapes were sampled at three time points from veraison to harvest from all canopy sides (i.e. E and W on North-South oriented rows and N and S on East-West oriented rows) and defoliation treatments and were analyzed for yield components and berry chemical composition. In addition, approximately 5 kg of grapes were collected from all plots and processed according to a standard winemaking protocol. Total yield and berry mass were lowest in the W-exposed and in the defoliated grapes. Must sugar and acid content, as well as wine alcohol and acidity did not depend on either factor. Skin anthocyanins and phenolics generally increased with defoliation. Skin anthocyanins and total berry phenolics were highest for north-south oriented rows, in the grapes of the cooler canopy side (E). Wines made from grapes of the E and W canopy sides showed the highest wine color, phenolic richness and the lowest pH. No interaction between row orientation and defoliation was detected for any of the measured parameters. According to the results, the E-facing grapes on the north-south oriented rows showed generally a superior grape and wine composition for Agiorgitiko variety, under the semiarid conditions of the Nemea area.

1 Introduction

Bunch zone defoliation is a very common viticultural practice applied to manipulate canopy microclimate. It modifies bunch light environment, inducing a positive effect on skin anthocyanins of grape berries [1]. It also promotes air circulation into the canopy, reducing humidity and subsequently the risks of fungal infection [2]. However, while in cool climates, increased bunch sunlight exposure is related to improved anthocyanin and phenolic profile [3], in warm to hot, semi-arid conditions, the increased berry temperature caused by the extended sunlight exposure may reduce flavonoid biosynthesis [4]. In fact, high berry temperature, caused by increased sunlight exposure, has been shown to inhibit color development [5].

Vineyard row orientation is an important factor determining canopy microclimate and thus grape berry composition. It has been suggested that photosynthetically active radiation (PAR) absorption by the grapevine canopy is higher in north–south and northeast–southwest than in east–west row orientations for latitudes between 30 and 50°N [6]. It has been also shown that east–west row orientations [7] can result in reduced growth, yield and total dry mass per vine compared to north–south and northeast–southwest orientations in Chardonnay grown in the Northern hemisphere, while juice total soluble solids and pH were not affected. However, little evidence exists on the effects

 https://doi.org/10.1051/e3sconf/20185001039

of canopy side on the chemical composition of grapes. Therefore, the aim of this study was to investigate the combined effects of row orientation and basal leaf removal on grape ripening of *Vitis vinifera* L. cv. Agiorgitiko under the semiarid climate of Southern Greece.

2 Materials and Methods

The experiment was conducted in 2017 in a 27-year old vineyard in Nemea (Greece), in two adjacent vineyard blocks situated on a flat site planted with two row orientations, North-South (N-S) and East-West (E-W). Both blocks were planted with *Vitis vinifera* L. cv. 'Agiorgitiko' grafted onto 41B rootstock and trained on a double cordon vertical trellis with uniform vineyard operations.

Defoliation was performed at berry set and included full leaf removal in the bunch zone and a non defoliated control, replicated three times in both blocks. Grapes were sampled at three time points from veraison to harvest from all defoliation treatments and canopy sides (i.e. E and W expositions on N-S oriented rows and N and S expositions on E-W oriented rows) and were analyzed for bunch and berry weight and berry chemical composition [8].

Approximately 5 kg of grapes were collected from all plots and processed according to a standard red winemaking protocol. The produced wines were analyzed for several classical analytical parameters [9] and total phenolic content by Folin–Ciocalteau method [10].

3 Results and discussion

3.1 Yield components

Defoliation did not affect total yield per vine and cluster weight at harvest (Table 1), but the defoliated vines produced denser clusters with smaller berries compared to non-defoliated ones.

Table 1. Defoliation (D) and Exposition (E) effects on vine production components at harvest (Y, yield per vine in kg; BW, berry weight in g; CW, cluster weight in g; CD, cluster density in berries per cm). Statistically significant differences within a factor are indicated by different letters ($p \leq 0.05$). * represents significance of the defoliation ×exposition (D×E) interaction at $p \leq 0.05$; ns, not significant.

		Y	BW	CW	CD
D	Non defoliated	2.42	1.80^{a}	201.0	7.72^{b}
	Defoliated	2.00	1.66^{b}	199.8	8.56^{a}
E	N	2.51^{a}	1.81	226.5^{a}	8.31^{ab}
	S	2.27^{a}	1.73	165.3^{b}	7.68^{b}
	E	2.54^{a}	1.74	227.8^{a}	8.88^{a}
	W	1.52^{b}	1.66	181.9^{b}	7.68^{b}
D×E		ns	ns	ns	*

Canopy side exposition exerted a significant impact on all yield components apart from berry mass. In particular, vines exposed W presented the lowest yield, and together with grapes of S exposition, presented the smaller berries and the lowest cluster density (Table 1). W and S canopy sides in the North hemisphere receive the highest heat loads and, although microclimate parameters were not measured, it can be suggested that increased cluster temperature under W and S expositions inhibited berry and bunch growth. There was no interaction between defoliation and exposition for the measured parameters, with the exception of cluster density (Table 1).

3.2 Grape composition

Neither defoliation nor canopy exposition affected juice analytical parameters at harvest [7] and no defoliation × exposition interaction was detected for the same parameters (Table 2).

Table 2. Defoliation (D) and Exposition (E) effects on must chemical composition at harvest (TSS, total soluble solids in °Brix; TA, titratable acidity in g of tartaric acid per L of juice); ns, not significant.

		TSS	TA	pH
D	Non defoliated	21.1	6.5	3.43
	Defoliated	22.3	5.9	3.55
E	N	21.6	6.7	3.40
	S	21.1	6.7	3.39
	E	21.9	5.9	3.52
	W	21.5	5.9	3.62
D×E		ns	ns	ns

Grape skins from the vines where defoliation was applied presented higher content (expressed per single berry) and concentration (expressed per berry weight) of total anthocyanins and total phenols (Table 3). High light incidence on grapes is generally considered to promote anthocyanin accumulation in the skins [1].

Table 3. Defoliation (D) and Exposition (E) effects on berry skin phenolic composition measured at harvest (TA/B, total anthocyanins in mg per berry; TP/B, total phenols in absorbance units (au) per berry; TA/W, total anthocyanins in mg per g of berry; TP/W, total phenols in au per g of berry;). Statistically significant differences within a factor are indicated by different letters ($p \leq 0.05$); ns, not significant.

		TA/B	TP/B	TA/W	TP/W
D	Non defoliated	1.46^{b}	3.11	0.85^{b}	1.79^{b}
	Defoliated	1.63^{a}	3.11	1.00^{a}	1.91^{a}
E	N	1.34^{c}	3.13^{ab}	0.77^{c}	1.78^{b}
	S	1.41^{bc}	2.83^{b}	0.86^{bc}	1.73^{b}
	E	1.81^{a}	3.40^{a}	1.08^{a}	2.02^{a}
	W	1.56^{b}	3.10^{ab}	0.93^{b}	1.85^{b}
D×E		ns	ns	ns	ns

Exposition also affected phenolic accumulation in the skins, the most advantageous one for phenolic compound accumulation being the E side on N-S oriented rows (Table 3), where the grapes were exposed to sunlight only during the cooler hours of the day. On the contrary, N exposed grapes had the lowest anthocyanin content, on average, possibly due to lower grape exposure to sunlight [1]. No interaction between defoliation and exposition was detected for skin polyphenolic parameters (Table 3).

A positive response of skin anthocyanin levels to light exposure on the coolest canopy side has been previously reported [4]. Another study [11] suggested that anthocyanin accumulation is usually higher in the berries on the bunch side facing outside the canopy rather than inside, with the exception of W exposed bunches, probably due to elevated temperature of the skins. It has also been suggested that the combined effect of solar radiation and elevated temperature on anthocyanin accumulation may be linked, among other factors, to genotype [12], allowing some cultivars to benefit from light exposure even at higher temperature levels.

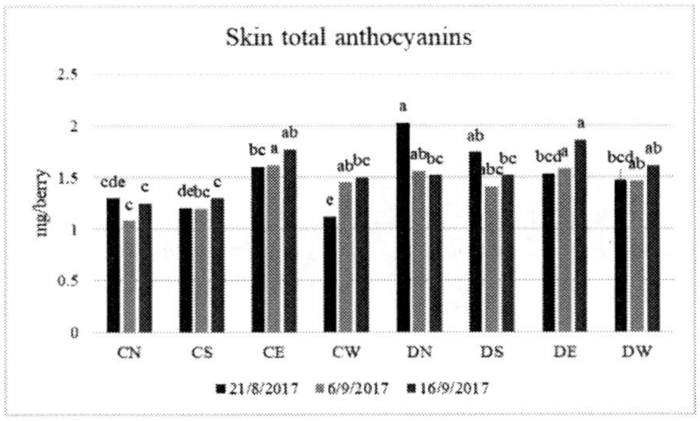

Figure 1. Skin anthocyanin accumulation in mg/berry during ripening (CN, non-defoliated/N exposition; CS, non-defoliated/S exposition; CE, non-defoliated/E exposition; CW, non-defoliated/W exposition; DN, defoliated/N exposition; DS, defoliated/S exposition; DE, defoliated/E exposition; DW, defoliated/W exposition). Statistically significant differences within a factor are indicated by different letters ($p \leq 0.05$).

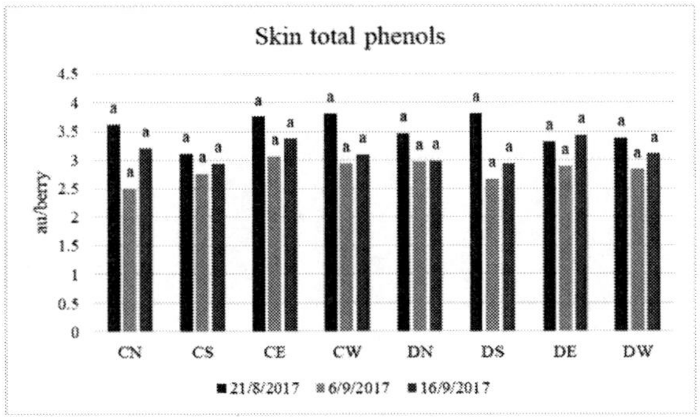

Figure 2. Skin total phenols evolution in au/berry during ripening (CN, non-defoliated/N exposition; CS, non-defoliated/S exposition; CE, non-defoliated/E exposition; CW, non-defoliated/W exposition; DN, defoliated/N exposition; DS, defoliated/S exposition; DE, defoliated/E exposition; DW, defoliated/W exposition). Statistically significant differences within a factor are indicated by different letters ($p \leq 0.05$).

Considering individual treatments, the defoliated treatment with E exposition (DE) presented the highest values of anthocyanins in the skins at harvest (followed by CE and DW) although the N-exposed defoliated grapes had the highest content at the beginning of the ripening period (Fig. 1). The non-defoliated treatments with N and S exposition (CN and CS) presented the lowest values at harvest. It is noteworthy that skin anthocyanins declined in most cases during harvest in the grapes of the N and S-exposed canopy sides whereas they increased reaching maximum values at harvest for the E and W-exposed sides. There were no differences in skin TP among the 8 treatments throughout ripening (Fig. 2).

3.3 Wine composition

Defoliation had no effect on wine alcohol content or titratable acidity (Table 4). S-exposed grapes produced the wines with the highest pH values, followed by vines exposed to N and E (Table 4). There was no interaction between defoliation and exposition, with the exception of wine pH. According to previous works, direct light exposure of berries led to higher ratio of tartaric to malic acid in Syrah grapes [13].

Table 4. Defoliation (D) application and Exposition (E) effects on wine composition. (TA, titratable acidity in g of tartaric acid per L of wine). Statistically significant differences within a factor are indicated by different letters ($p \leq 0.05$). * represents significance of the defoliation ×exposition (D×E) interaction at $p \leq 0.05$; ns, not significant.

		Alc. %	TA	pH
D	Non defoliated	12.91	6.27	3.77
	Defoliated	13.31	6.37	3.75
E	N	13.33	6.45	3.80^{ab}
	S	13.45	6.24	3.84^{a}
	E	12.83	6.24	3.71^{bc}
	W	12.85	6.34	3.69^{c}
D×E		ns	ns	*

Table 5. Defoliation (D) and Exposition (E) effects on wine phenolic composition. (CI, color intensity, OD 280, total polyphenolic index; TP, total phenols in mg of gallic acid per L of wine). Statistically significant differences within a factor are indicated by different letters ($p \leq 0.05$). * represents significance of the defoliation ×exposition (D×E) interaction at $p \leq 0.05$; ns, not significant.

		CI	OD 280	TP
D	Non defoliated	4.78	37.44^{b}	299.85
	Defoliated	5.01	41.2^{a}	310.68
E	N	4.33^{c}	37.3	286.03^{b}
	S	4.62^{bc}	37.81	289.84^{b}
	E	5.22^{ab}	40.96	309.24^{ab}
	W	5.40^{a}	41.21	335.96^{a}
D×E		ns	ns	ns

Wines produced from defoliated vines presented higher values of total polyphenolic index, but there was no difference in the concentration of total phenols (Table 5). Wines made from W-exposed grapes presented higher color (CI) and phenolic content (TP) values, followed by those from E exposition.

4 Conclusions

In warm, semiarid vine growing regions, achieving phenolic maturity when berry sugar content and acidity are at optimum levels can sometimes be an impossible task. As a result, growers either have to wait for phenolic maturity at the cost of producing wines with high alcohol content and unbalanced acidity, or harvest at technological maturity, risking to produce wines with poor color and unripe, astringent tannins.

Bunch zone defoliation at berry set promoted anthocyanin and phenolic accumulation in skins without advancing berry ripeness. As a result, the wines produced from grapes that had undergone defoliation presented a more intense color and higher total polyphenolic index compared to the non-defoliated ones, without presenting any signs of advanced or over ripeness: wine alcohol content and acidity were similar to those produced from the non-defoliated vines.

Regarding canopy side effect, E side exposition on N-S oriented rows was proved as the most beneficial for berry anthocyanin and phenolic compound biosynthesis. It is also very important that E exposition promoted phenolic maturity without advancing pulp maturity, as it had no effect on juice sugar content, acidity and pH. However, maximum wine color and lowest wine pH was achieved by the grapes of the W canopy side. E-W orientation, and particularly the N-facing side of the canopy, seems inappropriate for the variety and site in question, since it presented the lowest phenolic levels in both grapes and wines at harvest.

Since no defoliation × exposition interaction was detected for the most important quality parameters of grape and wine composition, it would be safe to conclude that in the conditions of the studied area, the best way to ensure maximum anthocyanin accumulation in the skins and optimize wine polyphenolic composition would be to establish vineyards with N-S row orientation and practice bunch zone defoliation as early as berry set.

References

1. N.K. Dokoozlian, W.M. Kliewer. J. Am. Soc. Hortic. Sci., 121, 869– 874 (1996).
2. R.W. Emmett, A.R. Harris, R.H. Taylor, J.K. McGechan. Grape diseases and vineyard protection. Viticulture Vol. 2. Practices. Coombe BG, Dry PR (Eds). Winetitles, Adelaide pp 232-278 (1992).
3. R.Ristic, M.O. Downey, P.G. Iland, K. Bindon, I.L. Francis, M.J, Herderich, S.P. Robinson, Aust. J. Grape Wine Res., 13, 53– 65 (2007).
4. J. Bergqvist, N. Dokoozlian, N. Ebisuda, Am. J. Enol. Vitic., 52, 1– 7 (2001).
5. J.M. Tarara, J. Lee, S.E. Spayd, C.F. Scagel. Am. J. Enol. Vitic. 59:235-247 (2008).
6. I. Campos, C.M.U. Neale, A. Calera, Aust. J. Grape Wine R., 23, 77–86 (2017)
7. C. Intrieri, O. Silvestroni, B. Rebucci, S. Poni, I. Filippetti. In: Kling, T.H., Wolf, T.E., Harkness, E.M. (Eds.), Proc. 4th Intern. Cool Climate Symp., 16–20 July, Rochester, USA. , pp. 10–15 (1996).
8. P. Iland, N. Bruer, G. Edwards, S. Weeks, E. Wilkes, Chemical analysis of grapes and wine: techniques and concepts (2004).
9. Compendium of International Methods of Analysis of Wines and Musts (2 vol.). Available from: http://www.oiv.int/en/technical-standards-anddocuments/methods-of-analysis/compendium-of-international-methodsof-analysis-of-wines-and-musts-2-vol.
10. P.G. Waterman, S. Mole S Analysis of phenolic plant metabolites, Oxford: Blackwell Scientific Publ pp 83-91 (1994).
11. P. Pieri, K. Zott, E. Gomès, G. Hilbert, *OENO One*, *50*(3), 23-33 (2016).
12. M.O. Downey, J.S. Harvey, S.P. Robinson. Aust. J. Grape Wine Res. 10:55-73 (2004).
13. S. DeBolt, R. Ristic, P.G. Iland, C.M. Ford, Hort.Science, 43, 957–961 (2008).

Productive response of Macabeo varietie in two locations of Spain

Amelia Montoro[1*], Irene Torija[1], Luis A. Mancha[2], Daniel Moreno[2] y David Uriarte[2]

[1]ITAP-FUNDESCAM, Polígono Campollano, Av. Segunda 61, 02007 Albacete, Spain
[2]CICYTEX, Finca La Orden, Ctra. A-V, km 372, Guadajira, Badajoz, Spain

Abstract. The work was carried out on Vitis vinifera, L. (cv. Macabeo) grafted onto 110 Ritcher rootstock and cultivated in two locations of the southern half of Spain: La Albuera (Badajoz) and Villarrobledo (Albacete), during 2016 and 2017. The soil of La Albuera is clay-loam texture, 1 m deep and 160 mm/m available water capacity (AWC); respect to climate is warm, cool nights and very dry according to the classification of [1]. The Villarrobledo soil is 0.75 m deep, silty-clay-loam texture and 135 mm/m AWC; the climate is very warm, cool nights and moderately dry climate. The study consisted in maintaining during two consecutive years and in both locations, the same hydric state of grapevine, which was evaluated by measurements of midday stem water potential, and comparing productive results obtained in the two places. Results derived from the analysis of yield components, as well as those derived of grape composition (total acidity, tartaric acid, malic acid, pH, potassium and soluble solids) show little variation between locations, instead great interannual variation among the parameters studied, so the "year-variety" effect seems to influence much more to the final result than the "locality-variety", for the case under study.

1 Introduction

Terroir is concerned with the relationship between the characteristics of an agricultural product (quality, taste, style) and its geographic origin, which might influence these characteristics. The concept of terroir has been described as an interactive ecosystem which include the climate, soil, the vine, and human factors [2, 3]. From another point of view, [4] defined *"Terroir"* as a ubiquitous term in the comercial wine world.

It is clear that the genotype is a major source of differences in fruit composition [5]. Fruit composition changes over time during berry development and ripening as part of the grapevine´s developmental program and is therefore under genetic control [6]. Although the effect of climate, soil and cultivar on berry development and berry composition cannot be separated [7], [8] and [9] found that the effect of climate was indeed the geatest, after which soil and cultivar followed.

Climate has been described by several authors as having a major influence on berry composition, highlighting temperature as one of the main factors [10, 11, 12]. In fact, weather differences among years, in addition to vineyard location rather than, say, within-vineyard differences in soil composition, are by far the strongest determinants of fruit composition and wine quality [6, 13].

Acids are among the main compounds that originate the grape berry, since they play a vital role in the production of quality wines as they ultimately determine the organoleptic and aesthetic character perceived in the wine. In addition, both the shelf-life and ageing potential is influenced since the physical, biochemical and microbial stability of the wine is determined by the levels of wine acidity [7, 14].

The aim of this study was to study the behavior of *Vitis vinifera* cv. Macabeo in two locations across two vintages.

2 Material and methods

2.1 Locations and soil

The study was carried out during 2016 and 2017 in two commercial vineyards (*Vitis vinifera* cv. Macabeo) located in different province:

1 Albacete (AB) (southeast of central area of Spain). The grapevine is located in Villarrobledo (39º15´N, 2º40´W, 714 m altitude). The silty-clay-loam texture soil and 0.75 m deep of the experimental site was characterized by a available water capacity (AWC) of 135 mm/m.

2 Badajoz (BA) (southwest of central area of Spain). The grapevine is located in La Albuera (38°42'N, 6°50'W, 253 m altitude). The clay-loam texture soil, 1 m deep and 160 mm/m AWC.

2.2 Plantations

The experiment was performed on grapevines of similar years old (5-6 for AB and 6-7 for BA). Both vineyards were grafted to 110 Richter rootstock. The spacing followed a 3.10 m x 2 m square pattern in AB and 2.9 m x 1.4 m in BA.

The experiment was realized in micro-plots in three replicate blocks, each block having three rows of 24 vines, in Albacete and six rows of 60 vines in Badajoz.

Vines were drip-irrigated. Both plots were managed according to commercial practices.

2.3 Climatic description

 https://doi.org/10.1051/e3sconf/20185001040

Meteorological variables during the experiment were measured with two automated weather station located over the trial plot in La Albuera and close to the experimental plot in Villarrobledo. All sensors were located between 1.5 and 2 m above the soil surface and weather data were registered in 15 min, hourly and daily time steps. Variables measured were: air temperature, relative humidity, wind speed, wind direction, shortwave radiation, longwave radiation and rainfall. Reference evapotranspiration (ET_0) values were calculated with the daily and hourly time step FAO56 Penman-Monteith (FAO56 P-M) equation [15] using the recorded meteorological variables.

The aridity index is semiarid (Mediterranean Type) in both grapevines according to [16]. The climate is very warm, cool nights and moderately dry climate in Villarrobledo; and warm, cool nights and very dry in La Albuera, according to the classification of [1].

2.4 Irrigation regime and measurements

The irrigation regime consisted in applying the 30% of ET_0 from veraison to harvest.

The main parameters measured in both vineyards were: phenological stage [17]; leaf area index (LAI), obtained with allometric measures of shoot length in Albacete and leaf nerve length in Badajoz related in both cases with leaf area; and midday stem water potential (Ψs), measured with a pressure chamber technique [18]. Stem water potential was determined around solar noon on exposed and bagged leaves covered with aluminum foil for at least 45 min before taking the measurements and stomata density were measured after veraison on fully developed leaves. Water stress integral was calculated from budburst to harvest according to [19].

It was tried that four vintages were realized at a similar level of total soluble solids (TSS) (close to 22°Brix). This level of technological ripeness was chosen to ensure that a potential unripe character of the must did not influence the rest of parameters and they could be comparable. Due to logistical reasons and sampling variability, the vineyards could not always be harvested at the desired TSS level. Table 5 provides an indication of the differences in the sugar content between vintages and between sites.

At harvest, the fruit of central lines were weighed in the field. The samples were transported from the field to the laboratory. In the laboratory, the bunches and 100 berry sample picked from the left, right, top and bottom of the bunch were counted and weighed. After weighing, 2000 g of each treatment and repetition they were refrigerated to 4°C and transported to laboratory for determining the total soluble solids (TSS), pH, titratable acidity, tartaric acid, malic acid and K.

Statistical analysis was performed using IBM SPSS Statistics v.19 computer software (IBM SPSS Statistics, 2012). Analysis of variance (ANOVA) was carried out to test the degree to which two or more groups vary or differ in the experiment. The significance level of the correlation coefficient was determinate at 5%.

3 Results and discussion

Table 1 shows the atmospheric evaporative demand and water applied from sprouting to harvest for two years and two locations. Highlights the ET_0 of 2016 that it was about 10% higher than 2017 for both locations and the rainfall so different (40%) in Badajoz for the two years of study.

Table 1. Reference evapotranspiration, rainfall, irrigation and total water applied.

Location	Year	ET0 (mm)(*)	Rainfall (mm)(*)	Irrigation (mm)(*)	Total water applied (mm)(*)
Albacete	2016	814	98	57	154
	2017	780	83	64	147
Badajoz	2016	835	154	56	210
	2017	721	61	53	114

(*) From sprouting to harvest

Figure 1 shows the average, máximum and mínimum temperatura during the main periods of crop in 2016 (a) and 2017 (b). Despite having a similar climate according to [16] and [1], differences in temperature between the two sites under study are observed, reaching the highest values Badajoz. These differences are much more marked on the day of the harvest (Figure 2).

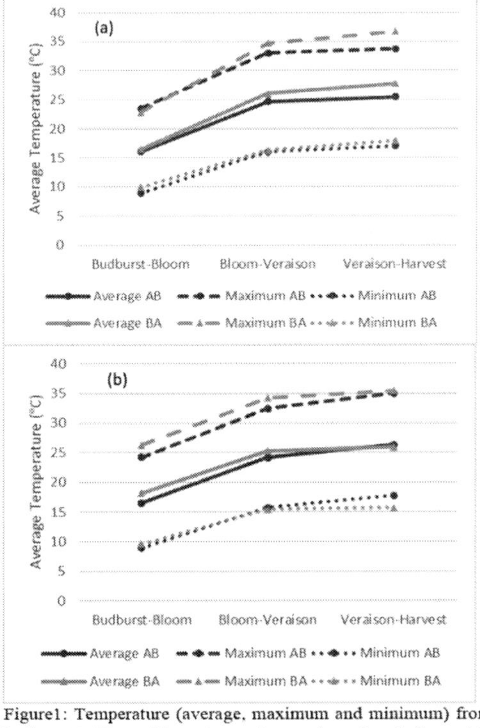

Figure1: Temperature (average, maximum and minimum) from budburst to bloom, bloom to veraison and veraison to harvest in Albacete (AB) and Badajoz (BA) for 2016 (a) and 2017 (b).

https://doi.org/10.1051/e3sconf/20185001040

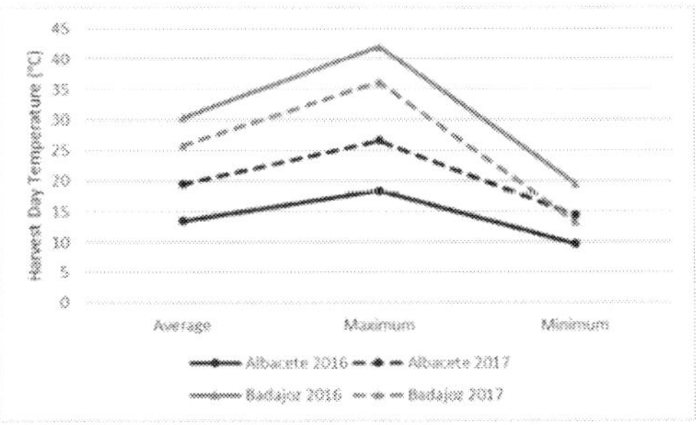

Figure 2. Harvest day temperature (average, maximum and minimum) in Albacete and Badajoz for 2016 and 2017.

Main phenological growth states are shown in table 2. Albacete and Badajoz presented an advance in 2017 with respect to 2016, in all its phenological stages from sprouting. Phenology is considered as the first biological indicator of climate [20], in this sense, the thermal integral analysis, that is used as one of the main criteria to assess the suitability of a cultivar to a location [21], is represented in figure 3. Budburst and flowering happened with the same accumulated temperature in both locations and the two years of study, instead the veraison was reached with higher temperature in Badajoz that in Albacete.

Table 2. Date of phenological growth stages

	Phenological Growth Stages Date			
	Albacete		Badajoz	
	2016	2017	2016	2017
Budburst	16-Apr	12-Apr	11-Apr	10-Apr
Flowering	12-Jun	25-May	6-Jun	22-May
Veraison	26-Jul	11-Jul	1-Aug	17-Jul
Harvest	14-Sep	29-Aug	5-Sep	16-Aug

An indication of the variability of the phenological dates per vintage and per cultivar is provided in Table 3. It is clear the great variation in the data, not only between the years, also between the locations, and within the year. The length of one interval per vintage does not dictate the length of the following interval in that vintage. Similar results were found by [7].

Table 3. Days between the major phenological stages in Albacete and Badajoz for 2016 and 2017.

	Albacete		Badajoz	
	2016	2017	2016	2017
Budburst-Bloom	57	43	56	42
Bloom-Veraison	44	47	56	56
Veraison-Harvest	50	49	35	30
Budburst-Harvest	151	139	147	128

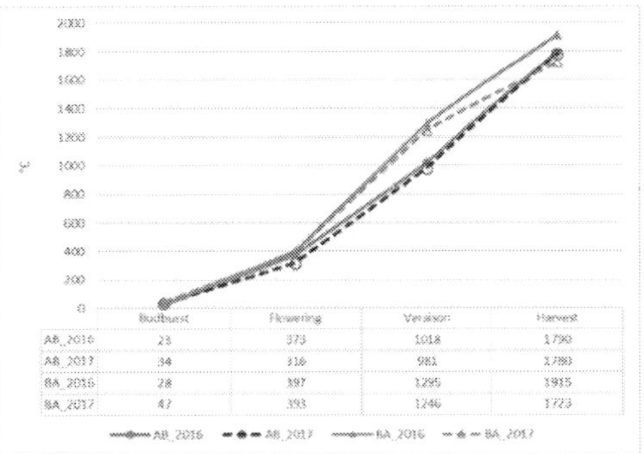

Figure 3. Thermal integral of grapevine for both locations in 2016 and 2017.

Leaf area index (LAI) in 2016 was higher than in 2017 and it was similar in both locations (Figure 4).

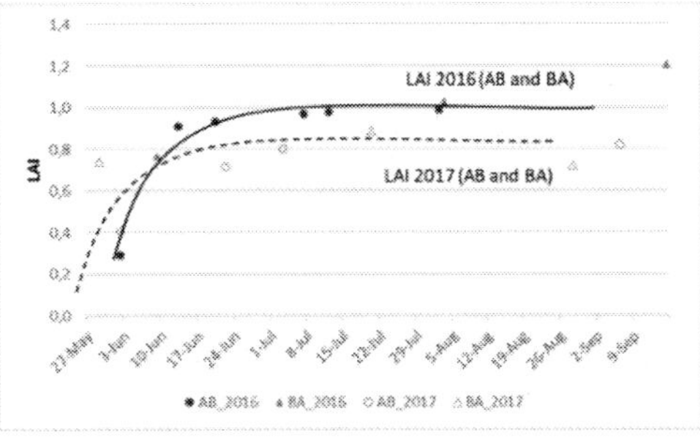

Figure 4. Leaf area index

The values of Ψs are shown in Figure 5 for 2016 (a) and 2017 (b) for both locations. In general, the beginning of the cycle in 2016 was less stressed than in 2017, however, the harvest occurred in 2016 with more negative potential values than in 2017, for both locations. Highlights the recovery of potential in Albacete after the first irrigation, not so in the case of Badajoz. Despite this, the water stress integral was similar in both locations and for two years of study (Table 4).

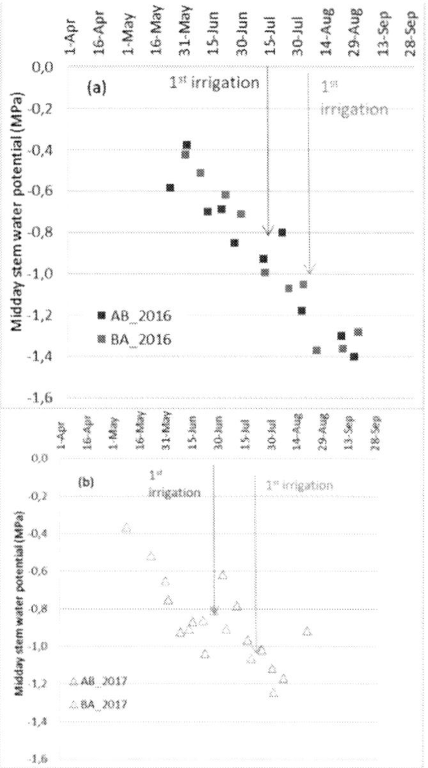

Figure 5. Midday stem water potential of grapevine for both locations in 2016 (a) and 2017 (b).

Table 4. Water stress integral from budburst to harvest.

Location	Year	Water stress integral
Albacete	2016	57,43
	2017	58,62
Badajoz	2016	57,83
	2017	59,91

Referent to yield, there were significant differences in yield in Albacete between 2016 and 2017 (Figure 6), not so in Badajoz and between Badajoz and 2017 in Albacete. This should be due to the production is defined in two years and 2015 (the year previous to experiment) the plot in Albacete was irrigated by the farmer without taking into account the water status of the plants. The yield difference was determined mainly by the bunches weight (Figure 7), and the number of berries since there was no significant difference in the weight of the berry (Figure 8) and the bunches number (Figure 9) was similar in both years. This agrees with that reported in previous studies [22] and it is true for both the variation between cultivars and the variation between growing seasons for the same cultivar [6].

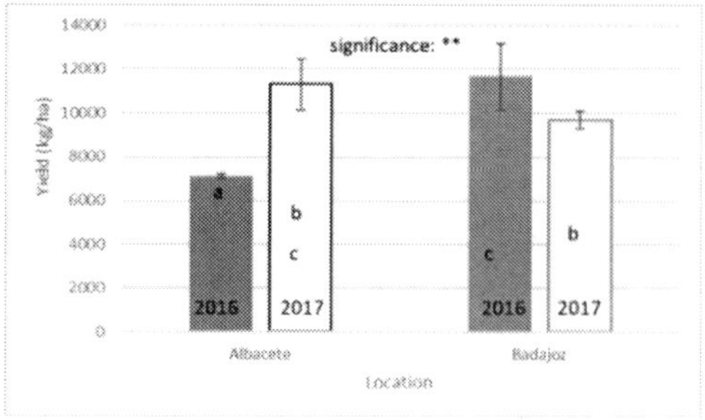

Figure 6. Yield of grapevine for both locations in 2016 and 2017.

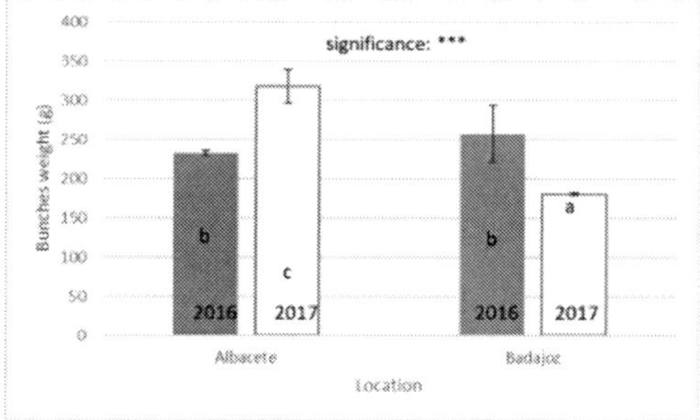

Figure 7. Bunches weight of grapevine for both locations in 2016 and 2017.

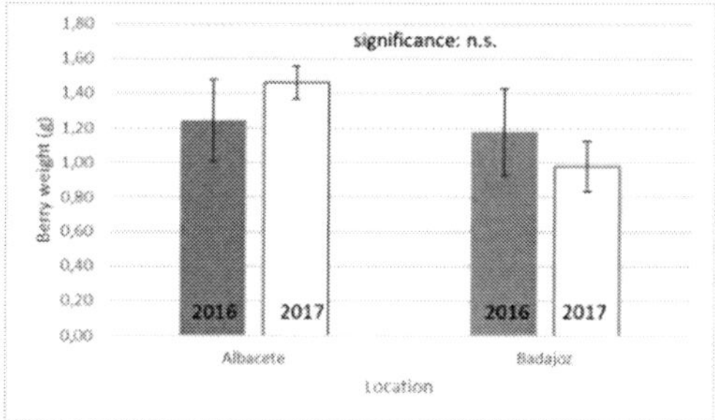

Figure 8. Berry weight of grapevine for both locations in 2016 and 2017.

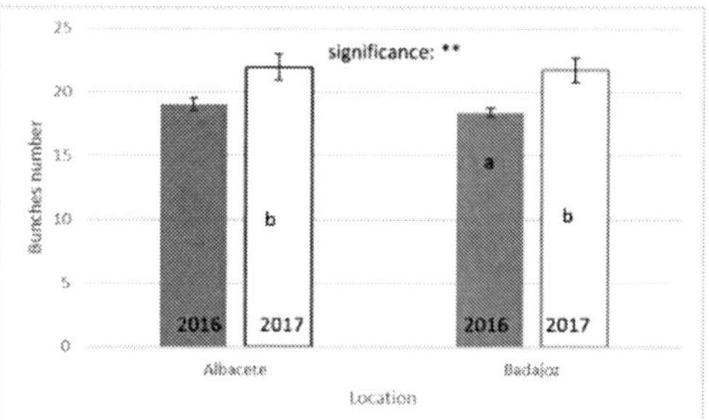

Figure 9. Bunches number of grapevine for both locations in 2016 and 2017.

Average technological maturity parameters at harvest are shown in Table 5. The location did not influence technological maturity except in malic acid (Figure 10 and Table 6), whereas the year effect seemed to have a greater influence in total acidity, tartaric acid and in the potassium content of the must (Table 6).

Table 5. Technological maturity parameters of must at at harvest for both locations in 2016 and 2017.

		Location	
	Year	Albacete	Badajoz
Total acidity	2016	4,17	4,22
(g/l Tart. Ac.)	2017	4,55	5,38
Tartaric acid	2016	6,18	6,23
(g/l)	2017	5,84	4,81
Malic acid	2016	1,27	0,65
(g/l)	2017	1,63	0,31
pH	2016	3,7	3,6
	2017	3,8	3,6
Potasium	2016	1,22	0,98
(g/l)	2017	1,23	1,52
°Brix	2016	22	20
	2017	26	23

The differences found in acids can be due to the different temperatures between locations and years. The effect of temperature on the organic acid content of the must is widely discussed with higher temperatures in general being associated with lower quantities of organic acids present in the juice, and lower temperatures during ripening associated with higher quantities, specifically in the case of malic acid [7].

Malic and tartaric acid has a definite synthesis period up until veraison, after which the content of tartaric acid remains constant in the berry and the content of malic acid decreases until harvest due to mainly respiration. The effect of climate, in particular temperature, on the metabolism of malic and tartaric acid, and the subsequent breakdown of malic acid after veraison, is well studied and understood. Although temperature has been found to have little or no effect on the tartaric acid content of the berries, the malic acid content reveal great changes according to the seasonal climatic differences [8, 10, 23]. Higher temperatures and higher respiration rates usually found in light exposed berries are in general attributed to lower malic acid levels [24].

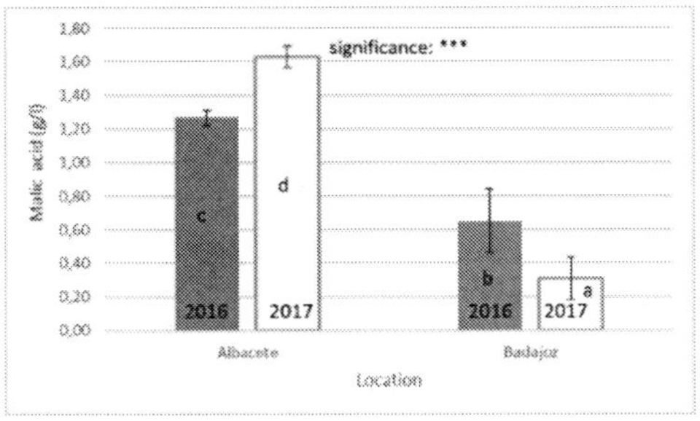

Figure 10. Malic acid of must for both locations in 2016 and 2017.

Table6. Results of analysis of variance for technological maturity parameters.

Parameter	Location	Year
Total acidity	n.s.	**
Tartaric acid	n.s.	*
Malic acid	***	n.s.
pH	n.s.	n.s.
Potasium	n.s.	*
Total soluble solids	n.s.	n.s.

***<0.05, **<0.01, ***<0.001, n.s.: non significance.**

4 Conclusion

Results show little variation between locations, instead great interannual variation among the parameters studied, so the "year-variety" effect seems to influence much more to the final result than the "locality-variety", for the case under study. This is mainly due to the variability of malic acid between vintages and according to the cultivar as the influence is not that noticeable for tartaric acid, glimpsing the effect of temperature on malic acid.

References

1. J. Tonietto, A. Carbonneau. Agricultural and Forest Meteorology, v.124, p.81-97 (2004)
2. M. Matthews. Terroir and other myths of winegrowing. University of California Press (2015).
3. G. Seguin. Experientia 42, 861-873 (1986)
4. G. Seguin. L. Fuller-Perrine (ed.) Proceedings of the Symposium on maritime Climate Winegrowing. Geneva, NY: Department of Horticultural Sciences, Cornell University (1988)
5. Z.W. Dai, N. Ollat, E. Gomes, S. Decroocq, J.P. Tandonnet, L. Bordenave et al. American Journal of Enology and Viticulture 62, 413-425 (2011)
6. M. Keller. The Science of Grapevines. USA (2015)
7. Z.A. Coetzee. Site and vintage response of malic and tartaric acid in Vitis vinifera L. cv's Cabernet Sauvignon and Sauvignon blanc. Theses in Stellenbosch University. Department of Viticulture and Oenology, Faculty of AgriSciences. (2013).
8. C. Van Leeuwen, P. Friant, X. Choné, O. Tregoat, S. Koundouras, D. Dubourdieu. American Journal of Enology and Viticulture 55 (3), 207-217 (2004)
9. G.Jones. https://www.guildsomm.com/public_content/features/articles/b/gregory_jones/posts/climate-grapes-and-wine (2015)
10. B. G. Coombe. Acta horticulturae 206, 23-35 (1987)
11. D.I. Jackson, P.B. Lombard. American Journal of Enology and Viticulture 44 (4), 409-430 (1993)
12. J.J. Hunter, V. Bonnardot. South African Journal of Enology and Viticulture 32 (1), 137-154 (2011)
13. A.G. Reynolds, G. Taylor, C. de Savigny. American Journal of Enology and Viticulture 64, 180-194 (2013)

14. R. Boulton. American Journal of Enology and Viticulture 31 (1), 76-80 (1980)
15. R.G. Allen, L.S. Pereira, D. Raes, M. Smith. FAO Irrigation and Drainage, paper no. 56 (1998)
16. E. De Martonne. La Meteorologie, vol. 2, pp. 449-58 (1926)
17. M. Baggiolini, M. Rev. Rom. Agr. Viticult. 8, 4-6 (1952)
18. P.F. Scholander, E.D. Bradstreet, E.A. Hemmingsen, H.T. Hammel . Science 148 (3668), 339-346 (1965)
19. B.J. Myers. Tree Physiology 4, 315-323 (1988)
20. A. Menzel, T.H. Sparks, N. Estrella,E. Koch, A. Aasa, R. Ahas, K. Alm-Kübler, P. Bissolli, O. Braslavská, A. Briede, F.M. Chmielewski, Z. Crepinsek, Y. Curnel, A. Dahl, C. Defila, A. Donnelly, Y. Filella, K. Jatczak, F. Måge, A. Mestre, O. Nordli, J. Peñuelas,P. Pirinen, V. Remišová, H. Scheifinger, M. Striz, A. Susnik, A.J.H. Van Vliet, F.E. Wielgolaski, S. Zach, A. Zust. Global Change Biology 12, 1969-1976. (2006)
21. J. Gladstones. Viticulture and Environment. Winetitles. Adelaide (1992)
22. O. Currle, O.Bauer, W. Hofäcker, F. Schumann, W. Frisch. Biologie der Rebe. Meininger, Germany (1983)
23. A.J. Winkler. General Viticulture. University of California Press (1962)
24. A.N. Lakso, W.M. Kliewer. American Journal of Enology and Viticulture 29 (3), 145-149 (1978)

 https://doi.org/10.1051/e3sconf/20185001041

Viticulture under climate change impact: future climate and irrigation modelling

Igor Sirnik[1,2], Hervé Quénol[1], Miguel Ángel Jiménez-Bello [2], Juan Manzano[3] and Renan Le Roux[1]

[1] UMR6554 LETG CNRS, Université Rennes 2, place du Recteur Henri le Moal 35043 Rennes Cedex, France
[2] Instituto de Ingeniería del Agua y Medio Ambiente, Universitat Politècnica de València, Camino de Vera s/n, 46022, Valencia, Spain.
[3] Centro Valenciano de Estudios sobre el Riego (CVER), Universitat Politècnica de València, Camino de Vera s/n, 46022, Valencia, Spain

Abstract. Vine is highly sensitive to climate changes, particularly temperature changes, which can be reflected in the quality of yield. We obtained meteorological data from weather station Llíria in viticultural site Valencia DO in Spain from the period 1961-2016 and elaborated the future modelling scenario Representative Concentration Pathways 4.5 (RCP4.5) and RCP8.5 for the period 1985-2100 within the Coupled Model Intercomparison, Project Phase 5 (CMIP5) for daily temperature, precipitation and evapotranspiration. The irrigation requirements (IR) future models for grape varieties Tempranillo and Bobal were elaborated. Temperature and evapotranspiration trends increased during observation period and are estimated to continue rising, according to the future model. Nevertheless, precipitation trend is estimated to decrease according to the model. The future scenarios show increase trend of temperature and evapotranspiration and decrease of precipitation. Total IR for the period 1985 – 2100 is expected to increase during growing season months according to the trendline for 16.6 mm (RCP4.5) and 40.0 mm (RCP8.5) for Tempranillo and 8.2 mm (RCP4.5) and 30.9 mm (RCP8.5) for Bobal grape variety. The outcome of this research is important to understand better the future climatic trends in Valencia DO and provides valuable data to face the future climate changes.

1 Introduction

Atmospheric conditions have strong influence in viticulture [1,2,3]. Due to recent climate change impact, irrigation practice in viticulture is becoming a reality in Mediterranean regions to control water condition of vine [4,5,6]. There are numerous methods to adjust irrigation with high precision in case of water shortage [7]. Shift of phenological stages of vine and changes of wine characteristics are facts [8], which clearly show recent impact of climate change in viticulture; indeed, the adaptation of viticulture is essential and should be based on future climate models [9]. Many studies were made addressing irrigation methodology in viticulture in Mediterranean environment [4,10,11]. The object of this paper is to create the future irrigation models for the period 2005 - 2100 for grape plant in viticultural site Valencia DO. There were several studies addressing irrigation and drought problems in viticulture [12,13]. Only few studies were made addressing the drought and irrigation subject in future climate change scenarios [1, 14]. This study was made on two representative grape varieties in viticultural site Valencia DO: Tempranillo and Bobal. Critical analysis was conducted for the reference period 2000 - 2016, to compare the ETo model calculated with FAO Blaney-Criddle formula with the ETo derived from Penman-Monteith formula [15]. Several studies addressed the evapotranspiration future modelling [16, 17] resulting positive trends of evapotranspiration in the future. According to the comparison statistical analysis, the correction calculations were performed, to correct the ETo model for the period 1985 – 2100. Using the precipitation model at the weather station Llíria, Euro-Cordex data set, corrected ETo model, single crop coefficient (Kc), crop evapotranspiration (ETc) and effective precipitation, the IR model was elaborated for the period 1985 – 2100 using two scenarios RCP4.5 and RCP8.5 [18]. The RCP8.5 scenario was chosen due to its highest radiative forcing increase until the year 2100 among the RCP scenarios, the most "pessimistic" scenario. The RCP4.5 the "optimistic" scenario ,was used due to its significantly lower increase for radiative forcing until the year 2100 comparing to the RCP8.5 scenario [19]. The Kc value for grape varieties Tempranillo and

Bobal, were determined using the results of studies conducted by Instituto Valenciano de Investigaciones Agrarias [20].

2 Methodology

We used the subset EUR-11 of the Euro-Cordex simulations, with spatial resolution 12.5 x 12.5 [21] and FAO Blaney-Criddle formula (ALLEN R. et al., 1986), to elaborate reference evapotranspiration model (ETo) for the period 1985 – 2100. By using climate models, we faced a certain level of uncertainty, due to simplified representation of future climate [23]. The Blaney-Criddle formula was used, due to available data set and its robust nature after calibration in comparison to Penman-Monteith formula, which requires higher number of parameters difficult to predict. Climate data were retrieved from the weather station Llíria, located inside of Valencia DO viticultural area, Spain [20]. At the weather station Llíria we retrieved the ETo data set from the period 2000 – 2016, calculated by Penman-Monteith formula (FAO-56 Method), commonly used formula in irrigation calculations [15]. The study assesses future IR by using two scenarios: RCP4.5 and RCP8.5 [19] for the period 1985 - 2100. The ETo model was calculated using the FAO Blaney-Criddle formula at the weather station Llíria for the period 2000 – 2016. From the same weather station we retrieved the ETo, calculated by Penman-Monteith formula [24]. We used the temperature and precipitation data set from Euro-Cordex data set [21], [25], retrieved from location of the meteorological station Llíria, located in the eastern part of the viticultural area Valencia DO (***Figure 1.*** *Location of the study site Valencia DO (left map) and location of the weather station Llíria, circled with the black ellipse (right map).* Figure 1). The methodology was elaborated in three steps:

1. Elaboration of temperature and precipitation model (1985 – 2100), using Regional Climate Model Euro-Cordex.

2. Calibration of FAO Blaney-Criddle formula (1) to create ETo model for the period 1985 – 2100. The calibration process is defined as calculation of the correction factors a and b.

$$ET_0 = \{a + b[P(0.46T + 8.13)]\}\left[1 + 0.1\left(\frac{Elev}{1000}\right)\right] \quad (1)$$

ET_0: estimated reference evapotranspiration, a and b: correction factors [22], P: daily percentage of annual daylight hours, T: daily mean temperature, $Elev$: elevation of the weather station Llíria.

3. Estimation of crop water requirement for grape variety Tempranillo and Bobal, using a simplified balance model, where soil moisture is assumed to be constant (2).

$$Necessity\ of\ irrigation = ETc - Pef \quad (2)$$

ETc: crop evapotranspiration, Pef : effective precipitation. Crop coefficient values (Table 1) were retrieved from IVIA [20].

Table 1. Monthly single crop coefficient (Kc) values for Tempranillo and Bobal variety, maximum quality during growing season.

Variety \ Month	April	May	June	July	August	September
Tempranillo, maximum quality	0	0	0	0.2	0.4	0.4
Bobal, maximum quality	0	0	0	0.15035	0.25059	0.25059

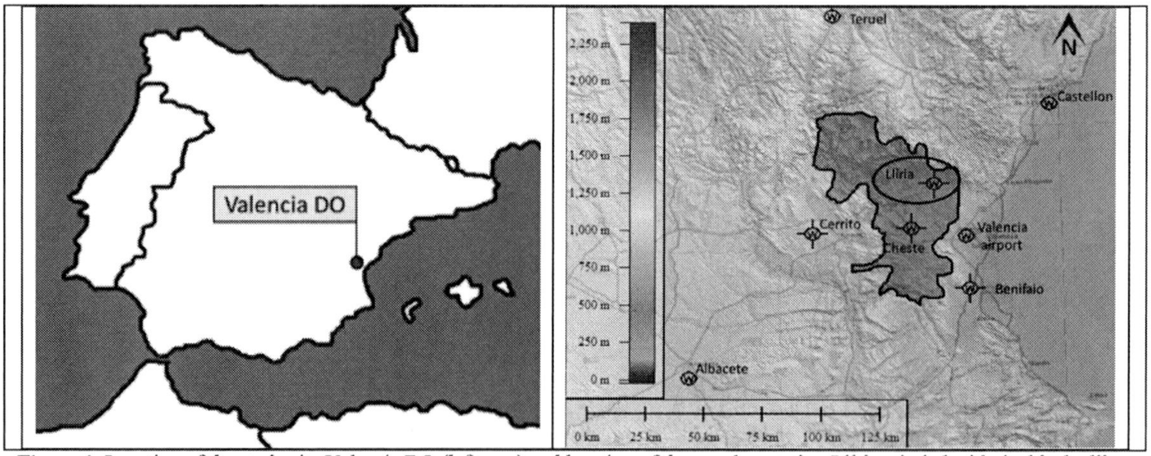

Figure 1. Location of the study site Valencia DO (left map) and location of the weather station Llíria, circled with the black ellipse (right map). Red area represents study site Valencia DO.

3 Results and discussion

Yearly average absolute difference between the ETo derived from Blaney-Criddle formula and ETo derived from observation data, using Penman-Monteith formula during the period 2000 - 2016 was 172.7 mm (Figure 2.**¡Error! No se encuentra el origen de la referencia.**). Difference between modelized and observation ETo is caused due to uncertainty of the model and its spatial resolution comparing to the observed data retrieved from one point, weather station Llíria. The absolute difference between the two ETo is notable, nevertheless, they share the same trend. Corrected ETo model for scenarios RCP4.5 and RCP8.5 show stronger correlation with ETo derived from observations during the period 2000 – 2016 (Figure 2.). The difference between the RCP4.5 and RCP8.5 ETo scenarios enhances from the year 2057, roughly estimated. Before the year 2057, both RCP scenarios show similar growing tendency. Using the ETo model from the period 1985 – 2100, the IR was calculated for Tempranillo and Bobal grape variety according to maximum quality of the grape. In this analysis we considered six months during the grape growing period (from April to end of September). The tendency of IR for Tempranillo and Bobal varieties shows a steady increase during the whole analysis period 1985 – 2100 (**¡Error! No se encuentra el origen de la referencia.** and **¡Error! No se encuentra el origen de la referencia.**), with an exception in September (Bobal), RCP4.5 scenario, where growth is negative. Stronger increase is noticed in RCP8.5 scenario comparing with RCP4.5 scenario in all six growing season months. Positive IR for Tempranillo was noticed in July, August and September during the whole analysis period and negative in April, May and June (Figure 3). The highest increase shows August (11.5 mm) for RCP8.5 and September (5.4 mm) for RCP4.5 scenario, according to the trend line. The lowest increase is noticed in April (2.4 mm) for RCP8.5 and in May (0.5 mm) for RCP4.5 scenario, according to the trend line. The most significant difference between the RCP4.5 and RCP8.5 scenarios was detected in August (6.2 mm) and the lest significant in April (1.5 mm), where RCP8.5 scenario is superior in both cases. The average increase of IR in the period 1985 – 2100, during the growing season months, is 16.6 mm for RCP4.5 and 40.0 mm for RCP8.5.

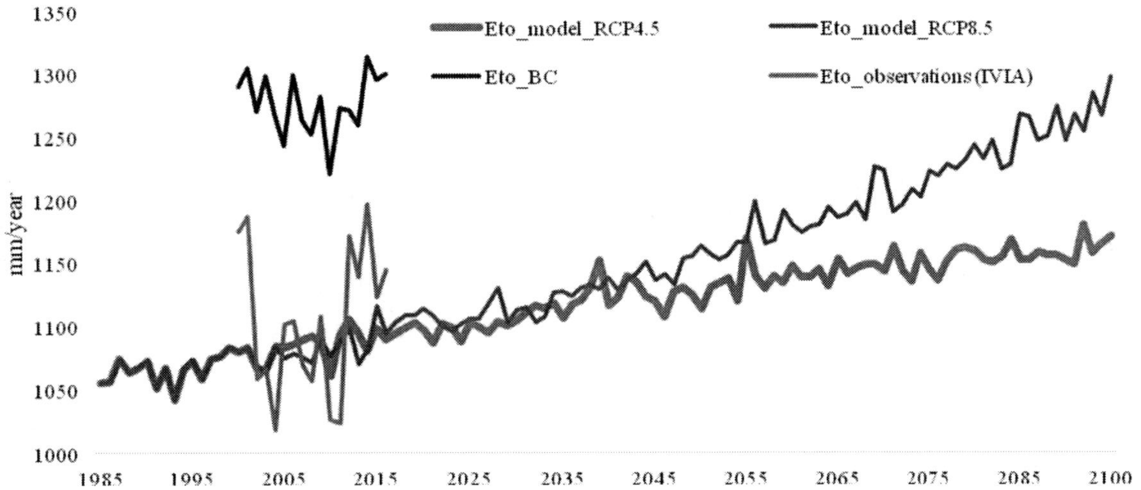

Figure 2. ETo models progress during the period 1985 - 2100, where ETo_model_RCP4.5 and ETo_model_RCP8.5 stand for ETo models using BC formula and Euro-Cordex data set, ETo_observations (IVIA) stands for ETo calculated by Penman-Monteith using observation data set and ETo_BC_2000_2016 stands for ETo model calculated by BC formula using observation data set.

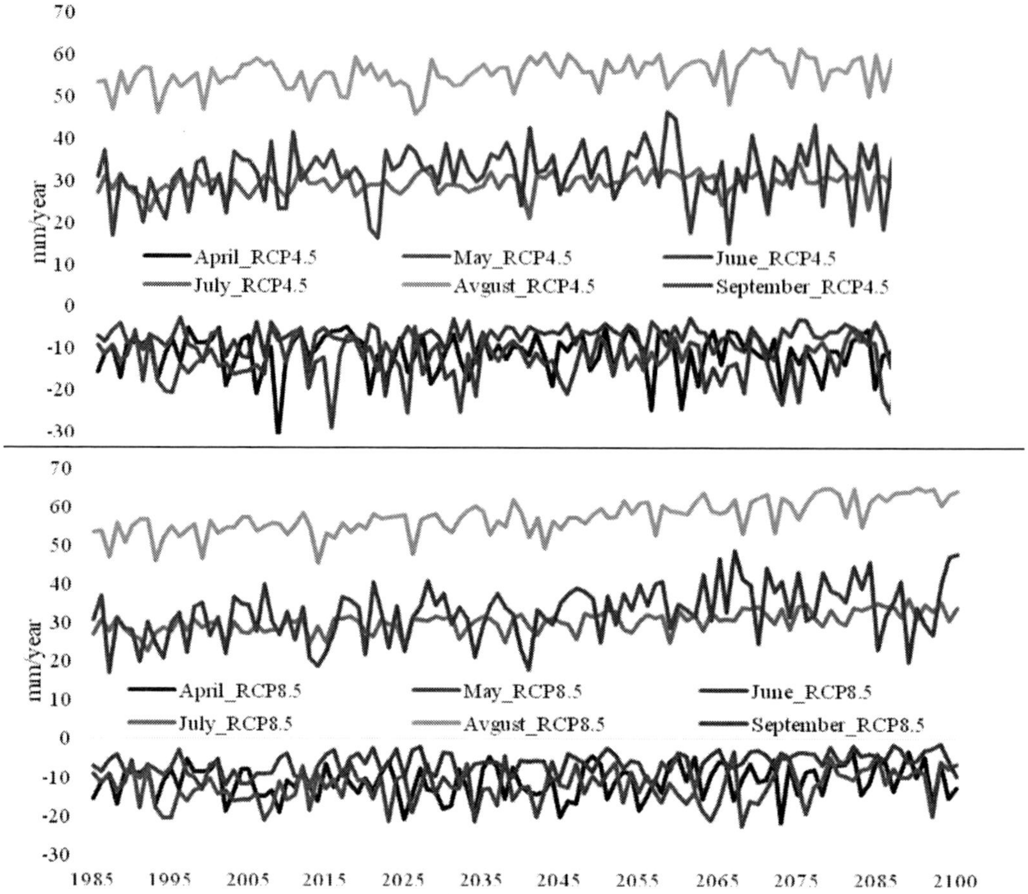

Figure 3. Progress of IR for grape variety Tempranillo, maximum quality, for growing season months. Upper graph represents IR progress according to RCP4.5 scenario and lower graph represents IR progress according to RCP8.5 scenario. Negative values of IR indicate the state, where precipitation is superior than crop evapotranspiration.

Positive IR for Bobal was noticed in July, August and September during the whole analysis period and negative in April, May and June (Figure 4). Exceptions were noticed in September in following years: 1987, 2022 and 2068. The highest increase shows August (37.8 mm) for RCP8.5 and 5.4 mm for RCP4.5 scenario, according to the trend line. The lowest increase was noticed in April (2.4 mm) for RCP8.5 and in September was noticed negative IR (-0.04 mm) for RCP4.5 scenario, according to the trend line. The most significant difference between the RCP4.5 and RCP8.5 scenarios was detected in September (7.8 mm) and the least significant in April (1.7 mm), where RCP8.5 scenario is superior in both cases. The average increase of IR in the period 1985 – 2100, during the growing season months, is 8.2 mm for RCP4.5 and 31.0 mm for RCP8.5. The progress of IR for Tempranillo and Bobal grape varieties during 1985 – 2100 showed similar tendency for all six growing season months. Highest increase of IR was detected in August (RCP8.5 scenario) for Bobal variety, 26.3 mm higher comparing to August for Tempranillo variety. Negative increase was noticed in September (RCP4.5 scenario) for Bobal variety, which is 0.5 mm lower than positive increase in May for Tempranillo variety. Total increase of IR for Tempranillo variety, according to RCP8.5 scenario, is 9.0 mm higher than Bobal variety. Tempranillo variety showed for 8.4 mm higher increase for RCP4.5 scenario comparing to Bobal variety.

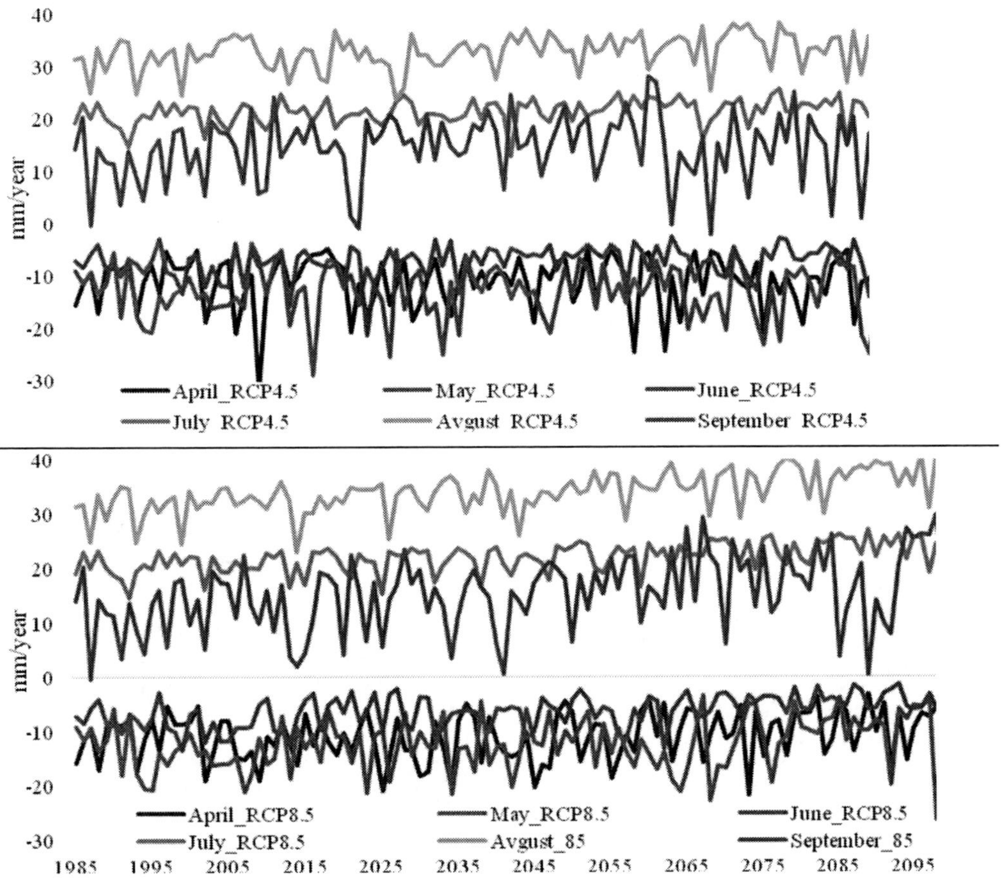

Figure 4. Progress of IR for grape variety Bobal, maximum quality, for growing season months. Upper graph represents IR progress according to RCP4.5 scenario and lower graph represents IR progress according to RCP8.5 scenario. Negative values of IR indicate the state, where precipitation is superior than crop evapotranspiration.

The created ETo models confirm increasing evapotranspiration demand, especially according to the RCP8.5 scenario (Figure 1), which consequentially influences higher future irrigation demand in viticulture. The highest IR is experienced in August for Tempranillo and Bobal variety (**¡Error! No se encuentra el origen de la referencia.** and **¡Error! No se encuentra el origen de la referencia.**). Yearly average of IR for the period 1985 – 2100, during growing season months for Tempranillo variety is 14.2 mm (RCP5.4) and 15.3 mm (RCP8.5); for Bobal variety is roughly 50% lower: 6.1 mm (RCP4.5) and 7.0 mm (RCP8.5). Negative increase of IR in September can be explained by heavy precipitation, which usually occurs in eastern part of Spain, after the summer, caused by mixing of cold autumn air and hot air above the Mediterranean Sea. The meteorological phenomenon is called "Gota fria" and it occurs on yearly bases [26].

4 Conclusion

The irrigation in agriculture is specifically important in eastern Spain, which is expected to experience higher evapotranspiration demands and more extensive droughts periods in the future [27], [28]. According to the created future models for IR for Bobal and Tempranillo grape varieties, IR will increase in the period 1985 – 2100 apart from September (Bobal verity, scenario RCP4.5). The increase of IR for Tempranillo variety during the growing season is 16.6 mm (RCP4.5) and 40.0 mm (RCP8.5). Bobal variety indicates lower increase of IR: 8.2 mm for RCP4.5 and 31.0 mm for RCP8.5. There are two important factors influencing future of irrigation state in viticulture region Valencia DO: temperature and precipitation. According to the future climate models, temperature will continue to rise, and precipitation will diminish, hence both factors do not favour to agriculture needs for viticulture in the future. Future precipitation model, according to the Euro-Cordex model [25], indicates decrease of precipitation in Valencia DO region in the future. Moreover, the study shows significant difference between IR between the Bobal and Tempranillo grape varieties according to the goal of obtaining the maximum quality of grape. Generally, the model can be used for other grape varieties in this viticulture region, however, for more accurate results, the single crop coefficient (Kc) should be modified according to the grape variety. The presented methodology can be used in any wine-growing site in to obtain the future irrigation scenarios for Tempranillo and Bobal grape varieties. With calibration of Blaney-Criddle formula and crop coefficient adjustment, this methodology can be used for any plant anywhere in the World. The elaborated IR models provide significant data, to deal with the future impact of climate change in irrigation field in Valencia DO region.

5 References

[1] H. Fraga, I. García de Cortázar Atauri, and J. A. Santos, Agric. Water Manag., **196**, 66–74, (2018)

[2] J. Tonietto and A. Carbonneau, Agric. For. Meteorol., **124**, no. 1–2, pp. 81–97, (2004)

[3] G. V. Jones and G. B. Goodrich, Clim. Res., **35**, no. 3, pp. 241–254, (2008)

[4] I. Buesa, D. Pérez, J. Castel, D. S. Intrigliolo, and J. R. Castel, Aust. J. Grape Wine Res., **23**, no. 2, pp. 251–259, (2017)

[5] G. V. Jones, Robert E. Davis, Am. J. Enol. Vitic, **51**, no. 3, (2000)

[6] P. R. Petrie and V. O. Sadras, Aust. J. Grape Wine Res., **14**, no. 1, pp. 33–45, (2008)

[7] G. Rallo, P. González-Altozano, J. Manzano-Juárez, and G. Provenzano, Agric. Water Manag., **180**, pp. 136–147, (2017)

[8] V. Guido, *Impact of climate change on vitis vinifera L. over Mediterranean area*, (2016)

[9] H. Quénol, Iñaki Garcia de Cortazar Atauri, Benjamin Bois, Andrew Sturman, Valérie Bonnardot and Renan Le Roux, J. Int. des Sci. la Vigne du Vin, **51**, no. 2, pp. 91–97, (2017)

[10] S. Vanino, G. Pulighe, P. Nino, C. de Michele, S. F. Bolognesi, and G. D'Urso, Remote Sens., **7**, no. 11, pp. 14708–14730, 2015

[11] J. M. Costa, M. Vazc, J. Escalonad, R. Egiptob, C. Lopesb, H. Medranod, M.M. Chaves, Agric. Water Manag., **164**, pp. 5–18, (2016)

[12] Z. Zsófi, E. Tóth, G. Váradi, D. Rusjan, and B. Bálo, Acta Biol. Szeged., **52**, no. 2, pp. 321–322, (2008)

[13] C. van Leeuwen and P. Darriet, J. Wine Econ., **11**, no. 1, pp. 150–167, (2016)

[14] P. Döll, Clim. Change, **54**, pp. 269–293, (2002)

[15] A. H. L. Penman, Natural Evaporation from Open Water, Bare Soil and Grass, **193**, no. 1032, pp. 120–145, (2013)

[16] Y.-P. Xu, S. Pan, G. Fu, Y. Tian, and X. Zhang, J. Geophys. Res. Atmos., **119**, no. 5, pp. 2174–2192, (2014)

[17] E. Obada, E. Alamou, A. Chabi, J. Zandagba, and A. Afouda, Hydrology, **4**, no. 3, p. 38, (2017)

[18] IPCC, "IPCC - Intergovernmental Panel on Climate Change," 2017. [Online]. Available: http://www.ipcc.ch/. [Accessed: 18-May-2017]

[19] G. P. Wayne, *The Beginner's Guide to Representative Concentration Pathways (RCPs),* Skept. Sciece, (2013)

[20] IVIA, "IVIA - Generalitat Valenciana, 2017. [Online]. Available: http://www.ivia.gva.es/. [Accessed: 18-May-2017]

[21] D. Jacob et al., EURO-CORDEX: New high-resolution climate change projections for European impact research, Reg. Environ. Chang., **14**, no. 2, pp. 563–578, (2014)

[22] R. Allen, W.O. Pruitt and members ASCE, J. Irrig. Drain. Eng., **112**, no. 2, pp. 139–155, (1986)

[23] J. Boé, *Changement global et cycle hydrologique: Une étude de régionalisation sur la France,* p. 278, (2007)

[24] L. G. de Carvalho, A. W. P. Evangelista, K. M. G. Oliveira, B. M. Silva, M. de Carvalho Alves, A. de Sá Júnior, W. L. Miranda, Idesia (Arica), **31**, no. 3, pp. 39–47, (2013)

[25] Euro-Cordex, "Euro-Cordex," 2017. [Online]. Available: http://www.euro-cordex.net/. [Accessed: 18-May-2017]

[26] CHJ, *Plan Hidrológico del Júcar 2015-2021,* Valencia, (2016)

[27] IPCC, *Climate Change 2014 Synthesis Report Summary Chapter for Policymakers*, (2014)

[28] M. V. O. Reig, *Collective management of irrigation in eastern Spain. Integration of new technologies and water resources,* Universitat Politècnica de València, (2015)

Experimental training systems in Champagne: an overview of agronomical and qualitative parameters

Julie Perry*[1], Sébastien Debuisson[1], Arnaud Descôtes[1]
[1]CIVC, Comité interprofessionnel du vin de Champagne, 5 rue Henri-Martin, CS 30 135, 51204 Epernay, France
*julie.perry@civc.fr

Abstract
Vine implantation in Champagne is strictly regulated. Row spacing is limited to 1,50 meter and the canopy height can not exceed 1,40 m. The traditional training system is therefore characterized by narrow spaced vines. From the late eighties, different vine training systems, such as lyres, have been tested in the Champagne area. The aim is to assess their interests in the terroir of Champagne, which is characterised by its cool climate, soil profile and its customs.
Whereas the lyre training showed its limits in the Champagne context, some other training systems have been implemented such as half-widely-spaced vines. These devices are characterised by a row spacing of two meters, a consistent cover crop and a canopy up to two meters. The plots are located in various places in the area and are strictly followed each year since 2006 (and 2000 for the first sites). Phenological, agronomical and ripening parameters are controlled and compared to the traditional training system plots. Experimental vinifications are done each year so that sensory analysis can be undertaken to assess the ability of these vines to produce wines with a Champagne typicality.
The results of this experimental device show interesting conclusions on the agronomical behaviour of experimental widely-spaced vines in a cool climate region. Spring frost resistance, cover crop management and ripening are some elements which show differences between the reference traditional system (REF) and the widely-spaced vines (VSL).

Introduction

The Appellation d'Origine Contrôlée Champagne has limited; distance of vine spacing (m) plus distance of row spacing (m) to 2.5m. This has been set in order to limit the vigor, and the yield, of the individual vine [Bonal, F. 1984]. These factors are considered favourable for the maturity of the grapes and their quality.

High density plantations however, are also associated to higher implantation, labour and specialised machinery costs, a greater demand for phytosanitary products, and in Champagne especially, an interrow management with elevated herbicide consumption and erosion susceptibility.

The appeal of VSL is therefore both financial and environmental. However, in order to satisfy both factors, the concept of vine balance is of paramount importance.

A fundamental principle of vineyard management is that total potential productivity is dependant on the quantity of intercepted photosynthetically active radiation (PAR), required for nutrient assimilation [Bates, S. 2010]. For vertical shoot positioned vines, canopy height (H) to row distance (E) is the primary consideration for light interception [Carbonneau, A. et al. 2007]. Optimal values of H/E are between 0.6 and 0.8 depending on viticultural practices and targets.

There have been few trials which have combined row spacing with an adaptation in vine architecture [Carbonneau, A. et al. 2007], and those studying the relationship of row spacing on vine physiology, yield, and berry composition often produce contradictory results [Archer, E. & Strauss, H.C. 1991]. This has been largely attributed to variances in vine vigor – influenced by scion and rootstock genetics, soil composition and texture, water availability, and cultivation practices such as; trellising, pruning, cover crops and fertilization [Steyn, J. et al. 2016] – all of which directly affects the results of vine spacing trials. As such, to understand the effects and feasibility of this mode of conduct in the Champagne appellation, 15 experimental parcels, approved by the INAO, have been established by a network of authorities and domains under the name « Mode de Conduite et Enherbement ».

Materials and Method

Experimental VSL parcels and their associated reference modalities, are scattered over the region and represent the 'experimental poles' of Champagne - reflecting the widest possible spectrum of climatic conditions and soils, whilst remaining true to the regions typical varieties, rootstocks and strict vitivinifcation procedures. The 15 parcels are comprised of 12 planted sites totalling 9 Ha and 1 Ha of 'transformed' vines in which traditional modalities have every second row removed and its canopy restructured accordingly.

Agronomic monitoring

The following data has been collected on sites at Essoyes and Plumecoq since 2000, on transformed sites at Chouilly, Verzernay and Mézy-Moulins as of 2007 and all sites as of 2012 :

- Load (number of buds / ha)
- Watch (number of clusters / m²)
- Fertility (number of clusters / bud)
- Foliar index: (exposed leaf area (m² / m²)
- Vigor : (dry matter g/m²)
- Average weight per cluster (g)
- Potential degree (% vol)
- Total acidity (gH2SO4 / L)
- Yield (kg / ha)
- Monitoring wood disease (esca)

In the case of climatic accidents, notations of frost or hail damage are made. Similarly, if there is a strong grey mould pressure, the affected cluster frequencies are indicated.

Vinification and tasting

Depending on the year and with the agreement of the domains, vinifications of 160 kg of grapes are carried out according to a strictly similar protocol at the CIVC's experimental winery. Complementary vinifications of 4000Kg are sometimes carried out by the domains and brought to the CIVC for tirage, aging and disgorgement. Triangular tests in black glasses are carried out on the base wine as well as 15 and 36 months after tirage to determine the existence or not of significant differences between VSL and REF. To date, 191 tasting comparisons have been conducted.

Additional studies

Production costs: An economic study was carried out comparing the production costs between VSL and REF. This study made it possible to apprehend all the factors (material investments, manpower, time of works, etc.) relating to the vegetative season and the production.

Life cycle analysis (LCA): LCA is a multi-criteria analysis that quantifies the environmental impacts of a product from the extraction of raw materials, to the disposal or recycling of the product by considering the manufacturing, logistics and use phases.

Results:

Charge: The bud load is, by definition, dependent on planting density, but also impacted by pruning systems. Differences measured are, on average, between 22 and 38% less in VSL than REF depending on the sites and years.

Fertility: Directly related to the vigor of the shoot, fertility is higher in VSL. This can also be explained by microclimatic conditions (notably sunshine) more favourable to floral initiation.

Foliar index: Exposed leaf area and the dry matter produced are always lower in VSL than REF. However, the difference between the two modalities is lower, on average, in the transformed sites (4% difference, all sites and all vintages combined) than in the planted sites (17% difference).

Average bunch weight: Cluster weights are generally higher in VSL, with all grape varieties, vintages and size systems combined - a difference of 11%.

Yields: Yields are 15-20% lower in VSL. This is explained by the density, and therefore the load in buds. VSL parcels pruned and trained as simple Guyot are, in general, producing the highest yields.

Frost sensitivity: VSL demonstrates a slightly lower sensitivity to spring frost, explained by an often later budburst, as well as by the higher average position of the buds.

Must composition: Of the entirety of the musts, all vintages combined, more differences occur according to the vintage and soils than by the mode of driving. However, on average, VSL musts contain a 5% increased total and ammonia nitrogen (mg N/L), 4 % increase in malic acid, 2 % in total acidity and a marginal elevation in pH.

Sensory results: In two out of three cases, the wines are not differentiated. There is no preference for one profile or the other of the wines. All the wines present a profile considered typical of the Champagne wines. Greatest differences are noted on Pinot Noir and Pinot Meunier.

Production costs: Results show that production costs for planting, seasonal work and mechanization are much lower in VSL. This is very logically explained by the density of plantation (reducing costs by 20%), minimizing the time of work (by 38%), and allowing above all, the use of vineyard tractors.

Life cycle analysis: For both VSL and REF systems, phytosanitary protection and soil maintenance are preponderant in terms of environmental impacts. VSL appears to be an interesting lever for reducing this impact with a 39% lower demand on resources than REF, and a 33% less impact on climate. These results are explained by reduced vineyard establishment materials, reduced tractor passing's and associated fuel consumption, and reduced phytosanitary requirements.

Conclusions and discussion

VSL offers an alternative mode of conduct to traditionally planted vines in the Champagne appellation, with both reduced production costs and impact on the environment. Furthermore, in the context of climate change and preserving Champagnes typicity, VSL demonstrates elevated levels of acidity, notably malic, without conclusive sensory differences (especially in Chardonnay).

While yields of VSL are typically lower, they do consistently meet the appellation's annually regulated and permitted yields, 'rendement de l'appellation'. However, additional yields that may be vinified for cellaring 'rendement reserve indivduelle' may not always be reached in VSL. It should be noted that differences in yield must not be simply filled by a higher load, a phenomenon that can be observed, especially in cases with double Guyot, which can compromise the vigor of the vine too much.

However, as the understanding of VSL management progresses, the economic viability of the system strengthens, and why this mode of viticulture is of continued interest. In 2017 (a spring frost year) for example, across all experimental plots observed, VSL yielded an average approximately the same as the traditional reference (-1%).

Due to large variances in performances between vineyard sites, further studies are required to better understand the characteristics of the vineyard (soil structure and fertility, water availability, etc.) where VSL performs most efficiently, allowing precise recommendations to be provided for the implementation of this mode of conduct in the Champagne appellation – subject to political approval.

References

ARCHER, E. and STRAUSS, H. C. 1991. *The effect of vine spacing on the vegetative and reproductive performance of Vitis vinifera L. (c.v. Pinot noir)*. South African Journal of Enology and Viticulture, 12:2:70–76.

BATES, T. 2010. *Resource Assimilation and Vineyard Productivity*. GRAPES 101. Viticulture & Enology. CornellCALS. 2010 Newsletters. Issue 2, April 2010.

BONAL, Cln. F.1984. *La définition et la législation de l'appellation*. Chapitre 6. L'appellation d'origine controlée champagne. Encylopédie 'Le Bonal'. Grandes Marques & Maisons de Champagne.

CARBONNEAU, A., DELOIRE, A., & JAILLARD, B. 2007. *System de conduite de la vigne.* La Vigne : Physiologie, terroir, culture. Dunod. p.303-13

STEYN, J. ALEXANDRE-TUDO, J. L. & ALEIXANDRE, J. L. 2016. *Grapevine Vigour and Within-Vineyard Variability: a Review*. International Journal of Scientific & Engineering Research, Volume 7, Issue 2, February-2016.

E3S Web of Conferences **50**, 01043 (2018)
XII Congreso Internacional Terroir
https://doi.org/10.1051/e3sconf/20185001043

Varietal responses to soil water deficit: first results from a common-garden vineyard near Bordeaux France

Mark Gowdy[1], Agnès Destrac-Irvine[1], Martina Haines[1], Gregory Gambetta[1], Philippe Pieri[1], Elisa Marguerit[1], Cornelis van Leeuwen[1]

[1]UMR EGFV, Bordeaux Sciences Agro, INRA, Univ. Bordeaux, ISVV, 33883 Villenave d'Ornon, France

Abstract. In wine producing regions around the world, climate change has the potential to decrease the frequency and amount of precipitation and increase average and extreme temperatures. This will both lower soil water availability and increase evaporative demand in vineyards, thereby increasing soil water deficits and associated vine stress. Grapevines control their water status by regulating stomatal closure and other changes to internal plant hydraulics. These responses are complex and have not been clearly characterized across a wide range of different *Vitis vinifera* varieties. Understanding how vine water status responds to changes in soil water deficits and other variables will help growers modify vineyard design and management practices to meet their quality and yield objectives. Carbon isotope discrimination measurements of certain plant tissues have been shown to provide effective characterization of stomatal closure, while water potential measurements provide a well-proven measure of overall vine water status. Using replicated data collected from an experimental common-garden vineyard at the *Institut des Sciences de la Vigne et du Vin* (ISVV) near Bordeaux, France, this project will analyze the effects on carbon isotope discrimination across 39 varieties and water potential across eight varieties against estimates of soil water deficits made using a water balance model running on local meteorology and considering the phenology of each variety. Similar to the literature, preliminary analysis finds as soil water deficit increases, carbon isotope data suggests greater stomatal closure and water potential measurements indicate greater vine stress. For both parameters, analysis will be performed to distinguish any difference in these responses between varieties.

1 Introduction

In wine producing regions around the world, climate change has the potential to decrease the frequency and amount of precipitation and increase average and extreme temperatures during the growing season. Understanding how vine water status responds to these changes across different *Vitis vinifera* varieties will help growers modify vineyard design and management practices to meet their quality and yield objectives [1]. Much has been published on how plants regulate water status, but not much to characterize the dynamics of this regulation across the wide range of *Vitis vinifera* varieties.

As soil water content decreases, the remaining water becomes more tightly bound with the soil matrix and the water potential in the plant must become more negative in order to remove the remaining water and maintain transpiration. In response to this, and other environmental factors, stomata begin to close to moderate water potential and protect the plant from hydraulic failure [2].

As stomata close there is a change in the carbon isotope composition of the carbohydrates and sugars generated by photosynthesis in the leaves at that time. Measurement of the isotope composition ($\delta^{13}C$) in berry juice then provides a useful indication of the level of stomatal closure that existed during the important berry-ripening period [3]. Water potential measurements collected pre-dawn and mid-day during the berry ripening also provides an assessment of soil water conditions and plant water status during this same period. For estimating the soil water content corresponding to the above measurements, a water balance model was used. Such models can be useful tools when other more costly and intensive methods may be prohibitive [4].

This paper presents preliminary data and analysis characterizing and comparing the stomatal closure response (as indicated by $\delta^{13}C$) for 39 *Vitis vinifera* varieties; and the plant water status response (as measured by water potential measurements) for eight varieties; both in response to changes in soil water content as estimated by a water balance model. The data presented in this paper was collected from an experimental common-garden vineyard near Bordeaux France, which allows for collection of replicate data [5].

2 Methods

2.1 Experimental vineyard

The experimental vineyard used for this study was planted in 2009 at the *Institute des Sciences de la Vigne et du Vin* (ISVV) situated at the *Institut National de la Recherche Agronomique* (INRA) *domain de la "Grande Ferrade"* (33883 Villenave d'Ornon) near Bordeaux, France (44°47'23.83"N 0°34'39.3"W). It consists of 52 varieties laid out in a randomized block replicate design, with 10 plants of each variety in each of five blocks. This design allows the ability to account for variability in soil or other conditions within the vineyard.

https://doi.org/10.1051/e3sconf/20185001043

Vines are pruned double Guyot style with 100 cm plant x 180 cm row spacing and trained by vertical shoot positioning with a canopy height of roughly 160 cm. Varieties planted are from the Bordeaux region, as well as other European regions, and some hybrids that might be suitable in Bordeaux in the future as the climate changes. All varieties were planted on SO4 rootstock clone 761 in gravelly soil typical of the region [5].

Many studies of drought adaptation of plants are carried out in controlled conditions, such as greenhouses, allowing strict control over experimental variables. Fundamental differences between plants in controlled versus field conditions include root distribution and water availability in the profiles of the root zone [6], along with climatic conditions. Hence, experimentation under field conditions is necessary to put results obtained in controlled conditions into perspective [7].

2.2 Carbon isotope discrimination measurement

The relative concentration of $^{13}CO_2$ to $^{12}CO_2$ inside the intercellular space of leaves changes relative to that of the atmosphere because of preferential reaction during photosynthesis with ^{12}C. These ratios, however, are modified when stomata close and cut off diffusion of CO_2 into the intercellular space of the leaves from the atmosphere, causing the sugars formed and incorporated into various plant tissues at that time to contain a modified ratio of carbon isotopes [8]. This makes the measurement of carbon isotope concentration a useful measure of the level of stomatal closure that existed at the time the glucose building blocks of sampled plant tissue or berry juice were photosynthesized [4].

Berries were sampled weekly in 2012 through 2016 from all varieties in all blocks of the vineyard and analyzed for numerous parameters, including sugar concentration, berry weights, and acid content. Afterwards, one sampling date from each season for each variety was selected as representative of full maturity based on sugar content and titratable acidity of the juice and cluster sanitary status. Previously frozen samples from the selected date were then defrosted and one sample per variety per block was prepared for analysis. The ratio of $^{13}CO_2$ to $^{12}CO_2$ was determined in the laboratory by elemental analysis coupled to an isotopic mass spectrometer and reported as a ratio $x10^{-3}$ (‰) with less negative values indicating more stomatal closure [9].

Of those originally planted in 2009, 48 varieties are now consistently producing enough fruit for analysis each year, however, only 39 varieties produced enough fruit to perform the various analyses across all five years (2012 through 2016) in this analysis. Data from measurements collected in 2017 are currently being analyzed and measurements will be collected again in 2018.

2.3 Water potential measurements

In order for a vine to continue transpiring, its water potential must become more negative as it overcomes decreasing soil water potential and conductivity in a drying soil [9]. Predawn leaf water potential is considered a good measure of the equilibrium reached in the plant with this water potential in the root zone [11].

Eight varieties in the experimental vineyard (Cabernet-Sauvignon, Merlot, Petit Verdot, Grenache, Tempranillo, Touriga nacional, Sauvignon blanc, and Semillon) were selected from the experimental vineyard for water potential measurement starting in 2017, with the intent of covering a range of varietal stomatal closure tendencies as described in various literature.

Midday leaf, midday stem, and predawn leaf water potential measurements were each made on leaves taken from two vines in each of four vineyard blocks for a total of eight vines per variety, totaling 64 for each water potential measurement. Measurements were performed using standard pressure chamber techniques [11].

2.4 Water balance modeling

A water balance model [12] was used to estimate daily soil water content in the vineyard for years 2012 through 2017 for comparison against $\delta^{13}C$ and water potential measurements. Inputs to the water balance model include phenology, canopy characteristics (dimensions, porosity, orientation), hedging dates, climate (solar radiation, reference evapotranspiration, temperature, precipitation), and an estimate of 160 mm for total transpirable soil water (TTSW) in the vine root zone. The model then outputs an estimate of soil water content as a fraction of total transpirable soil water (FTSW) on a daily time step. From this the average fraction of total transpirable soil water during the berry-ripening period ($FTSW_{BR}$) is calculated between dates of véraison and 75% of berry sugar accumulation for each variety. This averaging period was initially thought by the authors to correspond well with soil water content during berry sugar accumulation, but other averaging methods are being investigated as analysis continues.

3 Results

3.1 Water balance modeling

A box plot of average $FTSW_{BR}$ across all varieties for a given year shows 2012 and 2016 having significantly lower soil water content during the berry-ripening period than 2013 through 2015 (Figure 1). A plot of 2017 model output was not included as phenology and carbon isotope data are still under evaluation. FTSW estimates for 2017, however, were used for comparison against predawn water potential measurements. Absolute values of FTSW are likely underestimated due to difficulty estimating TTSW. Also, the model does not account for varietal differences in horizontal root system distribution. Overall, however, resulting relative differences between varieties remain valid.

3.2 Carbon isotope discrimination response

A box plot by year of the $\delta^{13}C$ measurements on berry juice across all varieties for a given year shows significant difference between both 2012 and 2016 when compared to 2013 through 2015 (Figure 2). Similarly, a comparison back with Figure 1 finds overall less negative $\delta^{13}C$ values (more stomatal closure) in the same years 2012 and 2016, which experienced lower $FTSW_{BR}$ (drier soil conditions during berry-ripening).

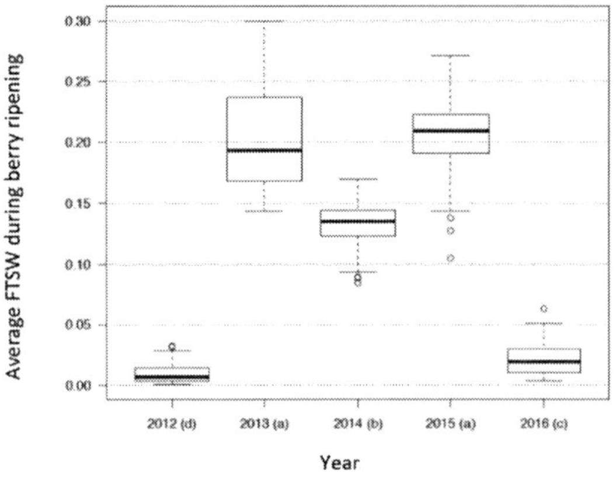

Figure 1. Box plot of $FTSW_{BR}$ (mm^3/mm^3) for all varieties that year for 2012 through 2016. Years grouped based Tukey Honest Significant Difference test at $\alpha = 0.05$. Dark bar is median; box ends are 75th / 25th percentiles; upper/lower horizontal lines are maximum/minimum respectively, with circles being outliers.

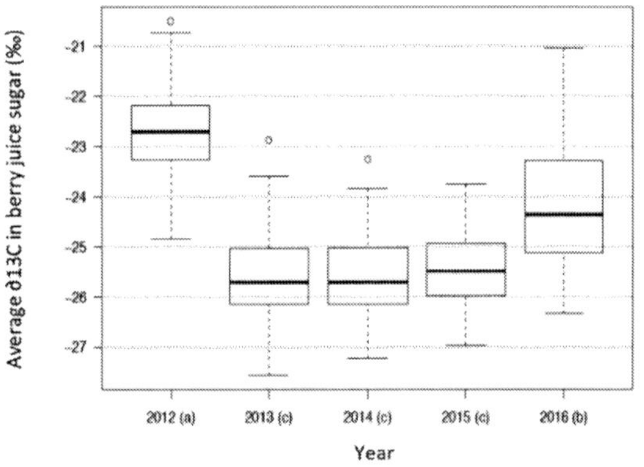

Figure 2. Box plot of $\delta^{13}C$ (‰) for all varieties measured for 2012 through 2016. Years grouped based Tukey Honest Significant Difference test at $\alpha = 0.05$. Dark bar is median; box ends are 75th / 25th percentiles; upper/lower horizontal lines are maximum/minimum respectively, with circles being outliers.

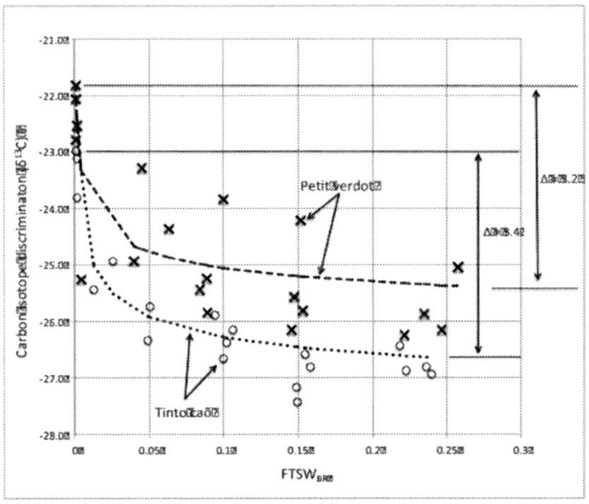

Figure 3. $\delta^{13}C$ (‰) versus $FTSW_{BR}$ (mm^3/mm^3) with power function regression lines for Tinto Cað and Petit Verdot and difference (Δ) between wet and dry years for the two varieties.

Statistically significant differences were also found between some varieties in the response of $\delta^{13}C$ to changes in soil water content ($FTSW_{BR}$). For example, a plot of $\delta^{13}C$ versus $FTSW_{BR}$ for both Tinto Cað and Petit Verdot shows similar response curves, but with consistently less negative values (more stomatal closure) seen across all $FTSW_{BR}$ values for Petit Verdot (Figure 3).

These curves also show the trend of less negative $\delta^{13}C$ overall in response to decreasing soil water content, with the different varieties demonstrating a good non-linear fit using a power function regression; with all varieties having similar power coefficients as a formula used for estimating soil matrix potential as a function of soil water content [13]. This suggests a the link between increasing soil water potential and associated stomatal closure (vine stress) in response to decreasing soil water content.

The difference in $\delta^{13}C$ between wet and dry $FTSW_{BR}$ conditions for the different varieties is also of interest as an indication of stomatal sensitivity to decreasing soil water content. In preliminary analysis many varieties demonstrated a difference in $\delta^{13}C$ of between 3 to 4 ‰, similar to that shown in Figure 3 for Petit Verdot and Tinto Cað. But before more definitive conclusions can be made about these varietal differences (and other responses described above), the authors are considering alternative calculations of $FTSW_{BR}$ and performing additional statistical analyses.

3.3 Water potential response

A plot of average predawn water potential measurements across blocks taken in 2017 versus corresponding water balance model estimates of FTSW shows low vine stress values (none less than -0.16 MPa) when FTSW was above 0.4 (Figure 4). After July 28th (moving right to left) predawn water potential dropped significantly as soils dried further past FTSW = 0.4. This threshold is similar to values reported in the literature [10,12]. There were also statistically significant differences between Semillon and three other varieties (Grenache, Petit Verdot, and Cabernet sauvignon) under drier soil conditions later in the year. The cause of this difference is uncertain and being investigated.

Also, a recently proposed method quantifies the relative stomatal closure tendency of different plant species by measuring the area covered by a plot of midday leaf water potential versus predawn leaf water potential across a range of water status, light, and VPD conditions. The relative size of these "hydroscapes" (as determined by a regression line using the most negative midday leaf water potential measurements) is indicative of the relative stomatal sensitivity of a given variety. A larger area indicating less stomatal closure and more variability in plant water potential and a smaller area indicating more stomatal closure and less variability of plant water potential [14]. Unlike other varieties, the water potential measurements for Semillon collected in 2017 began to fill out the hydroscape, although more data is still needed (Figure 5). More midday and predawn leaf water potential with more negative values are needed to fill out that important part of the hydroscape; although the results so far are promising.

 https://doi.org/10.1051/e3sconf/20185001043

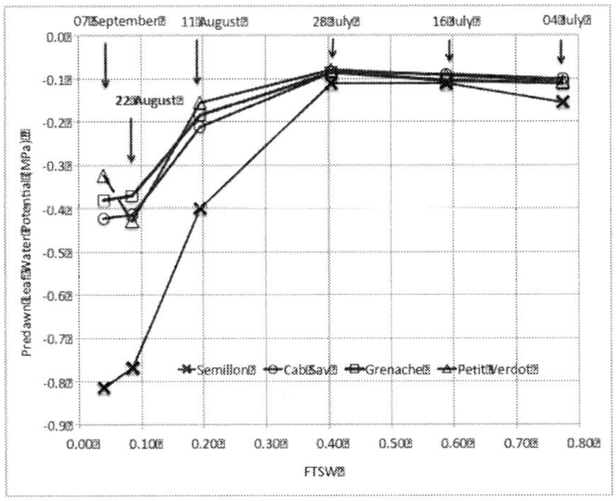

Figure 4. Average predawn leaf water potential (MPa) versus corresponding water balance model estimate of FTSW (mm^3/mm^3) for measurements taken in 2017 on Semillon, Cabernet sauvignon, Grenache, and Petit verdot.

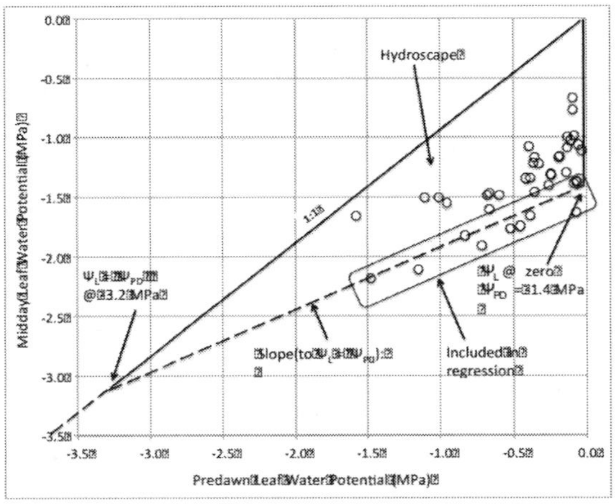

Figure 5. Plot of midday leaf versus predawn leaf water potential measurements (MPa) for Semillon in 2017, including hypothetical and calculated boundaries for the hydroscapes [14].

4 Discussion and conclusion

Preliminary analysis of $\delta^{13}C$ measurements from berry juice collected between 2012 and 2016 suggests greater stomatal closure for all varieties during years of lower average soil water content during the corresponding berry ripening period as estimated with a water balance model using phenology of the individual varieties. Some varieties appear to show significant differences from one another, with the response curves for individual varieties having a non-linear shape similar to those used for estimating soil matrix potential as a function of soil water content. Also, observed differences in $\delta^{13}C$ between years with wet and dry conditions during berry-ripening ranged between 3 to 4 ‰, but more consideration is being given by the authors to different soil water content averaging methods that might better represent soil water conditions during the time of berry sugar accumulation. Data from subsequent years should also improve the power of statistical determinations of differences between varieties.

Water potential data collected in 2017 from eight varieties finds predawn measurements correlated well with soil water content estimates from a water balance model using the phenology of the individual varieties. The reason for the significantly different response of Semillon is uncertain and being investigated. Also, plots of midday leaf versus predawn leaf water potential measurements show the early development of hydroscape plots that might be useful in further characterizing the stomatal closure behavior of the different varieties in response to changing soil water content. Comparison of the different water potential measurements may give additional insight into stomatal closure response to decreasing soil water content. The authors will also give consideration to effects of vapor pressure deficit as data continues to be collected in subsequent years.

5 Acknowledgments

These studies were carried out with support from the French National Research Agency (ANR) under the Investments for the Future Program, within Clusters of Excellence (ANR-10-LABX-45). The VitAdapt Project is also supported by the *Conseil Interprofessionnel des Vins de Bordeaux* and the *Conseil Régional d'Aquitaine*.

References

1. A. Pellegrino, E. Lebon, M. Voltz, J. Wery, Plant and Soil, **266** 129–42 (2004).
2. J.S. Sperry, U.G. Hacke, R. Oren, J.P. Comstock, Plant, Cell & Environment, **25** 251–63 (2002).
3. J.P. Gaudillère, C. van Leeuwen, N. Ollat, Journal of Experimental Botany, **53** 757–63 (2002).
4. C. van Leeuwen, P. Pieri, P. Vivin, *Methodologies and Results in Grapevine Research* (Springer: Dordrecht, 2010) 87–106.
5. A. Destrac-Irvine, C. van Leeuwen, CLIMWINE Proceedings, 165-171 (2016).
6. T. Améglio, P. Archer, M. Cohen, C. Valancogne, F-A. Daudet, S. Dayau, P. Cruiziat, Plant and Soil, **207** 155–67 (1999).
7. J. Bota, M. Tomás, J. Flexas, H. Medrano, J.M. Escalona, Agricultural Water Management, **164** 91–99 (2016).
8. A. Bchir, J.M. Escalona, A. Gallé, E. Hernández-Montes, I. Tortosa, M. Braham, and H. Medrano, Agricultural Water Management, **167** 11–20 (2016).
9. G. Farquhar, Annual Review of Plant Physiology and Plant Molecular Biology, **40** 503–37 (1989).
10. L. Krounbi, N. Lazarovitch, *Encyclopedia of Agrophysics*, (Springer: Dordrecht, 2011) 748–54.
11. X. Choné, C. van Leeuwen, D. Dubourdieu, J.P. Gaudillère, Annals of Botany, **87** 477–83 (2001).
12. E. Lebon, V. Dumas, P. Pieri, H.R, Schultz, Functional Plant Biology, **30** 699-710 (2003).
13. M.T. van Genuchten, Soil Science Society of America Journal, **44** 892–98 (1980).
14. F.C. Meinzer, D.R. Woodruff, D.E. Marias, D.D. Smith, K.A. McCulloh, A.R. Howard, A.L. Magedman, Ecology Letters, **19** 1343–52 (2016).

E3S Web of Conferences 50, 01044 (2018) https://doi.org/10.1051/e3sconf/20185001044
XII Congreso Internacional Terroir

Update in unified terroir zoning methodologies

Cornelis van Leeuwen[1] and Benjamin Bois[2]

[1]UMR EGFV, Bordeaux Sciences Agro, INRA, Univ. Bordeaux, ISVV, 33883 Villenave d'Ornon, France
[2] Centre de Recherches en Climatologie / Biogéosciences, UMR 6282, CNRS - Université de Bourgogne Franche-Comté, 6, Boulevard Gabriel, 21000 Dijon, France

Abstract. The concept of terroir is based on the assumption that the quality and the typicity of an agricultural product are linked to its origin. The precise definition of an origin requires zoning. Because terroir expression in viticulture is largely driven by interactions between the vine and its natural environment, soil and climate play a key role in terroir zoning. For clarity, soil-based and climate-based zoning are presented separately in this paper. They are, however, ideally carried out simultaneously, because of the existence of multiple interactions between these terroir factors. Prior to the implementation of zoning, the objectives need to be carefully defined. The appropriate scale at which the zoning is implemented depends on the objective and the available budget. The use of geomatics enables the production of maps at reduced cost and, possibly, increased accuracy. This paper is update of the unified zoning methodology which was adopted by the OIV in 2012 (OIV-VITI 423-2012).

1 Introduction

Wine style and quality are influenced by environmental factors, particularly soil and climate. Therefore, quality and pricing of wine are related to its origin. This feature is generally referred to as the "terroir" effect [1]. Because terroir has a spatial dimension, zoning is required to precisely delineate production areas. Many approaches for terroir zoning have been developed over the past decades, varying according to the objectives and scientific disciplines used. Comparing characteristics of terroirs worldwide then becomes difficult when zoning is carried out with such different approaches. In 2012 a unified methodology was developed by the International Organisation of Vine and Wine (OIV) with its *OIV Guidelines for vitivinicluture zoning methodologies on a soil and climate level* (Resolution OIV- VITI 423-2012 http://www.oiv.int/public/medias/400/viti-2012-1-en.pdf). Application of these methodologies enables the comparison of terroir zoning studies implemented in various winegrowing regions and countries. This paper presents an update of the OIV guidelines adopted in 2012.

Terroir zoning comprises soil and climate related aspects. Because soil and climate interact in terroir expression, their zoning should be ideally undertaken simultaneously. However, for the sake of clarity of presentation, soil and climate zoning are presented below separately. As a prerequisite to each zoning project, objectives and scale need to be clearly identified. Once the zoning is completed, several approaches allow validation of the zoning.

2 Terroir zoning objectives

Terroir zoning studies may have various objectives. It is important to define these objectives prior to the implementation of the zoning, as methodology and scale will vary accordingly. For instance, zoning of climatic risks requires focussing on specific climatic parameters, while landscape protection from urbanisation rather involves soil related zoning aspects. Optimization of technical vineyard management may require larger scale zoning (e.g. 1/5,000th), while landscape protection from urbanisation may be relevant at a smaller scale (e.g. 1/50,000th). Table 1 lists some potential terroir zoning objectives together with the relative importance of the role of soil, climate and soil/climate interactions. For some objectives only soil-based zoning is required, while others require only climate-based zoning. When soil-climate interactions are involved, it is necessary to implement soil- and climate-based zoning together.

Table 1. Objectives of terroir zoning with relative importance of the role of soil, climate and soil/climate interactions (modified from OIV Resolution OIV-VITI 423-2012).

Zoning objectives	Role of the soil	Role of the climat	Role of soil/climate interactions
Delineation of territories in accordance with their potential to produce wine of a certain quality and style	++	++	++
Zoning of potential precocity (phenology and grape ripening dynamics)	+	++	+
Optimisation of terroir expression by adaptation of plant material (variety and root stock) to soil and climate characteristics	++	++	+
Optimisation of technical and environmental management by adaptation of growing practices to soil and climate	++	+	+
Territorial management of crop protection risks	+	++	+
Plot selection	++	+	+
Territorial management of water resources	++	++	++
Zoning of climate risks and strong climatic constraints	0	++	0
Protection of terroirs and landscapes from various threats and in particular urbanisation	++	0	0

3 Soil based zoning

Soil based zoning is proposed to be implemented in three-steps. First, the approach needs to be defined, which is based on one or several scientific disciplines. Second, taking into account the objectives of the zoning and the available budget, a scale is chosen at which the zoning will be implemented. Finally, when available, complementary tools based on geomatic technologies are chosen to increase the accuracy of the zoning or to reduce the cost.

3.1 Selection of appropriate scientific basis

Geology enables a synthetic approach that is adapted to smaller-scale zoning (≤ 1/ 50,000th). Geological maps are quite cheap to produce and they do exist for most regions worldwide. However, the relation between geology and vine responses in terms of development, yield parameters and grape ripening is rather loose. Hence, geology can be used as a first approach in terroir zoning, but it needs to be augmented.

Geomorphology describes the surface landforms (plateau, slope, valley, terrace, etc.) resulting from the nature of rocks (especially the differences in hardiness), tectonic phenomena and erosion. Geomorphology maps can be easily established at reduced cost with the use of digital elevation models (DEMs), which are available at increasingly refined scales. This enables a synthetic approach well adapted to smaller-scale zoning (≤ 1/ 50,000th), as with geology, but explains vine behaviour only indirectly. Geomorphology helps to understand the distribution of soils and climatic parameters (in particular radiation and temperatures) over a given area.

Pedology (soil science) is an approach adapted to medium or larger-scale zoning (≥ 1/25,000th). The creation of soil maps requires auger sampling and soil pits studies. Pedology enables, to a certain extent, an explanation of soil effects on the physiological functioning of the vine. Many soil type classification systems exist worldwide but only a few are internationally validated; in particular the "Soil Taxonomy" (United States Department of Agriculture classification; USDA, [2]), the "World Reference Base for Soil Resources" (Food and Agriculture Organization; FAO ftp://ftp.fao.org/agl/agll/docs/wsrr103e.pdf) or the "Référentiel Pédologique" (Institut National de Recherche Agronomique, INRA; [3]). If a local classification is used, a match with one of the three above-mentioned classifications must be indicated. The advantages and drawbacks of the use of each of these three classifications are discussed in Appendix 1.

If only one discipline is chosen, pedology should be preferred. The combination, however, of geological, geomorphological and pedological approaches allows particularly accurate terroir zoning. Note that geological zoning can be mandatory to produce accurate pedology-based zoning, when subsoil is highly variable in space such as regions with a large number of faults. Certain disciplines, like the use of botanic indicators, can provide useful information to increase the accuracy of the zoning. They cannot, however, be considered an approach by themselves.

3.2 Selection of an appropriate scale

The scale of soil-based zoning must be defined beforehand. The scale depends on the zoning objectives (Table 1) and the scientific discipline chosen (Section 3.1). The larger the scale, the more precise the zoning, but the cost will also be higher. To produce soil maps at a given scale a certain density of observations are needed to provide acceptable resolution (Table 2). This table is based on a general rule of 0.5 to 1 observations per cm^2 of soil map, and the following:

for the scale 1/2,500 = 30 auger samples/soil pit
for the scale 1/10,000 = 20 auger samples/soil pit
for the scale 1/25,000 = 10 auger samples/soil pit
for the scale 1/100,000 = 5 auger samples/soil pit
for the scale 1/250,000 = 5 auger samples/soil pit

Table 2. Density of auger samples and soil pits, both as hectares (ha) per sample and number of samples per hectare needed to produce a soil map at a given scale (modified from OIV Resolution OIV-VITI 423-2012).

Scale	Number of ha per auger sample	Number of auger samples per hectare	Number of ha per soil pit	Total number of observations per ha
1/2,500	0.06 - 0.13	8 - 16	2 - 4	8 - 16
1/10,000	1 - 2	0.5 - 1.0	20 - 50	0.5 - 1
1/25,000	7 - 14	...	70 - 140	0.08 - 0.16
1/100,000	120 - 240	...	600 - 1,200	0.005 - 0.01
1/250,000	750 - 1,500	...	4,000 - 8,000	0.0008 - 0.0016

If soil distribution is locally complex, it may be necessary to increase the density of auger samples and / or soil pits, especially for the scales 1/25,000 and 1/100,000. For the scale 1/250,000 it is recommended to map one or more "reference areas" at a larger scale to establish rules for soil distribution in relation to the geology and geomorphology [4].

3.3 Spatial modelling, digital and remote sensing technologies

Various technologies, more or less recent, can be implemented to increase zoning accuracy and/or to reduce costs. Some of these new technologies have been developed in the framework of precision agriculture and are listed below. Although very useful, they cannot completely replace field work (auger samples, soil pits).

- Geographic Information Systems (GIS) are computer based technologies designed to manage spatial information. They are particularly useful to design maps and combine several layers of spatial information.
- Digital Elevation Models (DEM) can provide highly accurate geomorphological maps at reduced cost, through GIS modelling.
- Prior to soil mapping, the implementation of geophysical surveys such as electrical resistivity tomography (ERT) or electromagnetic induction (EMI) enables the production of very precise soil maps while limiting the number of auger sampling points. This approach is particularly adapted to large scale zoning studies (≥ 1/5,000th). In established vineyards, ERT provides best results, as the presence of metal wires interferes with EMI technology [5].

- Airborne remote sensing can help tanalyzing soil surface differences when no vegetation is present.
- Point-based information can be transformed into spatial information by means of geostatistics.

4 Climate-based zoning

Climate-based terroir zoning is generally implemented in three steps. First, appropriate climatic indicators are selected for the purpose. Second, high quality climatic data is sourced. Finally, homogeneous climatic zones are identified.

4.1 Selection of appropriate climatic indices

Climate-based terroir zoning uses different agro-climatic indices, depending on the objective of the zoning and the available data (Table 3).

Table 3. Accurate agro-climatic indices in relation to the objectives of climate-based terroir zoning (modified from resolution OIV-VITI 423-2012).

Zoning objectives	Appropriate agroclimatic indices and climate data	Timescale required
Precocity of phenological stages and grape ripening potential	HeatSum (GFV, HI, WI,...), AvGST	Month, day, hour
Delineation of territories in accordance with their potential to produce wine of a certain quality and style	WB, DI, RR (flowering - harvest), ET0, HeatSum, AvGST, MIN	Month, day, hour
Water management (vine water status and water resource management)	WB, DI, RR (vegetative period), ET0	Month, day, hour
Territorial management of crop protection risks	TM, RH, DH, GFV, ProtecModels	Day, hour
Frost threats	TN, TS, BudBurst, Hard, economic data, surveys	Day, hour
Hail threats	Hail pads, meteorological radar, economic data, surveys	Day, hour
Extreme heat threats	TX	Day, hour
Wind issues	W	Day, hour

Acromyms used:
AvGST: Average growing season temperature; DI: Dryness Index (climatic water balance); ET_0: Reference evapotranspiration; GDD: Growing degree days; GFV: Grapevine Flowering Veraison model; Hard: winter hardiness; HeatSum: indices based on heat summations; HI: Huglin Index; MIN: Indices based on temperature minimums in the ripening period; ProtecModels: crop protection models; RH: Relative humidity; RR: Cumulative rainfall; TM: Average air temperature; TN: Minimum temperature; TS: Surface temperature; TX: Maximum temperature, W: Wind speed; WB: Water Balance models based indices; WD: Wetness Duration; WI: Winkler Index.

For the purpose of comparison with other zoning operations performed at other sites or at other times, it is useful to work wherever feasible with commonly used and relevant indicators. In other words, producing a climate zoning with a non-conventional index has the drawback of preventing comparisons with previous work made in other regions. Most of climate zoning performed so far have been based on temperature, e.g. [6–8]. The average growing season temperature (AvGST) is a very popular and simple way to compare relative earliness in grapevine ripening in space and time. Indices derived from heat summation models are more precise, retaining only temperatures affecting grapevine growth rate and development. Amongst them, commonly used growing degree days accumulation indices are the "base 10°C" Winkler Index (WI) and the Huglin Index (HI), that can be computed either at a monthly or a daily time step. As these indices have been widely used, they allow appropriate time and space comparison of wine producing regions. The GFV model predicts flowering and veraison dates, using a "base 0°C" daily sum of temperatures [9]. As it has been calibrated for a wide range of grapevine cultivars, it is an accurate tool to perform zoning of phenological timing [10].

Several indices based on minimal temperature during the ripening period have been proposed such as the Cool Night Index or daily temperature amplitude. High night temperature has been shown to be detrimental to anthocyanin production, sugar and acidity in grapes [11]. These results were, however, obtained in controlled conditions with high night temperatures (over 25°C). Such conditions are seldom found in wine producing regions, except in some tropical viticulture areas. In contrast, no study, to our knowledge, validates the concept that low night temperatures (below 15°C) or considerable temperature ranges benefit to grape quality. Consequently, the use of such indices should be restricted to the potential need to cool the grape temperature prior to wine making.

Though water resource management is a crucial issue for many grape producing regions, few zoning attempts and indices regarding climate-related water issues have been proposed. Cumulative precipitation during the vegetative period or during the fruit development (RR) is a simple and useful way to depict the potential excess or deficit of water. Yet, it should not be considered alone for precise water requirement assessments. Vineyard potential water loss has to be addressed using potential evapotranspiration, ideally, through water balance modelling. Several water balance models are available, and they provide a large amount of inputs which can be synthetized through various indices. In many cases, grapevine water deficit during the growing season or during grape development is estimated by the actual/maximum daily transpiration ratio [12, 13], using rather complex water balance modelling. The dryness index (DI), a more easy-to-use alternative to estimate climate-related water resources for viticulture; it requires monthly climate data [14].

Vineyard pest and diseases management is a challenging issue for viticulture from economic, environmental and human health points if view. Yet, so far, terroir zoning do not account for pest related aspects. They are, however, to be considered for plant protection strategies. The zoning of phytosanitary risk can be achieved using climate-based models simulating a potential risk for diseases or pest population outbreaks,

 https://doi.org/10.1051/e3sconf/20185001044

combined with plant modelling to account for the sensitivity period of the grapevine [15].

Climate hazard zoning is challenging, because extreme events do not occur frequently. Thus, robustness of such type of zoning requires long series of climate data. For zoning of cold-related risks, comparisons based upon simple temperature threshold might be relevant at small scale only. At large scale (i.e. small territories), the risk for cold-related damage do not depend only on the occurrence of one or several cold events. Sensitivity of the grapevine to low temperatures also needs to be accounted for. The limit below which temperature damages grapevine tissues depends on various factors. For winter cold, grapevine sensitivity to low temperature differs according to the cultivar and changes dramatically during the season. This "winter hardiness" is estimated using specific models [16]. Concerning spring frost, budburst date has to be assessed, because, obviously, no frost damage will occur prior to this date. As budburst timing modelling is still subject to substantial error, zoning results can differ according to the choice of the budburst model [17].

Because long time climate data series and sufficiently accurate models might not be available, climate related hazard could be assessed through additional information, such as insurance pay-outs or surveys. This is specifically relevant for hail risk zoning [18], because hail measurement devices are usually not installed on weather stations.

4.2 Selection of high-quality climatic data

Three possible sources of climate data can be identified: (i) weather stations, (ii) remote sensing, and (iii) dynamic circulation models. Most of the relevant indicators needed for climate-based terroir zoning can be obtained from weather stations. For data sourced from weather stations, quality control is critical as invalid data from poorly located weather stations, or due to either sensor malfunctioning or database construction errors has to be identified and rejected. Optionally, such data can be replaced by estimates obtained from surrounding weather stations. Weather stations provide point-located data. Spatialization of point-data can be achieved by interpolation. Numerous methods are available to perform climate data spatial interpolation (see [19] for a review). An estimation of the uncertainty of the interpolation has to be carried out with an independent data set (e.g. "leave-one-out" cross validation). Spatial variability of a mapped climatic parameter needs to be greater than the uncertainty of the interpolation. Recently, small weather stations containing a temperature sensor, a shelter and a data logger have been increasingly used for terroir zoning studies [8]. Because of their low cost, they allow high density temperature assessment. This information, combined with the use of a DEM, allows temperature mapping at very high resolution through the implementation of spatial models [20]. When the weather stations are located inside the canopy, however, absolute temperatures recorded differ from those measured by standard weather stations. Hence, absolute values of climate indices obtained with small scale weather stations cannot be compared to those obtained with standard weather stations.

Remote sensing provides climate data over large areas and over continuous timescales. These data have to be pre-processed to eliminate artefacts (e.g. cloud cover; corrections for differences in land use, etc.). It is also important to check the quality of the data, especially with regard to the spatial and temporal uniformity of the signal.

Dynamic circulation models produce very large quantities of climate data, covering extensively large areas and manipulation of this type of data requires specific software packages, programming skills and considerable computational power. The spatial resolution of this type of data has considerably increased over the past few years and it is now currently available in grids of 8 km^2, and sometimes even 1 km^2 or lower. Quality assessment of data derived from dynamic circulation models, however, is challenging, because they provide climate data corresponding to an area, usually covering several km^2. Hence, comparing this data to climate observations from one weather station (point data) is not relevant. Spatial interpolation of climate station data should therefore be performed before.

Climatic zoning can be based on a selection of several relevant indices. For instance, the Multicriteria Climate Classification of Tonietto and Carbonneau [14] combines the use of the Huglin Index, a water balance model (DI) and a cool night index.

4.3 Identify climatically homogeneous zones

Unlike soil-based terroir-zoning, which is mainly based on qualitative data (soil types), climate zoning is based on quantitative data. Homogeneous zones therefore need to be delineated according to quantified climatic parameters. Spatial variability in the climatically homogeneous zones must be greater than or equal to the mapping error. It is also desirable that the areas should be delineated according to criteria that are relevant for viticulture.

Furthermore, since climate can vary considerably over time, climate-based terroir zoning must be based on descriptive statistics calculated over a sufficient number of years. The number of years required depends on the purpose of the zoning, the variable considered and the factors responsible for its variations in space (Appendix 2).

5 Validation of terroir zoning

The relevance of the terroir zoning can be validated by various methods. For example, ecophysiological studies charcaterize the response of the vines to environmental factors, including soil and climate. Major vine responses include phenology, vigor, yield parameters and grape composition. The interactions between the vine and the environment are frequently mediated by vine water and nitrogen status. This approach of validation can be carried out with point measurements across a network of reference blocks, or spatialized using maps of parameters such as vigor, timing of phenological stages, grape composition at ripeness, yield components, vine water status, and/or vine nitrogen status [21]. Terroir zoning can also be validated in other ways:

- Surveys among growers can make use of their empirical knowledge to validate terroir zoning [22].
- Risk levels of climatic threats or pest can be compared against maps of actual damage recorded on the vines.
- Sensory analyses implemented on wines obtained either by real scale or small-scale vinification can be used to validate the accuracy of terroir zoning
- Geographic information systems (GIS) are useful tools to overlay zoning maps with spatialized validation data.

5 Conclusions

Clear objectives have to be defined prior to choosing a terroir zoning methodology and the appropriate scale at which the study will be implemented. A scale of 1/5,000th is appropriate for zoning at the estate level, while 1/10,000th or 1/25,000th is appropriate for zoning at an appellation level. Highly relevant zoning at the soil level can be obtained by combining geology, geomorphology and pedology (soil science). When only one of these disciplines is chosen for the implementation of the study, however, pedology should be preferred. In climate zoning, relevant agro-climatic indicators must be selected and averaged over an appropriate number of years. Quality of source data is a critical issue. Differences among identified homogeneous zones must be greater than intra-zone uncertainty.

References

1. C. van Leeuwen, G. Seguin, J. Wine Res., **17** 1-10 (2006).
2. United States Department of Agriculture, Natural Ressources Conservation Services (USDA), Keys to Soil Taxonomy (2010).
3. D. Baize and M. Girard, INRA Eds. (1995).
4. C. Cam, P. Vital, J.-L. Fort, P. Lagacherie, R. Morlat, Etude et Gestion du Sol, **10** 35-42 (2003).
5. F. André, C. van Leeuwen, S. Saussez, R. van Durmen, P. Bogaert, D. Moghadas, L. de Rességuier, B. Delvaux, H. Vereecken, S. Lamblot, J. Appl. Geophys., **78** 113-122 (2012).
6. G. V. Jones, A. A. Duff, A. Hall, J. W. Myers, Am. J. Enol. Vitic., **61** 313–326 (2010).
7. M. A. Amerine, A. J. Winkler, Hilgardia. **15**, 493–673 (1944).
8. H. Quénol, *Changement climatique et terroirs viticoles* (Lavoisier, Paris, 2014), *Tec & Doc.*
9. A. K. Parker, I. García de Cortázar-Atauri, C. van Leeuwen, I. Chuine, Aust. J. Grape Wine Res., **17** 206–216 (2011).
10. A.K. Parker *et al.*, Agric. Forest Meteorol., **180** 249-264 (2013).
11. K. Mori *et al.*., Vitis, **44** 161-165 (2005).
12. C. van Leeuwen *et al.*, J. Int. Sci. Vigne Vin., **43** 121–134 (2009)
13. H. Fraga, I. Garcia de Cortázar Atauri, A. C. Malheiro, J. A. Santos, Glob. Change Biol. **22** 3774–3788 (2016).
14. J. Tonietto, A. Carbonneau, Agric. Forest Meterol., **124** 81-97 (2004).
15. D. Molitor *et al.*, Eur. J. Plant Pathol. **144**, 785–798 (2016).
16. J. C. Ferguson, M. M. Moyer, L. J. Mills, G. Hoogenboom, M. Keller, Am. J. Enol. Vitic., **65** 59–71 (2014).
17. G. Sgubin *et al.*, Agric. For. Meteorol., **250–251** 226–242 (2018).
18. F. Vinet, Atmospheric Res., **56** 309–323 (2001).
19. J. Li, A. D. Heap, Ecol. Inform., **6** 228–241 (2011).
20. R. Le Roux *et al.*, Agric. For. Meteorol., **247** 159–169 (2017).
21. C. van Leeuwen, in *Managing Wine Quality*, A. G. Reynolds, Ed. (Woodhead Publishing, 2010), *Woodhead Publishing Series in Food Science, Technology and Nutrition*, pp. 273–315.
22. R. Morlat, *Collection Avenir Œnologie, Œnoplurimedia, Chaintré* (2001).

Acknowledgement

We are grateful to Mark Gowdy for proofreading and of the paper.

Appendix 1: Soil classifications for soil-based terroir zoning

There are many soil classifications. For standardisation, the OIV recommends that its members use one of the following three classifications for soil-based terroir zoning works: the "Soil Taxonomy" (American classification [2]), the "World Reference Base for Soil Resources" (FAO Classification ftp://ftp.fao.org/agl/agll/docs/wsrr103e.pdf) or the "Référentiel Pédologique" (French classification [3]). Each of these classifications has interests and limits of use.

The "Soil Taxonomy" [2] is the classification that allows the most accurate definition of the soil types encountered. It is used in many countries. Its complexity, however, makes it a tool for soil science experts rather than for soil-based terroir zoning objectives.

The "World Reference Base for Soil Resources" (FAO classification ftp://ftp.fao.org/agl/agll/docs/wsrr103e.pdf), also called the "FAO classification", is an internationally recognised classification that is easy to use, however, it is limited to only 32 proposed references. Furthermore, this classification does not recognise the predominant role of

 https://doi.org/10.1051/e3sconf/20185001044

the rock type in pedogenesis (soil formation). Consequently, there is no group of carbonated soils, which clearly limits its interest for vitiviniculture zoning.

The Référentiel Pédologique, established by the French Institut National de Recherche Agronomique [3] is a relatively comprehensive and easy reference to use. It is based on both morphological criteria (diagnosis horizon) and pedogenetic criteria (type of parent rock in particular). Even though this classification is used in several countries, its national origin (French) is a limit.

Appendix 2: Temporal sampling requirements for agro-climatic indices used in climate-based terroir zoning

The climate differs from the soil in particular due to its temporal variability. Also its characterization, for vitiviniculture requires a study over several years. The size of this temporal sample, hereafter referred to as the "study duration", is highly dependent on the objectives defined. There are 2 cases, among others:

- The zoning objective is limited only to the identification of areas considered as climatically homogeneous (in terms of one or more agroclimatic indices) within the study area.
- The zoning objectives are (1) to distinguish areas considered climatically homogeneous within the study area, (2) to compare the climatic characteristics of areas identified in the study area with other wine producing regions (intra- and extra-regional comparison).

In the first case, the study duration can be variable, depending on the spatial scale and the atmospheric and environmental factors that govern the spatial variability of the climate. Thus, for large-scale zoning (study area of a size less than 100 km), certain variables such as air temperature may be affected, primarily, by lasting geographical features or those that are only slightly variable over time, such as the relief or land use. Thus, a study period of several years (5 years minimum) may be sufficient to demonstrate redundant spatial structures over the years. However, for variables where the spatial distribution depends largely on weather conditions, such as rainfall, substantial study duration is required. It is recommended that the times given for the calculation of climate normals, as defined by the World Meteorological Organization (WMO) are used, which is 30 years.
In the second case, it is also recommended that a study period of 30 years is used. It is clear that the comparison of the climatic characteristics of the areas identified in the study area with other wine regions requires identical periods of study, due to climate change over the long term.

Parrel, vinífera aragonesa de la depresión del Ebro. Adaptación a terroir semiáridos de cultivo.

Ernesto Franco[1], Alberto Pavon[2], Yolanda Gogorcena[3], Ricardo López[4], Javier Andreu[5], Jesús J. Usón[2], y Eva Herrero[6]

[1] Centro de Transferencia Agroalimentaria - Gobierno Aragón. Unidad de Enología. Avda. de Movera, s/n, 50194, Zaragoza (España).
[2] Centro de Transferencia Agroalimentaria - Gobierno Aragón. Unidad de Tecnología y Mejora de la Vid. Avda. de Movera, s/n, 50194, Zaragoza (España).
[3] Estación Experimental Aula Dei - Consejo Superior de Investigaciones Científicas. Laboratorio de Genómica de frutales y vid. Avda. de Montañana, 1005, 50059, Zaragoza (España).
[4] Laboratorio de Análisis de Aromas y Enología - Instituto Agroalimentario de Aragón. Departamento de Química Analítica de la Facultad de Ciencias de la Universidad de Zaragoza, 50009, Zaragoza (España).
[5] Centro de Transferencia Agroalimentaria - Gobierno Aragón. Unidad de Cultivos Leñosos. Avda. de Movera, s/n, 50194, Zaragoza (España).
[6] Centro de Investigación y Tecnología Agroalimentaria de Aragón. Avda. de Montañana, 930, 50059, Zaragoza (España).

Resumen. En el marco del proyecto VALOVITIS (Programa POCTEFA 2014-2020) de identificación y evaluación de variedades minoritarias en el Piedemonte Pirenaico, se está trabajando y sometiendo a estudio la variedad PARREL. Tras los resultados obtenidos en sucesivas campañas de vendimia y vinificación semi-industrial, la variedad Parrel destaca como una de las variedades autóctonas de la región de Aragón con mayor potencial para su posterior desarrollo comercial atendiendo a sus propiedades agronómicas, enológicas a su resistencia a enfermedades como el oídio y a su capacidad de adaptación a condiciones climáticas continentales caracterizadas por elevadas oscilaciones térmicas a lo largo del año.

1 Proyecto VALOVITIS

El trabajo aquí presentado forma parte del Proyecto VALOVITIS (www.valovitis.eu, Programa POCTEFA 2014-2020), que tiene como objetivo la identificación y evaluación agronómica, tecnológica y aromática de variedades desconocidas de la zona de Piedemonte Pirenaica. Los resultados obtenidos en este proyecto pretenden contribuir a la implantación de variedades desconocidas y proporcionar, a través de la innovación, ventajas competitivas en el mercado.

VALOVITIS trabaja en la localización de variedades de vid locales que poseen perfiles sensoriales diferentes, y que además presenten capacidad de adaptación a las nuevas condiciones climáticas y no muestren vulnerabilidad significativa a plagas y enfermedades.

2 Banco de germoplasma del Centro de Transferencia Agroalimentaria del Gobierno de Aragón

Desde 1989 el Centro de Transferencia Agroalimentaria (CTA) del Gobierno de Aragón, en su Banco de Germoplasma, ha implantado, conservado e identificado el material vegetal de vid prospectado. La función principal de este banco es evitar la pérdida de biodiversidad. Para ello se trabaja en la prospección y recuperación de variedades antiguas a punto de desaparecer.

En el marco de VALOVITIS y durante los años 2016 y 2017 se ha profundizado en el estudio de más de 40 accesiones de esta colección, a través de la evaluación agronómica, vinificación y análisis sensorial de los vinos producidos.

 https://doi.org/10.1051/e3sconf/20185001045

3 La variedad Parrel

La PARREL es una de las variedades sometidas a estudio en este trabajo. Se trata de una variedad minoritaria localizada exclusivamente en Aragón y que muestra características agronómicas y enológicas peculiares que merecen un estudio en profundidad de su potencial.

Esta variedad, siendo "desconocida", no está incluida en la relación de variedades de viñedo de vinificación autorizadas en Aragón.

3.1 La vinífera

La uva de Parrel es de hollejo tierno y color intenso casi negro.

El origen del material vegetal introducido en el conservatorio del CTA procede de diferentes puntos del valle del Ebro, principalmente en la provincia de Zaragoza, como se muestra en la **Figura 1**. Los datos climáticos se describen en la **Tabla 1**.

Tabla 1. Datos climáticos de los lugares de origen del material vegetal de la variedad Parrel evaluados en el proyecto VALOVITIS.

	Origen	H (m)	P (mm)	ET_0 (mm)	T (°C)
1	Ainzón (Z)	489	421	1.133	14,5
2	La Almolda (Z)	355	329	1.277	14,3
3	Lécera (Z)	541	361	1.090	13,8
4	Híjar (TE)	268	335	1.208	14,6
5	Daroca (Z)	894	429	1.102	12,2
6	Zuera (Z)	310	419	1.361	14,1
7	Sástago (Z)	287	339	1.324	14,9

H: altitud; P: precipitación anual; ET_0: Potencial de evapotranspiración; T: temperatura anual media; Z: Zaragoza; TE: Teruel. Fuente: Atlas Climático de Aragón.

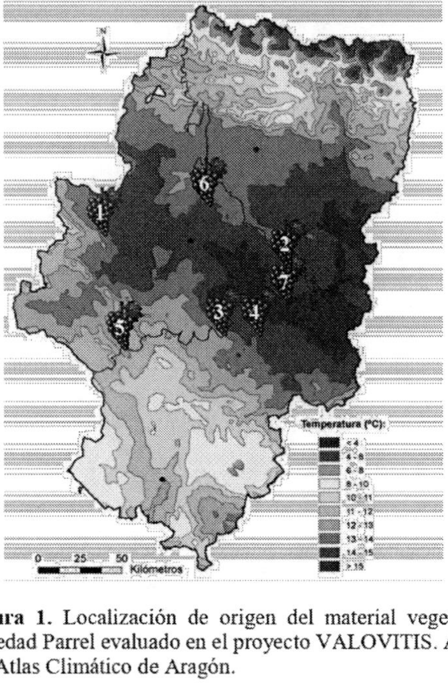

Figura 1. Localización de origen del material vegetal de la variedad Parrel evaluado en el proyecto VALOVITIS. Adaptado del Atlas Climático de Aragón.

La depresión aragonesa del Ebro, es una zona semiárida con precipitaciones anuales no superiores a 400 mm y temperaturas medias anuales que oscilan entre los 4-6 °C de enero y los 24-26 °C en julio. La elevada oscilación térmica anual en los lugares en los que ha aparecido esta variedad, induce a pensar que se trata de una variedad con gran capacidad de adaptación a diferentes situaciones climáticas y terroir.

3.2 Caracterización ampelográfica

En la Tabla 2 y en la Figura 2 se muestran las características más distintivas de esta variedad.

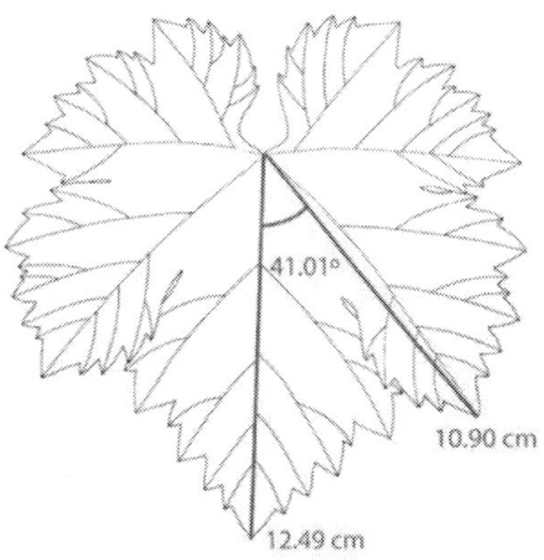

Figura 2. Hoja media reconstruida de Parrel. Análisis ampelométrico. Fuente: Misión Biológica de Galicia-CSIC.

Tabla 2. Descriptores ampelográficos.

Pámpano joven	Distribución de la pigmentación antociánica no ribeteada y de intensidad media.
Hoja adulta	Limbo en forma pentagonal con cinco lóbulos y dientes rectilíneos. Densidad de pelos entre los nervios del envés: media-alta. Seno peciolar poco abierto y con baja o nula densidad de pelos.
Racimo	Mediano y de baja compacidad media.
Baya	Esférica de tamaño medio con una epidermis azul-negra.
Pulpa	Sin coloración y con pepitas.

Figura 3. Imagen de los racimos de Parrel.

3.3 Caracterización molecular

Los análisis moleculares realizados en proyectos anteriores con seis marcadores microsatélites, demostraron que se trataba de una variedad no descrita anteriormente y con un perfil molecular no coincidente con los de las bases de datos de décadas pasadas [1].

Tabla 3. Análisis moleculares del material vegetal según su procedencia.

VMC4F3-1		VVIN16		VVIV37		VVIV67	
179	179	153	159	165	171	364	366

VVMD27		VVIP31		VVS2		ZAG79	
179	189	180	196	131	151	243	261

Fuente: IMIDRA 2014-15 [2].

La caracterización molecular de varias de las accesiones recuperadas de las zonas de origen, incluidas en la Tabla 1, se llevó a cabo en la Estación Experimental de Aula Dei mediante 8 microsatélites: VMC4F3-1 VVIN16, VVIV37, VVIV67, VVMD27, VVIP31, VVS2 y ZAG79, en el ámbito del proyecto RF2012-00027-C05-02 [2]. El perfil molecular obtenido para todas las accesiones fue idéntico independientemente de la zona de origen (Tabla 3). La identificación llevada a cabo en el IMIDRA indica que se trata de la variedad Parrel (GEN 0905) según consta en su base de datos.

En las accesiones que resulten destacadas de este proyecto se deberán llevar a cabo nuevos análisis para completar la caracterización ampelográfica y molecular, con el fin de identificar los biotipos de la variedad que nos permita salvaguardar la variabilidad genética recuperada.

3.4 Caracterización agronómica

Durante las campañas de 2016 y 2017, se han realizado vinificaciones de rosado y de tinto con uvas de esta variedad.

Los datos mostrados en las tablas 4 y 5 corresponden a las campañas de 2016 y 2017 vendimiadas en las parcelas del Banco de Germoplasma del Gobierno de Aragón ubicadas en la finca de La Alfranca (Pastriz, Zaragoza). En estas parcelas la conducción se realiza en espaldera y poseen una instalación de riego por goteo. Su ciclo es de secano con algún riego de apoyo si el verano es muy seco.

La variedad Parrel es de ciclo largo y de madurez más tardía que la Garnacha Tinta. Destaca su elevada producción, que se sustenta en el gran número de racimos por cepa y en el tamaño medio de la baya.

Tabla 4. Datos de vendimia de las accesiones utilizadas para la minivinificación de rosados.

Rosado	**2016**	**2017**
Fecha de vendimia	4/10	16/10
Producción (*kg/cepa*)	8,1	5,2
Peso del racimo (*g*)	396	310
Peso de 100 bayas (*g*)	266	205
N. de bayas por racimo	148	151
N. de racimos por cepa	20,5	16,9

Tabla 5. Datos de vendimia de las accesiones utilizadas para la minivinificación de tintos.

Tinto	**2016**	**2017**
Fecha de vendimia	26/9	22/9
Producción por cepa (*kg*)	11.7	8.2
Peso del racimo *g*	586	395
Peso de 100 bayas *g*	171	168
N. de bayas por racimo	342	235
N. de racimos por cepa	20.0	20.8

Se observa que tiene un buen potencial productivo, si las condiciones son favorables, influyendo bastante las condiciones climáticas en el momento del cuajado. Si estas son desfavorables, como en 2016 cuando el cuajado fue irregular, el número de bayas por racimo es menor, lo que provoca una importante minoración productiva de la variedad. El peso de las bayas y el número de racimos por cepa suelen ser valores más constantes que el número de bayas por racimo.

Agronómicamente la variedad Parrel presenta una producción elevada y mantenida en el tiempo, así como una buena resistencia a oídio.

3.5 Caracterización química - Análisis clásicos

Durante el proceso de embotellado, un volumen representativo de vino se tomó para analizar, entre otros parámetros, el grado alcohólico adquirido a 20 °C por absorción de infrarrojos, la acidez total y el pH a través de una potenciometría automática, el ácido málico por método enzimático (límite de detección: 0,0; límite de cuantificación: 0,1), la intensidad de color por espectroscopia UV-VIS y el índice de polifenoles totales por absorbancia a 280 nm. La inyección en flujo se utilizó para analizar el contenido en azúcares reductores, acidez volátil y el dióxido de azufre total y libre.

Los vinos obtenidos de esta variedad, presentan una analítica adecuada y acorde a las características descritas para vinos rosados y tintos en las Denominaciones de Origen Protegidas (DOP) de Aragón (tabla 6).

 https://doi.org/10.1051/e3sconf/20185001045

Tabla 6. Análisis moleculares del material vegetal según su procedencia.

	Rosado		Tinto	
	2016	2017	2016	2017
Azucares reductores *(g/l)*	0,90	0,80	0,80	0,90
Grado alcohólico adquirido a 20 °C *(%v/v)*	12,95	10,79	13,29	13,11
Extracto seco total	21,50	23,00	24,00	26,40
Acidez total (ác. Tartárico) *(g/l)*	5,90	5,40	5,90	5,80
Acidez volátil (ác. Acético) *(g/l)*	0,36	0,20	0,32	0,41
pH	3,57	3,48	3,60	3,64
Ácido málico *(g/l)*	1,40	1,50	1,30	1,70
Dióxido de azufre total (mg/l)	17,00	44,00	33,00	102,00
Dióxido de azufre libre (mg/l)	5,00	5,00	5,00	7,00
Intensidad de color	2,50	1,55	12,43	8,26
Índice de polifenoles totales	12,20	10,40	39,60	30,00

Análisis: Laboratorio Agroambiental del Gobierno de Aragón.

La graduación alcohólica es moderada, lo que aporta valor añadido a esta variedad si se tiene en cuenta el aumento de grado observado en los vinos aragoneses debido al cambio climático. La acidez y el pH son normales para los vinos de la región. Destaca la intensidad de color adecuada y suficiente incluso presentando polifenoles totales bajos, circunstancia similar a lo observado en los vinos tintos de Parraleta, variedad minoritaria localizada en la DOP Somontano.

3.6 Caracterización sensorial

Los vinos se someten a análisis sensorial, que lo realiza un panel compuesto por 11 jurados expertos de los Centros participantes en el proyecto VALOVITIS (CTA, CITA y LAAE). Se usa para ello una ficha descriptiva donde se valoran descriptores del aroma y del gusto.

En las figuras 4 y 5, se muestra el perfil sensorial de los vinos rosados y tintos para los dos años de estudio. Es de gran importancia comprobar como los perfiles, que se corresponden con los descriptores valorados en las fichas de cata, son prácticamente iguales con independencia de la cosecha.

En el caso de los rosados, presentan una intensidad media alta, destacando los aromas de carácter fermentativo, mientras que en el gusto destaca la intensidad aromática y la acidez. Los tintos presentan también una intensidad aromática media-alta, destacando por igual el carácter terpénico, especiado y fermentativo. En el gusto destaca la intensidad aromática y la acidez, observando ausencia casi total de los descriptores tánicos.

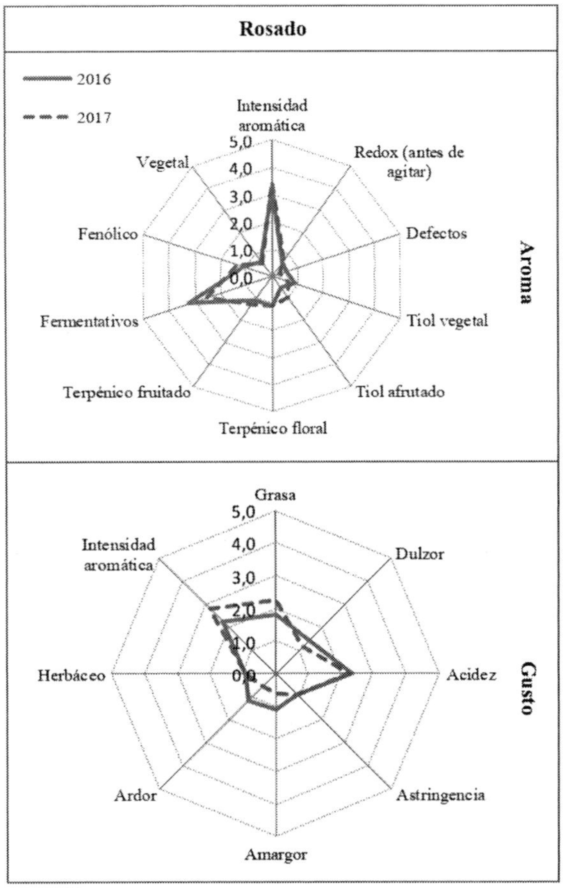

Figura 4. Perfiles aromáticos de la vinificación de rosado obtenidos a partir de la variedad Parrel.

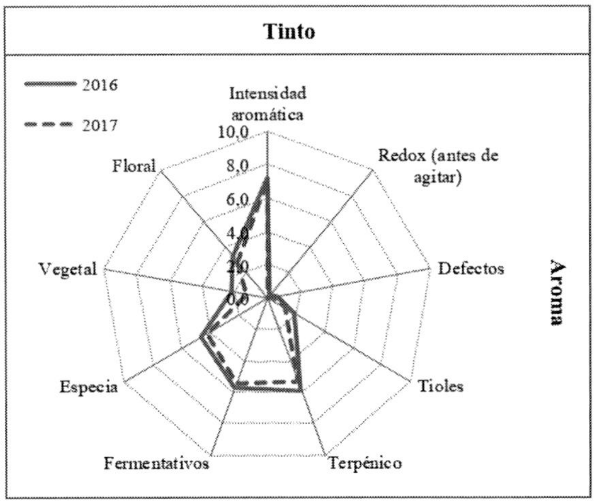

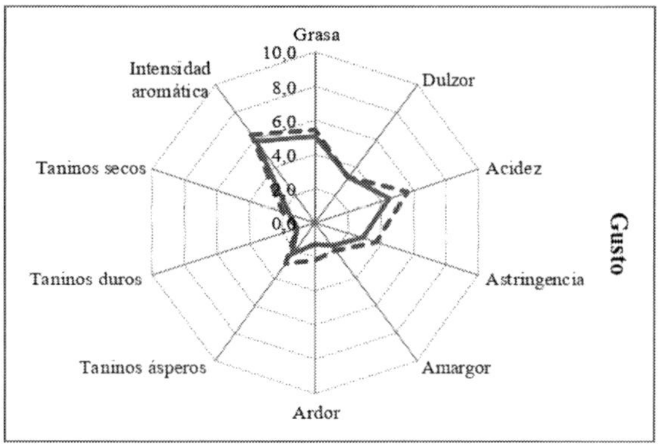

Figura 5. Perfiles aromáticos de la vinificación de tinto obtenidos a partir de la variedad Parrel.

Tanto el grado alcohólico como el contenido fenólico está relacionado con la producción del viñedo. En este caso la elevada producción condiciona que la concentración fenólica sea limitada, incidiendo en que los descriptores de la tanicidad obtengan bajas puntuaciones.

El peso de variedad es muy superior al del año, y mantiene la tipicidad varietal de los vinos, en cuanto a características sensoriales como analíticas.

4 Conclusiones

A partir de los resultados de los análisis químicos y sensoriales realizados, se ha observado que los vinos tintos obtenidos presentan un grado alcohólico suficiente y, si bien los componentes fenólicos son bajos, esto no condiciona el potencial de la variedad para producir tanto tintos como rosados de calidad, estos últimos con predominio de aromas terpénicos.

El terroir ha mostrado un efecto menor que la propia caracterización varietal en las producciones obtenidas.

La variedad PARREL, de acuerdo con sus características agronómicas y enológicas, merece poner en valor su potencial a través de estudios posteriores en profundidad sobre su comportamiento agronómico y enológico, en especial en situaciones de producciones más moderadas.

5 Agradecimientos

Este proyecto está cofinanciado por el Fondo Europeo de Desarrollo Regional (FEDER).

Referencias

1. T. Buhner-Zaharieva, S. Moussaoui, M. Lorente, J. Andreu, R. Núñez, J.M. Ortiz, Y. Gogorcena. *Am J Enol Vitic*, **61**, 557-562 (2010).

2. Proyecto RF2012-00027-C5-02. Documentación, caracterización y racionalización del germoplasma de vid prospectado y conservado en España. Creación de una colección nuclear.

E3S Web of Conferences **50**, 02001 (2018)
XII Congreso Internacional Terroir
https://doi.org/10.1051/e3sconf/20185002001

What rules to frame the development of new vineyards plots in Bandol DOP?

Paul Minvielle[1], Loïc Villier[2] and Jacques Daligaux[1], Stéphane Angles[3]

[1]Aix Marseille Univ, CNRS, TELEMME, Aix-en-Provence, France
[2]Pierre and Marie Curie University - Paris 6, Paris UPMC, France
[3]Université de Lorraine, EA LOTERR, Nancy, France

Resumen. The vineyard of Bandol is confronted with an important land pressure in a very urbanized and very touristic area. The new parcels that appear at the northern edge of the vineyard must comply with a certain number of rules in order to maintain the quality and typicality of the wines.

1. Introduction

Since 1935, l'INAO, Institut National de l'Origine et de la qualité, delimits *terroirs*, in France, at the plots fine scale, relying on geological and pedological criteria. The boundaries of the Bandol vineyard date from 1941 and considers the territory of 8 communes of Var department, except the lowest areas, covered with alluvial formations, (figure 1). Diversity and mapping of wine *terroirs* have been analyzed in a precedent study, under the prism of the expressions of soil and geological substrate properties in the Bandol red wines. [1].

The paper proposes first to define Bandol terroir using Warren Moran approach [2], then to deal with the question of the valorization of soils from the point of view of its legal zoning, its *delimitation parcellaire*. We consider the development of agricultural land-use planning and the durability of the strategies of a qualitative wine production with strong identity. Plots planted with vines cover about 1500 hectares, preferentially implanted on sandy-clay soils rich in limestone.

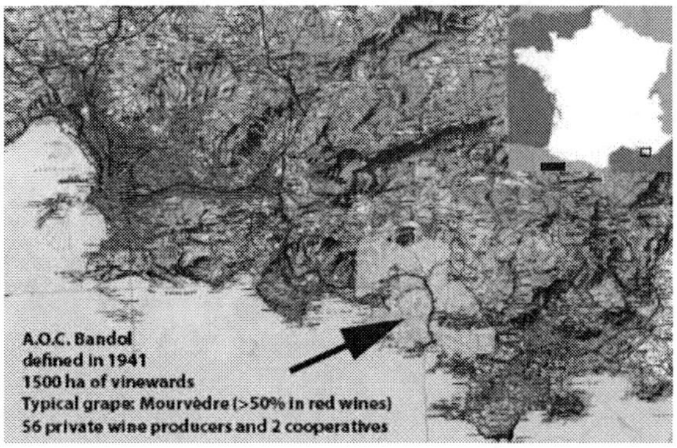

Figure 1. Bandol DOP localization

Large areas not planted, mainly forested, extend in the northern part of the area. This territory is generally on medium to steep slopes, on compact limestones of the Upper Cretaceous. For several decades, viticulture practices have consisted in developing these forested areas, often to compensate for losses suffered by the vine in the lower parts, gradually consumed by urbanization. These developments took the form of land clearing and then development of plots, crushing and grinding of the hardest surface horizons and terracing. These practices that were long tolerated, while they modified the original *terroir,* are today problematic to professionals and authorities in charge of vineyard management.

2. Bandol an archetypal terroir

Bandol DOP is a quiet archetypal territory to illustrate a terroir in the meaning of the French National Institute of Origin and Quality (INAO): that is to say: a "*terroir is a delimited geographical space, in which a human community built, in the course of its history, a collective knowledge of production, based on a system of interactions between a physical and biological environment and a set of human factors*". [3]. We are going to precise first the agro/vini terroir dimension, then complete with social terroir aspects, according to Warren Moran approach. It is important to reaffirm the need for an interdisciplinary approach for terroir studies, including pedologists, climatologists, agronomists, oenologists, but also sociologists, lawyers, geographers… (figure 2)

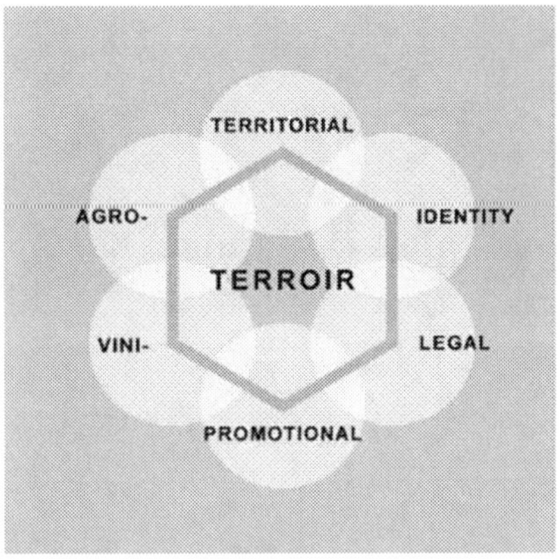

Figure 2. Warren Moran approach of terroir (2006, op.cit.)

2.1. The agro/vini terroir

Bandol is famous for red wines that differentiate from other French appellations using the Mourvèdre grape at a minimum of 50% of the blends. Historically, the strategy developed by the local wine producers and regulation authorities was too emphasize the high quality and typicality of the red wines. The average high quality relies on high constraints imposed for the AOC label. The initial delimitation in 1941 is justified by the history of wine production, the current practices and spatial partitioning of high and poor-quality products, and the zonation was subdued to a geological study. The definition of the appellation considered geology only for exclusion of the low land vineyards that grow on recent alluvial deposits, and that were dedicated to low quality, highly productive grapes. The statutory order of the A.O.C. revised in 2011 cites geological criteria only in an historical perspective, although the limits initially placed according to the borders of geological units are retained.

The question arises of how the geological substrate and soil properties may deserve justification of the Appellation boundaries. The geological substrate and soil properties vary spatially within the appellation and their incidence on wine typicality remains difficult to demonstrate. We hereby present a preliminary assessment of a potential terroir effect but dependent to human processing. However, an analytical description of the spatial variations of the soil and geological properties, allow to test their relevance for the zonation of wine productions and the justification of the Appellation boundaries

Within the 8 communes of the appellation, vineyards are grown on varied geological substrates, but their distribution is non-random. The mixed carbonate siliciclastic of Coniacian-Santonian age that form the basement of the Beausset syncline are highly favoured for vineyard settlements. Middle and Upper Triassic substrates also seem to be appreciated (figure 3).

	Communes of the AOC 229 km² Percentage	Vineyards 48 km² Percentage
Coniacian-Santonian	47	77
Cenomanian-Coniacian	16	0
Aptian-Albian	4	1
Berriasian-Barremian	6	0
Upper Jurassic	7	1
Middle Jurassic	6	4
Lower Jurassic	4	2
Triassic	10	14
Permian/Basement	1	1

Figure 3 Distribution of geological substrates according to communes and vineyards

The landscape of the area is highly contrasted, from low alluvial plains to steep canyons. The vineyards occupy preferentially the slopes between 2° and 10° (figure 4).

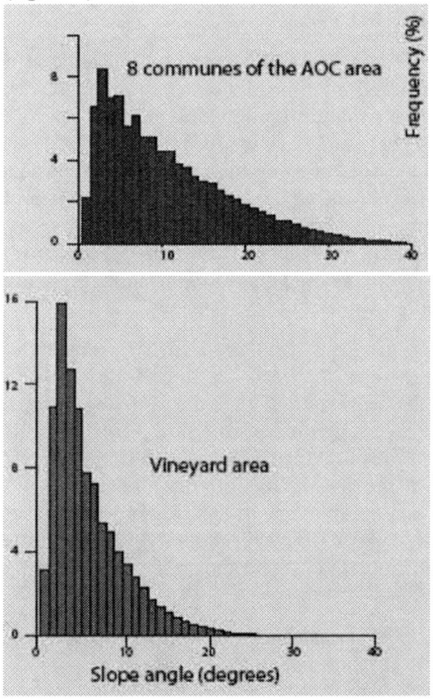

Figure 4. The slopes in the communes of the AOC area and in the vineyard area

Unpublished pedological maps, the 1/50 000 geological maps, complimented with eld-work were used to investigate the link between geology, soil properties and Terroir effect. Statistical analysis of the pedological dataset suggests:

- The lowlands are associated to rich soils, with a high clay component and low level of carbonates. These soils usually fall in the area discarded for production of Bandol wines.

- The highly permeable, quartz-rich soils derived from alteration of the Coniacian-Santonian calcareous silt- and sandstone house most of the Bandol vineyards. From place to place, the soil properties vary according to grain size (silt to sand), and the input of clay through superficial formations (alluvial and slope deposits).
- The Rudist-rich carbonates produces valuable soils for wine growing on slopes, where soils are thicker and rich in carbonate pebbles and clays.
- In the vineyards, the substrates made of Triassic and Lower Jurassic limestones are usually associated with soils rich in pebbles and fine elements (silt, clay).

The analyses of the geological, soil, and topographic variables allows to the discrimination of potential "Terroir Units" within the Bandol appellation. The "Terroir Units" can be defined from a set of objective criteria and are therefore operational for zonation. The Terroir effect is potentially significant at the scale of a commercial product and the "Terroir Units" could help for further revision of the AOC zonation. (figure 5).

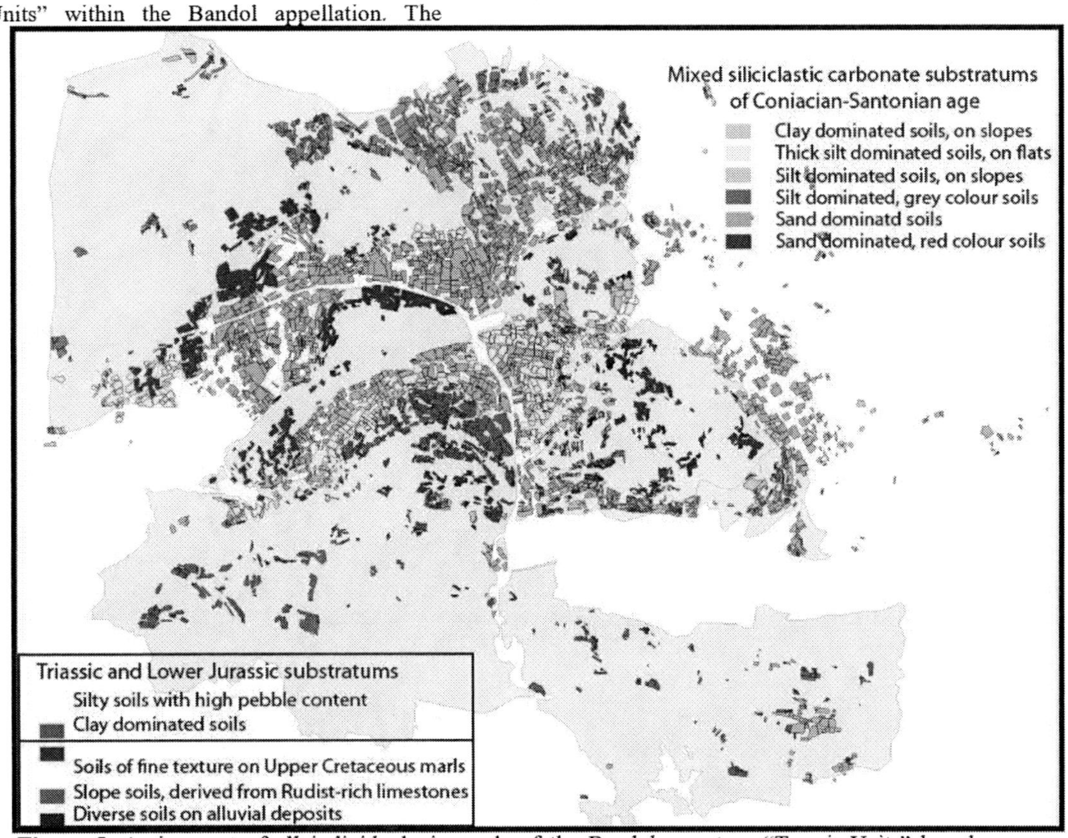

Figure 5. Assignment of all individual vineyards of the Bandol area to a "Terroir Units" based on geological substrate and soil properties

2.2. The social terroir

To this relative agro-pedological homogeneity of the vineyard, we can add favorable human factors which allowed the emergence of an appellation of elite from 1941. While the Provence wine was resolutely turned towards the production of mass wines , some visionaries promote the vineyard of Bandol by giving him in his naming decree particularly elitist rules: yields not exceeding 40 hl / ha, choice of a local grape variety, the *mourvèdre*, in assemblages, obligation of aging in cask of 18 months for the red wines ...

A few decades later the same requirements prevail. The leaders of the vineyard of Bandol systematically refuse any mutation likely to lower the quality of their wines. To cite only one revealing example of a state of mind, they refuse any extension or revision of the bounded Bandol area. They defend every hectare of Bandol in local planning commissions. Finally the technical committee of the appellation makes an important work of improvement and harmonization of viticultural practices. Despite the disappearance of tastings during the 2006 reform, the entire production of the 56 cellars and two cooperative cellars is tasted several times a year.

https://doi.org/10.1051/e3sconf/20185002001

3. Territorial regulations in Bandol DOP

Since 2016, several plots cleared in the forest, have been downgraded by the INAO experts, alerted by the ODG, the organization and defense of the vineyard of Bandol. A *terroir* study, based on concrete cases, was conducted as part of the downgrading requested by the INAO [4].

It must first be specified that all the settlements of new parcels in the forest are not covered by these procedures. The phenomenon that aims to transfer the vineyards of the plains or hillsides that urbanize towards the wooded margins of the vineyard is not in itself necessarily problematic. For some years the supervisory authorities have even encouraged these practices in certain vineyards particularly affected by land pressure. This "move of viticulture" that has been observed in many vineyards of Provence has sometimes completely transformed some appellations. To cite just one example, mention is made of the Cassis AOC, which no longer corresponds at all to the location of the historic vineyard, classified in 1936.

In the vineyards of Bandol, the implantation of these new parcels has remained limited for the last twenty years. We conducted a landscape diagnostic study of the Bandol vineyard by digitizing the entire plots at different reference dates between 1998 and 2017, as part of the PATERMED scientific program, Mediterranean landscapes and terroirs. [5]. We were able to accurately measure the area of these new parcels planted in the forest. At the scale of the entire vineyard this corresponds to less than 5% of the planted area of the vineyard in twenty years.

Establishments that are problematic, although rare, must be treated with firmness so as not to harm the image and quality of the wines. At the scale of the whole of Provence three practices appeared these last years, practices sometimes cumulated on the same parcel:

The first problem concerns the spreading of toxic inert products on the lands classified in appellation. This situation results from the importance of major infrastructure works in the region (new roads, tunnels, real estate operations). These spreading of rubble or waste can sometimes reach several meters on a plot of several hectares. The regional press has echoed the problem, viticulture professionals have had to react vigorously to reverse the trend and put in place sanctions within the framework of the DOP (Figure 6).

Var-matin

Des vins varois au bon goût de gravats? L'affaire risque d'éclabousser les AOC

Figure 6. Var matin newspaper (2016/05/03): "*Var wines with a good taste of rubble. The case risks splashing the DOP*"

The second point that needs to be highlighted is the contribution of exogenous land in parcels classified as AOC. This practice was long tolerated as the land came from the area of appellation. Today the contributions of exogenous lands are no longer accepted by the INAO and the various specifications of the defense and management organizations, ODG are now clear on this point. AOC Bandol is no exception.

Finally, the third and final point is equally worrying is the appearance of quarries in classified wooded areas. Private companies exploit the limestone that is very popular in the construction of villas. The layers are cut several meters thick, sometimes up to 10 or 15 meters (Figure 7).

Figure 7. Extraction of limestone in the vineyard of Bandol

Except for environmental and landscape issues, the development of these quarries poses a real problem for ODG Bandol. When all the rock has been discharged, a plot of vines is laid out in large terraces, using fine particles and crushed rock mobilized on site. New soils appear with agro-pedological conditions very different from the area classified in production. (Figure 8)

To try to find solutions to these three problems in the vineyard, the following points were discussed to redefine the rules for the development of new vineyards plots. A working group including ODG Bandol and INAO experts is working on the initiative of Olivier Colombano, technical and administrative head of ODG Bandol.

From what stage of transformation will we consider that the pedological sequence has been too strongly upset? The vineyard soils of Bandol are generally not very thick and very little structured. The development of any new parcels on the slopes involves disturbances in the structure and texture of the soil.

Should hard limestone slabs from plots be crushed or removed? Most soils of the PDO have high Ph, related to the omnipresence of limestone. On-site crushing of the calcareous bedrock has a lasting impact on the soil's chemical properties and is probably not desirable in areas where soil pH is already high.

In the current state of thinking, the proposal is not to accept soil transformations that include crushing and grinding the hard bedrock, whether it is directly flush, or that it is at a certain depth. Practices that have been tolerated so far include crushing on few decimeters, without disruption of the soil sequence. The topographic profile of the parcel must not be significantly modified.

What are the heights and widths of terraces to remember? The new developments are very spectacular because of their landscape impact compared to traditional practices. Extending the vineyard by destroying the forest is also problematic from the point of view of environmental regulation and sustainable development.

Figure 8: new large terraces in Bandol vineyard

These spectacular landscape transformations in the terraced vineyards have already been observed in many vineyards. For example in the Cinque Terre, one can see new parcels of vines with modernized landscape far from the historical and patrimonial area. At Bandol, the traditional terraces like the sector of Cabassaou or the Tourtine, are separated by walls having a maximum of three meters of height. The width of the terrace is generally less than three meters (three rows or four rows of vines on average). The maximum slope is of the order of 30%. Many terraced plantations, which have emerged for two decades, no longer fit this pattern at all. The proposal of the working group is to accept walls from a height of up to eight meters on slopes that do not exceed 30%. Remember that the Bandol vineyard does not accept the use of the machine to pick up the grapes. It does not need to transform its traditional vine landscape.

The fact of limiting the slope of the wooded parcels that one wishes to plant in vines, is to avoid the cornices having high thicknesses of limestones. They can be developed only with very important means like the quarries already mentioned. Limiting the height of the walls is a way to maintain the traditional landscape character of the vineyard, while allowing some modernization.

In those news plots in forest,the use of irrigation-is unavoidable. the water reserves of the soils are much lower than the average of the soils currently exploited. Recall that the use of irrigation in a AOP is tolerated only derogatory in summer by prefectural decree. And still there must be a water resource nearby. We have already exclude the use of allogenic inputs of land. These practices, tolerated in other territories, would be necessary for optimal exploitation of the northernmost PDO territories. These recently invested soils are very little compatible with maintaining the expression of wines according to the current quality standards of AOP.

4. Conclusion

The central question related to these new developments is the maintenance of the typicality of wines. One of the strong points of the Bandol vineyard is the great homogeneity of its products. Homogeneity linked certainly to a welded winegrowing community, but also to a particularly coherent agro-terroir. It therefore seems desirable that the vineyard can be maintained in its current terroirs, even if punctual transformations remain acceptable to compensate for inevitable losses related to urbanization. And these new parcels in the forest must finally develop with a minimum of coherence with the existing vineyard. Indeed besides the question of the rules for the new plots in forest that we posed in this paper, one must also think about the structure of the vineyard exploitations in order to avoid too much fragmentation. This fragmentation of the plot of land causes inevitable movements of agricultural machinery, not compatible with sustainable development of the vineyard.

References

1. L. Villier, P. Minvielle, A. Fournillon, P.-O. Bruna, *Expressions of soil and geological substrate properties in the red wines of Bandol?* IXe International Terroirs Congress, Dijon, poster (2012)

2. W. Moran, *Crafting terroir: people in cool climates, soils, and markets*, Sixth International Cool Climate Symposium for Viticulture and Oenology, Christchurch, 27 p. (2006).

3. G., Flutet et al., Pour une définition opérationnelle au service du développement durable, *Actes du colloque international sur les terroirs,* Aix-en-Provence, p. 27-32. (2007)

4 F. Gillot, P Minvielle, Ph. Moustier, *Rapport d'expertises : procédure de déclassement parcellaire, AOP Bandol*, INAO, centre de La Valette, Toulon, 2017, 43 p.

5 P. Minvielle, S. Angles, J. Daligaux, La gouvernance territoriale agricole : l'exemple de la viticulture et de l'oléiculture dans l'agglomération de Toulon (France), *Revue Environnement Urbain*, Montréal, 6 : 74-99. (2012)

E3S Web of Conferences 50, 02002 (2018) https://doi.org/10.1051/e3sconf/20185002002
XII Congreso Internacional Terroir

Influencia del tipo de suelo en la características de los mostos, vinos y destilados del cv. L Palomino fino

Influence of soil on the characteristics of musts, wines and distillates from cv. L Palomino fino

Mª Victoria Jimenez-Povedano[1], Emma Cantos-Villar[1], Mª Jesús Jimenez-Hierro[1], José Alberto Casas[2], Luis Miguel Trillo[2], Salvador Guimera[2] y Belén Puertas[1*]

[1]Instituto de Investigación y Formación Agraria y Pesquera (IFAPA), Centro Rancho de la Merced. Consejería de Agricultura, Pesca y Desarrollo Rural. Cañada de la Loba (CA-3102) PK 3.1. 11471 Jerez de la Frontera. Spain.

[2]Centro de Investigación, Desarrollo, Innovación y Medio Ambiente (CIDIMA). Bodegas González Byass Jerez, SLU. C/ Manuel María González nº 12. 11470 Jerez de la Frontera. Spain.

Resumen. Tradicionalmente los Brandis de Jerez se elaboran con destilados o aguardientes de vinos de la variedad Airén. Sin embargo, algunas bodegas del marco de Jerez están apostando por la elaboración de Brandis a partir de la variedad Palomino fino, mayoritaria en el marco de Jerez. En este trabajo se ha estudiado la composición química y se ha realizado el análisis sensorial de mostos, vinos y destilados procedentes de uva de la variedad Palomino fino cultivada en tres tipos de suelos diferentes: albarizas, barros y arenas. Los mostos de la uva cultivada en albarizas presentaron mayor grado sacarimétrico, más acidez total y menor pH. Consecuentemente con los mostos de partida, fueron los vinos procedentes de la uva cultivada en suelo de albarizas los que presentaron mayor grado alcohólico. Los vinos procedentes de la uva cultivada en suelo de barros presentaron más concentración de acetato de etilo y de n-propanol. Sensorialmente los vinos mejor valorados fueron los procedentes de la uva cultivada en suelo de barros. Respecto a los destilados, los mejor valorados sensorialmente fueron los procedentes de uva cultivada en suelo de albarizas.

Abstract. Brandies are usually elaborated from white grapevine Airen. However, wineries are currently trying to wide the brandies production testing other varieties. Palomino fino variety is mainly cultivated in albarizas soil, and in minor proportion in mud and sandy soils. In the current work, the analytical composition of Palomino fino musts, wines and distillates, which have been cultivated in albarizas, mud and sands soils, has been studied. The must from grape cultivated in albarizas soil highlighted for showing high sugar content and total acidity, meanwhile the pH the lowest value. The wine come from grape cultivated in albarizas showed the highest ethanol content, in agreement with sugar content in the must. The wines from the grapes grown in mud soil had a higher concentration of ethyl acetate and n-propanol. In tasting, the wines come from grape cultivated in mud soil obtaining the highest score in fruit aroma intensity. Regarding distillates, those come from albarizas soil showed the highest scores.

1. Introducción

La calidad de los aguardientes de vino de baja graduación, denominados tradicionalmente holandas, viene determinada, principalmente, por las materias primas de partida y por cómo se realice el proceso de elaboración. Así, la calidad del Brandy de Jerez viene determinada por los destilados de vino utilizados para su obtención.

Al producirse alguna modificación en el proceso de elaboración o en la misma materia prima se podrían obtener nuevos productos con peculiaridades y características diferentes a los ya existentes diversificando así la producción [1].

Estas modificaciones pueden ser, entre otras, la variedad de uva, la temperatura de fermentación de los vinos, la fecha de vendimia, la zona de cultivo o la presencia de lías en el proceso de destilación. Todos, o algunos de ellos, podrían conducir a la obtención de productos de calidad diferenciada o bien nuevos productos para su posterior comercialización tras su envejecimiento en madera.

En este trabajo, se muestran los resultados de los análisis químico-enológicos y sensoriales de los vinos y destilados de la variedad Palomino fino, de la cosecha de 2017, procedentes de uva cultivada en distintos tipos de suelos, como son: albarizas, barros y arenas.

En la cosecha de 2016 se realizó un estudio

 https://doi.org/10.1051/e3sconf/20185002002

preliminar, a escala de laboratorio, en la bodega González Byass.

2. Material y Métodos

2.1 Características del ensayo y material vegetal

Los tres tipos de suelos de donde procede la uva para este estudio son: albarizas, barros y arenas.

El suelo de albarizas se caracteriza por su elevado contenido en carbonato cálcico, con entre un 25 y un 40 % de caliza activa, pobre en materia orgánica y en nitrógeno y alto nivel de porosidad, lo que ayuda a retener la humedad. Los suelos de barros se ubican en valles y vaguadas, se componen de arcilla, arena, caliza y materia orgánica descompuesta, tienen un color más oscuro y mayor fertilidad. Los de arenas se localizan en las zonas costeras, se componen de arena, arcilla y menos del 20% de caliza activa [2].

En todos los casos la variedad cultivada es la Palomino fino, que ocupa más del 95 % de la superficie de vid cultivada en el marco de Jerez. Esta variedad es idónea para la elaboración de los vinos generosos de la zona. Actualmente, se está estudiando su aptitud para obtener holandas de calidad aptas para envejecer como brandis de Jerez.

En la Tabla 1 se muestra, para cada tipo de suelo el municipio, el pago, la finca, la parcela, el tipo de poda y la edad del viñedo.

Tabla 1. Producción de uva y parámetros de los mostos.

	Albarizas	Barros	Arenas
Municipio	Jerez de la Frontera	Jerez de la Frontera	Sanlúcar de Barrameda
Pago	Carrascal	Carrascal	La Loma alta
Finca	La Canariera	La Canariera	Las Lomas
Parcela	Esteve	Canariera	Las Lomas
Poda	Vara y pulgar	Cordón doble	Vara y Pulgar
Edad de la viña (años)	31	26	30

2.2 Controles enológicos

2.2.1 Vinificación

La vendimia se realizó manualmente a finales de agosto, en fechas similares para los tres tipos de suelos, con el fin de poder comparar los resultados.

La uva se trasladó a la bodega experimental del Rancho de la Merced, en cajas de plástico perforadas, de unos 18 Kg de capacidad. Se despalilló y se prensó en prensa neumática. Al mosto, a la salida de la prensa, se le adicionaron 2,5 ml/Hl de enzimas pectolíticas (Blanco L, Sepsa-Enartis) y se introdujo en una cámara frigorífica a 4°C para su desfangado, donde se mantuvo 24 horas. No se adicionó anhídrido sulfuroso ya que el destino de estos vinos era la destilación. En los mostos desfangados se determinaron los siguientes parámetros: grado Brix, acidez total, pH, ácido tartárico, ácido málico e índice de Folin-Ciocalteu (IFC).

Tras el desfangado, el mosto se encubó en 2 depósitos de 50 L de capacidad, de acero inoxidable, provistos de sistema de refrigeración por ducha. Se corrigió la acidez con ácido tartárico hasta llevar el pH a 3,3 y se adicionaron 25 g/Hl de levaduras comerciales previamente hidratadas (Viniferm BY, Agrovin S.A.).

La temperatura de fermentación se estableció en 18 °C. Diariamente se midió el grado Baumé, la densidad y la temperatura de los mostos. Al alcanzar densidad inferior a 1000 g/ml se desliaron y se mantuvieron en cámara frigorífica a 4°C hasta su destilación.

Tras el deslío se determinaron a los vinos los parámetros químico-enológicos de calidad. Los análisis de los mostos y vinos han sido realizados siguiendo los métodos internacionales de análisis de la OIV [3].

2.2.2 Destilación

Los vinos se destilaron de forma discontinua simple (un único proceso de destilación para cada vino) en alquitara de cobre. De cada vino se realizaron dos destilaciones.

El calentamiento se llevó a cabo por fuego directo utilizando como combustible gas propano y la refrigeración se realizó por circulación de agua fría. De esta manera, la temperatura de salida del destilado fue inferior a 15°C. Se destilaron por duplicado en lotes de 12 L. Se recogieron distintas fracciones a lo largo del proceso de destilación. En base al análisis por cromatografía de gases (GC), se definieron las fracciones de cabeza y corazón también llamadas "holanda". Por otra parte, la fracción de cola se definió en base al grado alcohólico. En todos los casos, el porcentaje de la fracción de cabezas respecto al volumen de vino inicial de partida fue de un 1%. La destilación se consideró terminada cuando el grado alcohólico del destilado obtenido fue inferior a 20% vol.

Las diferentes fracciones obtenidas de los destilados fueron almacenadas en frio hasta su posterior análisis en el laboratorio. Los análisis de los destilados han sido realizados siguiendo los métodos internacionales de análisis de la OIV [4].

Las determinaciones analíticas realizadas a las fracciones de cabezas, corazones y colas de los destilados fueron: grado alcohólico, pH, sulfuroso total y cobre. Asimismo, se identificaron y cuantificaron por CG a los corazones u holandas los siguientes compuestos volátiles: acetaldehído, acetona, acetato de metilo, acetato de etilo, 1-1-dietoxietano, metanol, 2-butanol, butirato de etilo, n-propanol, isobutanol, alilico, acetato de isoamilo, n-butanol, isoamílicos, 2-hexanol, hexanoato de etilo, heptanoato de etilo, lactato de etilo, n-hexanol, cis-3-hexen-1-ol, octanoato de etilo, ácido acético, nonanoato de etilo, linalol, decanoato de etilo, succinato de dietilo, α-terpineol, undecanoato de etilo, acetato 2-fenil-etilo, laurato de etilo, 2-fenil-etanol y miristato de etilo.

2.2.3 Análisis organoléptico de los vinos

Primeramente se realizaron catas triangulares, en copas negras, para ver si se detectaban, olfativamente, diferencias entre las réplicas de cada ensayo. Para los vinos procedentes de cada tipo de suelo se pusieron 6 puestos y se contó con 12 jueces. En cada puesto había dos copas iguales y una distinta.

El análisis sensorial descriptivo de los vinos se realizó mediante un panel de 7 catadores entrenados en nuestro Centro y 5 catadores de la Bodega González Byass, siendo los jueces los que reflejaron en las correspondientes fichas de cata la percepción de los parámetros encontrados en los vinos en cuanto a la sensación visual, olfativa y gustativa. La valoración de cada atributo se realizó en una escala del 0 al 10, considerando las puntuaciones cercanas al 0 ausentes y las cercanas al 10 muy presentes.

2.2.4 Análisis organoléptico de los destilados

El análisis sensorial de los destilados se realizó con el mismo panel de catadores que para los vinos. Los jueces reflejaron en las correspondientes fichas de cata la percepción de los parámetros encontrados en los destilados en cuanto a limpidez en la fase visual y a intensidad, punzante, fragancia, afrutado, vinoso, floral y calidad, en la fase olfativa. La valoración de cada atributo se realizó en una escala del 0 al 8, considerando las puntuaciones cercanas al 0 ausentes y las cercanas al 8 muy presentes.

2.3 Análisis estadístico

Se realizo un análisis de la varianza (ANOVA) para cada determinación analítica, utilizando el programa Statistix software (v. 9.0 Analitycal software, FL, USA).

3. Resultados y Discusión

3.1 Mostos

En la Tabla 2 se muestra, para cada tipo de suelo, la producción de uva y los parámetros analizados a los mostos desfangados.

Tabla 2. Producción de uva y parámetros de los mostos desfangados.

	Albarizas	Barros	Arenas
Producción (Kg/ha)	8.000	10.000	10.000
Grado Brix	22,5	20,2	18,7
pH	3,72	3,87	3,83
Acidez total (g/L TH_2)	2,94	2,60	2,72
Ácido tartárico (g/L)	4,92	5,32	5,12
Ácido málico (g/L)	0,41	0,45	0,62
Índice Folin-Ciocalteu	7,24	4,55	5,90

La menor producción de uva y la mayor graduación Brix se obtuvo cuando se cultivó en suelo de albarizas. Asimismo, los mostos procedentes de uva cultivada en albarizas fueron los que presentaron menor valor de pH y mayor acidez total. El menor grado Brix fue el de la uva cultivada en suelo de arenas, con la misma producción de uva que la cultivada en suelo de barros.

La variedad Palomino fino se caracteriza por valores bajos de ácido málico [5], se observó que cuando se cultiva en suelo de arenas la cantidad este ácido es superior que cuando se cultiva en suelo de albarizas o de barros.

Respecto al índice de Folin-Ciocalteu los valores más altos se obtuvieron cuando la uva se cultivó en suelo de albarizas y los más bajos cuando la uva procedía de suelo de barros.

3.2 Vinos

En la Figura 1 se muestra la cinética de la fermentación alcohólica (FA). La FA de los mostos procedentes de suelo de albarizas duró 15 días, mientras que los procedentes de suelo de barros y de arenas duraron 8 y 10 días respectivamente. Hay que tener en cuenta que el grado Brix de los mostos de suelo de albarizas era más elevado. Todas las fermentaciones se desarrollaron correctamente y los vinos quedaron con contenido en azucares reductores inferiores a 2 g/L.

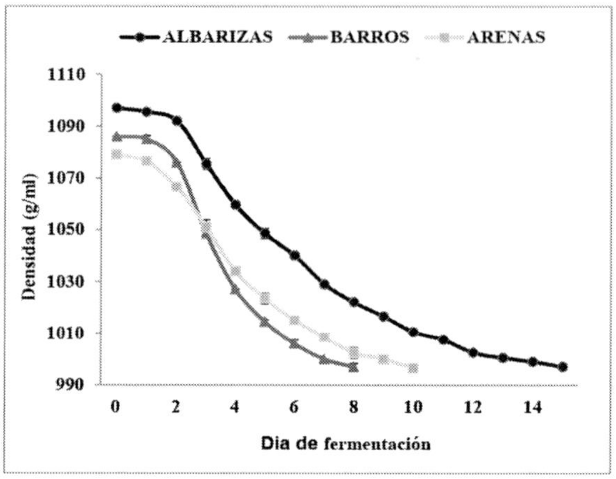

Figura 1. Evolución de la fermentación.

Los resultados de las determinaciones analíticas de los vinos se muestran en la Tabla 3.

Los grados alcohólicos de los vinos están en concordancia con los grados Brix de los mostos de partida, siendo significativamente más altos los de los vinos que procedían de uva cultivada en suelo de albarizas. En concordancia está la densidad relativa, es decir, valores más bajos a mayor concentración de etanol. Generalmente, los vinos de mayor graduación alcohólica presentan valores más altos de glicerol, por lo que los vinos procedentes de uva cultivada en suelo de albarizas presentaron valores más altos de dicho compuesto.

No se encontraron diferencias significativas ni en azúcares ni en el extracto seco. Todos ellos presentaron valores bajos de acidez volátil, lo que indicaba ausencia e desviaciones en la fermentación alcohólica.

Respecto a los ácidos orgánicos, no hubo diferencias significativas en los valores del ácido acético ni del ácido cítrico. Los vinos que presentaron valores más bajos de ácido succínico fueron los procedentes de uva cultivada en suelo de barros. Los vinos procedentes de uva cultivada en suelo de arenas realizaron la fermentación maloláctica, por lo que presentaron los valores más altos de ácido láctico y más bajos de ácido málico.

En cuanto a compuestos volátiles mayoritarios no se encontraron diferencias significativas ni en acetaldehído, ni en isoamílicos, ni en el sumatorio de alcoholes superiores. Los vinos procedentes de uva cultivada en suelo de barros fueron los que presentaron valores más altos de acetato de etilo, metanol y n-propanol y los procedentes de uva cultivada en suelo de albarizas los que presentaron mayor concentración de isobutanol.

Los vinos con tonalidades más verdosas, es decir valores más bajos de a* fueron los procedentes de uva cultivada en suelo de barros y los de mayor tonalidad amarilla fueron los procedentes de uva cultivada en albarizas, mayor valor de b* y de absorbancia a 420 nm.

Las concentraciones de los metales están dentro de las permitidas en vinos blancos jóvenes. Cabe destacar que los vinos procedentes de uva cultivada en suelo de arenas presentaron valores más altos de sodio y de hierro.

Tabla 3. Determinaciones analíticas de los vinos.

	Albarizas	Barros	Arenas	G.S.
Parámetros convencionales				
Etanol (% v/v)	14,69 a	12,51 b	11,48 c	***
Densidad relativa	0,9871 c	0,9895 b	0,9906 a	***
Azúcares (g/L)	1,97	1,36	1,12	n.s.
Ac. total (g/L TH_2)	3,91 a	3,27 b	3,75 a	*
Ac. volátil (g/L AcH)	0,22 a	0,12 b	0,13 b	***
pH	3,25 b	3,36 a	3,29 b	**
Extracto seco (g/L)	12,55	13,05	12,23	n.s.
Glicerol (g/L)	7,10 a	5,98 b	6,06 b	*
Ácidos orgánicos (g/L)				
Acético	0,19	0,04	0,10	n.s.
Cítrico	0,05	0,09	0,03	n.s.
Succínico	0,73 a	0,29 b	0,75 a	*
Tartárico	1,96 b	2,17 a	2,10 a	*
Málico	0,41 a	0,31 a	0,06 b	**
Láctico	0,03 c	0,12 b	0,54 a	***
Compuestos volátiles mayoritarios (ppm)				
Acetaldehído	24,5	22,9	25,2	n.s.
Acetato de etilo	41,0 b	71,0 a	35,5 b	*
Metanol	28,4 b	35,4 a	30,9 b	*
N-propanol	14,7 c	26,2 a	16,4 b	***
Isobutanol	13,8 a	11,2 b	10,6 b	*
Isoamílicos	169,5	131,8	137,8	n.s.
Alcoholes superiores	198,0	169,1	164,7	n.s.
Parámetros relacionados con el color				
L	96,49	97,98	97,26	n.s.
a*	-1,34 a	-1,65 b	-1,40 a	*
b*	13,64 a	9,96 b	10,78 b	*
C^*_{ab}	13,71 a	9,99 b	10,87 b	*
h	95,61 c	99,49 a	97,36 b	**
D.O. 420 nm	0,096 a	0,064 b	0,075 b	**
D.O. 520 nm	0,020	0,011	0,015	n.s.
Metales (ppm)				
Calcio	26	24	30	n.s.
Sodio	14,5 b	11,0 c	17,5 a	**
Potasio	419 b	511 a	522 a	***
Hierro	0,27 b	0,28 b	0,59 a	**
Cobre	0,52	0,58	0,44	n.s.
Zinc	0,18	0,18	0,20	n.s.

Dentro de cada fila, letras diferentes indican diferencias significativas a $P \leq 0,05$ según el test LSD. G.S.: grado de significación; n.s.: no significativo; *: significativo a $P \leq 0,05$; **: significativo a $P \leq 0,01$; ***: significativo a $P \leq 0,001$.

3.3 Destilados

En las Tablas 4, 5 y 6 se muestran los parámetros físico-químicos de las tres fracciones de los destilados: cabezas, corazones y colas. Como se ha comentado anteriormente los cortes de cabeza se realizaron en el mismo punto para todos los destilados (1% del vino inicial), es decir 120 ml. El % de etanol más alto, en las fracciones de cabeza, fue el de los destilados procedentes de uva cultivada en suelo de albariza, al igual que los correspondientes vinos de partida.

En todas las fracciones (cabezas, corazones y colas) de cada destilado las determinaciones de pH, cobre y SO_2 total estuvieron dentro de los rangos esperados. En todos los casos, el cobre tuvo el mismo comportamiento, aumentó de cabeza a corazón y de corazón a cola siendo el último aumento más pronunciado que el primero.

Los caudales de destilación, como es lógico, disminuyeron de las fracciones de cabezas a las de corazones y de las de corazones a las de colas, manteniendo la temperatura de salida del destilado prácticamente constante a lo largo de cada destilación.

Tabla 4. Parámetros físico-químicos de las cabezas.

	Albarizas	Barros	Arenas
Etanol (% v/v)	77,45	75,47	76,50
pH	5,95	5,93	5,27
Cobre (ppm)	2,63	2,11	5,60
SO_2 total (ppm)	4	4	7
Volumen (ml)	120	120	120
Caudal (ml/min)	32	30	29
Tª salida destilado (°C)	11	11	11

Tabla 5. Parámetros físico-químicos de los corazones.

	Albarizas	Barros	Arenas
Etanol (% v/v)	66,50	65,57	68,00
pH	5,95	5,92	5,13
Cobre (ppm)	2,98	2,52	8,13
SO_2 total (ppm)	3	4	8
Volumen (ml)	2000	1250	1250
Caudal (ml/min)	27	25	26
Tª salida destilado (°C)	10	10	10

Tabla 6. Parámetros físico-químicos de las colas.

	Albarizas	Barros	Arenas
Etanol (% v/v)	39,1	40,5	41,5
pH	5,27	5,41	5,33
Cobre (ppm)	9,93	7,53	9,16
SO_2 total (ppm)	4	5	4
Volumen (ml)	450	650	700
Caudal (ml/min)	19	19	20
Tª salida destilado (°C)	10	10	10

Para una "holanda" de calidad el grado alcohólico de los corazones debe estar comprendido entre 65 y 70 % v/v. Al unir las diferentes fracciones de los destilados para obtener el grado alcohólico deseado en los corazones u holandas (65 70 % v/v), los volúmenes finales fueron mucho mayores en los destilados de vinos procedentes de uva cultivada en suelo de albariza, respecto a los destilados de vinos procedentes de uva cultivada en suelos de barros o de arenas (2000 ml frente a 1250 ml), debido a la mayor graduación alcohólica de estos vinos.

El grado alcohólico de las fracciones de colas estuvo en torno al 40% v/v, alcanzando un volumen de éstas menor en los destilados procedentes de la uva cultivada en suelo de albarizas.

En la Tabla 7 se muestran las concentraciones de aldehídos, cetonas y alcoholes de las fracciones corazón de los destilados. No se encontraron diferencias significativas en acetaldehído, cuyo descriptor es oxidación, estando todos correctos con valores inferiores a 70 g/Hl a.a.

Cuando la relación n-propanol/isobutanol de los aguardientes es ≥ 1 es síntoma de alguna alteración bacteriana en los vinos de partida. En este trabajo los destilados que presentan una relación mayor entre estos compuestos son los procedentes de suelo de arenas.

Los destilados de uva cultivada en suelo de arenas presentaron valores más altos de metanol, 2-butanol, n-propanol, n-butanol, 2-hexanol y cis-3-hexen-1-ol.

Tabla 7. Aldehídos, cetonas y alcoholes de los corazones (g/Hl a.a.).

	Albarizas	Barros	Arenas	G.S.
Acetaldehído	17,73	34,40	17,43	n.s.
Acetona	0,385 a	0,281 c	0,344 b	***
Metanol	15,33 c	19,45 b	20,68 a	***
2-Butanol	0,235 b	0,253 b	0,318 a	**
n-Propanol	10,88 c	16,66 b	23,80 a	***
Isobutanol	10,25	11,10	10,50	n.s.
n-Butanol	0,459 b	0,375 c	0,575 a	***
Isoamílicos	134 b	157 a	135 b	**
2-Hexanol	0,369 b	0,378 b	0,500 a	**
n-Hexanol	0,467 b	1,080 a	0,478 b	***
Cis-3-hexen-1-ol	0,126 b	0,114 b	0,151 a	**
2-Feniletanol	3,21 a	2,66 a	1,20 b	***

Dentro de cada fila, letras diferentes indican diferencias significativas a $P \leq 0{,}05$ según el test LSD. G.S.: grado de significación; n.s.: no significativo; **: significativo a $P \leq 0{,}01$; ***: significativo a $P \leq 0{,}001$.

El metanol puede provenir de un prensado excesivo, se considera correcto cuando se encuentra en concentración inferior a 50 g/Hl a.a. y en este trabajo están siempre por debajo de dicho valor.

El 2-butanol, cuyo descriptor es fermentación butírica, en el destilado procedente de uva cultivada en suelo de arenas está por encima del valor correcto (≥ 0,3 g/Hl a.a.) aunque por debajo del inaceptable (≥0,6 g/Hl a.a.). El descriptor del n-butanol también es la fermentación butírica, considerándose correcto cuando se encuentra en concentraciones inferiores a 1.0 g/Hl a.a., encontrándose este compuesto, en todos los destilados, por debajo de este valor.

El 2-hexanol y el cis-3-hexen-1-ol tienen como descriptores los herbáceos, considerándose inaceptables cuando superan 0,6 g/Hl a.a.

Los destilados procedentes de uva cultivada en suelo de barros fueron los que presentaron valores superiores de n-hexanol, cuyo descriptor es resina o hierba verde cortada.

En cuanto al 2-feniletanol, cuyo descriptor es rosa [5], fueron los destilados procedentes de uva cultivada en suelos de albarizas y barros los que presentaron valores más altos.

En las Tablas 8 y 9 se muestran los ésteres (acetatos) y ésteres etílicos superiores.

No se encontraron diferencias significativas en el acetato de metilo cuyo descriptor es disolvente o quitaesmalte, ni en acetato de 2-feniletilo, cuyo descriptor es floral y sotobosque.

El acetato de etilo se encontró en mayor concentración en los destilados procedentes de uva cultivada en suelo de arenas, tiene como descriptor el picado acético y se considera correcto cuando su concentración está por debajo de 455 mg/L [6].

El descriptor del acetato de isoamilo es fruta madura, plátano y pera, su umbral de percepción es de 1,6 mg/L [7]. Al igual que el acetato de etilo, se encontró en mayor cantidad en los destilados procedentes de uva cultivada en suelo de arenas.

Tabla 8. Ésteres (acetatos) de las fracciones de corazón (mg/L).

	Albarizas	Barros	Arenas	G.S.
Acetato de metilo	1,95	2,01	1,95	n.s.
Acetato de etilo	159,8 b	185,5 b	431,5 a	***
Acetato de isoamilo	7,57 b	6,82 b	26,62 a	***
Acetato 2-feniletilo	1,53	1,24	1,75	n.s.
Total acetatos	**170,9 b**	**195,7 b**	**461,8 a**	***

Dentro de cada fila, letras diferentes indican diferencias significativas a $P \leq 0{,}05$ según el test LSD. G.S.: grado de significación; n.s.: no significativo; ***: significativo a $P \leq 0{,}001$.

Respecto a los ésteres etílicos no hubo diferencias significativas ni en el hexanoato (caproato), ni en el nonanoato, ni en el decanoato (caprato).

Los que presentaron valores más altos de octanoato (caprilato) fueron los procedentes de uva cultivada en suelo de arenas, cuyos descriptores son fruta madura, manzana, piña, plátano y pera [1].

Tabla 9. Ésteres etílicos superiores de las fracciones de corazón (mg/L).

	Albarizas	Barros	Arenas	G.S.
Hexanoato de etilo	0,06	0,10	0,11	n.s.
Heptanoato de etilo	0,56 a	0,51 a	0,27 b	***
Octanoato de etilo	2,44 b	2,79 b	4,05 a	**
Nonanoato de etilo	1,97	1,30	1,77	n.s.
Decanoato de etilo	2,16	2,35	2,57	n.s.
Total esteres etílicos superiores	**7,19 b**	**7,07 b**	**8,77 a**	**

Dentro de cada fila, letras diferentes indican diferencias significativas a $P \leq 0{,}05$ según el test LSD. G.S.: grado de significación; n.s.: no significativo; **: significativo a $P \leq 0{,}01$; ***: significativo a $P \leq 0{,}001$.

3.3 Análisis organoléptico de los vinos

En las catas triangulares se detectaron diferencias significativas ($p \leq 0.05$) entre las repeticiones de los vinos procedentes de la uva cultivada en suelo de albarizas y de la cultivada en suelo de arenas y no hubo diferencias significativas entre los vinos procedentes de uva cultivada en suelo de barros. Al encontrar diferencias significativas entre las repeticiones de los vinos procedentes de uva cultivada en el mismo tipo de suelo se realizaron las catas descriptivas de cada uno de los vinos.

En la Figura 2 se representan los resultados del análisis sensorial de los vinos.

En la fase visual los vinos procedentes de uva cultivada en suelo de arenas estaban algo más turbios que los procedentes de uva cultivada en suelo de albarizas y en suelo de barros. Los puntuados con valores más altos en color fueron los procedentes de uva cultivada en suelo de albarizas, lo que coincide con los datos analíticos mostrados en la Tabla 3.

En la fase olfativa los vinos mejor valorados fueron los procedentes de uva cultivada en suelo de barros, que fueron puntuados con valores más altos en las notas cítricas, fruta de árbol (manzana y pera), fruta de hueso (albaricoque y melocotón) y fruta madura.

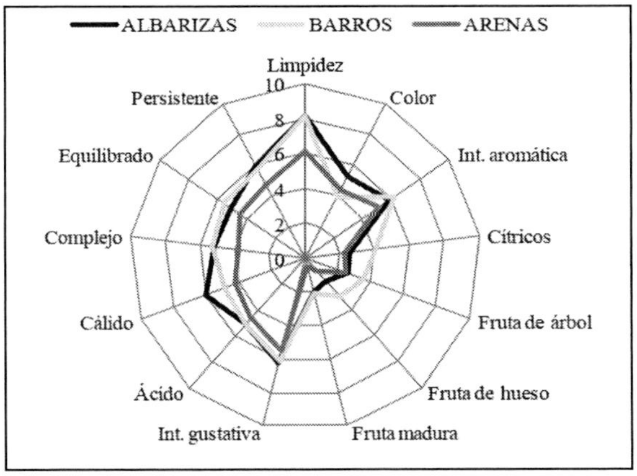

Figura 2. Análisis sensorial descriptivo de los vinos.

En la fase gustativa destacaron los vinos procedentes de uva cultivada en suelo de albarizas por ser más cálidos (mayor grado alcohólico). En cuanto a complejidad, equilibrio y persistencia, los vinos procedentes de uva cultivada en suelo de albarizas y barros alcanzaron puntuaciones similares y más altas que los vinos procedentes de uva cultivada en suelo de arenas.

3.4 Análisis organoléptico de los destilados

En la Figura 3 se representan los resultados medios, por tipo de suelo, del análisis sensorial de los destilados. La cata descriptiva se realizó, en las fracciones corazón, de los destilados procedentes de uva cultivada en cada tipo de suelo. Para la cata, los destilados se hidrataron con agua osmotizada, para que tuvieran un 30 % v/v de etanol.

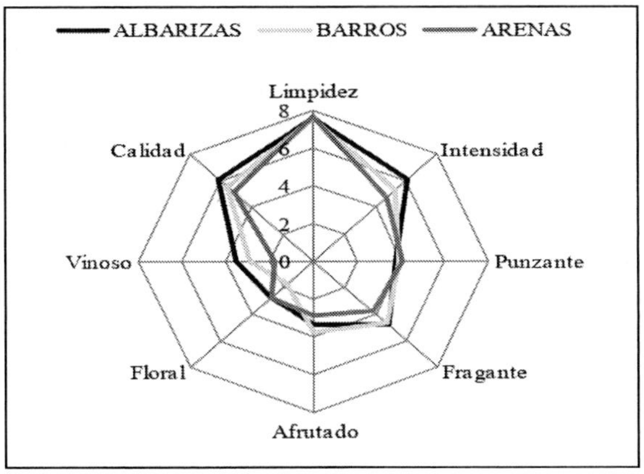

Figura 3. Análisis sensorial descriptivo de los destilados.

Todos ellos alcanzaron idénticas puntuaciones en limpidez.

Los destilados procedentes de uva cultivada en suelo de albarizas fueron los mejor valorados en los atributos: intensidad, vinoso y calidad. Los procedentes de uva cultivada en suelo de barros fueron mejor puntuados en el atributo afrutado.

En general, los que alcanzaron puntuaciones más bajas, fueron los procedentes de uva cultivada en suelo de arenas. Posiblemente las notas afrutadas del acetato de isoamilo, muy presente en los destilados procedentes de uva cultivada en suelo de arenas, se han visto enmascaradas por el acetato de etilo, que se encontraba en concentraciones elevadas en dichos destilados.

4. Conclusiones

Los vinos procedentes de uva cultivada en suelos de albarizas fueron los de más color y mayor grado alcohólico y los procedentes de suelos de barros los mejor valorados organolépticamente.

Las holandas mejor valoradas sensorialmente fueron las que procedían de suelos de albarizas y con las que se obtuvo mayor volumen de la fracción corazón, sin embargo la producción de uva fue menor que en suelos de arenas, con la misma poda y edad.

Comparando los resultados de los trabajos realizados en las cosechas de 2016 y 2017 se puede decir que, en ambas cosechas, los vinos peor valorados fueron los procedentes de uva cultivada en suelo de arenas. Asimismo, las holandas mejor valoradas fueron las procedentes de uva cultivada en suelo de albarizas.

En cualquier caso, son necesarios más años de estudio para poder obtener resultados más sólidos.

5. Agradecimientos

Los resultados presentados han sido obtenidos en el marco del proyecto BESTBRANDY, cuyo título completo es "Factores que influyen en la calidad del Brandy y nuevos sistemas de elaboración del mismo, desde el viñedo al envasado", que cuenta con el apoyo financiero del Centro para el Desarrollo Tecnológico Industrial (CDTI) a través de la convocatoria Interconecta, cofinanciada con fondos FEDER de la Unión Europea.

6. Bibliografía

1. M.S. Jurado, *Potencial agronómico y enológico de diez variedades blancas de vid como materia prima en la obtención de aguardientes para Brandy de Jerez. Estudio comparativo de las holandas producidas*. Tesis Doctoral, Universidad Cádiz (2016)
2. A. García de Luján. *La Viticultura de Jerez*. ISBN 97-8847- 1146-786, (1997)
3. O.I.V. *Recueil des méthodes internationales d'analyse des vins et des moûts*. Paris. Office International de la Vigne et du Vin (2015)
4. O.I.V. *Recueil des méthodes internationales d'analyse des boissons spiritueuses d'origine vitivinicole*. Paris. Office International de la Vigne et du Vin (2014)
5. B. Puertas, *Estudio del potencial enológico de 15 variedades blancas de vid en la zona del Jerez*. Tesis Doctoral, Universidad de Cádiz ISBN 84-7498-399-1(1989)
6. J. Muñoz-Ramos, *Enciclopedia de los alcoholes. Características y elaboración de licores y aguardientes de todo el mundo*. ISBN 84-08-01842-6 (1996)
7. S.C. Diéguez, M.L.G. de la Peña, E.F. Gómez. *Volatile composition and sensory characters of commercial Galicia orujo spirits*. J. Agric. Food Chem. 53, 679-6765 (2005)

Influencia del terroir en el perfil aromático de Tempranillo Blanco en la D.O.Ca. Rioja

Juana Martínez*[1], Pilar Rubio-Bretón[1], Mª Eva Vicente[1] y Enrique García-Escudero[1]

[1]Instituto de Ciencias de la Vid y del Vino- ICVV (Gobierno de La Rioja, CSIC, Universidad de La Rioja), 26071 Logroño, España

Resumen. En los vinos blancos el aroma es uno de los principales atributos de calidad, siendo los compuestos aportados por la uva los que determinan el perfil varietal. Tempranillo Blanco es una variedad derivada de Tempranillo Tinto por mutación natural, que desde el año 2008 está autorizada únicamente en la D.O.Ca. Rioja. Sus vinos manifiestan características organolépticas de alta calidad, con notas afrutadas y florales intensas. En este trabajo se estudió el perfil aromático varietal de Tempranillo Blanco en siete localizaciones geográficas de la D.O.Ca. Rioja durante la campaña 2016. Las parcelas se distribuyeron en las tres subzonas que integran la denominación, con características vitivinícolas diferenciadas debido a las condiciones de clima y suelo. Los resultados obtenidos mostraron diferencias notables en el contenido aromático de la uva procedente de las distintas ubicaciones del viñedo. Los compuestos más abundantes fueron los norisoprenoides (48-70%), seguidos por los bencenoides (8-29%) y los aldehídos y cetonas (3-18%); otros muchos volátiles fueron detectados en menores porcentajes (compuestos C6, alcoholes, terpenos, ésteres, ácidos...). A pesar de las diferencias observadas, el perfil aromático varietal se mantuvo, por lo que se puede concluir que Tempranillo Blanco es una variedad con amplias posibilidades de adaptación a diferentes entornos vitícolas.

Influence of terroir on the aromatic profile of Tempranillo Blanco in the D.O.Ca. Rioja

Summary. Aroma is one of the main attributes of white wines quality and, in particular, prefermentative aromas provided by the grape are the main responsible for the varietal characteristics. Tempranillo Blanco variety was originated from Tempranillo Tinto by natural mutation, being authorized since 2008 just in the D.O.Ca. Rioja. Its wines show high quality organoleptic characteristics, with intense fruity and floral notes. In this work, the varietal aromatic profile of Tempranillo Blanco was studied in seven geographical locations of the D.O.Ca. Rioja during the 2016 vintage. The plots were distributed in the three sub-areas that make up the Denomination, with distinct vitivinicultural characteristics due to climate and soil conditions. The results obtained showed notable differences in the aromatic content of the grape in the different vineyard plots studied. Norisoprenoids were the most abundant compounds (48-70%), followed by benzenoids (8-29%) and aldehydes and ketones (3-18%), other groups of volatiles were also detected in lower percentage (compounds C6, alcohols, terpenes, esters, acids…). Despite the observed differences, the varietal aromatic profile was maintained stable, so that it could be concluded that Tempranillo Blanco is a variety with wide possibilities of adaptation to different growing conditions.

Introducción

El aroma es uno de los atributos determinantes de la calidad en los vinos blancos, siendo los volátiles aportados por la uva los responsables de su tipicidad [1-2]. Estos compuestos proceden del metabolismo secundario de las plantas, y su presencia está influenciada por muchos factores, tales como la variedad, las prácticas culturales, el clima, el suelo y la localización geográfica [3-4]. Se encuentran principalmente en el hollejo y la pulpa de las bayas, tanto en forma libre (odorantes) como en forma de precursores, ligados a moléculas de glucosa y otros compuestos [5-8]. En la uva se han identificado gran número de compuestos aromáticos pertenecientes a diferentes familias químicas, entre los que destacan monoterpenos, C13 norisoprenoides, bencenoides y compuestos C6 [9-12]. La familia de los monoterpenos presenta un poder odorífero muy elevado, relacionado con notas florales y afrutadas, son compuestos característicos de las variedades muy aromáticas (Muscat, Riesling, Gewürztraminer...), aunque también se encuentran en menor cantidad en otras consideradas neutras.

El impacto de las condiciones ambientales puede modificar de forma importante la composición volátil de la uva [13]. Numerosos factores (exposición a la luz, temperatura, estado hídrico...) pueden afectar a la evolución de la maduración y a otros procesos que se desarrollan paralelamente, entre los que destaca la síntesis, translocación y acumulación de los aromas varietales [14].

La caracterización aromática de las variedades blancas de difusión internacional ha sido objeto de gran número de estudios [15-17]. Por el contrario, la bibliografía existente sobre las variedades blancas españolas es más escasa [18-20], y prácticamente inexistente en algunas de reciente incorporación al viñedo, como es Tempranillo Blanco.

La variedad Tempranillo Blanco se originó a partir de Tempranillo Tinto por mutación natural, siendo detectada en un viñedo de La Rioja en 1988 [21]. Esta variedad se cultiva desde el año 2008 únicamente en la D.O.Ca. Rioja, y en la actualidad es la segunda variedad blanca en cuanto a superficie. Se trata de una variedad de ciclo corto y producción media, que permite elaborar vinos con una composición química equilibrada en grado alcohólico, acidez y contenido polifenólico, por lo que puede ser adecuada para diversificar la elaboración de vinos blancos. Aporta características organolépticas de alta calidad, destacando por un elevado potencial aromático con notas afrutadas y florales intensas [22].

El estudio de la composición aromática, así como de la influencia de las condiciones del entorno, en las variedades blancas poco conocidas, pueden ser de gran interés para la obtención de vinos con mayor complejidad aromática, que reflejen el carácter del terroir y presenten características de diferenciación en el mercado.

El objetivo de este trabajo fue el estudio de la influencia del entorno geográfico en el perfil aromático de la variedad Tempranillo Blanco en la D.O.Ca. Rioja (España).

Material y métodos

El estudio se llevó a cabo con la variedad Tempranillo Blanco durante la campaña 2016, en diferentes viñedos de la D.O.Ca. Rioja (España). Se seleccionaron siete parcelas experimentales distribuidas en las tres subzonas que integran la denominación, con características vitivinícolas diferenciadas debido a las condiciones de clima y suelo. En la subzona de Rioja Baja se localizaron dos parcelas (Alfaro y Corera), tres en Rioja Media (Logroño, Nalda y Albelda) y dos en Rioja Alta (Cenicero y Fonzaleche). Las condiciones climáticas recogidas en las estaciones meteorológicas más próximas (desde 1 de octubre de 2015 a 30 de septiembre de 2016) se muestran en la Figura 1.

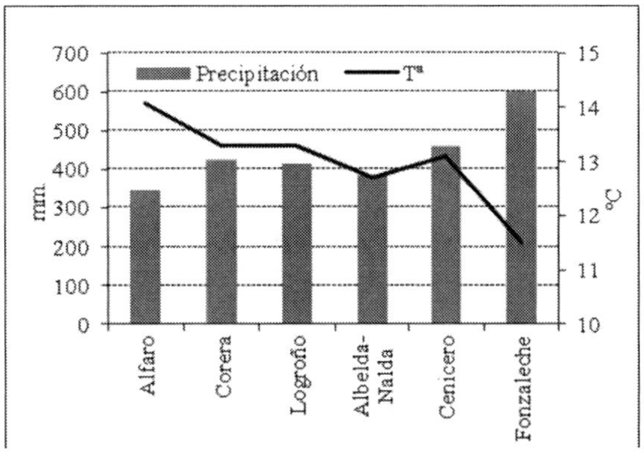

Figura 1. Condiciones climáticas en parcelas de Tempranillo Blanco.

Los viñedos fueron plantados entre los años 2008-2011, conducidos en espaldera, doble cordón Royat, con un marco de plantación de 2.90 x 1.10 m. Su altitud media osciló entre 345 y 555 m, con la excepción de una parcela en Rioja Media (Albelda) situada a 720 m de altitud.

La vendimia de la uva se efectuó al alcanzar la madurez tecnológica, entre los días 31 de agosto y 10 de octubre. En ese momento se evaluaron los parámetros siguientes: producción (kg uva/cepa), peso medio del racimo (g) y de la baya (g). Previamente, se realizó el muestreo de 500 bayas, por triplicado, en cada una de las parcelas, que se separaron en dos lotes. 200 bayas se destinaron al análisis de la composición físico-química (grado probable, pH, acidez total, ácidos tartárico y málico y potasio) según los métodos oficiales [23]. Las 300 bayas restantes se procesaron con una trituradora Thermomix® TM5 durante 15 segundos (velocidad 4) y el mosto obtenido se congeló a -20°C para el posterior análisis de aromas varietales.

Los compuestos volátiles libres fueron determinados por microextracción en fase sólida y posterior cromatografía Gases-Masas (HS-SPME-GC-MS) Las muestras de uva fueron extraídas según el método descrito por [24] y modificado por [25]. La fibra SPME utilizada fue DVB/CAR/PDMS, 50/30 µm. Las extracciones se realizaron en viales de vidrio de 20 ml, a los que se adicionó 9 ml de muestra, 2.5 g de NaCl y 10 µl de patrón interno (2-Octanol) en una concentración de 24.33 mg/l. Después de agregar un imán de agitación el vial se colocó en el GC-MS. El acondicionamiento de la muestra se hizo a 60°C, durante 15 minutos con agitación de 500 rpm. A continuación, la fibra SPME se insertó automáticamente en el espacio superior del vial de muestra, y se mantuvo durante 105 min con agitación. Una vez concluida la extracción, la fibra se introdujo en el puerto de inyección del GC a 250°C y se mantuvo durante 15 minutos para la desorción. Los compuestos desorbidos se separaron en un cromatógrafo de gases 7890B con detector de masas de triple cuadrupolo (MS/MS) 7000C de Agilent Technologies (Palo Alto, CA, USA) conectado al programa Agilent MassHunter Qualitative Analysis B.07.00. Se usó helio como gas portador (1,2 ml/min). Las inyecciones se realizaron en modo splitless (1 min). Las condiciones cromatográficas fueron las descritas por [25]. El análisis de los compuestos C6 se llevó a cabo por el mismo procedimiento, pero con la muestra diluida 1/30 con agua milliQ, ya que se encuentran en concentración muy elevada. Para la cuantificación de los compuestos se empleó el método semicuantitativo, en relación al área y la concentración del patrón interno empleado.

El estudio estadístico de los resultados se realizó mediante el análisis de varianza ANOVA y en caso de existir diferencias significativas, $p < 0.05$, se utilizó el test de Tukey para la separación de medias (programa estadístico SPSS versión 20.0). También se aplicó un análisis multivariante de componentes principales (programa InfoStat).

Resultados y discusión

Los parámetros productivos obtenidos para la variedad Tempranillo Blanco en las diferentes parcelas se muestran en la Tabla 1. Como puede observarse, se encontraron diferencias entre Logroño, donde se alcanzó la producción mayor, y Alfaro, Albelda y Nalda, con rendimientos más bajos. Los racimos de mayor tamaño se obtuvieron en Corera y Logroño, y los más pequeños en Alfaro y Nalda. En cuanto al tamaño de las bayas, las más grandes se encontraron en Nalda y las de menor tamaño en Alfaro. Este parámetro estuvo menos influenciado por las condiciones de cultivo, lo que parece indicar que se trata de una característica varietal.

Tabla 1. Parámetros productivos en parcelas de Tempranillo Blanco (2016). Media ± desviación estándar (n= 3).

Subzona	Localidad	Producción (kg/cepa)	Peso racimo (g)	Peso baya (g)
Rioja Baja	**Alfaro**	3.10±0.28**b**	150±0.90**b**	1.37±0.08**c**
	Corera	4.61±0.87**ab**	300±5.30**a**	2.03±0.29**ab**
Rioja Media	**Logroño**	6.81±1.83**a**	290±4.10**a**	1.79±0.20**bc**
	Albelda	3.67±0.58**b**	240±4.90**ab**	1.85±0.72**bc**
	Nalda	3.53±0.15**b**	180±1.70**b**	2.38±0.17**a**
Rioja Alta	**Cenicero**	4.32±1.15**ab**	190±5.10**ab**	1.63±0.25**bc**
	Fonzaleche	4.97±1.13**ab**	220±2.20**ab**	2.13±0.09**ab**

Letras diferentes en la misma columna indican diferencias según el test de Tukey p <0.05.

La composición físico-química de la uva de Tempranillo blanco de las diferentes parcelas (Tabla 2) estuvo claramente influenciada por las condiciones climáticas, con la excepción del grado probable. No obstante, este parámetro mostró valores medios más bajos en la parcela de Albelda, en la que no alcanzó un nivel correcto de madurez, probablemente a causa de su elevada altitud. El pH más alto se obtuvo en la uva procedente de Corera debido a su elevado contenido de potasio, que podría estar relacionado con el tipo de suelo del viñedo. La acidez total fue superior en las parcelas de Rioja Alta (Cenicero y Fonzaleche) y más baja en Alfaro, debido a las condiciones climáticas en que se desarrolló la maduración.

Tabla 2. Composición físico-química de la uva de Tempranillo Blanco (2016). Media ± desviación estándar (n= 3).

Subzona	Localidad	Grado Prob. (% v/v)	pH	Acidez Total (g/l)	Ac. Tartárico (g/l)	Ac. Málico (g/l)	Potasio (mg/l)
Rioja Baja	**Alfaro**	12.9±0.10**a**	3.48± 0.04**ab**	4.99±0.06**c**	6.77±0.08**bc**	2.17±0.06**b**	1318±95**bc**
	Corera	13.1±0.25**a**	3.58±0.05**a**	6.28±0.48**ab**	8.32±0.08**ab**	3.39±0.33**a**	1786±83**a**
Rioja Media	**Logroño**	13.0±0.41**a**	3.43±0.06**ab**	5.23±0.36**bc**	7.32±0.34**bc**	2.19±0.33**ab**	1255±44**bc**
	Albelda	11.5±0.04**a**	3.23±0.04**b**	5.78±0.33**bc**	6.87±0.30**bc**	1.91±0.17**b**	832±46**d**
	Nalda	13.9±0.35**a**	3.30±0.06**b**	5.48±0.11**bc**	7.07±0.39**bc**	2.19±0.18**b**	1184± 44**c**
Rioja Alta	**Cenicero**	13.4±0.90**a**	3.37±0.04**b**	6.64±0.14**a**	8.63±0.33**a**	2.69±0.24**ab**	1491±177**b**
	Fonzaleche	13.7±0.14**a**	3.17±0.02**b**	6.89±0.02**a**	6.76±0.24**c**	2.93±0.04**ab**	1193±12**c**

Letras diferentes en la misma columna indican diferencias según el test de Tukey p <0.05

En el análisis de la composición volátil varietal de Tempranillo Blanco se detectaron 43 compuestos, que pueden agruparse en 8 familias químicas (alcoholes, aldehídos y cetonas, terpenos, norisoprenoides, ésteres, ácidos, compuestos C6 y bencenoides). En la Tabla 3 se muestra la abundancia relativa de cada uno de los grupos de compuestos identificados en las diferentes parcelas. En todos los casos, los norisoprenoides fueron los compuestos más abundantes (48-70%), seguidos por los bencenoides (8-29%) y los aldehídos y cetonas (3-18%). En menor proporción se encontraron los compuestos C6 (5-16 %), alcoholes (2-15 %), terpenos (2-5 %) y otros (ácidos, ésteres…). En Alfaro los norisoprenoides fueron mucho más abundantes que en el resto de parcelas, probablemente porque la temperatura fue más alta, tal como ha sido observado por otros autores [13]. En Nalda y Albelda (parcelas muy próximas, pero con diferente altitud) se encontró el mayor porcentaje de bencenoides; y en Corera y Fonzaleche destacaron los aldehídos y cetonas y los alcoholes. En el viñedo de Logroño se obtuvieron más compuestos C6, lo que podría deberse a la elevada producción obtenida. Las diferencias en el porcentaje de terpenos relacionadas con la ubicación del viñedo fueron muy escasas, lo que indicaría que se trata de un carácter varietal.

Tabla 3. Composición volátil (%) en parcelas de Tempranillo Blanco (2016).

Familia química	Rioja Baja		Rioja Media			Rioja Alta	
	Alfaro	Corera	Logroño	Albelda	Nalda	Cenicero	Fonzaleche
Norisoprenoides	70.3	49.1	51.1	53.3	50.7	50.4	47.6
Bencenoides	8.53	8.29	13.7	28.5	20.2	16.7	13.3
Aldehídos y Cetonas	5.84	17.4	10.1	3.27	11.5	12.8	17.9
Compuestos C6	8.41	4.94	16.0	9.90	5.34	6.56	5.69
Alcoholes	2.89	15.1	3.94	1.90	8.41	8.43	11.2
Terpenos	3.23	4.27	4.75	2.26	3.29	3.30	3.78
Ácidos grasos	0.32	0.63	0.23	0.56	0.30	1.11	0.19
Ésteres	0.50	0.38	0.21	0.30	0.30	0.70	0.24

El contenido total de aromas varietales (Figura 2), así como el de la mayor parte de las familias químicas analizadas (Tabla 3), fue superior en la uva procedente de la parcela de Nalda, mientras que los valores más bajos se encontraron en Logroño, Alfaro y Albelda. Los resultados obtenidos no mostraron una clara relación entre la localización geográfica del viñedo según la clasificación por subzonas de la D.O.Ca. Rioja y el contenido de aromas varietales de la uva. En Rioja Baja el contenido aromático fue mucho más bajo en Alfaro que en Corera, probablemente debido a las diferencias climáticas entre ambas parcelas. En Rioja Media las diferencias entre las parcelas se acentuaron más, a pesar de su proximidad geográfica y climatología similar, y podrían estar relacionadas con la producción y la altitud, además de la interacción con otros muchos factores no considerados en este estudio (tipo de suelo, exposición…) [13]. En Rioja Alta el contenido aromático alcanzó valores medios, algo más bajos en Fonzaleche, donde las precipitaciones fueron las más elevadas y las temperaturas más bajas.

Los norisoprenoides y los terpenos son los compuestos más odoríferos, presentan aromas florales y cítricos que juegan un papel clave en el aroma varietal de los vinos [2]. En Tempranillo Blanco se detectaron 8 norisoprenoides, entre los que destacaron por su mayor concentración los dos isómeros de β-damascenona, β-ionona y dehidro-β-ionona (Tabla 4). Los valores más elevados correspondieron al viñedo de Nalda, aunque con pocas diferencias respecto a Corera y Cenicero.

El contenido de compuestos terpénicos fue en general bajo, se cuantificaron 6 compuestos pertenecientes a este grupo, entre los que destacaron acetona de geranilo, p-cymenene, p-cymene y linalol. Todos ellos, junto con α-terpineol y limoneno tuvieron mayor concentración en la uva procedente de las parcelas de Nalda y Corera.

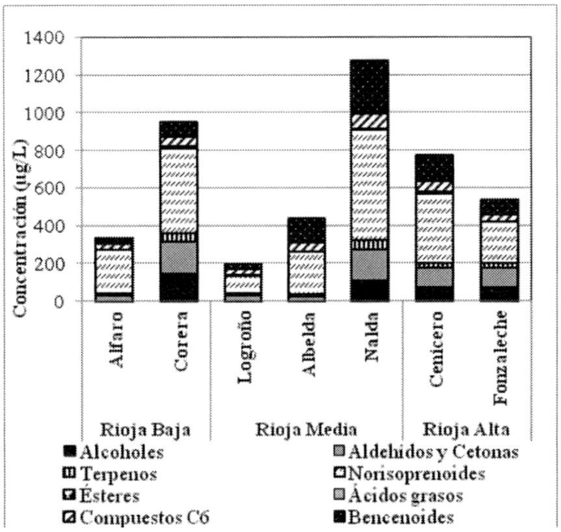

Figura 2. Composición volátil total (µg/l) de Tempranillo Blanco en diferentes parcelas (año 2016).

En el grupo de los bencenoides se detectaron 3 compuestos, 2-feniletanal, benzaldehído y 2-feniletanol. Estos últimos puede contribuir positivamente al aroma del vino aportando notas dulces y florales (rosa), y se forman principalmente durante la fermentación alcohólica [7]. Estos compuestos fueron significativamente más elevados en el viñedo de Nalda, y apenas se observaron diferencias entre el resto de parcelas.

Entre los volátiles clasificados como C6 se detectaron 6 compuestos, dos de los cuales (hexanal y E,E-2,4-hexadienal) tuvieron concentraciones mayores al resto. Este grupo de sustancias son responsables de aromas verdes y herbáceos, y dependiendo de su concentración pueden afectar negativamente a la calidad del vino [7]. Su formación se produce principalmente durante las etapas prefermentativas de la vinificación. Los resultados observados por [19] en diferentes variedades blancas mostraron que estos compuestos fueron los más abundantes (70-93%) en los mostos obtenidos después del estrujado y prensado de la uva.

En el análisis de componentes principales (Figura 3) las dos primeras funciones explicaron el 98% de la varianza, y permitieron la clasificación de las parcelas de Tempranillo Blanco en función de su composición volátil. La función 1 explicó el 83.7% de la varianza, y se correlacionó positivamente con los terpenos y norisoprenoides. En la parte positiva del eje X se posicionaron las parcelas de Nalda y Corera, en la negativa Alfaro, Logroño y Albelda.

Tabla 4. Composición volátil (μg/l) en parcelas de Tempranillo Blanco (2016). Media ± desviación estándar (n= 3)

Compuesto	Rioja Baja		Rioja Media			Rioja Alta	
	Alfaro	Corera	Logroño	Nalda	Albelda	Cenicero	Fonzaleche
β-Ciclocitral	2,89±0,95a	5,86±1,02a	3,55±1,41a	5,68±1,36a	2,57±0,54a	5,25±1,95a	5,21±2,16a
(Z)-β- Damascenona	12,03±1,56bc	23,83±2,24ab	4,91±0,74c	30,93±12,40a	14,95±5,68abc	21,01±4,49abc	13,99±5,35bc
(E)-β- Damascenona	201±7,87bc	387±23,00ab	76,6±13,96c	495±173a	190±45bc	300±26abc	169±116bc
β -Ionona	7,67±2,71b	15,72±3,15ab	8,85±3,37ab	17,45±4,57a	6,36±1,14b	14,45±3,82ab	10,46±4,15ab
Dehidro-β-ionona	3,91±0,33b	16,29±3,08ab	2,38±0,40b	35,48±16,30a	7,20±2,96b	18,92±4,39ab	19,99±13,64ab
Vianol	0,42±0,03a	2,63±2,26a	0,14±0,07a	1,69±1,32a	0,56±0,41a	0,99±0,44a	0,63±0,15a
β-Maaliene	0,26±0,09a	0,35±0,15a	0,12±0,04a	0,20±0,03a	0,18±0,15a	0,44±0,20a	0,17±0,09a
Metil-dihidroepijasmonato	1,39±0,33a	1,46±0,25a	0,10±0,02b	1,44±0,73a	0,16±0,04b	1,03±0,26ab	0,63±0,44ab
Σ Norisoprenoides	**229±70bc**	**453±134ab**	**97±26,2c**	**588±171a**	**222±66bc**	**363±103b**	**220±57bc**
Benzaldehído	5,96±6,17c	29,52±5,15bc	5,73±2,36c	107,89±18,60a	29,90±7,69bc	52,35±13,45b	30,76±10,93bc
2-Feniletanal	21,59±7,86d	48,72±0,91cd	20,21±10,57d	163,94±28,75 a	99,03±23,21b	82,51±20,49bc	45,08±8,75cd
2-Feniletanol	1,43±0,34b	3,72±1,43b	1,02±0,37b	11,96±1,84a	3,46±0,46b	2,15±0,68b	3,91±1,28b
Σ Bencenoides	**28,98±10,6c**	**81,96±22,6bc**	**26,97±10,0c**	**283,79±77a**	**132,39±49b**	**137,01±41b**	**79,75±21bc**
Limoneno	0,57±0,17b	1,96±0,72a	0,61±0,19b	2,02±0,48a	1,03±0,26ab	1,15±0,39ab	0,96±0,44ab
p-Cymene	1,19±0,69b	5,28±0,76a	0,82±0,22b	4,90±1,28a	1,16±0,37b	3,57±1,84ab	3,00±2,44ab
p-Cymenene	0,86±0,06bc	4,73±1,11ab	0,58±0,08c	7,23±2,08a	1,27±0,10bc	2,99±0,76bc	4,36±2,77abc
Linalol	2,27±0,41cd	3,98±0,59ab	1,45±0,26d	4,55±0,51a	1,66±0,30d	3,19±0,48bc	2,17±0,52cd
α-Terpineol	1,51±0,08b	3,01±0,68ab	0,63±0,22b	6,25±3,00a	1,20±0,21b	3,20±0,56ab	3,05±1,81ab
Acetona de geranilo	4,56±2,57b	23,24±8,5a	5,28±0,80b	21,35±9,87a	4,19±0,61b	11,94±2,21ab	9,14±5,02ab
Σ Terpenos	**10,96±1,46b**	**42,20±8,0a**	**9,36±1,85b**	**46,30±6,91a**	**10,52±1,21b**	**26,05±3,82ab**	**22,69±2,85ab**
Hexanal	8,04±1,14a	9,46±3,56a	10,76±1,90a	13,63±1,51a	19,16±16,13a	12,54±6,39a	6,16±1,52a
(Z)-3-Hexen-1-ol	0,87±0,13a	1,20±0,04a	0,82±0,10a	1,51±0,63a	1,49±1,25a	1,66±0,59a	0,56±0,14a
1-Hexanol	2,98±0,50a	5,63±0,99a	2,71±0,34a	9,89±1,63a	9,07±8,83a	2,80±0,44a	2,00±0,39a
(E)-2-Hexen-1-ol	1,56±0,37cd	3,30±0,30bc	1,53±0,63cd	5,20±0,91a	1,26±1,16d	4,15±0,43ab	1,50±0,19cd
(E)-2-Hexenal	4,94±0,95b	8,33±0,78ab	7,81±2,07ab	8,98±5,62ab	4,11±3,58b	13,23±1,2a	4,62±1,21b
E,E-2,4-Hexadienal	10,16±0,63b	20,97±3,18b	7,82±2,68b	35,96±8,19a	10,88±1,26b	19,66±3,7b	19,31±7,75b
Σ Compuestos C6	**28,55±3,70b**	**48,89±6,99ab**	**31,45±4,09b**	**75,17±12,20a**	**45,97±6,87b**	**54,04±7,21ab**	**34,16±6,99b**

Letras diferentes en la misma fila indican diferencias según el test de Tukey p <0.05

La función 2 explicó un 14.31% de la varianza y se correlacionó positivamente con el contenido en bencenoides. Según esta función en la parte positiva del eje Y se posicionaron las parcelas de Albelda y Nalda, y en la negativa Corera. Las parcelas de Cenicero y Fonzaleche se situaron en la zona intermedia par ambas funciones.

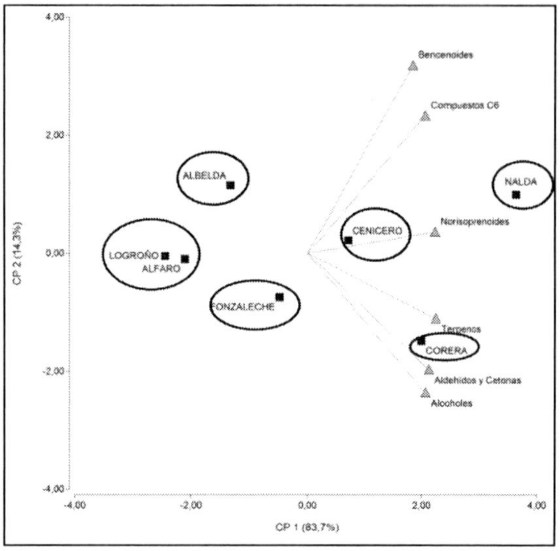

Figura 3. Componentes principales de la composición volátil de Tempranillo Blanco en diferentes parcelas (año 2016).

Conclusiones

En este trabajo se cuantificó por primera vez la composición volátil de la variedad Tempranillo Blanco en siete localizaciones geográficas de la D.O.Ca. Rioja, durante la campaña 2016. Los resultados obtenidos mostraron diferencias notables en el contenido aromático de la uva procedente de las distintas ubicaciones del viñedo, pero no se pudo establecer ninguna relación con la clasificación por subzonas de la D.O.Ca. Rioja. A pesar de las diferencias observadas, el perfil aromático varietal se mantuvo, lo que parece indicar que Tempranillo Blanco es una variedad con amplias posibilidades de adaptación a diferentes entornos vitícolas.

Finalmente, es preciso remarcar la necesidad de continuar estos estudios en otras campañas, con condiciones climáticas muy diferentes, con la finalidad de ampliar los conocimientos acerca de la influencia del entorno sobre el perfil volátil de Tempranillo Blanco, lo que permitirá confirmar sus posibilidades para la elaboración de vinos blancos con características diferenciadas que reflejen el carácter del terroir.

Bibliografía

1. Z. Gunata, C. Bayonove, R. Baumes, R Cordonnier, J. Chromatogr. **331**, 83-90 (1985)
2. P. Ribéreau-Gayon, Y. Glories, A. Maujean, D. Dubourdieu, *Handbook of Enology. The Chemistry of Wine*, **Vol 2** (Chichester: John Wiley and Sons Ltd, 2006)
3. M.J. Cabrita, A.M. Freitas, O. Laureano, R. di Stefano, J. Cromatogr. A **778**, 219-224 (1997)
4. S.M. Bureau, A.L. Rzungles, R.L. Baumes, J. Sci. Food Agric. **80**, 2012-2020 (2000)
5. M.A. Sefton, I.L. Francis, P.J. Williams, Am. J. Enol. Vitic. **44,** 359–370 (1993)
6. M.A. Sefton, I.L. Francis, P.J. Williams, J. Food Sci. **59**, 142–147 (1994)
7. A. Pedroza, A. Zalacain, J.F. Lar, M.R. Salinas, Food Res Int. **43**, 1003-1008 (2010)
8. J.M. Oliveira, I. Araujo, O. Pereira, S. Maia, A.J. Amaral, M.O. Maia, Anal. Chim. Acta **513**, 269–275 (2004)
9. J.M. Oliveira, M. Faria, F. Sa, F. Barroa, I. Araujo, Anal. Chim. Acta **563**, 300–309A (2006)
10. A. Panighel, R. Flamini, Molecules 40 (19), 21291-21309 (2014)
11. E. Sánchez-Palomo, R. Alonso-Villegas, M.A. González Viñas, Food Chem. **173**, 1195-1202 (2015)
12. A. Zalacain, J. Marín, G.L. Alonso, M.R. Salinas, Talanta **71(4),** 1610-1615 (2007)
13. J. Marais, G. Versini, C. Wyk, A. Rapp, S. Afr. J. Enol. Vitic. **13**, 71-77 (1992)
14. A.L. Robinson, P.K. Boss, P.S. Solomon, R.D. Trengove, H. Heymann, S.E. Ebeler, Am. J. Enol. Vitic. **65,** 1-23 (2014)
15. C. Peyrot des Gachons, C. van Leeuwen, T. Tominaga, G.P. Soyer, J.P. Gaudillère, D. Dubourdieu, J. Sci. Food Agric. **85,** 73-85 (2005)
16. T. Tominaga, C. Peyrot des Gachons, D. Dubourdieu, J. Agric. Food Chem. **46**, 5215-5219 (1998)
17. M. Vilanova, Z. Genisheva, M. Graña, J.M. Oliveira, S. Afr. J. Enol. Vitic. **34 (2)**, 212-222 (2013)
18. J.E. Bueno, R. Peinado, J. Moreno, M. Medina, L. Moyano, L. Zea, J. Food Sci. **68(1)**, 158-163 (2003)
19. A. Gonzalo-Diago, E. Terroba –Pérez, E. Baroja, E. García-Escudero, J. Martínez, *Abstracts Macrowine*, 200 (2016)
20. E. López-Tamames, N. Carro-Mariño, Y. Gunata, C. Sapis, R. Baumes, C. Bayonove. J. Agric. Food Chem. **45(5),** 1729-1735 (1997)
21. J. Martínez, T. Vicente, T. Martínez, J.B. Chávarri, E. García-Escudero. *Actas XXIX Congreso Mundial de la Viña y el Vino*,**1**, 41-43 (2006)
22. J. Martínez, A. Gonzalo, E. Baroja, E. García-Escudero, Enoviticultura **47**, 2-10 (2017)
23. ECC, Official Journal of the European Comunity, **L272 (3),** 1-192 (1990)
24. R. Perestrelo, A.S. Barros, S.M. Rocha, J.S. Câmara. Talanta **85,**1483-1493 (2011)
25. T. Garde-Cerdán, L. González-Arenzana, N. López, R. López, P. Santamaría, I. López-Alfaro. Innov. Food Sci. Emerg. Technol. **20**, 91-99 (2013)

Agradecimientos

Este trabajo ha sido realizado mediante un proyecto de investigación cofinanciado al 50% por el Gobierno de La Rioja y Fondos FEDER durante 2016 (Programa Operativo FEDER de La Rioja 2014-2020)

Role of natural terroir attributes on Barbera grapevine performance and grape quality in Piemonte region (Italy)

Gonzalez Perez Laura Alexandra[1], Bavaresco Luigi[2] and Etienne Neethling[1]

[1]ESA, International Vintage Master, Department of Viticulture and Oenology, Ecole Supérieure d'Agricultures, Angers 49007, France
[2]UCSC, Department of Sustainable Crop Production, Pomology and Viticulture section, Università Cattolica del Sacro Cuore, Piacenza 29211, Italy

Abstract. The trial was carried out in the Bersano estate (Nizza Monferrato, Piemonte, Italy) in different vineyards (Generala, Cremosina and Monteolivo) devoted to the production of DOCG Nizza, the newest and highly reputed Barbera denomination. For Bersano, this research provided access to scientific information and tools to valorise its Nizza wines and to characterize and select some sub-zones. The study investigated the effect of natural terroir attributes on vine behaviour and grape composition of cv. Barbera. Conducted in 2017, the tests included phenological phases, vegetative growth, chlorophyll and gas exchange measures. Maturation curves were also developed to better understand the time course of berry ripening, from fruit-set to harvest and statistical relationships were studied. The vineyards presented different climatic and soil conditions which influenced bud burst, shoot growth and plant photosynthesis; concerning soil properties, texture was the main differentiating factor. During the growing season, the berries followed the normal growth pattern, but an extreme drop occurred at harvest due to high diurnal temperatures. Ripening curves showed dissimilarities between the three studied vineyards, i.e. TSS, berry mass, total anthocyanins and total phenols. La Generala presented the best viticulture features regarding the vine performance and grape composition.

1 Introduction

A terroir can be defined as an agro-ecosystem whose natural attributes - soil, subsoil, relief and climate - constitute a unique set of factors that, through plants and animals, give the food product specific characteristics [1]. According to the current and official definition of the OIV, a viticulture terroir is: ''A concept which refers to an area in which collective knowledge of the interactions between the identifiable physical and biological environment and applied vitivinicultural practices develops, providing distinctive characteristics for the products originating from this area. "Terroir" includes specific soil, topography, climate, landscape characteristics and biodiversity features'' [2]. It can be said that the structural character and potential wine quality is derived from the environment features of the planting site and the wine's personality from the vintage conditions, whilst the final balance and style are shaped by the viticulturist. When these three influences meet in perfect harmony, a great and unique wine will result [3].

Within this perspective, the "zoning" technique initiated, allowing the division of a territory in function of comparable and divergent features of environmental (and human) characteristics and checking the qualitative result of different species and varieties [4]. Viticultural zoning can be defined as the spatial characterization of zones or units that produce grapes or wines of similar performance and composition, while enabling operational decisions to be implemented more effectively [5]. The aims of viticultural zoning are related to the delimitation of protected viticultural areas, which serve to give added value to wines according to their origin and typicality, or the delimitation of homogeneous areas for vineyard management, which can be used to optimize the grape quality [6]. With strong national and international competition, as well as challenging conditions such as a changing climate, studies to determine suitability for the cultivation of grapevines and wine management are becoming more and more widespread [7].

In this context, the trial was carried out in the Bersano estate (Nizza Monferrato, Piemonte, Italy) devoted to the production of Nizza DOCG, the newest and highly reputed Barbera denomination. For Bersano, this research provide access to scientific information and tools to valorise its Nizza wines and to characterize and select some sub-zones. The study investigated the

 https://doi.org/10.1051/e3sconf/20185002004

effect of natural terroir attributes on Barbera grapevine performance and grape quality.

2 Material and Methods

2.1 Natural environmental features

The research project considered three vineyard areas, which are namely La Generala, Cremosina and Monteolivo, located in the heart of the Barbera production area (Monferrato, Asti province, Piedmont region), where Bersano winery owns 10 ha devoted to the production of DOCG Nizza wines. The climatic conditions of the three studied areas were characterized by two automatic meteorological stations, one placed near Agliano Terme (covering La Generala area) and the other in Nizza Monferrato (covering La Cremosina and Monteolivo areas). Bioclimatic indices such as the Huglin index and Winkler index were calculated for the growing season of 2017.

The physical and chemical soil properties of the three estates were also determined. The training system for the three areas was simple Guyot with a fruit cane of 7 or 8 buds with a renewal spur (two buds) from the other side to get an average yield of 9 tons per hectare. The following plant material was utilized:

- La Generala: Barbera standard vines, grafted on Rupestris du Lot.
- Cremosina: Barbera clone AT84 grafted on 420 A.
- Monteolivo: Barbera clone AT84 grafted on SO4

2.2 Vine phenology and growth

The control of phenological phases was done according to the modified Eichhorn and Lorenz system [8]. The control of phenological stages begun in April and ended with harvest. Starting mid-May, shoot length was measured weekly on two shoots per vine (20 shoots per vineyard) until shoot trimming. When inflorescences were clearly visible (stage 15), the total number of buds and inflorescences per plant were counted and bud fertility determined [9].

Gas exchange readings were performed on 15 vines per vineyard, by checking one leaf/shoot inserted at nodes 7–10. Leaves were sampled on 30 June (fruit set) and 08 August 2017 (veraison). Readings were taken with a ADC BioScientific LCi Analyser Serial No. 31581 within the 10h to 13h time windows. Two field measurements of greenness index were made with the Soil-Plant Analysis Development SPAD-502 chlorophyll meter throughout the season. The measurements were made on 30 June (fruit set) and 08 August 2017 (veraison) on 1 leave per vine (15 vines per plot), located two nodes above the second cluster on a middle vigorous shoot.

2.3 Berry growth and ripening

In order to calculate berry growth (on a weight basis) according to Gray and Coombe [10] a 20-berry sample for each plant tagged (ten per vineyard) was collected, at 10-day intervals beginning from fruit set.

Ripening curves of berry mass, must TSS (Total Soluble Solids), TA (Titratable Acidity) and anthocyanins and phenolics concentrations were calculated in each vineyard [8] at 10-day intervals beginning from veraison, by analyzing a 300-berry sample from each vineyard.

At harvest, (04/09 in Monteolivo, 08/09 in Cremosina and 12/09 in La Generala), yield and cluster number per vine were recorded, mean cluster mass was then calculated. A two basal-cluster sample was collected from each tagged vine. A first 50-berry sample was used for determining total anthocyanins and phenolics concentrations (mg/g) according to the method reported [9]. A second 20-berry sample was used for identification and quantification of phenolic compounds in grape [10]. The remaining amount of each two-cluster sample was used for individual berry mass assessment and, upon crushing, must was processed for total soluble solids concentration (°Brix), pH and titratable acidity (g/L) with standard methods.

3 Results and Discussion

3.1 Influence of climate and soil on vine phenology and growth

The year 2017 is considered one of the warmest and driest, with a Winkler index of 1988°C (Generala) and 1835°C (Cremosina and Monteolivo) and a seasonal rainfall of 359.2 mm (Generala) and 330.6 mm (Cremosina and Monteolivo), resulting in a phenological earliness of 10 to 20 days. According to the recorded data, Generala vineyards are warmer and wetter than Cremosina and Monteolivo (Figures 1, 2).

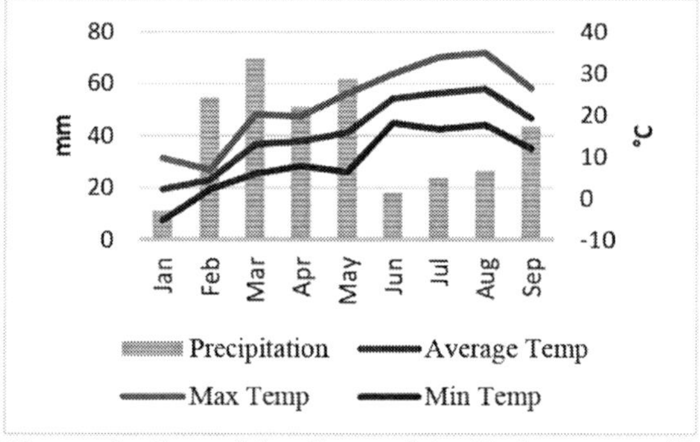

Figure 1. Generala's climate data (1 January to 21 September 2017). Data source: Coldiretti Grower Union.

Except for bud burst, phenological stages occurred at the same time. According to Lee [11], higher temperatures in the latter part of winter anticipate the time of budburst. In March, Generala had a higher average temperature than the other vineyards, resulting in an earlier bud burst in Generala as compared to Cremosina and Monteolivo.

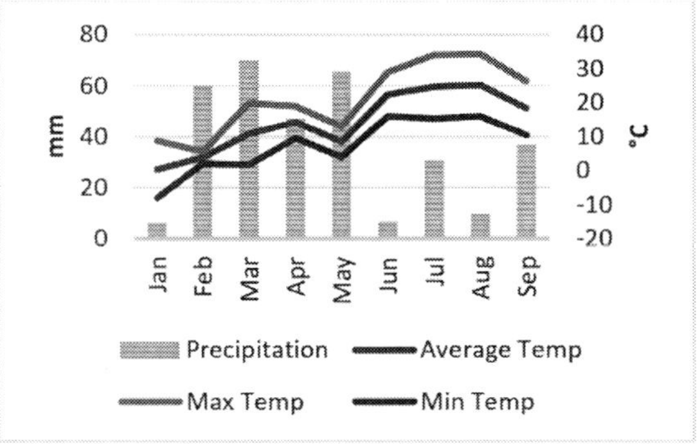

Figure 2. Cremosina and Monteolivo climate data (1 January to 21 September 2017). Data source: Coldiretti Grower Union.

In the three vineyards, the rapid shoot elongation almost stopped at the end of April (132 DOY) (Figure 3), due to unexpected low temperatures (-3 °C in Cremosina and Monteolivo, and 0.2 °C in Generala), as shoot elongation is positively related to Heat Degree Days (HDD). In addition of this event, the difference in shoot length between Cremosina, Monteolivo and Generala was also due to soil temperature. We can speculate that the lower shoot length recorded in Monteolivo during all the measures is due to its loamy soil, while as Cremosina has less clay and more sand that warms up faster (Figure 4), resulting in a higher shoot length.

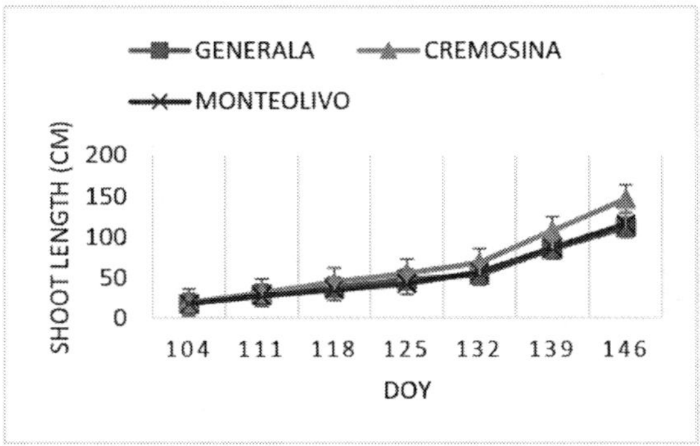

Figure 3. Pre-trimming shoot length measured at weekly intervals from 4 April to 26 May 2017 (DOY 104 to 145). Vertical bars indicate standard errors, n=7.

Grapevine photosynthesis is affected by environmental factors, one of them is the relative humidity, which stomatal conductance. According to Stoev & Dobreva [12], an optimal hydrometric is 60 to 70%. During the leaf gas exchange measures (June and July), Cremosina and Monteolivo vineyards had an average relative humidity around 70% and in the case of Generala it was around 60%. These values compared to the stomatal conductance and photosynthesis data recorded (Table 1) show a positive interaction, the increase in relative air humidity is reflected in the stomatal conductance.

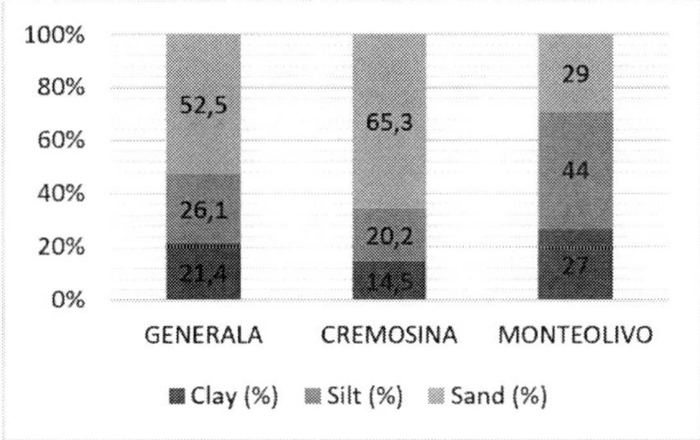

Figure 4. Proportion of sand, silt and clay in Generala, Cremosina and Monteolivo soils. Data source: Coldiretti Grower Union and Impresa Verde Cuneo.

Table 1. Leaf gas exchange measured at pea size (30 June 2017) and veraison (8 July 2017) growth stages.

Site	**Pea-size**		**Veraison**	
	g_s (mol/m^2·s)	A (µmol/m^2·s)	g_s (mol/m^2·s)	A (µmol/m^2·s)
Generala	0.07	9.20	0.12	9.73
Cremosina	0.10	11.28	0.11	10.80
Monteolivo	0.15	12.78	0.09	8.36
Sig.	ns	ns	ns	n.s

*, $P \leq 0.05$; Within each column, means followed by different letters are significantly different ($P < 0.05$) according to the Student–Newman–Keuls test. ns, not significant. (gs) Stomatal conductance (A) Photosynthesis rate.

3.2 Influence of climate and soil on berry growth and ripening

During the growing season of 2017, the berries in the three vineyards followed the normal growth pattern, according to a double sigmoidal curve [8]. However, an extreme drop at the harvest occurred (Figure 5) due to higher temperatures (>40°C) and lack of rain which caused the dehydration of the grapes.

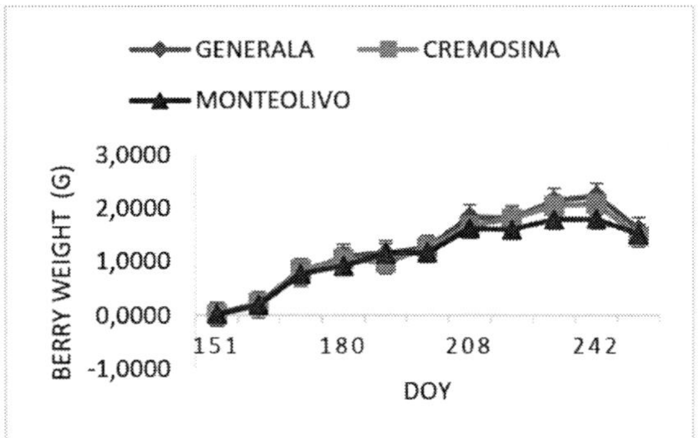

Figure 5. Berry growth measured at intervals of 10 days within the day of the year (doy) 151–harvest period (31 May to 4, 8 and 14 September) in 2017. Vertical bars indicate standard error, n = 11

https://doi.org/10.1051/e3sconf/20185002004

Among the different parameters (Figure 6), four of them showed differences between the three vineyards (TSS, berry mass, total anthocyanins and total phenols). Also Carlomagno et al. [13] demonstrated a role of the location on Barbera volatiles during ripening.

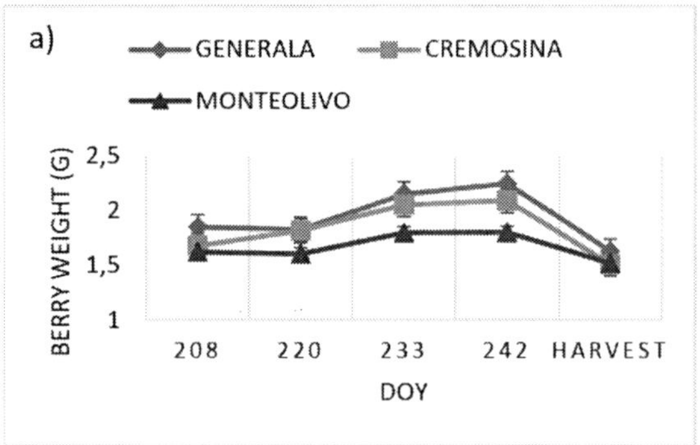

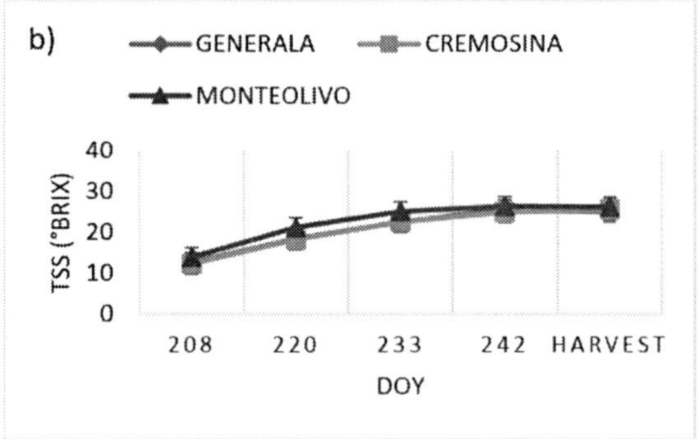

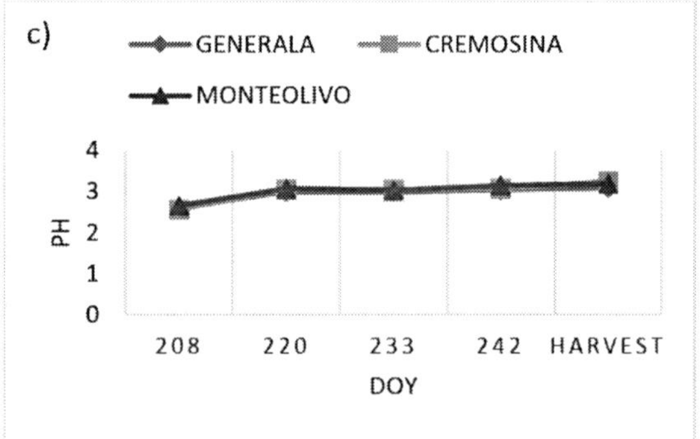

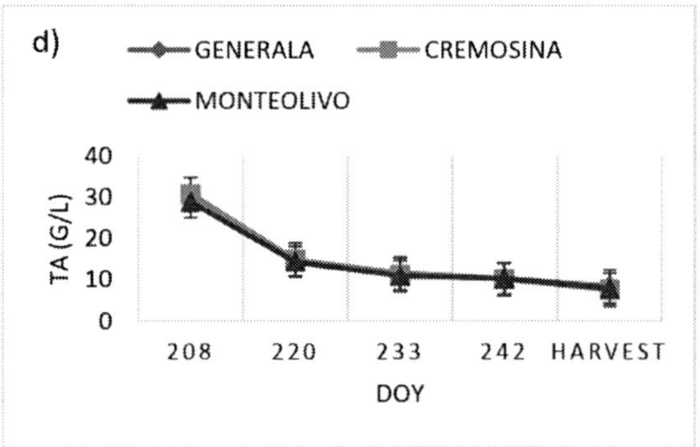

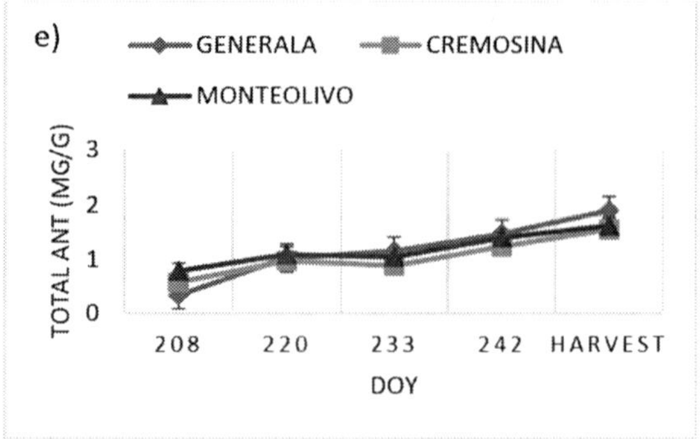

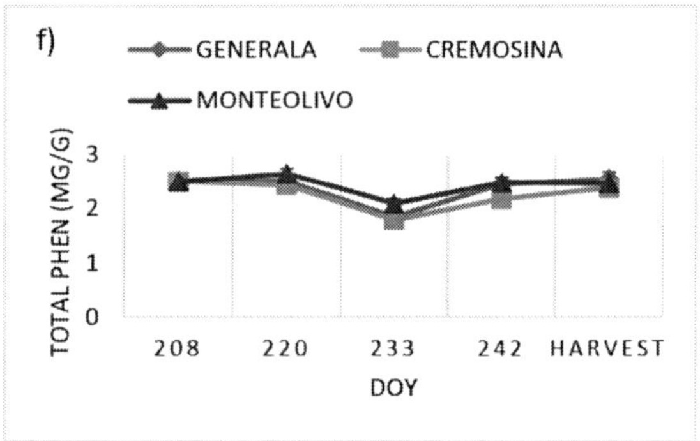

Figure 6. Growth and ripening curves for (a) berry weight, (b) TSS, (c) pH, (d) total acidity, (e) total anthocyanins and (f) total phenols of Barbera grapes. Vertical bars indicate standard error, n = 5; day of the year (DOY).

In addition to climatic conditions, soil has a strong influence on grape composition during ripening, as being the main water resource for the plant, and soil moisture-holding capacity is related to the soil texture. The effect of water deficit correlates with the decrease in the size of berries and changes in the ratio between skin and berry weight, which in turn affects the concentration of phenolic compounds [15]. These principles explain the rapid concentration of total anthocyanins and phenols from the 233 DOY till harvest (Figures 6e and 6f), period in which there was a combination of lack of rain and high average temperatures that reached 34.5°C (Generala) and 34.4°C (Cremosina and Monteolivo).

3.3 Grape quality and vine performance at harvest

There was no significant difference in almost all the parameters of vine performance and grape composition. Generala presented higher concentrations of total anthocyanins (1.90 mg/g) and total phenolics (2.55 mg/g) than Cremosina (1.55 mg/g and 2.39 mg/g) and Monteolivo (1.62 mg/g and 2.49 mg/g) (Table 2). However, the three vineyards presented better levels of berry anthocyanins in comparison to the general average levels recorded by Regione Piemonte during the last years across the region [14].

Table 2. Grape composition at 2017 harvest in Generala, Cremosina and Monteolivo vineyards.

Vineyard	TSS (°Brix)	TA (g/L)	Total Ant (mg/g)	Total Phen (mg/g)
Generala	24.9	8.4	1.90	2.55
Cremosina	25.5	7.6	1.55	2.39
Monteolivo	26.1	7.8	1.62	2.49
Sig	n.s.	n.s.	n.s.	n.s.

*, $P \leq 0.05$; Within each column, means followed by different letters are significantly different ($P < 0.05$) according to the Student–Newman– Keuls test. ns, not significant.

4 Conclusions

The effectiveness of the results achieved lies in the methodology applied on the same grape variety (vitis vinifera L cv. Barbera) and under the same cultural practices (except for the clone and rootstock). We believe that, under these conditions, the viticulture profiles reported represent a very good approach to vineyard potentialities. The three tested vineyards have the following viticulture profile:

Generala: sandy-clay-loamy soils, alkaline, medium levels of total nitrogen and organic matter, annual growing cycle of 167 days (31 March-14 September), with a seasonal rainfall of 359.2 mm and heat accumulation of 1998 GDD (growing degree days); it is listed under the region IV in the Winkler scale. Due to its climate and soil characteristics, it presents an excellent concentration of total anthocyanins (1.90 mg/g) and phenolics (2.55 mg/g).

Cremosina: sandy-loamy soil, alkaline, with poor levels of total nitrogen and organic matter, annual growing cycle of 159 days (2 April-8 September) with a seasonal rainfall of 331 mm and heat accumulation of 1835 GDD (growing degree days); it is listed under the region III in the Winkler scale. Due to its climate and soil characteristics, it presents good levels of total anthocyanins (1.55 mg/g) and total phenolics (2.39 mg/g).

Monteolivo: loamy soil, alkaline, with poor levels of total nitrogen and organic matter, annual growing cycle of 155 days (2 April-4 September) with a seasonal rainfall of 331 mm and heat accumulation of 1835 GDD (growing degree days); it is listed under the region III in the Winkler scale. Due to its climate and soil characteristics, it present good levels of total anthocyanins (1.62 mg/g) and total phenolics (2.49 mg/g).

The three terroirs studied and classified as Nizza DOCG present differences in the vegetative growth, vine performance and grape quality related to the variation in the main terroir environmental factors (soil properties and meteorological conditions). On the basis of the results, we can conclude that Generala terroir in comparison with Cremosina and Monteolivo, presented the best viticulture features regarding the vine performance and grape composition. However, the interaction of the three terroirs with the variety needs to be deepened by further research, as well as correlation with wines (chemical and sensory profile). This study is the first step for Bersano winery to characterize its three terroirs inside Nizza DOCG.

5 References

1. C. Van Leeuwen, & G. Seguin, J Wine Res. *The Concept of Terroir in Viticulture*, **17**, 1-10 (2006)
2. OIV, Resolution OIV/Viti 333/2010, *Definition of vitivinicultural "Terroir"* (2010)
3. M. Fregoni, Phytoline, *Terroir, zonazione, viticoltura*, (2003)
4. E. Constantini, R. Barbetti, P. Bucelli, A. Cimato, E. Franchini, G. L'Abate, & P. Storchi, *Zonazione viticola ed olivicola della provincia di Siena. Colle val d'Elsa*, 224 (2006)
5. E. Vaudour, *Les terroirs viticoles. Définitions, caractérisation, protection*, (2003)
6. E. Vaudour & A. Shaw, SAJEV, *Worldwide Perspective on Viticultural Zoning*, **26(2)**, 106-115 (2005)
7. E. Vaudour, E. Costantini, G.V. Jones, S. Mocali. *An overview of the recent approaches to terroir functional modelling, footprinting and zoning*. Soil 1, 287-312 (2015)
8. D. Lorenz, K. Eichorn, H. Bleiholder, R. Klose, U. Meier & E. Weber, Aust. J. Grape Wine Res., *Phenological growth stages of the grapevine (Vitis vinifera L. spp. vinifera). Codes and descriptions according to the extended BBCH scale*, **1**, 100-103 (1995)
9. P. Iland, P. Dry, T. Proffitt & S. Tyerman, *The Grapevine from the science to the practice of growing vines for wine*, 32-62 (2011)
10. J. Gray & B.G. Coombe, Aust. J. Grape Wine Res. *Variation in Shiraz berry size originates before fruitset but harvest is a point of resynchronisation for berry development after flowering*, **15**, 156-165 (2009)
11. I. Lee, & B. Coombe, *Viticulture*, 2, 150-166 (2004)
12. K. Stoev & S. Dobreva, *Connaisance de la vigne et du vin*, 2(10), 125-139 (1976)
13. A. Carlomagno, A. Schubert & A. Ferrandino, *Volatiles in Vitis vinifera L. cv Barbera during ripening as influenced by growing location. IXe Congrès International des Terroirs vitivinicoles* 7, 29-32 (2012)
14. Regione Piemonte, *BARBERA Studio per la caratterizzazione del territorio, delle uve e dei vini dell'area di produzione del barbera d'Asti*, (2001)

Efecto combinado de cepa de levadura y *Terroir* en vinos Malbec de Mendoza

Combined effect of yeast strains and *Terroir* on Malbec wines from Mendoza

Dolores Pérez[1], Carolina Capaldi[2], Laura Mercado[3], Adriana Malizia[4] y Santiago Sari[1]

[1]Centro de Estudios de Enología, Estación Experimental Agropecuaria Mendoza, Instituto Nacional de Tecnología Agropecuaria (INTA), San Martín 3853,
5507 Luján de Cuyo, Mendoza, Argentina.
[2]Facultad de Ciencias Agrarias, Universidad Nacional de Cuyo, Almte. Brown 500, 5507, Luján de Cuyo, Mendoza, Argentina
[3]Laboratorio de Microbiología Enológica Estación Experimental Agropecuaria Mendoza, Instituto Nacional de Tecnología Agropecuaria (INTA), San Martín 3853, 5507 Luján de Cuyo, Mendoza, Argentina.
[4] LALLFERM S.A., Lateral Este de Acc. Sur 1751, Luján de Cuyo, Mendoza, Argentina.

Resumen. Las cepas de levaduras producen atributos distintivos en los vinos, como así también, la uva proveniente de diferentes zonas puede modificar el *flavour* del vino. El objetivo del presente trabajo fue evaluar el efecto producido por levaduras comerciales (EC1118, Tango Malbec, Flavia e Ionys) e indígenas sobre vinos Malbec provenientes de Ugarteche, El Cepillo y Gualtallary (Mendoza, Argentina). Se determinaron perfiles fermentativos, controles de implantación, parámetros físico-químicos y cromáticos generales y análisis sensorial de los vinos. Se aplicó ANAVA de dos vías con test LSD Fisher sobre los parámetros y se graficó un ACP. Las levaduras comerciales se implantaron en un 100%, mientras que los perfiles genéticos de las fermentaciones espontáneas fueron diferentes entre sí. El factor región, evidenció un importante efecto, destacándose los vinos de Ugarteche con altos niveles de pH y bajos valores de alcohol, intensidad colorante, IPT y antoncianos totales. El factor levadura reveló efectos significativos respecto a la levadura Ionys, la cual en las tres regiones, finalizó su fermentación dos días después que el resto de las levaduras y en sus vinos se destacó la nota a fruta fresca y la baja intensidad de astringencia y amargo. Sumado a esto, produjo 0.5 % menos grados de alcohol, 0.2 menos de pH.

Abstract. Yeast strains may impact on wine characteristics, producing distinct attributes. In addition, it is generally accepted, that grape must from different agro-ecological zones can impart on wine flavour. The aim of the present research was to evaluate the effects of commercial (EC1118, Tango Malbec, Flavia and Ionys) and indigenous yeast strains on chemical Malbec wines made from Ugarteche, El Cepillo and Gualtallary vineyards (Mendoza, Argentina). Fermentation profile, implantation controls, general chemical and chromatic wine parameters and sensory characterization were determined. Two-way ANOVA with LSD Fisher test and PCA were applied. In all commercial inoculated fermentations, implantation showed almost 100 % of similarity, while spontaneous fermentations exhibited different profiles. The region had a significant individual effect, where wines from Ugarteche region were characterized by lower alcohol degree, spectrophotometric detection at 280 nm, colour intensity and total anthocyanidins and huge pH values, in wines. The yeast factor presented significant effects mainly due to wines fermented by Ionys yeast strain, which its fermentation duration lasted two more days than the others yeast, but it produced outstanding effects on sensorial aspects of wines, favouring fresh-fruit notes and lowering bitterness and astringency. Furthermore, this yeast produced a reducing effect on pH and alcohol degree of about 0.2 points and 0.5 %, respectively.

1. Introducción

El cultivar Malbec (*Vitis vinifera* L), oriundo de Burdeos-Francia es considerado una variedad emblemática de la producción vitivinícola argentina; ya que permite lograr excelentes vinos apreciados nacional e internacionalmente. En la actualidad existen 40.401 ha de esta variedad en Argentina, de las cuales el 85,8% se encuentra en Mendoza [1]. La Zona Alta del Río Mendoza es la región vitícola más tradicional en el cultivo de este cepaje, ubicada en el pedemonte de la Cordillera de Los Andes, a 33° de latitud sur. Organolépticamente el Malbec se caracteriza por una alta intensidad colorante, con matices violetas y negros mientras que a nivel aromático, se pueden apreciar notas a ciruelas pasas, moras, violetas y anís como también algunas notas herbáceas [2].

La composición fenólica en las variedades de uva tinta es un parámetro cada vez más importante en la calidad del vino. Los compuestos fenólicos se clasifican en diferentes grupos, entre los cuales los antocianos y los flavonoides son los que tienen mayor influencia sobre el color, su evolución y otras características organolépticas relacionadas con su calidad [3].

Existen diversos factores que influyen sobre la acumulación de compuestos fenólicos en la uva y por lo tanto en el vino, uno de ellos es el factor edafoclimático. Factores ambientales como la luz y la temperatura, que caracterizan diferentes zonas de cultivo, inciden en forma directa en la actividad fotosintética de la planta y sobre las enzimas de la ruta fenilpropanoide modificando la biosíntesis de estos compuestos [4–6].

El término francés "*Terroir*" se refiere a la relación entre un determinado vino y una zona específica de producción del mismo, la cual le provee características y atributos únicos con respecto a otras zonas [7]. Dentro de este concepto, también se encuentra la microbiota nativa, la cual se ha demostrado que puede ser característica en una determinada zona vitícola [8].

Por otro lado, el pH, ya sea en el mosto o en el vino, es otro aspecto importante a tener en cuenta en relación al efecto sobre los antocianos. Estos pigmentos principales responsables del color de los vinos se encuentran en forma libre y en equilibrio constante entre diferentes formas químicas en función del pH, intensificándose las tonalidades rojas y aumentando el color total cuando los valores de pH son bajos [3]. Sumado a esto, la influencia de las levaduras que intervienen en la fermentación de los mostos sobre el color final del vino, es uno de los factores que ha sido estudiado más recientemente. Algunas cepas de levaduras debido a una desviación en su ruta metabólica durante la fermentación alcohólica, pueden producir un efecto acidificante sobre el mosto en fermentación [9, 10] produciendo sobre el color del vino el efecto arriba mencionado. Además, los bajos pH favorecen el fenómeno de copigmentación, típico en vinos jóvenes, en el cuál a través de uniones no covalentes tipo "sándwich" entre antocianos y copigmentos se favorece la estabilidad del color en el tiempo como también se produce una intensificación del mismo [11]. Sin embargo, se ha postulado que las levaduras vínicas también pueden actuar eliminando antocianos, adsorbiéndolos sobre su pared y también hidrolizando los heterósidos de antocianos por la actividad β-glucosidasa, dando lugar a agliconas más inestables [3]. En resumen, según la cepa de levadura utilizada pueden variar tanto el color del vino como la concentración de antocianos, por lo que, la influencia sobre el color es uno de los criterios primordiales en la selección de levaduras para la vinificación en tinto.

El aroma es otro importante parámetro de calidad en el vino. Debido a que es el resultado de diversos procesos biológicos, químicos y tecnológicos que ocurren desde la uva hasta el vino, se lo clasifica según su origen en: aroma primario o varietal, secundario o fermentativo y terciario o de crianza [12]. El aroma de origen varietal está constituido por sustancias volátiles (aromáticas) y no volátiles (no aromáticas) generadas durante el desarrollo de la baya y que contribuyen a la diferenciación varietal. Por lo tanto, dependen de las rutas metabólicas de la vid y consecuentemente están mayormente influenciados por factores que afecten a la planta [13]. Sin embargo, los procesos tecnológicos y biológicos aplicados durante la obtención del vino también pueden influenciar su contenido, por ejemplo, ciertas levaduras hidrolizan las formas ligadas de estos compuestos y los transforman en aromáticos [14], por otro lado, se ha determinado que las bajas temperaturas de fermentación favorecen la retención de estos aromas [15].

Los compuestos aromáticos de origen fermentativo provienen de dos procesos biológicos, la fermentación alcohólica, llevada a cabo principalmente por levaduras del género *Saccharomyces cerevisiae* y la fermentación maloláctica, llevada a cabo por bacterias lácticas. Es el grupo aromático más abundante y es el responsable, entre otras, de la nota vinosa. Su composición y concentración dependen de factores que impactan sobre el metabolismo de los microorganismos implicados en el proceso fermentativo, tales como temperatura de fermentación, nutrición, cepa de levadura, composición del mosto, etc.[15]. Se ha demostrado también, que existen diferencias en la producción de metabolitos secundarios entre cepas de levaduras, lo que influye en las características químicas y organolépticas del vino [9, 16].

A partir de lo anteriormente expuesto, resulta fundamental el contribuir al conocimiento del efecto individual y de interacción entre levaduras y zonas vitícolas de Mendoza sobre los parámetros relacionados con la calidad de los vinos Malbec.

2. Materiales y métodos

2.1. Materia Prima

Se utilizaron uvas de viñedos pertenecientes a la bodega Doña Paula ubicados en tres zonas diferentes de la provincia de Mendoza, cuyas características se detallan en la siguiente **Tabla 1**.

2.2. Microorganismos

Para llevar a cabo la fermentación alcohólica se utilizaron las siguientes levaduras secas activas provistas por la empresa Lalferm S.A.: Lalvin EC1118 (*Saccharomyces cerevisiae bayanus*), Ionys (*Saccharomyces cerevisiae),* Tango Malbec (*Saccharomyces cerevisiae)* y Flavia (*Metschnikowia pulcherrima)*. Los tratamientos con Flavia, 24 horas después de su inoculación, fueron secuencialmente implantados con EC1118. Paralelamente, como testigo, se realizaron fermentaciones espontaneas a partir de la flora nativa contenida en la uva de cada zona.

Tabla 1. Características agro-ecológicas de los viñedos.

Localidad	Sistema de Conducción	Latitud	Longitud	Altitud (m)	Suelo	Riego	Rendimiento (Kg/ha)
Ugarteche, Luján de Cuyo	Espaldero	33° 15' 00"	68° 57' 38"	1012	Franco-Arenoso	Goteo	8000
El Cepillo, San Carlos	Espaldero	33° 49' 13,6"	69° 11' 51,7"	1117	Franco-Arenoso	Goteo	6500
Gualtallary, Tupungato	Espaldero	33° 23' 13,07"	69° 13' 52"	1293	Franco-Arenoso	Goteo	6000

2.3. Protocolo de vinificación

Por cada región, se descobajaron y molieron 20 kg de uva colocados en recipientes plásticos de 25 L de capacidad, esto se realizó por triplicado. Se analizaron los parámetros básicos de rutina en todos los mostos (°Brix, pH, acidez total) y luego fueron adicionados con solución acuosa de 3 g/hL de metabisulfito de potasio y de ser necesario, se corrigió la acidez a 6 g/L con ácido tartárico.

Se aplicó un diseño multifactorial con dos factores (Zonas y Levaduras), con 3 niveles para el primer factor y 5 niveles para el segundo factor, basados en la cantidad de zonas y las levaduras utilizadas.

La hidratación y aclimatación de las levaduras comerciales se realizó según protocolo brindado por el fabricante. Las fermentaciones se llevaron a cabo a una temperatura controlada de 25°C y el seguimiento diario de las mismas se realizó a través del control de la densidad del mosto (g/cm^3) con un periodo de maceración para todos igual correspondiente a 10 días. La nutrición se aplicó en dos momentos, al principio de la FA y al llegar a una densidad igual a 1,070 g/cm^3.

Finalizando la fermentación alcohólica se tomó una muestra de cada repetición para llevar a cabo un control de implantación de las mismas, seguidamente, se inocularon bacterias *Oenococcus oeni* Lalvin VP41 (Lalferm S.A). Finalizado el período de maceración se realizó el descube de los tratamientos continuando con la fermentación maloláctica (20 °C). Una vez finalizados los procesos fermentativos, se adicionó anhídrido sulfuroso hasta 35 mg/L y se procedió con las estabilizaciones pertinentes a una temperatura aproximada de 4-8 °C, durante 2 meses. El envasado se realizó en botellas de vidrio de 750mL de capacidad con tapa a rosca, previo trasiego.

2.4. Determinaciones analíticas

Las siguientes determinaciones analíticas se realizaron en mosto y vino según protocolos de la Organización Internacional de la Viña y el Vino [17, 18]: sólidos solubles por refractrometría, acidez total, acidez volátil, pH, azúcares reductores, grado alcohólico por método aerométrico. Los parámetros relacionados con el color se determinaron siguiendo lo descripto por Zamora, 2003: índice de polifenoles totales a 280nm, antocianos totales, intensidad colorante. El control de implantación de las cepas de levaduras se determinó mediante la técnica de diferenciación intraespecífica, PCR Interdelta (Lallemand, Canadá). La caracterización sensorial se realizó mediante un análisis descriptivo no estructurado mediante panel entrenado (15 personas) perteneciente al Centro de Estudios en Enología, EEA Mendoza INTA.

2.5. Tratamiento de datos y análisis estadístico

Primeramente se verificó la normalidad de los datos, procesados por el software estadístico Infostat®. Se aplicó análisis de la varianza multifactorial y Test LSD Fisher para obtener una comparación de las medias buscando diferencias significativas entre los diferentes tratamientos y evaluar el efecto de los factores con un nivel de significancia de $p \leq 0,05$. Además, se les aplicó un análisis multivariado, utilizando como método de reducción de información, el Análisis de Componentes Principales (ACP).

Tabla 2. Datos iniciales de los mostos de las diferentes zonas vitícolas

Variable *	El Cepillo	Gualtallary	Ugarteche
°Brix	$23,4 \pm 0,26^a$	$24,65 \pm 0,56^c$	$24,3 \pm 0,1^b$
pH	$3,52 \pm 0,02^b$	$3,66 \pm 0,02^c$	$3,42 \pm 0,02^a$
Acidez Total	$5,25 \pm 0,36^b$	$4,35 \pm 0,17^a$	$5,86 \pm 0,1^c$

* valores promedios ± desviación estándar, letras diferentes indican diferencias estadísticas con respecto al test LSD Fisher para un p-valor<0,05.

3. Resultados y discusión

Con respecto a los datos iniciales de los mostos de uva Malbec pertenecientes a cada zona (**Tabla 2**) se pudo observar que se diferenciaron ampliamente mostrando una marcada influencia de la zona de origen sobre la uva. Gualtallary alcanzó un mayor grado de madurez y mayor nivel de pH con un consecuente nivel de acidez total inferior al resto. En cuanto a las características de El Cepillo, se destacó su menor nivel azucarino pero niveles intermedios de pH y acidez respecto al resto, siendo Ugarteche el mosto con mayores niveles de acidez y pH pero de madurez intermedia. Este último resultado llama la atención si se considera que este viñedo se encuentra en una zona de menor altura y más cálida entre las estudiadas en este trabajo, por lo que se esperaban mostos con menor acidez [5]. Sin embargo, a pesar de los resultados del análisis estadístico, estas diferencias entre zonas pueden ser consideradas de poca magnitud e impacto a nivel enológico.

Tabla 3. Duración promedio de las fermentaciones y valores promedios ± desviación estándar de los análisis finales de los vinos obtenidos de cada región geográfica y levadura.

Zona	Levaduras	Duración de la FA (días)	pH	Ac.Total (g/L)	Ac.Volátil (g/L)	Alcohol (%v/v)
El Cepillo	EC1118	8	$3,72 \pm 0,01^{bcd}$	$4,84 \pm 0,05^{cd}$	$0,33 \pm 0,07^{bcd}$	$13,95 \pm 0,00^{e}$
	Flavia	8	$3,68 \pm 0,01^{bc}$	$4,48 \pm 0,56^{abc}$	$0,25 \pm 0,00^{abc}$	$13,73 \pm 0,18^{cd}$
	Tango Malbec	8	$3,73 \pm 0,01^{cde}$	$4,54 \pm 0,05^{bcd}$	$0,35 \pm 0,10^{cde}$	$13,9 \pm 0,00^{de}$
	Ionys	10	$3,59 \pm 0,02^{a}$	$5,55 \pm 0,21^{e}$	$0,18 \pm 0,02^{a}$	$13,4 \pm 0,07^{ab}$
Gualtallary	EC1118	8	$3,86 \pm 0,02^{g}$	$4,50 \pm 0,11^{abc}$	$0,34 \pm 0,03^{bcd}$	$14,83 \pm 0,11^{g}$
	Flavia	9	$3,76 \pm 0,04^{def}$	$4,87 \pm 0,00^{cd}$	$0,35 \pm 0,04^{cde}$	$14,78 \pm 0,04^{g}$
	Tango Malbec	8	$3,80 \pm 0,01^{efg}$	$4,76 \pm 0,37^{cd}$	$0,45 \pm 0,06^{e}$	$14,75 \pm 0,00^{g}$
	Ionys	11	$3,65 \pm 0,01^{ab}$	$5,55 \pm 0,21^{e}$	$0,24 \pm 0,05^{ab}$	$14,23 \pm 0,11^{f}$
Ugarteche	EC1118	8	$4,02 \pm 0,03^{hi}$	$4,05 \pm 0,00^{a}$	$0,40 \pm 0,00^{de}$	$14,00 \pm 0,07^{e}$
	Flavia	8	$3,97 \pm 0,04^{h}$	$4,47 \pm 0,05^{abc}$	$0,25 \pm 0,03^{abc}$	$13,58 \pm 0,04^{bc}$
	Tango Malbec	7	$4,07 \pm 0,04^{i}$	$4,13 \pm 0,11^{ab}$	$0,28 \pm 0,07^{abc}$	$13,98 \pm 0,11^{e}$
	Ionys	11	$3,83 \pm 0,07^{fg}$	$5,00 \pm 0,04^{d}$	$0,18 \pm 0,02^{a}$	$13,38 \pm 0,04^{a}$
	Zona (Z)		***	**	**	***
	Levadura (L)		***	***	**	***
	Z x L		*	*	*	*

Letras diferentes indican diferencias estadísticas de la aplicación del análisis de la varianza con test LSD Fisher, p-valor<0,05. *** p<0,001; ** p<0,05; * p>0,05.

El control de implantación reveló que en cada tratamiento las fermentaciones se realizaron exitosamente con las cepas inoculadas, resultando un nivel de implantación cercano al 100% en todos los casos (datos no mostrados). Con respecto a las fermentaciones espontáneas se observaron diferentes perfiles genéticos demostrando que las cepas nativas difieren entre y dentro de una misma zona [19] y que en ningún caso estos perfiles coincidieron con las cepas comerciales. A partir de este resultado se decidió no tomar en cuenta los tratamientos de fermentaciones espontáneas en el análisis factorial debido a que el factor levadura no era el mismo entre repeticiones y zonas.

Tabla 4. Intensidad colorante (IC), índice de polifenoles totales a 280nm (IPT) y antocianos totales (AT) de los vinos de tres zonas geográficas fermentados por diferentes levadaduras.

Zona	Levaduras	IC	IPT	AT (mg/L)
El Cepillo	EC1118	$1,44 \pm 0,04^{ab}$	$86,85 \pm 1,63^{d}$	$918,09 \pm 32,48^{bc}$
	Flavia	$1,65 \pm 0,15^{cd}$	$85,2 \pm 13,86^{d}$	$985,69 \pm 64,97^{cde}$
	Tango Malbec	$1,85 \pm 0,12^{e}$	$94,3 \pm 5,94^{d}$	$1051,09 \pm 45,48^{def}$
	Ionys	$1,99 \pm 0,03^{ef}$	$83,9 \pm 7,78^{cd}$	$924,66 \pm 31,86^{bcd}$
Gualtallary	EC1118	$1,58 \pm 0,11^{bc}$	$112,6 \pm 0,57^{e}$	$1052,19 \pm 9,28^{ef}$
	Flavia	$1,65 \pm 0,03^{cd}$	$92,85 \pm 9,4^{d}$	$1170,75 \pm 37,12^{f}$
	Tango Malbec	$1,83 \pm 0,08^{de}$	$113,65 \pm 10,11^{e}$	$1173,59 \pm 20,73^{f}$
	Ionys	$2,1 \pm 0,08^{f}$	$82,26 \pm 5,06^{bcd}$	$1029 \pm 158,39^{cde}$
Ugarteche	EC1118	$1,43 \pm 0,07^{ab}$	$69,35 \pm 5,87^{abc}$	$833,74 \pm 8,85^{ab}$
	Flavia	$1,37 \pm 0,12^{a}$	$55,65 \pm 0,49^{a}$	$903 \pm 63,73^{bc}$
	Tango Malbec	$1,47 \pm 0,09^{abc}$	$68,05 \pm 6,15^{ab}$	$913,72 \pm 33,72^{bc}$
	Ionys	$1,5 \pm 0,08^{abc}$	$54,6 \pm 5,09^{a}$	$764,97 \pm 16,4^{a}$
	Zona (Z)	***	***	***
	Levadura (L)	***	*	**
	Z x L	**	**	*

Valores promedios ± desviación estandar, letras diferentes indican diferencias estadísticas de la aplicación del análisis de la varianza con test LSD Fisher, p-valor<0,05. ***p<0,001; **p<0,05; *p>0,05.

La levadura Ionys se destacó primeramente con respecto a la duración promedio de la fermentación produciendo fermentaciones más lentas [10], las cuales finalizaron dos o tres días después de las demás cepas (**Tabla 3**). El análisis de dos vías reveló que no hubo efecto interactivo entre levadura y zona (L x Z; p>0,05) sobre ninguno de los parámetros medidos. Sin embargo, se observó un significativo efecto individual de ambos

 https://doi.org/10.1051/e3sconf/20185002005

factores.

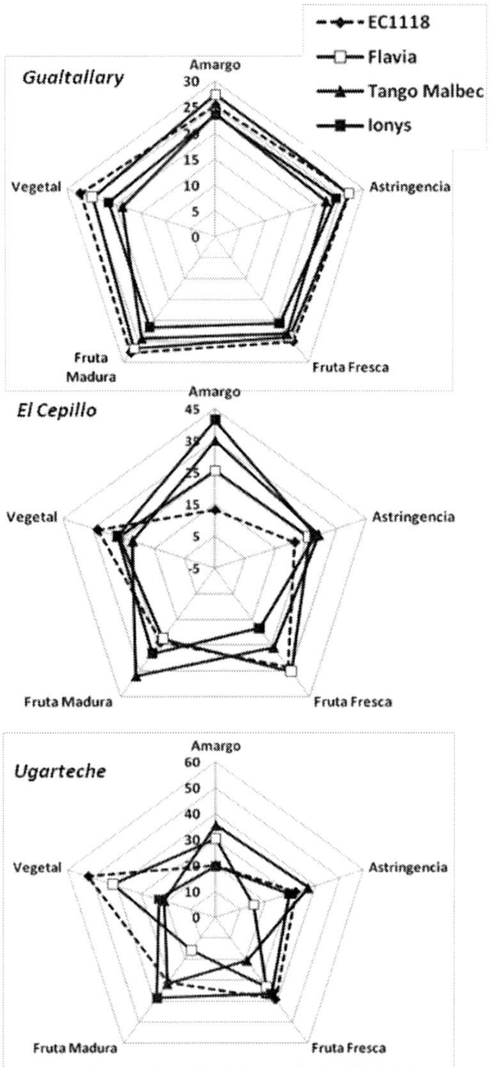

Figura 1. Media de las intensidades percibidas por los degustadores en el análisis sensorial descriptivo de los vinos Malbec de tres zonas fermentados con diferentes levaduras.

En el caso de la levadura, pH, la acidez total y el grado alcohólico fueron fuertemente afectados por las diferentes cepas [9, 10]. En mayor medida este efecto se observó en la cepa Ionys, y en menor medida en la cepa Flavia, las cuales produjeron un aumento de la acidez total, disminución del pH y del grado alcohólico. Por otro lado, debido a las diferencias detalladas en los mostos al principio de esta sección, los vinos de El Cepillo alcanzaron menores niveles de alcohol debido a su menor grado de madurez, sin embargo, los vinos de Ugarteche presentaron mayores niveles de pH a pesar de que sus mostos presentaban menores valores de este parámetro. Esto puede deberse al mayor contenido de potasio que poseían los mostos de Ugarteche (dato no mostrado). En cuanto a los valores de acidez volátil, Ionys produjo los menores niveles.

En la **Tabla 4** se detallan las características cromáticas y de composición polifenólica determinadas en los vinos. En primer lugar, el efecto de la levadura (L) fue muy significativo ($p<0,001$) en IC donde los mayores niveles fueron alcanzados por Ionys seguida de Tango Malbec. Este dato es interesante debido a que se sabe que la mayor intensidad colorante en el caso de Ionys es debido a su bajo pH [3], sin embargo no fue el caso de Tango Malbec, cuya mayor intensidad se debió al mayor contenido de antocianos libres y consecuente mayor nivel de IPT, aunque no siempre significativo, en cada zona. La levadura EC1118 alcanzó los menores niveles de estos parámetros y Flavia con un nivel un poco superior a ésta, no logró diferenciarse de la misma. En el caso del efecto producido por las zonas sobre estos parámetros se observó un efecto significativo ($p<0,001$), destacándose la zona de Ugarteche la cual se caracterizó por menores niveles de color, antocianos e IPT, debido principalmente sus condiciones agroecológicas, menor altitud y mayor rendimiento del viñedo. La levadura Ionys mostró en general un menor contenido de antocianos e IPT debido a que por el bajo pH y menor grado alcohólico de los vinos se ve favorecido el fenómeno de copigmentación eliminando del medio compuestos antocianos en forma libre y brindandole al vino mayor estabilidad en el tiempo [11].

Sensorialmente el panel de degustadores evaluó durante 3 sesiones 3 atributos aromáticos (vegetal, fruta fresca y fruta madura) y 2 atributos en boca (astringencia y amargo). El análisis estadístico no arrojó diferencias estadísticas como tampoco se observa una tipificación de las cepas en las tres zonas. Sin embargo, la levadura EC1118 se caracterizó a nivel aromático por la mayor intensidad de nota vegetal en todas las zonas (**Fig. 1**). Los vinos de Gualtallary, resultaron con perfiles muy similares entre sí con respecto a las levaduras, mientras que en Altamira y Ugarteche se observó un mayor efecto sensorial producido por éstas.

Con el objetivo de observar claramente la incidencia de los factores levadura y zona sobre todos los parámetros medidos, se realizó un análisis de componentes principales (ACP, **Fig. 2**), permitiendo explicar el 73% de la varianza existente entre los datos. En principio se observa un fuerte efecto de la zona, permitiendo una agrupación de los tratamientos según el origen de las uvas. Con respecto a la CP1, en el lado positivo de la misma se ubicaron los tratamientos correspondientes a las uvas de Gualtallary, caracterizados por la mayor intensidad de color, grado alcohólico, nota vegetal, fruta madura y astringencia. El tratamiento Ionys-G se separa con respecto a las demás levaduras acercándose a la mayor acidez total, mayor intensidad de color. Este comportamiento de Ionys se observa en las demás zonas, en las cuales se separa de las otras levaduras marcando un efecto sobre el vino, relacionándose con la nota de fruta fresca, acidez total, IPT e mayor intensidad colorante.

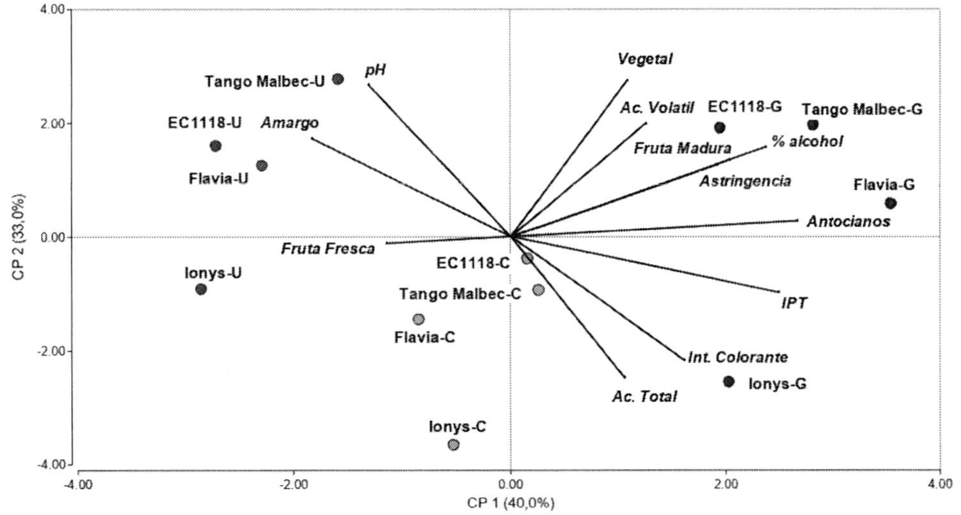

Figura 2. Análisis de componentes principales de los parámetros medidos en los vinos de la variedad Malbec provenientes de 3 regiones (C: El Cepillo, U: Ugarteche, G: Gualtallary) fermentados con 4 levaduras (EC1118, Ionys, Tango Malbec, Flavia).

4. Conclusiones

El objetivos del presente trabajo fue observar el efecto de las levaduras y de las zonas sobre el vino Malbec. Como conclusión, se puede afirmar que algunos aspectos del vino, principalmente los relacionados con los polifenoles, se encuentran principalmente influenciados por la región de origen de la uva (*Terroir*), mientras que ciertos parámetros químicos y sensoriales son determinados por la cepa de levadura utilizada.

5. Bibliografía

[1] DPTO. ESTADÍSTICA Y ESTUDIOS DE MERCADO SUBGERENCIA DE ESTADÍSTICA Y ASUNTOS TÉCNICOS INTERNACIONALES. *Informe variedad Malbec* [online]. 2017. Dostupné z: http://www.inv.gov.ar/inv_contenidos/pdf/estadisticas/tespeciales/2017/Malbec_2017.pdf

[2] CATANIA, C. a S. AVAGNINA. Malbec. In: *Curso Superior de Degustación de Vinos*. 2007, s. 1–11.

[3] ZAMORA, Fernando. *Elaboración y crianza del vino tinto: Aspectos científicos y prácticos*. Madrid: Mundi-Prensa, 2003.

[4] BERLI, F, J D'ANGELO, B CAVAGNARO, R BOTTINI, R WUILLOUD a M F SILVA. Phenolic composition in grape (Vitis vinifera L. cv. Malbec) ripened with different solar UV-B radiation levels by capillary zone electrophoresis. *Journal of Agricultural and Food Chemistry* [online]. 2008, **56**(9), 2892–2898. ISSN 00218561. Dostupné z: doi:10.1021/jf073421+

[5] FANZONE, Martín. *Caracterización de la composición fenólica de uvas y vinos de la variedad Malbec (Vitis vinifera L.): su relación con el origen geográfico, factores vitivinícolas y valor comercial*. B.m., 2012. Universitat Rovira I Virgili, Tarragona.

[6] COSTA, Elisa, João F. DA SILVA, Fernanda COSME a António M. JORDÃO. Adaptability of some French red grape varieties cultivated at two different Portuguese terroirs: Comparative analysis with two Portuguese red grape varieties using physicochemical and phenolic parameters. *Food Research International* [online]. 2015, **78**, 302–312. ISSN 09639969. Dostupné z: doi:10.1016/j.foodres.2015.09.029

[7] FORONI, Francesco, Miriam VIGNANDO, Marilena AIELLO, Valentina PARMA, Maurizio Guido PAOLETTI, Andrea SQUARTINI a Raffaella I. RUMIATI. The smell of terroir! Olfactory discrimination between wines of different grape variety and different terroir. *Food Quality and Preference* [online]. 2017, **58**, 18–23. ISSN 09503293. Dostupné z: doi:10.1016/j.foodqual.2016.12.012

[8] DEL MÓNACO, Silvana M., María E. RODRÍGUEZ a Christian A. LOPES. Pichia kudriavzevii as a representative yeast of North Patagonian winemaking terroir. *International Journal of Food Microbiology* [online]. 2016, **230**, 31–39. ISSN 18793460. Dostupné z: doi:10.1016/j.ijfoodmicro.2016.04.017

[9] TILLOY, Valentin, Anne ORTIZ-JULIEN a Sylvie DEQUIN. Reduction of Ethanol Yield and Improvement of Glycerol Formation by Adaptive Evolution of the Wine Yeast Saccharomyces cerevisiae under Hyperosmotic Conditions. *Applied and Environmental Microbiology* [online]. 2014, **80**(8), 2623–2632. Dostupné z: doi:10.1128/AEM.03710-13

[10] ENGLEZOS, Vasileios, Francesco CRAVERO, Fabrizio TORCHIO, Kalliopi RANTSIOU, Anne ORTIZ-JULIEN, Milena LAMBRI, Vincenzo GERBI, Luca ROLLE a Luca

COCOLIN. Oxygen availability and strain combination modulate yeast growth dynamics in mixed culture fermentations of grape must with Starmerella bacillaris and Saccharomyces cerevisiae. *Food Microbiology* [online]. 2018, **69**, 179–188. ISSN 0740-0020. Dostupné z: doi:10.1016/j.fm.2017.08.007

[11] BOULTON, Roger B. the Copigmentation of Anthocynins and Its Role in the Color of Red Wine: Comments on a Critical Review. *American Journal of Enology and Viticulture*. 2001, **52**(2), 67–87. ISSN 00029254.

[12] BAYONOVE, C., R. BAUMES, J. CROUZET a Z. GÜNATA. Aromas. In: A. MADRID VICENTE, ed. *Enología : fundamentos científicos y tecnológicos / coord. por Claude Flanzy*. 2° Ed. Madrid: Mundiprensa, 2003, s. 137–176.

[13] ROBINSON, Anthony L., Paul K. BOSS, Peter S. SOLOMON, Robert D. TRENGOVE, Hildegarde HEYMANN a Susan E. EBELER. Origins of grape and wine aroma. Part 1. Chemical components and viticultural impacts. *American Journal of Enology and Viticulture*. 2014, **65**, 1–24.

[14] HERNÁNDEZ-ORTE, P., M. CERSOSIMO, N. LOSCOS, J. CACHO, E. GARCIA-MORUNO a V. FERREIRA. The development of varietal aroma from non-floral grapes by yeasts of different genera. *Food Chemistry*. 2008, **107**(3), 1064–1077.

[15] PÉREZ, Dolores. *Efecto de la temperatura de fermentación y de la nutrición con amonio sobre la cinética de fermentación y el perfil aromático de vinos cv . Torrontés Riojano*. Mendoza, Argentina, 2016. Facultad de Ciencias Agrarias, UNCuyo.

[16] GAMERO, A. *Study of the production and release of aromas during winemaking carried out by different Saccharomyces species and hybrids*. Valencia, España, 2011. Universidad Politécnica de Valencia.

[17] OIV. *Compendium of international methods of wine and must analysis. Vol. 1*. Paris: OIV, 2013.

[18] OIV. *Compendium of international methods of wine and must analysis. Vol. 2*. Paris: OIV, 2013.

[19] MERCADO, L., A. DALCERO, R. MASUELLI a M. COMBINA. Diversity of Saccharomyces strains on grapes and winery surfaces: Analysis of their contribution to fermentative flora of Malbec wine from Mendoza (Argentina) during two consecutive years. *Food Microbiology* [online]. 2007, **24**(4), 403–412. ISSN 07400020. Dostupné z: doi:10.1016/j.fm.2006.06.005

Upward shifts in elevation – a winning strategy for mountain viticulture in the context of climate change?

Lukas Egarter Vigl[1,a], Arno Schmid[2], Franz Moser[3], Andrea Balotti[1], Erwin Gartner[4], Hermann Katz[3], Siegfried Quendler[4], Stefania Ventura[2] and Barbara Raifer[2]

[1]Eurac Research, Viale Druso 1, I-39100 Bolzano, Italy
[2]Laimburg Research Center, Laimburg 6, I-39040 Ora, Italy
[3]Joanneum Research, Leonhardstraße 59, A-8010 Graz, Austria
[4]Obst u. Weinbauzentrum Kärnten, Schulstraße 9, A-9433 St. Andrä, Austria

Abstract. The advent of global climate change has major impacts upon viticultural production. Changes in the spatial limits of wine production are already being observed around the globe; vineyards are now viable at higher elevations and more polar latitudes. Climatic conditions are also threatening production in existing appellations. Therefore, sound management strategies are vital to maintain high-quality wines and varietal typicity, and to respond to changing market conditions. In mountainous regions such as the European Alps, new production areas at higher elevations are increasingly considered to be a promising solution. However, the suitability of viticulture in general, and even specific varieties of wine grapes, can change drastically across short distances in complex mountain terrain. Variations in temperature and radiation accumulation directly influence plant suitability, yield quantity, and quality. This paper shares initial findings from the REBECKA Project, a transnational research initiative designed to assess the impacts of climate change on mountain viticulture and wine quality in South Tyrol (Italy) and Carinthia (Austria). A three-part approach is utilized to better assess these dynamics: (1) historical crop yield data from local vineyards are assessed, (2) plant phenology stages and polyphenolic compounds of the Pinot Noir variety are analyzed along an elevation gradient and related to bioclimatic indices, and (3) a suitability map is developed that considers small-scale topographic and agro-environmental conditions. Taken together, these components contribute in clarifying many of the opportunities and threats facing high altitude viticulture in a changing world and provide new insights for sound decision-making in alpine vineyards.

1 Introduction

Global climate change is proceeding at an unprecedented rate. According to the Intergovernmental Panel on Climate Change (IPCC), the past century has seen the global average temperature rise by more then 1 °C and all climatic models agree that drastic temperature increases will continue to occur through 2100 [1]. These temperature changes are likely to affect agro-economic activities, including grape growing, both temporally and spatially [2]. In fact, over the past years the land area devoted to viticulture at the cooler end of the suitability spectrum has increased globally [3], reflecting the trend towards a warmer climate. While most research on the subject focuses on the latitudinal shifts towards the poles (i.e. in southern UK, Scandinavia, northern Germany) and its consequences for wine production, only little attention has been given so far to the emerging wine producing areas in mountainous terrain.

Recent vineyard plantings in the Alpine wine growing region of Alto Adige, however, are more frequently located at higher altitudes than in the past (Fig. 1), confronting traditional viticulture practices with a number of new challenges: harsher environmental conditions, shortened growing seasons, reduced yields and poorer soil structures. Understanding and assessing the opportunities and risks

 https://doi.org/10.1051/e3sconf/20185002006

grape growers are likely to face in these regions is thus crucial for sustainable, quality-oriented viticulture [4].

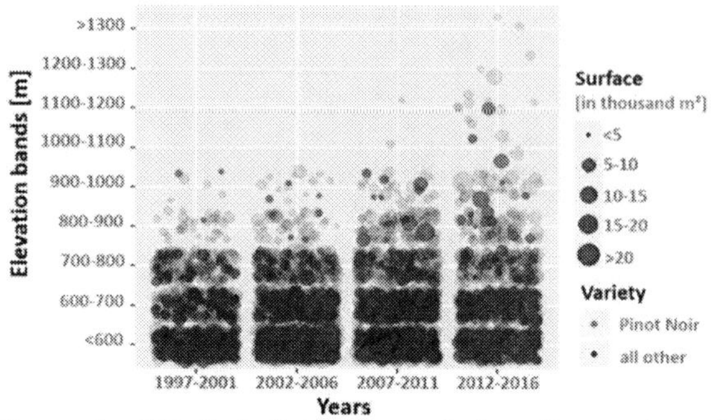

Figure 1. Altitudinal distribution and size of new vineyard plantings in the Alto Adige wine growing region (IT) since 1997. Pinot Noir variety superimposed in red.

In this work, we present some preliminary results of the transnational EU-funded "REBECKA" project (Interreg V-A IT-AT: ITAT1002, duration: 2017-2019) that has been specifically designed to address the impacts of climate change on the mountain wine growing regions of Alto Adige (Italy) and Carinthia (Austria). The project is composed of three main parts. First, historic yield data from different wineries are analyzed. Herein, the main focus is given to the assessment of spatiotemporal shifts in the dates of harvest, and in the yield quantity and quality (mainly sugar and acidic content). Second, plant phenology stages (bud burst, flowering, ripening, harvest) and polyphenolic compounds are analyzed across different elevation ranges, covering warm, intermediate, cold and extreme sites. Finally, a spatially explicit land suitability map is developed, based on bioclimatic indices that characterize the wine growing potential of the study sites. The work presented in this paper, however, is limited to the Alto Adige study site (Fig.2).

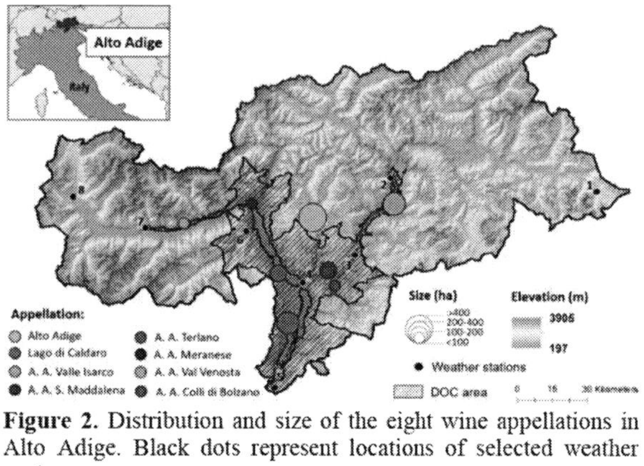

Figure 2. Distribution and size of the eight wine appellations in Alto Adige. Black dots represent locations of selected weather stations.

2 Material and Methods

2.1 Study area

The Alto Adige wine region is located in the Central Alps and covers an area of approximately 5400 ha (Fig. 2). It is situated between an Alpine climatic zone and a Mediterranean one, with vineyards usually growing between 200 m and 1100 m above sea level (a.s.l.). Some extreme sites reach 1300 m a.s.l. The current viticultural area includes both red and white cultivars (40% and 60%, respectively). The main varieties are Pinot Grigio, Pinot Blanc, Chardonnay and Pinot Noir, along with a few local varieties such as Schiava (Vernatsch), Lagrein and Gewürztraminer. Currently, the South Tyrolean wine growing region is divided into eight appellations (federally designated regions, called "controlled designations of origin – DOC"): Kalterersee, Eisacktaler, St. Magdalener, Terlaner, Meraner, Vinschgau, Bozner Leiten and Südtiroler.

2.2 Yield data from wineries

Historic yield records from five important local wineries (Tramin, Nals, Eppan, Kurtatsch, and Eisacktal) across Alto Adige have been collected and statistically analyzed for quantitative and qualitative parameters at the land parcel level. The main analyses included the assessment of spatiotemporal changes in yield quantity, sugar and acidic content and harvest dates in relation to topo-climatic site conditions. In total, the past 20 years of wine production have been analyzed, covering approximately 1000 ha of land (20% of total Alto Adige wine growing surface).

2.3 Phenological data and grape maturity records

At thirty different vineyard plots, the phenological stages (bud burst, flowering, ripening and harvest) were assessed over the vegetation period (April 1st to October 31st). These sites have been selected in order to cover all eight wine appellations and most of the variety in topo-climatic characteristics (i.e. different altitudinal and topographical patterns) of the study area. Pinot Noir was selected as the reference variety for the field surveys as it is cultivated over the entire altitudinal gradient, including vineyards located between 250 m and 1100 m a.s.l. Moreover, at each site, maturity tests were performed that analyzed the polyphenolic compounds (i.e. pH-value, sugar and acidic content) of the grapes at specific points in time. Finally, both plant-growth data and polyphenolic compounds were related to bioclimatic indices, in order to estimate their reliability for predicting plant growth and grape quality compounds.

2.4 Climate data and zoning approach

A 20-year time series of daily climatological observations from a network of 80 different stations has been used to calculate bioclimatic indices. These indices are commonly used to characterize wine-growing regions worldwide in a standardized way and as a suitability-zoning tool, as there is a significant relationship between plant growth and heat accumulation [5]. A multiple linear regression model has been applied to spatialize the climatological data into a high-resolution grid, using topographic parameters such as elevation, slope, aspect and potential solar radiation as predicting variables. Finally, climate maturity groupings were built based on growing degree days [6] to characterize wine-growing suitability at a parcel scale.

3 Results and discussion

The Alto Adige wine-growing region presents great diversity, both in terms of topo-climatic conditions and cultivated varieties. The cultivars include both white and red varieties and are spread across an altitudinal gradient of approximately 1000m. This results in large differences between single production zones, including cool, intermediate and warm regions [7]. Moreover, over the past 50 years, the South Tyrol wine region has registered a temperature increase of more than 1°C, directly affecting crop growth by advancing phenological stages and by lengthening the overall growing season (Tab. 1).

Table 1. Climate records of selected weather stations in Alto Adige: locations, elevation, time periods, trend over the period of record (POR) and current growing season average temperature (GST, for April-October).

Location	Elevation [m a.s.l.]	Time Period	Trend [°C, POR]	Current GST [°C]
Sesto	1310	1956-2016	1.0	12.4
Varna	590	1956-2016	1.2	17.4
Barbiano	490	1977-2016	0.8	17.5
Bolzano	254	1984-2016	1.5	19.4
Salorno	212	1977-2016	1.5	19.6
Pavicolo	1200	1972-2016	1.4	12.8
Silandro	698	1956-2016	1.0	16.6
Monte Maria	1310	1967-2016	0.9	13.4
Average		48 years	1.2	16.1

For some crops, like wine grapes, a warmer climate can have a positive impact by extending their climatic niche into areas formerly considered to be too cold. As an increase in surface temperature often directly triggers an advancement of plant phenology, earlier dates of fruit pick would be expected. The historic yield records from five different wineries, however, depict that harvest dates have not considerably advanced during the overall period of study (1995-2016). In the first decade of the time series, an advancement in harvest dates is visible that coincides with increased grape maturity levels (sugar content). This trend, however, is counterbalanced by the second decade of records, where fruits are picked later and sugar levels flatten out (Fig. 3). Producers most likely take advantage of warmer climate conditions to harvest fruits at a greater level of ripeness. Thus, changes in plant phenology stages due to warmer climatic conditions do not necessarily entail a change in harvest dates.

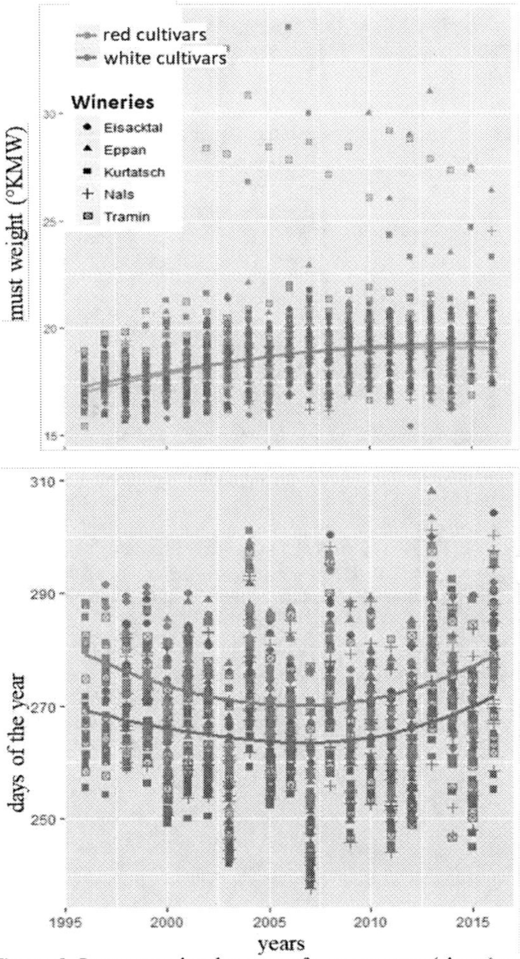

Figure 3. Long-term development of sugar content (above) and harvesting dates (below). Colors represent red and white cultivars, while the symbols depict the five different wineries under study.

The preliminary analysis of phenological data at the 30 different vineyard plots confirmed a strong relationship between the vegetation cycle and climate properties, especially in the beginning and in the end of the growing season. In fact, budburst events of 2017 could be directly related to altitude, which is used as a proxy for the different climatic site conditions in this study (Fig. 4). This strong dependency, however, decreases once budburst occurs and the growing season progresses. Favored temperature regimes after budburst most likely allowed Pinot Noir grapes to develop similarly along the entire altitudinal gradient, as only minor variations in the length of flowering and ripening stages across plots could be observed. Towards the end of the growing season, elevation is decisive in regulating grape maturity levels and the correlation coefficients between plant growth and elevation become significant once again. Thus, our first field observations suggest that vineyards at the cooler limits of suitability experience an increased exposure to climate extremes (i.e. spring or autumn frosts) because decisive phenological stages for quality wine production (i.e. harvesting) are directly related to climatic site conditions during the early- and late-stages of the growing season, exactly when these extremes are most likely to occur,

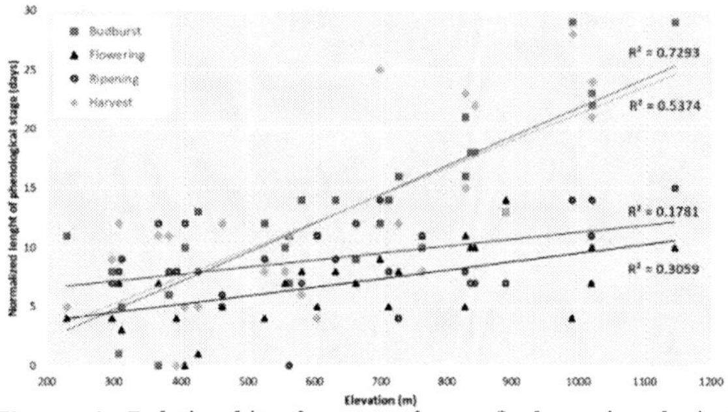

Figure 4. Relationships between dates of plant phenological stages and elevation. Dates are normalized to the first vineyard plot entering the respective phenological stage. Elevation is used as a proxy for climate site condition.

Moreover, for each vineyard block, grape compounds were compared to different bioclimatic indices to estimate index performance in predicting chemical properties (Fig. 5). Specifically, the Winkler index, Huglin Index, Gladstone index, Cool Night Index and Growing Season Average Temperature were compared with grape compounds such as sugar content, pH-value, potassium concentration, and malic, tartaric and total acid. Overall, no single index could be identified that explains all variables in a statistically significant manner. However, except for the sugar levels, heat-summation-based indices performed best in predicating chemical grape compounds. Of these, the Winkler index most consistently fits the data. Thus, it has been selected as the most representative index for use in climate suitability mapping.

The mountainous Alto Adige region contains many vineyards near the cooler limits of wine grape cultivation. On the Winkler index (WI), these areas mostly range between classes I and II (Fig. 6), making this region especially suited to white and early ripening varieties (i.e. Pinot Noir). However, the warmest locations in the region, such as the lowlands around Bolzano, receive temperature sums in excess of 1700 °C, qualifying them as region III areas and for the production of high-quality red cultivars. These different classes occur in relatively close proximity due to the huge altitudinal gradients in the region. The coldest climate zones (WI Ia) in Alto Adige theoretically reach up to 1300 m. However, the current limit for quality wine production actually lies at approximately 1100 m, due to other environmental factors such as more frequent weather extremes [8]. In such extreme conditions, mostly hybrid varietals such as Solaris are cultivated. The cool climate region WI Ib is characterized by elevations up to 1100 m and early ripening cultivars such as Müller Thurgau, as well as Chardonnay and Pinot Noir for sparkling wine production.

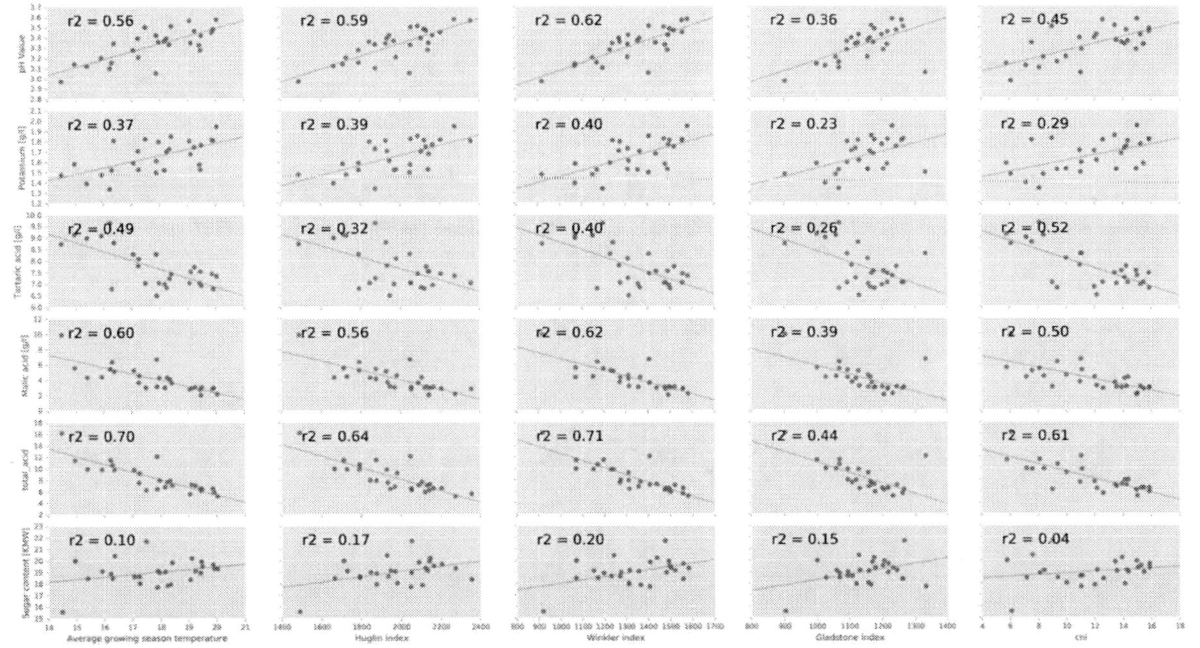

Figure 5. Relationships between different bioclimatic indices and wine grape compounds of Pinot Noir cultivar for the year 2017. Regression lines and correlation coefficient superimposed.

Winkler index region II areas are suited for varieties like Pinot Gris, Pinot Noir, Gewürztraminer and Chardonnay.The final Winkler region found in South Tyrol is classified as region III and extends upward to approximately 450 m in the sunniest and warmest areas. These sites are characterized by varieties like Schiava, Lagrein, Merlot and Sauvignon Blanc.

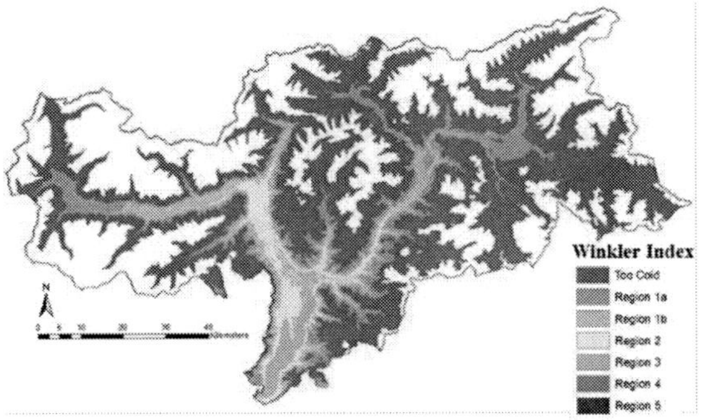

Figure 6. Heat accumulation groupings using the Winkler bioclimatic index based on an averaged temperature map of the period 1991-2010.

4 Conclusion

As increasing temperatures associated with global climate change continue to be felt across the landscape, opportunities for wine grape cultivation are expanding into areas that were previously unsuitable. In particular, areas at extreme latitudes and elevations (i.e. mountainous regions) are experiencing increases in wine grape production. There is therefore a need to analyze changing viticulture suitability for growers and decision makers. Increasing temperatures alone, however, cannot be directly linked to enhanced suitability of mountain viticulture. Such trends suggest enhanced potentials of cool climate viticulture, but also critically mask the additional impact of the more highly variable climate regimes predicted by climate models. In fact, the inter-annual variability and extreme weather events that already threaten productivity in mountain areas are likely to become more frequent and intense. Thus, those who choose to cultivate these newly suitable areas face an increased risk of reduced overall production and catastrophic weather events. While warmer climates might open new opportunities for some varieties, for others, especially existing autochthonous cultivars, higher temperatures come along with insuperable challenges for physiological plant growth. Farmers must therefore be aware of these changing environmental conditions in order to meet market demands and consumer expectations for specific products. Moreover, the economic investments associated with opening new vineyard land must be carefully considered in this context. This is especially difficult in newly suitable viticulture areas, where farmers may lack expertise and information about wine cultivation, and take unnecessary monetary risks. As climate change alters mountainous regions, understanding the risks and opportunities it presents is vital to maintaining a socio-ecologically sustainable viticulture system.

5 References

1. IPCC. Fifth Assessment Report of the Intergovernmental Panel on Climate Change. Synthesis Report – Summary for Policymakers. Geneva, Switzerland, 151 pp, (2014).
2. A. Nesbitt, B. Kemp, C. Steele, A. Lovett, S. Dorling. Impact of recent climate change and weather variability on the viability of UK viticulture – combining weather and climate records with producers' perspectives. *Aust J Grape Wine Res* **22**, 324-335 (2016).
3. G.V. Jones. Climate, grapes, and wine: structure and suitability in a changing climate. *Acta Hort* **931**,19-28 (2012).
4. L. Egarter Vigl, E. Tasser. Die Weinbau-Zonierung in Südtirol. *SZOW* **8**, 6-9 (2017).
5. J. Tonietto, A. Carbonneau. A multicriteria climatic classification system for grape-growing regions worldwide. *Agric For Meteorol* **124**, 81–97 (2004).
6. M. Amerine, A.J. Winkler. Composition and quality of musts and wines of California grapes. *Hilgardia* **15**: 493–675 (1944).
7. L. Egarter Vigl, E. Tasser, S. Williams, U. Tappeiner Defining suitable zones for viticulture on the basis of landform and environmental characteristics: a case study from the South Tyrolean Alps. *Fifth International Congress on Mountain and Steep Slope Viticulture, Congegliano. Italy* (2017).
8. R. Zorer, D. Rocchini, M. Metz, L. Delucchi, F. Zottele, F. Meggio, M Neteler. Daily MODIS land surface temperature data for the analysis of the heat requirements of grapevine varieties. *IEEE Transactions on Geos* (2013).

How can remote sensing techniques help monitoring the vine and maximize the terroir potential?

Cecile Tondriaux[1], Anne Costard[1], Corinne Bertin[1], Sylvie Duthoit[1], Jérôme Hourdel[2] and Jacques Rousseau[2]

[1]TerraNIS SAS, 10 Avenue de l'Europe, 31 520 Ramonville, France
[2]ICV, La Jasse de Maurin, 34970 LATTES, France

Abstract. In each winegrowing region, the winegrower tries to value its terroir and the oenologists do their best to produce the best wine. Thanks to new remote sensing techniques, it is possible to implement a segmentation of the vineyard according to the qualitative potential of the vine stocks and make the most of each terroir to improve wine quality. High resolution satellite images are processed in several spectral bands and algorithms set-up specifically for the Oenoview service allow to estimate vine vigour and a heterogeneity index that, used together, directly reflect the vineyard oenological potential. This service is used in different terroirs in France (Burgundy, Languedoc, Bordeaux, Anjou) and in other countries (Chile, Spain, Hungary and China). From this experience, we will show how remote sensing can help managing vine and wine production in all covered terroirs. Depending on the winegrowing region and its specificities, its use and results present some differences and similarities that we will highlight. We will give an overview of the method used, the advantage of implementing field intra-or inter-selection and how to optimize the use of amendment and sampling strategy as well as how to anticipate the whole vineyard management.

1 Precision viticulture: why and for whom

In a context of growing competition on international markets, it becomes of utmost importance to help winegrowers improve profitability and maximize wine quality. The use of precision viticulture is a step in this direction. Multi-spectral satellite and aerial images are frequently used to estimate vegetation indices or biophysical parameters linked to vegetation status (biomass, nutrient deficiencies, water stress status, or health status) [1], thus allowing to map intra and inter-field spatial variability [2]. The consideration of this particularly high variability within vineyards is crucial for a better management on all aspects: technical to maximize agronomic performances, economical to optimize profitability and environmental to limit the impact of cultural practices.

Nowadays, four concrete main uses of precision viticulture can be highlighted; all answering wineries or winemakers needs:

- Selective harvesting : an allotment based on vine vigor and heterogeneity indicators is performed to gather parcels or part of parcels with similar organoleptic properties. The efficiency of the vine field selection is improved to achieve specific production goals.
- Reduction of fertilization costs : fertilizer inputs are adjusted according to the vigor areas delineated inside each vine plot, leading to optimized and rational fertilization scheme.
- Adaptation of cultural practices to each vintage with the aim of homogenizing vine plots.
- Optimize sampling strategy to guide field observations such as maturity controls, yield estimation or sampling to ease the technician work.

In France, several operational solutions exist and are based on multispectral UAV, aerial or satellite imagery. Most of these solutions use vegetation indices such as NDVI (Normalize difference vegetation index) to reflect vine vigor (such as Fruition Analytics, EarthLab & Millesime), excepted Œnoview® which generates vine biophysical parameters such as FCover (Fraction of green cover vegetation).

 https://doi.org/10.1051/e3sconf/20185002007

2 The Œnoview® service

2.1 Purpose and scope of the Œnoview® service

Œnoview® is a comprehensive precision viticulture service, available worldwide, helping winegrowers improve profitability and wine quality. The Œnoview® service is based on the analyses of satellite imagery associated with an expertise in viticulture and winemaking. Its development has been initialized in 2007 by Airbus Defence & Space and the ICV group; Since 2014, Œnoview® is commercialized by TerraNIS, which brings expertise in satellite imagery analyze and supply of value-added products, in partnership with the ICV group that brings its knowledge in viticulture and winemaking.

The Œnoview® cartography measures the variability of vine vigor through satellite imagery and the level of internal variability of each vineyard and vine plot. This information is then analyzed and formatted in order to help users make the good decisions. Zoned maps are used to improve agricultural practices such as field selection, fertilization or harvest. On a large scale, groups of vineyards presenting the same characteristics can be created to improve harvest organization.

Œnoview® has been sold since 2008 mainly in France but also in Morocco, Japan, China, Chile, Hungary and Spain. For the year 2016 for example, Œnoview® was sold in France for a total of 5500 hectares concerning around 290 wineries. In the perspective of addressing other countries, which is a key strategic goal for TerraNIS development, demonstrations are still going on in Italy and Greece particularly.

2.2 Methodology

The main steps of the process can be seen in Figure 1. Once information on the area(s) of interest has been provided by the customer (fields contours, grape variety, practices etc), at least one satellite image is acquired at the veraison stage, but more acquisitions can be envisaged if a vegetation monitoring is required. Œnoview® is based on optical satellite imagery with, most of the time, a spatial resolution less or equal to 1m50 (Spot6/7, Pleiades, …). Images -which embed several spectral bands from visible to near infra-red - are processed by the Overland software suite (developed and maintained by Airbus Defence & Space). It retrieves physical characteristics from spectral information (biophysical processing) using state-of-the-art reflectance models of vegetation (PROSPECT/SAIL model [3]), integrating the atmospheric correction of the satellite image (based on LOWTRAN model [4]) and using advanced numerical techniques to perform model inversion.

The retrieved biophysical parameters such as FCover (Fraction of green cover vegetation) which directly reflects vine vigor is known to be robust, without any bias due to agricultural practice or atmospheric conditions, and reproducible enough to allow the absolute characterization of a parcel, and thus the spacialization of agronomic advices.

Three to four weeks before harvest, vine vigor maps and corresponding statistical indicators are delivered in different kinds of formats according to user needs (GIS compatible format, kml, pdf, posters, …). A Smartphone application allows to display Œnoview® products anywhere and to enrich them with in-field reports. All along these steps, appropriate and customized recommendations and on-site visits are proposed.

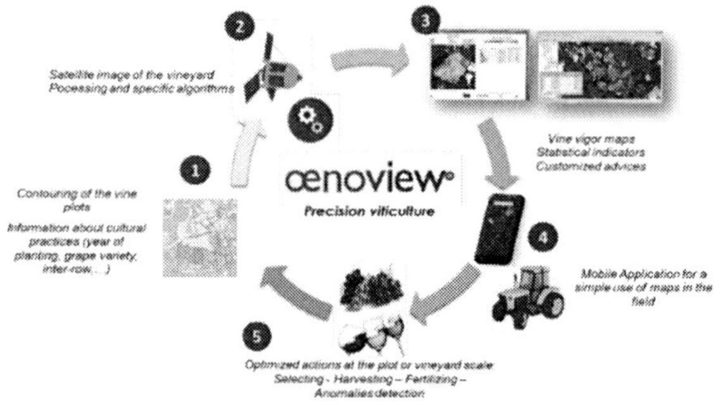

Figure 1 : Main steps of the Œnoview® service.

3.1 Main benefits for the wine industry

Œnoview® has a threefold benefit: decrease production costs by optimizing farming practices, reduce environmental impact by limiting chemical inputs and improve both quality and productivity of the vineyard.

If you are a winegrower, Œnoview® offers you the opportunity to improve the profitability of your vineyard thanks to an accurate mapping of each field. Image processing algorithms combined with the oenological know-how make possible to extract useful indicators for a better management of the harvest and an optimization of agricultural practices (fertilization, irrigation, health protection etc). Areas with anomalies (diseases or nutrient deficiencies for instance) can be early detected to make appropriate decisions.

If you manage a winery, Œnoview® provides you with an instantaneous and large-scale characterization of the qualitative potential of the vineyard parcels. An allotment based on oenological quality indicators is performed to gather parcels or part of parcels with similar organoleptic properties. You can then improve the efficiency of your vine parcel selection to achieve specific production goals.

The following section show examples of uses of Œnoview®.

https://doi.org/10.1051/e3sconf/20185002007

3 Concrete examples of uses

3.1 For a simple and reliable field selection

In the Languedoc winemaking region, where several big wineries or winemakers (*i.e* Vinovalie, Canet winery, Ouveillan winery, Les Grands Chais de France, ...) use Œnoview® since several years, one of its main uses is the characterization of the qualitative potential of the vineyard parcels for parcels selection. According to the winery production goals, an allotment based on vigor and heterogeneity indicators is performed, leading to an improvement of the efficiency of the field selection. In the example shown in Figure 2 :, fields with high vigor were selected to make Rosé wine; with adapted winemaking methods, homogeneous parcels with moderate vigor were selected to give full-bodied red wine and parcels with moderate vigor and high heterogeneity gave a fruity red wine. Two complementary red wine styles are produced and a regularity from one year to another is thus reached.

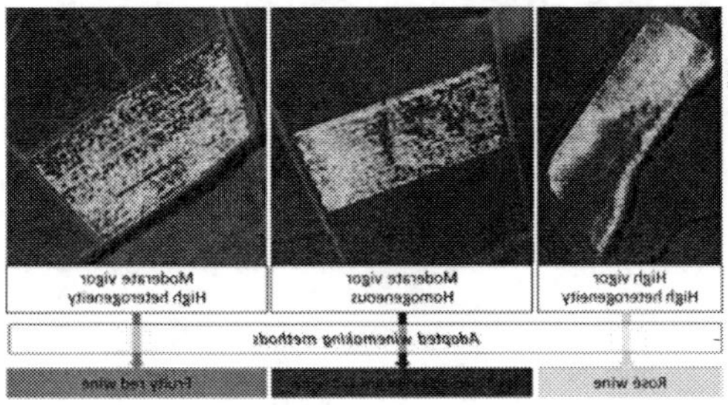

Figure 2 : Illustration of the use of the precision viticulture service Œnoview® in the Languedoc winemaking region (France) for field selection.

3.2 To reduce fertilization costs

The Grands Chais de France (GCF) group is one of the largest vineyard owners in the world. The group currently owns 33 wineries in many wine-growing areas of France and 1,500 hectares of vines. One of the main objective of GCF when choosing to map vine vigor with Œnoview® in 2016 and 2017 for an area of 600 hectares was to be thrifty and eco-aware when applying fertilizers and phytosanitary products, improving their profitability in the meantime. Adapted zoned vine vigor products with 4 to 8 zones per vine plot, associated with fertilizer spreaders capable of variable rate application, allowed them to set up a rational fertilization schemes (Figure 3). Even without such a sophisticated equipment attached to a tractor, it is possible to target fertilization input differentially, for example by speeding up or slowing down tractor speed locally. The profitability or the fertilization scheme set up in 2017 still has to be clearly evaluated but previous uses of such maps in the Bordeaux winemaking region (Châteaux Castel and Château Fayau for example) disclosed that 10 to 30% saving has been made on the fertilization costs.

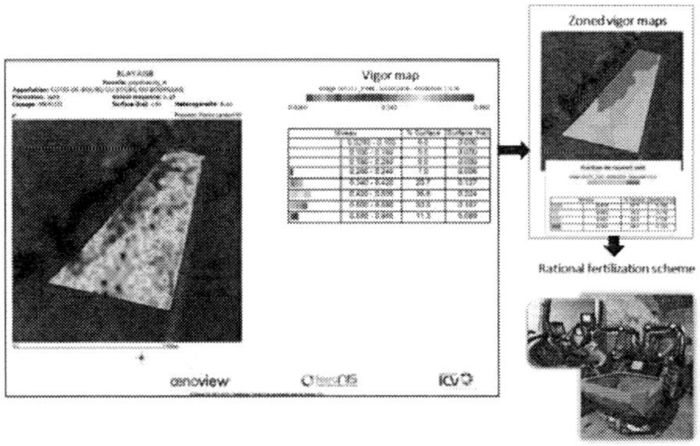

Figure 3 : Use of Œnoview®zoned vine vigor maps obtained from satellite image acquired at veraison stage in 2016 to set up a rational fertilization scheme (Grand Chais de France).

3.3 To monitor vineyard evolution

The « Château Fortia» is one of the oldest winery of Châteauneuf-du-Pape in the Rhone winemaking region. Over its 30 hectares, Fortia has always developed integrated vineyard practices and Œnoview® is used since several years as a decision support tool for diverse applications: parcel selection, vineyard restructuration scheme or reduction of intra-field variability for example. Moreover, the monitoring available (

Figure *4*) allowed them to understand the vintage effect, to quantify the impacts of changes in the agricultural practices and to adjust practices to each vintage.

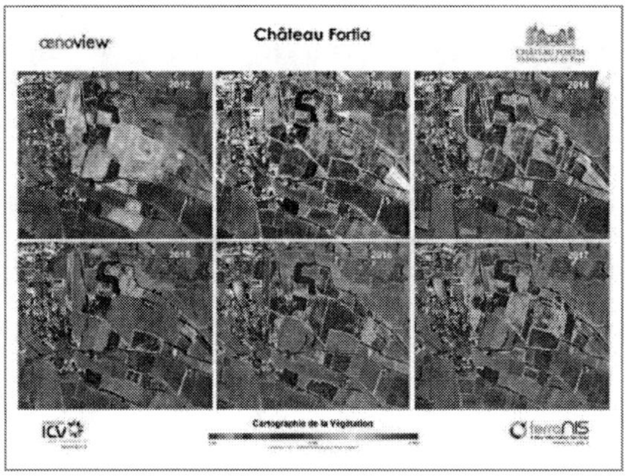

Figure 4 : Example of vineyard evolution monitoring with the precison viticulture service Œnoview® between 2012 and 2017 in the Rhone winemaking region (France).

3.4 As a global decision tool

In Chile, the use of such vigor maps is quite diverse. When big wineries (from 5.000 to 11.000 hectares) harvest all the vine mechanically and look for a tool to both detect anomalies and help them plan the harvest calendar at a plot level, others, smaller and dedicated to high quality wine, use the Oenoview maps combined with other measurement systems to detect at an early stage any kind of growth problem and early signs of water stress. They monitor the vegetation development thanks to several images, carefully analyse each area of heterogeneity at the pixel level, to then harvest all by hand, following a meticulous planning. Then, the maps are also used as a decision-making tool to plan their fertilization strategy, to define the quantity and spot the needier areas.

3.5 To improve management of sub-field and inter-field variability

With its high positive progression, the Apulia wine production has to cope with economic and environmental constraints. Improving wine quality and quantity while reducing environmental impacts and the use of chemical inputs is a key issue for the sustainability of Apulia vineyards. In this context, Œnoview®was proposed in Apulia by Planetek, thanks to the Eugenius platforms network implemented in the frame of the Eugenius H2020 Copernicus project. In 2017, a first experimental pilot service was conducted for Tormaresca vineyard in Salento, an Apulian vinery property of the Antinori family. More than 200 hectares were monitored using very high-resolution satellite images and agronomical in-situ dataset (grape variety and cultural practices). Several critical stages of the grape development were monitored, with a special focus during veraison stage (a few weeks before harvesting). Zoned maps were also produced to define homogeneous areas on which specific agricultural practices such as fertilization can be implemented. Finally, Œnoview® supported Tormaresca wine-growers in managing sub-field and inter-field variability of the vine vigor (Figure). Œnoview® customer found satellite monitoring the most accurate and affordable survey method to support the production of high quality wine.

Figure 5: Vine vigor maps displayed on the Smartphone application (left) and homogeneous areas for harvest optimization and wine quality improvement defined by Œnoview® (right) - Tormaresca Bocca di Lupo Vineyard in Minervino Murge (Apulia).

4 Conclusion and evolutions to come

Precision viticulture allows to estimate and map vine vegetation development through vegetation indices or biophysical parameters linked to vine vigor, highlighting then spatial variability within fields, but also at the entire vineyard scale. Concrete uses for wine cooperatives or wine growers are diverse but field selection, fertilization and vineyard monitoring are the main uses observed. The Oenoview® service, which was mainly proposed in France vineyards so far, is now addressing other countries. Customers' feedbacks, in France but also worldwide with the first conducted experimentations, are very positive (Figure), proving that Oenoview® can be fully operational even in a new climate and soil environment.

TerraNIS also tries to constantly improve its portfolio of services to answer vineyards needs. TerraNIS is then currently working with partners to detect and characterize water stress in the vineyard, which is an important issue for the wine industry in France but also worldwide. The use of remote sensing imagery is coupled with in-situ measurements using the IoT technology. The objective is to optimize water resources, bringing the right amount of water at the right place, thus ensuring vine health and wine quality.

https://doi.org/10.1051/e3sconf/20185002007

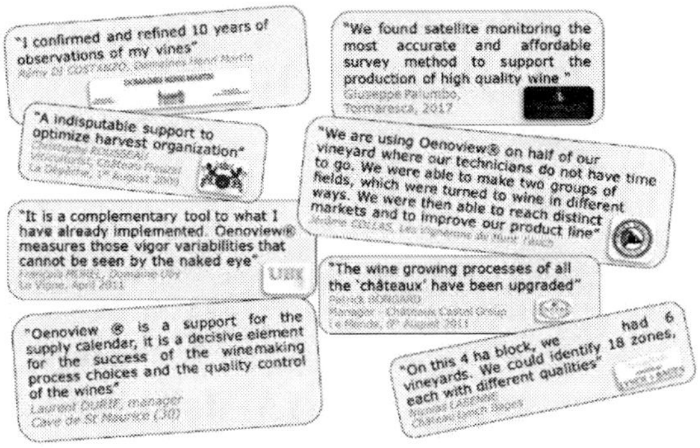

Figure 6: Examples of Oenoview Users' feedback

5 References

1. L.F. Johnson and al., 2003, J. Rousseau and al., 2008; C. Acevedo and al., 2007; D. Lamb and al., 2004
2. A. Hall et al., 2002, B.Tisseyre and al., 2007
3. PROSPECT/SAIL: S. Jacquemond and F. Baret, 1990; J.B. Feret et al, 2008; W. Verhoef, 1984
4. LOWTRAN : F. Kneizys and al, 1988

Una base de datos espacial integrada en un Sistema de Información Geográfica para la gestión del terroir: un nuevo sistema consistente e interactivo

Alberto Lázaro-López[1], María L. González-SanJosé[1] y Vicente D. Gómez-Miguel[2]

[1]*Dpto. Biotecnología y Ciencia de los Alimentos, Facultad de Ciencias, Universidad de Burgos. Burgos, España.*
[2]*Dpto. Producción Agraria (Edafología), ETSIAAB, Universidad Politécnica de Madrid, Madrid. España*

Abstract. Methodology of the Integrated Zoning of the Terroir involve the generation and management of a high volume of big size data that are interrelated conceptually and spatially, of which its analysis and storage is necessary. These actions are critical in the investigation process and, therefore, there is a special interest in an effectively and efficiently data management. The databases are widely used because of their capabilities on integrity, consistency, redundancy and independence of data management as a method for optimizing resources, so the main goal is to develop a spatial database that facilitates the access, management and analysis of the data of the ZIT. The result is TEZISdb, an acronym for Terroir Zoning Information Service Database, a multilocal and multiscalar modeling that allows to collect thematic and spatial data of all environmental factors from terroir as well as functions for its analysis.

Resumen. La Zonificación Integrada del Terroir (ZIT) y su proceso metodológico implican la generación y el manejo de un elevado volumen de datos de gran peso que se encuentran interrelacionados conceptual y espacialmente y el cuál es preciso almacenar y analizar. Estas acciones son críticas en la investigación y, por ello, existe especial interés en manejar los datos eficaz y eficientemente. Las bases de datos (BBDD) se emplean ampliamente por sus capacidades en integridad, consistencia, redundancia e independencia en la gestión de datos como método para la optimización de recursos, por lo que se plantea como objetivo el desarrollo de una BD espacial que facilite el acceso, manejo y análisis de los datos procedentes de la ZIT. El resultado es TEZISdb, acrónimo de Terroir Zoning Information Service Database, un modelado multilocalización y multiescalar que permite recoger datos temáticos y espaciales de todos los factores ambientales que conforman el terroir así como funciones para su tratamiento y análisis.

1. Introducción

La técnica de zonificación se utiliza frecuentemente, y cada vez con mayor precisión, como una herramienta adecuada para el conocimiento, descripción, delimitación y cartografía de los terroir. La zonificación fue definida en 2003 por Fregoni y col. [1] como "la investigación para el estudio del territorio con el fin de repartirlo en zonas relativamente homogéneas como interacción entre el viñedo y el ambiente" y permite trabajar sobre los elementos con influencia en el desarrollo de los viñedos y en la calidad del vino de forma conjunta.

De los distintos sistemas desarrollados para zonificar, actualmente se utilizan métodos que promueven el empleo de modelos complejos [2], que combinan mapas temáticos e información espacial de diferente índole, y que han sido utilizados con éxito en distintas DO españolas [3].

A través de esta metodología se estudia, cuantifica y cartografía un amplio rango de parámetros de los diferentes factores implicados en la definición del medio (clima, litología, geomorfología, suelo, vegetación y usos del suelo) junto con datos asociados al cultivo (variedad, portainjerto, distribución del viñedo, densidad de plantación, producción...). Su análisis permite seleccionar los parámetros y variables que se consideran de mayor influencia en la diferenciación zonal y constituye la Zonificación Ambiental que proporciona el mapa de las "Unidades Homogéneas del Medio" (UHM). Posteriormente, la relación de estas unidades con los productos (uva, mosto, vino) procedentes de ellas da lugar a la Zonificación Integrada del Terroir (ZIT) [4].

Durante todo este proceso se utilizan intensivamente datos procedentes de teledetección, como el Modelo Digital del Terreno (MDT o DEM en inglés) o imágenes multi-espectrales, de los que son derivados nuevos índices en formato de imágenes ráster de alta resolución espacial y espectral, así como datos georreferenciados expresados en geometrías (puntos, líneas o polígonos) con atributos asociados. Esto supone la generación y el manejo de un elevado volumen de datos de gran peso que se encuentran interrelacionados conceptual y espacialmente, de los cuales es preciso su análisis atendiendo a dicho aspecto para generar nueva información, así como su almacenamiento, con el objetivo de preservarlos y permitir que estén disponibles para futuras aplicaciones que permitan derivar nuevos usos y obtener nuevas conclusiones. Estas acciones son críticas y, por ello, existe especial interés en manejar los datos eficaz y eficientemente.

De forma generalizada, la manera de abordar esta situación ha sido mediante una gestión basada en un sistema de ficheros con archivos individuales y de diferentes tipos organizados manualmente. Sin embargo, esto lleva asociadas ciertas contrapartidas: la carencia de integridad y consistencia en los datos, al no poder asegurarse que un valor es coherente con el atributo al que representa ni que un mismo valor sea representado regularmente de la misma forma; la existencia de redundancia, al generarse duplicidad cuando se establecen relaciones; la definición de estructuras rígidas, que dificultan su actualización; la falta de independencia entre la estructura de los datos y las consultas que se quiera hacer sobre ellos; y la falta de optimización de estas consultas. Además, la interoperabilidad entre diversos tipos de archivos no siempre está garantizada, supone una capa añadida de complejidad y dificulta la sistematización de los flujos.

Alternativamente, a partir de la búsqueda de superación de estas dificultades surgen las bases de datos (BBDD) y los sistemas gestores de BBDD (SGBD), que aportan ventajas en integridad, consistencia y no redundancia de los datos; una gestión con una organización independiente del sistema de ficheros; y en el análisis, con búsquedas optimizadas y desligadas de la estructura [5]. Estas capacidades han hecho que su uso se haya extendido a multitud de campos y, en concreto, han sido ampliamente utilizadas para gestionar datos pertenecientes a recursos naturales y del medio como el suelo, la vegetación, meteorología y clima, hidrología, sanidad vegetal, etc [6]. Aun así, no ha sido posible reconocer ninguna referencia bibliográfica de BBDD específicas para el terroir.

2. Antecedentes

El concepto de BD se refiere a una colección de datos especialmente organizados para lograr búsquedas o disposiciones con posterioridad [7] y, para cumplir con estos requisitos, se desarrollan modelos de BBDD que pretenden alcanzarlos desde diferentes perspectivas. En unos sistemas prima la eficiencia, creando estructuras orientadas al modo de consulta. En 1970 se introdujo un nuevo sistema, el modelo relacional de BBDD, que centró su atención en la consistencia y la integridad de los datos sobre la eficiencia.

Según este concepto, las entidades físicas reales podían asimilarse a tablas relacionadas entre sí, para sostener, de esa manera, datos relativos a ellas. Tomando como base el modelo de BD relacional se desarrolló posteriormente el modelo objeto-relacional como una evolución que incorporaba conceptos del modelo orientado a objetos, como herencias y tipos de datos complejos y definidos por el usuario, aportando aún mayor flexibilidad en la construcción de estructuras y robustez [8].

Ejemplo de datos complejos son los datos espaciales, que reúnen en un único atributo información de geometría, sistemas de referencias y sistemas de coordenadas. Estos datos cuentan con un estándar de almacenamiento y acceso recogido en la norma ISO 19125, denominada *Simple Feature Access* (SFA) [9].

La definición de los datos espaciales como atributo abrió la posibilidad de diseñar bases de datos espaciales que reúnen en un mismo espacio de trabajo datos temáticos y georreferenciados y permiten su tratamiento conjunto. Esto supuso un nuevo paradigma que superaba a los modelos previos de separación de BD, para datos exclusivamente temáticos, y herramientas SIG, encargadas del tratamiento de los georreferenciados, que habían predominado desde el origen de las herramientas SIG en la década de 1960. Este hecho se ve reforzado por la mejora de los motores de los SGBD que soportan el modelo objeto-relacional, con la incorporación también de funciones con las que manipular los datos georreferenciados.

Esta transición queda patente, por ejemplo, en el crecimiento de lo que hoy día es el National Soil Information System (NASIS) [10] y los productos que incluye desarrollados por la National Soil Survey Cooperative en los EUA. Tiene sus inicios en unas herramientas SIG junto a una BBDD relacional y, en la actualidad, ha adoptado como herramienta para la gestión de datos una BBDD con un sistema de modelado objeto-relacional con soporte de datos georreferenciados [11]. De forma similar, se ha producido el desarrollo de otras BBDD espaciales para la gestión de recursos naturales [6].

Tomando en cuenta las consideraciones previas sobre la zonificación del terroir, que implica un elevado volumen de datos temáticos y georreferenciados combinados, y las ventajas asociadas a la utilización de BBDD, el empleo de una BBDD espacial se intuye como una herramienta crítica para el avance cualitativo en la investigación del terroir. Sin embargo, no se ha identificado ninguna referencia bibliográfica relacionada.

 https://doi.org/10.1051/e3sconf/20185002008

Por ello, se establece como objetivo general, facilitar la gestión y el acceso a los datos de la ZIT, mediante los objetivos particulares de desarrollar un modelado para todos los datos ligados a la zonificación del terroir, implementarlo en una BBDD y generar funciones de análisis para el terroir. Como método de comprobación y validación, el diseño será aplicado a la zonificación del terroir de la DO Campo de Borja.

3. Material y métodos

3.1. Fuente de los datos

La DO Campo de Borja comprende 65.296,20 ha localizadas en la provincia de Zaragoza, al noreste de la Península Ibérica y cuenta con 6.242,89 ha de viñedo, de las que 3.442,69 ha (2015) se dedican a Garnacha. Los datos utilizados proceden del proyecto de zonificación de la DO [4] y se centra en las UHM de la variedad Garnacha caracterizadas y cartografiadas durante el estudio [12].

3.2. La estructura de los datos

Para el desarrollo de una BD se ha seguido el estándar de arquitectura ANSI, que contempla tres fases progresivas: Conceptual, Lógica y Física.

Los requisitos generales establecidos durante la fase Conceptual contemplan que la BD sea interconectable con diversas aplicaciones, posibilitando tanto el acceso como la edición; extensible, permitiendo incorporar nuevos módulos de funcionalidad en el futuro; y de código libre, facilitando que los avances realizados puedan ser compartidos con la comunidad científica. Desde el punto de vista del contenido específico, el modelado para la zonificación del terroir debe aglutinar datos de parámetros provenientes de los diferentes factores ambientales del terroir, a saber, geología y geomorfología, suelo, clima, vegetación, uso del suelo.

De estos factores existen estructuras de datos ya conceptualizadas y en funcionamiento adaptadas a diferentes áreas, lo que haría de una BD para la zonificación del terroir una suerte de agrupación que integrase las BBDD de los factores. Esto ha servido de apoyo para la elaboración del diagrama ER y su posterior transformación a modelo lógico empleando un sistema objeto-relacional.

Se ha tomado como referencia para el suelo, geología y geomorfología la teoría expuesta en el Soil Survey Manual [13] [14] y las estructuras de SOTER [15] y NASIS [10]. Para la vegetación y uso del suelo se ha adoptado la nomenclatura y organización del Mapa Forestal de España y SIGPAC y el Catastro respectivamente [16].

Para la implementación física se ha utilizado PostgreSQL en su versión 10, un SGBD de tipo servidor-cliente, de código libre, que cumple con las propiedades ACID y que se integra con múltiples extensiones. Entre ellas PostGIS, una extensión para datos espaciales tanto vectorial como ráster que cumple con la norma SFA y cuenta con numerosas funciones de manipulación y consulta, empleada en su versión 2.4. Como apoyo para la visualización se ha empleado la aplicación SIG QGIS, también de código libre, con un amplio soporte para PostgreSQL.

Al encontrarse el desarrollo de la BD en fase de prototipo, no se ha establecido perfiles ni grupos de acceso específicos al no estar planteada actualmente su distribución.

4. Resultados

El resultado es una BD para la zonificación del terroir que ha sido denominado TEZISdb, acrónimo de Terroir Zoning Information Service Database. Se trata de un modelado multilocalización y multiescalar, extensible a cualquier región de interés y a cualquier escala que aloja datos temáticos y espaciales de todos los factores ambientales que conforman el terroir.

Se ha optado por diferenciar tantos esquemas como factores se han descrito, resultando en los esquemas de geología y geomorfología (geo) (Figura 1), suelo (suelos) (Figura 2), clima (clima), vegetación (vege) y usos (usos); esquemas auxiliares como base cartográfica (base) y un esquema propio para los resultados del estudio (terroir). Cada uno de estos esquemas agrupan tablas principales con soporte en tablas auxiliares e índices y vistas sobre dichas tablas.

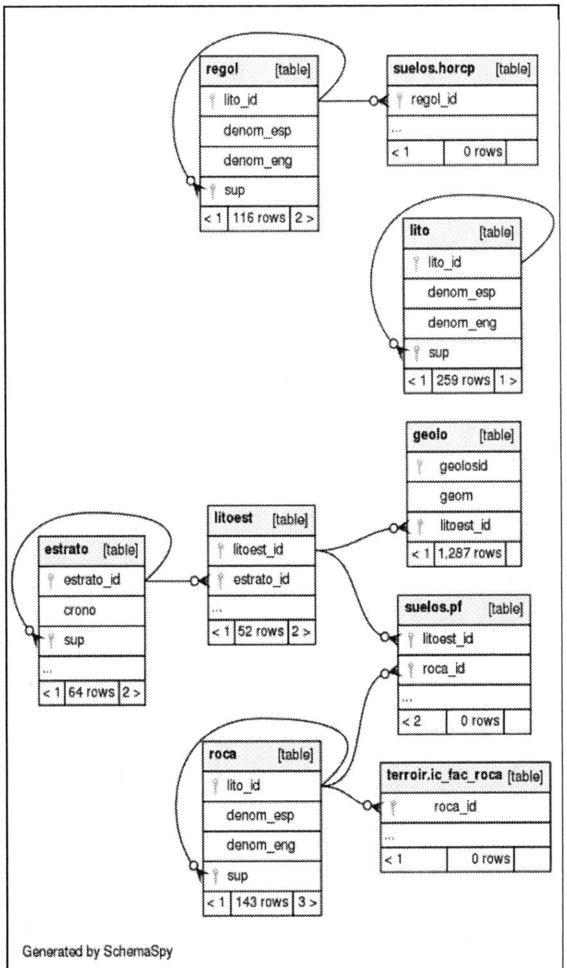

Figura 1. Modelado de las tablas principales del esquema para geología y geomorfología, "geo" (Generado mediante SchemaSpy)

Las principales funciones incluidas están relacionadas con el cálculo de índices, ya sean índices bioclimáticos, edáficos o varios que intervienen en la caracterización de las UHM o el Índice de Calidad del terroir. Dichas funciones de consulta y análisis de datos están desarrolladas principalmente bajo SQL, estando integradas en los esquemas.

La validación del modelado desarrollado a través de su aplicación a la zonificación del terroir de en la DO Campo de Borja ha resultado satisfactoria. Ha conseguido alojar todos los datos que componen el proyecto de zonificación, con una distribución de 11 tablas para el esquema "geo"; 26 en "clima"; 6 para "vege"; 8 en "usos"; 10 en "base", 9 en "terroir" y, por último, 58 tablas para "suelos", además de diferentes vistas y vistas materializadas. Es relevante que el tamaño final de la BD supone menos del 10% del total del conjunto de archivos que formaban el proyecto de zonificación original, considerando versiones, subversiones y uniones pertenecientes al flujo de trabajo.

También, durante la carga de los datos ha sido posible identificar y corregir ciertos fallos con valores que no cumplían las restricciones establecidas en el diseño, permitiendo verificar la integridad y consistencia de los datos de la zonificación en la DO Campo de Borja al final del proceso. Los fallos encontrados se centraban fundamentalmente en errores de transcripción y nomenclatura.

El uso de esta nueva herramienta y la metodología asociada ha conseguido reducir los tiempos de cálculos y procesos en el flujo de trabajo, lo que establece una nueva marca de eficiencia para la técnica de zonificación del terroir, dejando abierta la posibilidad de nuevas mejoras en el código que los redujeran aún más.

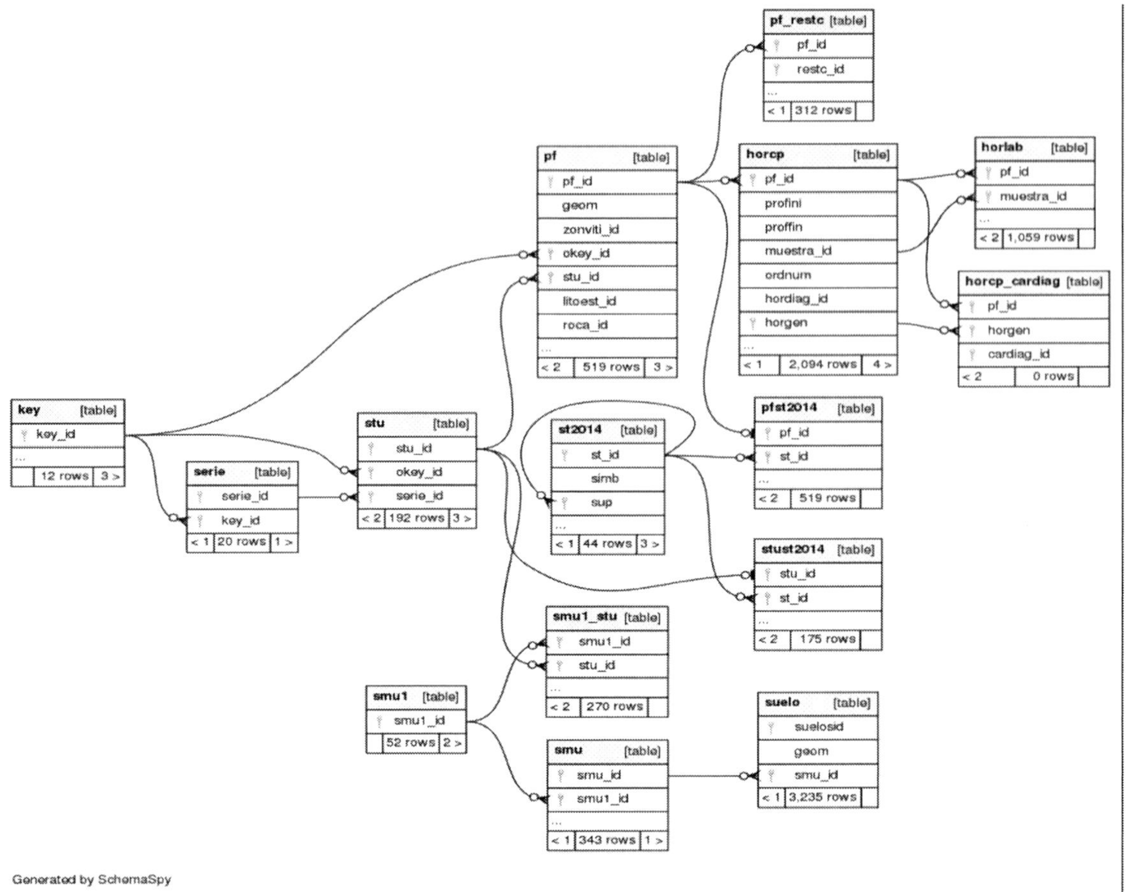

Figura 2. Modelado de las tablas principales del esquema "suelos" (Generado mediante SchemaSpy)

5. Conclusiones

Se ha diseñado un modelado de BD espacial para la ZIT denominado TEZISdb, acrónimo de Terroir Zoning Information Service Database, que permite recoger datos temáticos y espaciales de todos los factores ambientales que conforman el terroir, extensible a cualquier región de interés y escala. Ha sido desarrollado bajo un modelo objeto-relacional para el que se ha elegido como SGBD PostgreSQL con su extensión PostGIS, que lo habilita para la gestión y el análisis de datos espaciales.

Esta herramienta permite implantar un nuevo paradigma de flujo de trabajo con implicaciones sobre:

La integridad, consistencia y redundancia de los datos, al almacenar sólo los datos originales mientras se aplican restricciones y comprobaciones automáticas con lo que se evitan múltiples copias dispersas y se optimiza el espacio empleado.

La gestión y análisis de los datos, al centralizar el mantenimiento y control de acceso a los datos sobre el SGBD y desligando las consultas de las estructuras. Esto posibilita la conexión de múltiples aplicaciones que realizan consultas complejas y la definición de funciones de análisis de datos en diversos lenguajes que automatizan cálculos de la ZIT.

El modelo ha sido aplicado a la zonificación de la DO Campo de Borja como método de comprobación y validación, dando como resultado una BD TEZISdb funcional para esa región.

Si bien con la BD se logra una gestión y análisis de los datos más robusto y sistemático, la interacción con ella queda limitada a aquellos técnicos y científicos con conocimientos específicos en manejo de bases de datos y SIG, por lo que resulta interesante disponer de un sistema visual, accesible e intuitivo, que facilitase información actualizada y adaptada para cada perfil de usuario.

Una línea de desarrollo futuro contemplada es el diseño de una Infraestructura de Datos Espaciales (IDE), que sume, a la gestión de los datos con TEZISdb, una interfaz de interacción con usuarios (UI) y diferentes canales de comunicación vía telemática, creando una experiencia de acceso unificada a los datos del terroir. Actualmente se trabaja sobre una UI que se apoya en tecnologías disponibles en visualización de datos SIG, personalizándola para los datos del terroir: como consultas directas sobre las delineaciones del mapa para obtener información de las SMU, la composición de STU, la descripción de cada STU y los perfiles característicos definidos para cada STU. También la información conjunta de todos los factores para cada UHM.

6. Bibliografía

[1] M. Fregoni, D. Schuster, y A. Paoletti, *Terroir, zonazione, viticoltura : trattato internazionale*. Verona: Phytoline, 2003.

[2] OIV, «Resolución 423/2012: OIV Guidelines for vitiviniculture zoning methodologies on a soil and climate level». OIV/VITI, 22-jun-2012.

[3] V. D. Gómez-Miguel, «Inventarios del Recurso Suelo de las DO de Castilla y León. Parte I: Importancia, metodología y aplicaciones», *Innovación y Tecnología Agroalimentaria*, vol. 2, p. 14, nov-2007.

[4] V. D. Gómez-Miguel y V. Sotés, «Zonificación del Terroir: Estudio de Suelos y Ordenación del Cultivo de la Vid en la DO Campo de Borja (Zaragoza)», Universidad Politécnica de Madrid (UPM), Madrid, 2015.

[5] N. Matthew y R. Stones, *Beginning Databases with PostgreSQL: From Novice to Professional*, Second edition. 2005.

[6] P. M. Huang, Y. Li, y Sumner, Eds., *Handbook of Soil Sciences: Resource Management and Environmental Impacts*. CRC Press, 2011.

[7] C. J. Date, *The new relational database dictionary: a comprehensive glossary of the concepts arising in connection with the relational model of data, with definitions and illustrative examples*. Sebastopol, CA: O'Reilly Media, 2015.

[8] R. O. Obe y L. S. Hsu, *PostgreSQL: Up and Running*, Second edition. Beijing: O'Reilly, 2015.

[9] R. O. Obe y L. S. Hsu, *PostGIS in Action*, Second edition. Shelter Island, NY: Manning, 2015.

[10] Soil Survey Division Staff, «Chapter 1: Understanding NASIS», en *NASIS User Guide*, Version 6.1., USDA-Natural Resources Conservation Service, 2011, p. 1:19.

[11] D. E. Beaudette y A. T. O'Geen, «SoilWeb: 13 Years Later», ISRIC Spring School 2016.

[12] A. Lázaro-López, J. Cámara, Á. Martínez, V. D. Gómez-Miguel, y V. Sotés Ruiz, «Terroir Zoning in Appellation Campo de Borja (Northeast Spain): Homogeneous Terroir Units (HTU) and Terroir», presentado en International Terroir Congress, 11th, McMinnville, OR (USA), 2016, p. 452:457.

[13] Soil Survey Division Staff, *Soil Survey Manual*. Washington, D.C.: U.S. Government Printing Office, 2017.

[14] A. Van Wambeke y T. Forbes, *Guidelines for using Soil Taxonomy in the names of soil map units*. Soil Conservation Service. US Department of Agriculture (USDA), 1986.

[15] V. W. P. van Engelen y J. A. Dijkshoorn, *Global and National Soils and Terrain Digital Databases (SOTER)*, 2.0. ISRIC. World Soil Infomation, 2013.

[16] Instituto Geográfico Nacional, «Geoportal IDEE». [En línea]. Disponible en: http://idee.es/. [Accedido: 18-abr-2018].

Modelling the Terroir of the Douro Demarcated Region, Portugal

Helder Fraga[1], Ricardo Costa[1] and João Santos[1]

[1]Centre for the Research and Technology of Agro-Environmental and Biological Sciences, CITAB, Universidade de Trás-os-Montes e Alto Douro, UTAD, 5000-801 Vila Real, Portugal, hfraga@utad.pt

Abstract: The Douro Demarcated Region (DDR) is one of the main winemaking regions in Portugal. Viticulture is tied to the Douro region, having a strong impact on the local economy. This mountainous region, characterized by the slopes of the Douro Valley, provides a vast range of environmental characteristics for winegrapes. Different climatic conditions, soil characteristics, topography, varieties and practices comprise the Douro terroirs, resulting in the uniqueness of the wines produced. In the current study, an assessment of the homogeneous viticultural zones is proposed for the Douro, by integrating different terroir elements (thermal, hydric, soils, topography and vegetation), using state-of-the-art very-high resolution datasets (1-km) and a large number of variables. A multivariate zoning was carried out using a principal component analysis (PCA) and a subsequent clustering. A geospatial assessment of the terroir elements was also performed separately for each sub-region in the DDR. The connections of the different terroirs prevailing at a given location were also innovatively assessed. This zoning may promote a more appropriate selection of vineyard sites, the selection of more locally-adapted varieties and rootstocks and the adoption of appropriate viticultural practices. The present study may be a standard methodology that can be applied to other winemaking regions worldwide.

1. Introduction

Detecting the unique characteristics within a given viticultural region is key to assessing the winemaking suitability and potential. This zoning assesses the most relevant viticultural characteristics in each winegrowing region, providing a basis for strategic vintage planning (Costantini et al. 2016). The regional climate, soil properties, topography, varieties (biodiversity) and management practices of any given winemaking regions are usually described by the *terroir* concept (OIV 2010).

Climate is the main driver for winegrape development and growth (Keller 2010). Temperature and water availability influence winegrape yields and are responsible for a balanced grape ripening (Fraga et al. 2015). Soil is another key *terroir* element, crucial for water and nutrients, which influences grape quality attributes (Morlat and Jacquet 2003). The topographic elements are another important factor that influences viticultural characteristics of a given region, such aselevation, slope and solar exposure (Yau et al. 2013). Additionally, the vegetation: winegrape cultivars; cover crops and management practices are also important *terroir* elements (Böhm 2010).

The Douro Demarcated Region (DDR) is famous for its Port Wine and other high-quality wines. The DDR encompasses over 250,000ha in the Douro valley (Figure 1b). It is one of the most important winemaking regions in Portugal, accounting for approximately 20% of the country's vineyard area and wine production (IVV 2015).

https://doi.org/10.1051/e3sconf/20185002009

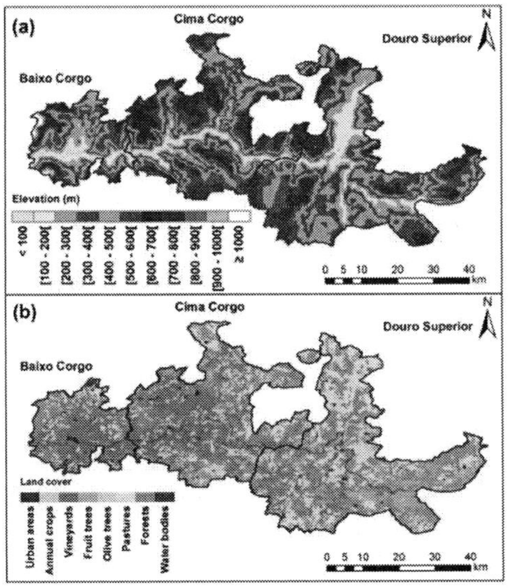

Figure 1 – (a) Digital elevation model of the Douro Demarcated Region (DDR). (b) Land cover classification in the DDR, according to the CORINE land cover map. The three sub-regions of the DDR, Baixo Corgo (BC), Cima Corgo (CC) and Douro Superior (DS), are also outlined.

Given the distinctness of the DDR, it is important to define its *terroir* characteristics in a thorough integrated approach. Given the unique Mediterranean-type climatic characteristics and the complex orography of the DDR, very high resolution zoning data is required. Similar *terroir* conditions across the regions should produce wines with similar characteristics. The present study aims to develop a multivariate zoning approach, through a holistic integration of the main *terroir* elements, herein illustratively applied to the DDR. A similar methodology can be conducted in other viticultural regions worldwide.

2. Material and Methods

2.1 Terroir elements

In the current study, five *terroir* elements were considered: (i) thermal conditions; (ii) hydric conditions, (iii) soils, (iv) topography and (v) vegetation dynamics.

To assess (i) and (ii) terroir elements, high-resolution datasets, produced by the WorldClim project (Hijmans et al. 2005) at ca. 1 km spatial resolution, were used. For the thermal conditions, the following variables were assessed: annual, seasonal (DJF: December–January–February, MAM: March–April–May, JJA: June–July–August, SON: September–October–November) and growing season (April–October) minimum, maximum and mean temperatures; Cool night index (CI; Tonietto 1999) from Fraga et al. (2014); Growing degree-day (GDD; Winkler 1974) from Fraga et al. (2014); Thermicity index (TI; Rivas-Martínez and Rivas-Saenz 2011) from Costa et al. (2017). For the hydric conditions, the following variables were selected: annual, seasonal (DJF, MAM, JJA, SON) and growing season (April–October) precipitation totals; Aridity index (AI; Rivas-Martínez and Rivas-Saenz 2011) from Costa et al. (2017); Dryness index (DI; Tonietto and Carbonneau 2004) from Fraga et al. (2014).

Soil properties were obtained from the Harmonized World Soil Database (HWSD; FAO/IIASA/ISRIC/ISSCAS/JRC 2012). HWSD soil properties were originally mapped at a 1:1,000,000 scale and are available at a resolution of ~1 km. The selected variables were: Bulk density (kg dm^{-3}); Clay content (%); pH (–log(H+)); Salinity (dS m^{-1}); Sand content (%); Gravel content (%); Silt content (%); Organic content (%); Total exchangeable bases (cmol kg^{-1}) and Sodicity (%). Additionally, water table depth, retrieved from Fan et al. (2013), was also included in the analysis.

A fourth set of variables was chosen to characterize the topography of the DDR, including elevation (m), slope (°) and solar exposure (number of hours of sun, excluding clouds). Elevation was retrieved from a digital elevation model, DEM (GTOPO30). Slope and solar exposure (photoperiod) were calculated from the DEM dataset using geographical information systems - ArcGIS (version 10.3.1) Spatial Analyst Tool.

A fifth set of variables was included to take into account the effect of vegetation growth and winegrape phenology. The Normalized Difference Vegetation Index (NDVI) was used for the analysis of the vegetative growth. NDVI takes into account the difference between near-infrared and red reflectances, providing a measure of vegetation greenness, where higher NDVI corresponds to higher vegetation greenness (Huete et al. 2002). NDVI from the Moderate Resolution Imaging Spectroradiometer (MODIS) is obtained from the National Aeronautics and Space Administration (NASA) Land Processes Distributed Active Archive Center (LP DAAC; https://lpdaac.usgs.gov/). Two tiles covering all of Portugal (h17v04, h17v05) are retrieved, at a 1 km spatial resolution and on a 16-day temporal resolution, from 2000 to 2015. Furthermore, a set of maps containing the Julian dates of the main winegrape phenological timings (budburst, flowering, veraison and harvest) were also included as variables in the vegetation dynamics set. This data, which has been previously validated, was obtained from Fraga et al. (2016a).

2.2 Principal component analysis and clustering methodology

The delineation of homogeneous zones in the DDR was carried out using a principal component analysis (PCA) and a subsequent *k*-means clustering methodology, using MATLAB (version R2017a). Variables of each of the 5 *terroir* elements (thermal, hydric, soils, topography and vegetation) were first normalized by the corresponding mean prior to PCA, as variables have different physical units, also warranting equal weighting in clustering. PCA was used herein as a method to reduce the number of variables (over 40 variables for all *terroir* elements) to retain for subsequent clustering (*k*-means on the subspace of the leading orthogonal modes). After a preliminary sensitivity analysis, the leading two components (PC1 and PC2) of each *terroir* element were retained, in all cases cumulatively explaining much of the variance. The number of resulting clusters was tested and four clusters were eventually selected for each element. Finally, the resulting clusters were mapped, corresponding to the viticultural *terroirs* in the DDR at ca. 1 km spatial resolution (total of 3910 1km grid-cells).

2.3 Relationships between the different terroir elements

The highest occurrence between clusters of different *terroir* elements at the same 1 km grid-cell is determined. This assessment was done separately for each of the sub-regions within the DDR (Baixo Corgo: BC; Cima Corgo: CC and Douro Superior: DS). The frequencies of occurrence of each clusters/*terroir* element were also assessed separately by each land cover class, using the CORINE land cover classification (EEA 2002) shown in Figure 1b.

3. Results

3.1 Clustering approach

Cumulatively, the explained variance of the first two PCs was over 93% for the thermal, hydric and vegetation groups and over 66% for the soil and topography groups. The first principal component represents more than 90% of the total variance for the thermal and hydric groups, while the second component represents much lower variance (3.5% and 1.5%, respectively). For the other three *terroir* elements, the first PC, while still explaining large fractions of variance, showed comparatively low values (soil: 44.9%; topography; 34.7%; vegetation: 62.8%), while the second PC now acquires higher relative relevance. For these *terroir* elements, the spatial organization of the region in near-homogeneous areas is much less marked than for the thermal, hydric and vegetation groups, thus explaining the overall lower fractions of represented variance.

A description of the four resulting clusters of each *terroir* element is provided in Table I. While the description of the thermal, hydric, topography and vegetation clusters are self-explanatory, the soil clusters correspond to the underlining 'soil unit name' from FAO (Food and Agriculture Organization of the United Nations) in the original soil dataset (FAO/IIASA/ISRIC/ISSCAS/JRC 2012). Therefore, the nomenclature used in the HSWD was kept. Regarding the thermal clusters (Fig.2a), a clear distinction is evident in all the corresponding variables, with minimum temperature in the winter playing a key role for the zoning. Also for the hydric conditions (Figure 2c), a strong variable differentiation is apparent, being DI the leading variable. For the soil clustering (Figure 2e), some of the base variables have a more decisive role in zoning. Luvisols have higher gravel levels, Humic Cambisols have higher total exchangeable bases, Eutric/Dystric Cambisols show sand content and higher exchangeable sodium and Dystric Regosols have higher levels of organic matter. Regarding topography (Figure 2g), the description of the clusters is directly based on the weight of each of the three base variables. For the vegetation clustering (Figure 2i), NDVI acquires a more important role for zoning, particularly its summer values.

Table I – Description of the clusters for each *terroir* element.

***Terroir* elements**	**Clusters**			
	1	**2**	**3**	**4**
Thermal	Cool	Moderate cool	Moderate warm	Warm
Hydric	Wet	Moderate wet	Moderate dry	Dry
Soil	Luvisol	Humic cambisol	Eutric/Dystric cambisol	Dystric regosol
Topography	High elev. Low solar exp.	High elev. Low slopes	Low elev. High slopes	Low elev. High solar exp.
Vegetation	Early phen. High green.	Early phen. Low green.	Late phen. Low green.	Late phen. High green.

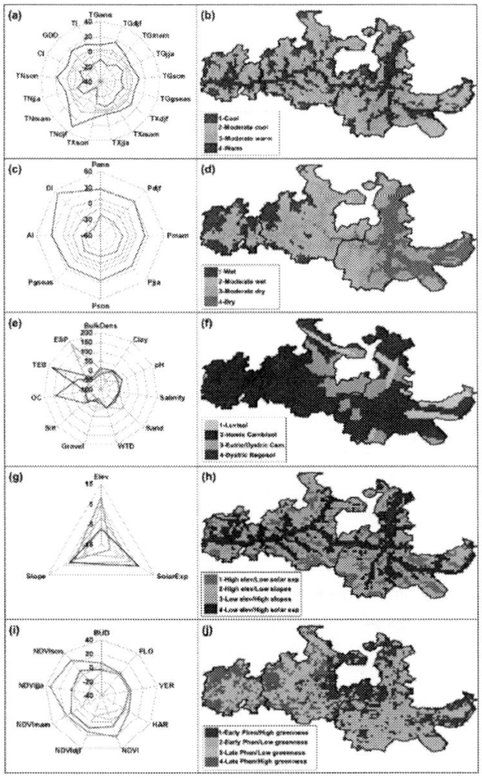

Figure 2 – Radar chart with the normalized values of the variables used in the clustering methodology: (a, b) thermal; (c, d) hydric; (e, f) soils; (g, h) topography; (i, j) vegetation dynamics. Geographical representation of the clusters of each of the five *terroir* elements in the Douro Demarcated Region.

3.2 Spatial homogeneity in the DDR

After processing data from each of the *terroir* elements, a preliminary zoning of the DDR was achieved (Figure 2). Different homogeneous areas were established for each of the five *terroir* elements. The cooler (thermal cluster 1; henceforth Thermal-1) and moderate cooler zones (Thermal-2) are located in the outer/higher elevation areas of the DDR, while the moderate warm (Thermal-3) and warm (Thermal-4) zones are located in the inner/lower elevation areas (Figure 2b). For the whole DDR, the cluster with the highest frequency is Thermal-2, followed by Thermal-3 (Table II). The DDR is primarily characterized by moderately wet/wet (Hydric-2/Hydric-1) conditions in the western and central areas (BC and CC), and moderately dry/dry (Hydric-3/Hydric-4) conditions in the east (DS) (Figure 2d). The Hydric-2 cluster has the highest occurrence (34%; Table II). Regarding the soils of the DDR (Figure 2f), most of the region corresponds to the Soil-3 cluster: Dystric Regosols (63%), while other clusters occur significantly less. Concerning topography, there is a strong association between high elevations, lower solar exposures and lower slopes (Topography-1 and -2, respectively) and between low elevations, high solar exposures and high slopes (Topography-3 and -4, respectively). The predominant cluster in the DDR is Topography-4 (33%) (Table II). For the vegetation clusters, a clear distinction between early phenology (Vegetation-1 and -2), in the eastern areas, and late phenology (Vegetation-3 and -4), in the west, is apparent. Higher greenness tends to occur in the higher elevation outer areas, while low greenness occurs in the inner areas of the DDR. The predominant cluster is Vegetation-2: early phenology and low greenness. The corresponding full dataset of these *terroir* classes is available as Supporting Information (S1 Appendix) so as to potentiate their application by researchers, decision-makers and stakeholders of the DDR winemaking sector.

Table II – Percentage of each cluster occurrence in the Douro Demarcated Region (DDR). Grey shaded values indicate the highest cluster occurrence for each class.

Terroir elements	Clusters			
	1	2	3	4
Thermal	15%	**33%**	30%	22%
Hydric	15%	**34%**	29%	22%
Soil	9%	12%	16%	**63%**
Topography	18%	28%	21%	**33%**
Vegetation	26%	**35%**	22%	17%

3.3 Integrated sub-regional terroir assessment

The relationships between clusters of the different *terroir* elements was also performed on a sub-regional level (for BC, CC and DS). The highest association between clusters is shown in Figure 3. In all sub-regions Dystric Regosols dominate the associations between soils and other *terroir* elements. For the BC sub-region (Figure 3a), the moderately cool and the wet areas are strongly associated. Warm areas are associated with low elevations/high solar exposures and late phenology/low greenness. Wet areas are also tied to late phenology/high greenness and humic cambisols. In this sub-region, moderate wet areas are associated with low elevations/high solar exposures. In the CC sub-region, moderate cool and moderate wet areas are highly associated, being also connected to high elevation areas/low slopes and late phenology/low greenness. In the DS sub-region, moderate warm areas show links to early phenology/low greenness, while warm dry areas with low elevations and high solar exposures largely overlap.

https://doi.org/10.1051/e3sconf/20185002009

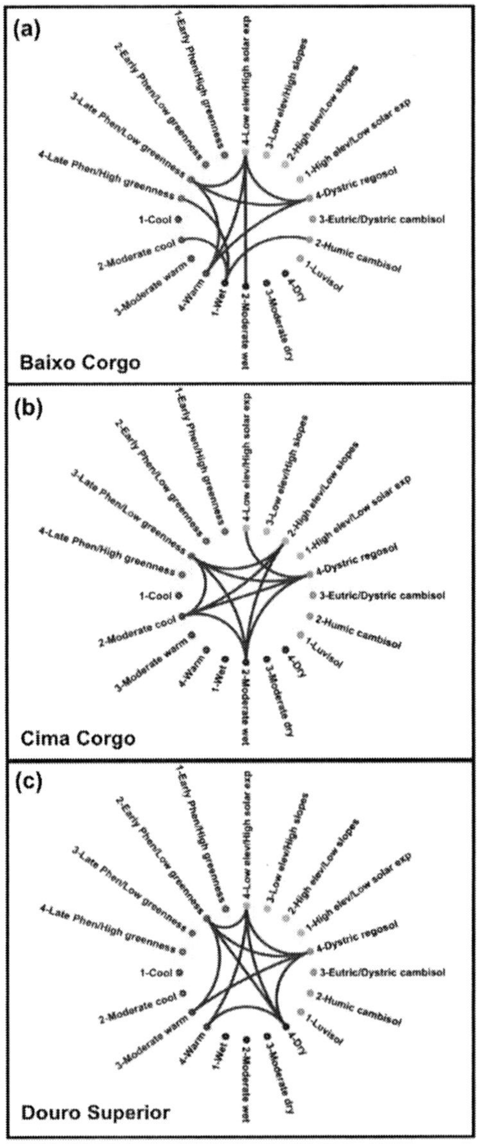

Figure 3 – Chart representing the relationships between clusters of the different *terroir* elements. Only the highest association inside each sub-region is plotted: (a) Baixo Corgo (BC); (b) Cima Corgo (CC) and (c) Douro Superior (DS).

4. Discussion and conclusions

In the current study, homogeneous zones of interest for viticultural activities has been proposed for the DDR, by integrating different *terroir* elements, using state-of-the-art high resolution datasets. The integration of all these factors, quantified by a large number of variables, into a high resolution viticultural zoning (~1km) in the DDR was not previously performed. The present study thereby significantly improves our understanding of the *terroir* spatial variability in the DDR.

Overall, four classes of each *terroir* element (thermal, hydric, soils, topography and vegetation dynamics) have been established, by applying a *k*-means clustering on the subspace of the leading orthogonal modes. The delimited thermal classes may allow growers to select and grow winegrape varieties based on their optimum thermal requirements. The hydric classes may also assist the selection of suitable rootstocks to improve plant water status. These issues are particularly relevant taking into account the projected climate change impacts for the region, mainly the drying trend over the next decades (Fraga et al. 2016b; Santos et al. 2016). Furthermore, other *terroir* elements considered herein, such as slope degree, solar exposure, soil type and vegetation greenness, may assist growers in their management activities and vintage planning throughout the year. The inter-connections of the different *terroir* elements that exist at a given location was innovatively assessed. Although similar methodologies were already applied by previous studies to other winemaking regions, allowing the identification of the existing *terroirs*, an integration of all these factors was still incipient. These interactions may allow identifying the different environmental processes that account for the winemaking quality potential of the different zones in the DDR.

The multivariate zoning of viticultural *terroirs* provided in the present study may also be of great importance to policymakers and stakeholders operating in the DDR, acting as an additional decision support tool. In order to promote the usage of this new information, the corresponding dataset is provided in Supporting Information (S1 Appendix). Currently, in this region, a ranking system for vineyard plots is in force, with a classification scheme ranging from the best sites (class A) to the worst (class F) (*Instituto dos Vinhos do Douro e Porto* – IVDP). This ranking system may also be revised based on the current findings. In addition, taking into account the most recent climate change projections for the region (Fraga et al. 2016b; Santos et al. 2017), this ranking may require substantial adaptations in the future. Lastly, the present study can also be used as an archetypal methodology that can be applied to other winemaking regions in the country and even worldwide.

Acknowledgements

This work was supported by: the ModelVitiDouro project (PA 53774), funded by the Agricultural and Rural Development Fund (EAFRD) and the Portuguese Government (Measure 4.1 – Cooperation for Innovation PRODER Programme – Rural Development Programme); the INNOVINE&WINE project (NORTE-01-0145-FEDER-000038), co-funded by the European Regional Development Fund through NORTE 2020 Programme; European Investment Funds (FEDER/COMPETE/POCI), POCI-01-0145-FEDER-006958, and Portuguese Foundation for Science and Technology (FCT), UID/AGR/04033/2013. The postdoctoral fellowship (SFRH/BPD/119461/2016) awarded to the leading author is also acknowledged.

References

Böhm, J., 2010: *Portugal vitícola: o grande livro das castas.* C. Ferreira, 234 pp.

Costa, R., H. Fraga, P. M. Fernandes, and J. A. Santos, 2017: Implications of future bioclimatic shifts on Portuguese forests. *Reg Environ Change*, **17**, 117-127,10.1007/s10113-016-0980-9

Costantini, E. A. C., R. Lorenzetti, and G. Malorgio, 2016: A multivariate approach for the study of environmental drivers of wine economic structure. *Land Use Policy*, **57,** 53-63,10.1016/j.landusepol.2016.05.015

EEA, 2002: CORINE Land Cover update, I&CLC2000 project, Technical Guidelines.

Fan, Y., H. Li, and G. Miguez-Macho, 2013: Global Patterns of Groundwater Table Depth. *Science, N.Y.*, **339,** 940-943,10.1126/science.1229881

FAO/IIASA/ISRIC/ISSCAS/JRC, 2012: Harmonized World Soil Database (version 1.2). FAO, Rome, Italy and IIASA, Laxenburg, Austria.

Fraga, H., I. Garcia de Cortazar Atauri, A. C. Malheiro, and J. A. Santos, 2016a: Modelling climate change impacts on viticultural yield, phenology and stress conditions in Europe. *Glob Chang Biol*, **22,** 3774-3788,10.1111/gcb.13382

Fraga, H., A. C. Malheiro, J. Moutinho-Pereira, G. V. Jones, F. Alves, J. G. Pinto, and J. A. Santos, 2014: Very high resolution bioclimatic zoning of Portuguese wine regions: present and future scenarios. *Reg Environ Change*, **14,** 295-306,10.1007/s10113-013-0490-y

Fraga, H., J. A. Santos, J. Moutinho-Pereira, C. Carlos, J. Silvestre, J. Eiras-Dias, T. Mota, and A. C. Malheiro, 2016b: Statistical modelling of grapevine phenology in Portuguese wine regions: observed trends and climate change projections. *J Agric Sci*, **154,** 795-811,10.1017/S0021859615000933

Fraga, H., R. Costa, J. Moutinho-Pereira, C. M. Correia, L.-T. Dinis, I. Gonçalves, J. Silvestre, J. Eiras-Dias, A. C. Malheiro, and J. A. Santos, 2015: Modeling Phenology, Water Status, and Yield Components of Three Portuguese Grapevines Using the STICS Crop Model. *Am J Enol Vitic*, **66,** 482-491,10.5344/ajev.2015.15031

GTOPO30: Global 30 Arc-Second Elevation (GTOPO30), USGS Products. Data available from the U.S. Geological Survey.,

Hijmans, R. J., S. E. Cameron, J. L. Parra, P. G. Jones, and A. Jarvis, 2005: Very high resolution interpolated climate surfaces for global land areas. *Int. J. Clim.*, **25,** 1965-1978,10.1002/joc.1276

Huete, A., K. Didan, T. Miura, E. P. Rodriguez, X. Gao, and L. G. Ferreira, 2002: Overview of the radiometric and biophysical performance of the MODIS vegetation indices. *Remote Sens Environ*, **83,** 195-213,10.1016/S0034-4257(02)00096-2

IVV, 2015: Vinhos e Aguardentes de Portugal, Anuário 2015. *Ministério da Agricultura, do Desenvolvimento Rural e das Pescas: Instituto da Vinha e do Vinho, Lisboa*, 236,

Keller, M., 2010: The Science of Grapevines: Anatomy and Physiology. Elsevier, Inc. 400pp,

Morlat, R., and A. Jacquet, 2003: Grapevine root system and soil characteristics in a vineyard maintained long-term with or without interrow sward. *Am J Enol Vitic*, **54,** 1-7,

OIV, 2010: Resolution OIV/VITI 333/2010, Definition of vitivinicultural "TERROIR", Tbilisi, 25th June 2010.

Rivas-Martínez, S., and S. Rivas-Saenz, 2011: Worldwide Bioclimatic Classification System. Phytosociological Research Center, Spain, http://www.globalbioclimatics.org.

Santos, J. A., R. Costa, and H. Fraga, 2017: Climate change impacts on thermal growing conditions of main fruit species in Portugal. *Clim Change*, **140,** 273-286,10.1007/s10584-016-1835-6

Santos, J. A., M. Belo-Pereira, H. Fraga, and J. G. Pinto, 2016: Understanding climate change projections for precipitation over western Europe with a weather typing approach. *J Geophys Res-Atmos*, **121,** 1170-1189,10.1002/2015JD024399

Tonietto, J., 1999: Les macroclimats viticoles mondiaux et l'influence du mésoclimat sur la typicité de la Syrah et du Muscat de Hambourg dans le sud de la France: méthodologie de caractérisation. PhD Dissertation, 233p, Ecole Nationale Supérieure Agronomique, Montpellier, France.

Tonietto, J., and A. Carbonneau, 2004: A multicriteria climatic classification system for grape-growing regions worldwide. *Agric For Meteorol*, **124,** 81-97,10.1016/j.agrformet.2003.06.001

Winkler, A. J., 1974: *General viticulture.* University of California Press.

Yau, I. H., J. R. Davenport, and R. A. Rupp, 2013: Characterizing Inland Pacific Northwest American Viticultural Areas with Geospatial Data. *Plos One*, **8**,10.1371/journal.pone.0061994

Mapping Cabernet Franc vineyards by unmanned aerial vehicles (UAVs) for variability in vegetation indices, water status, and virus titer

Andrew G. Reynolds[1]*, Hyun-Suk Lee[1], Briann Dorin[1], Ralph Brown[2], Marilyne Jollineau[3], Adam Shemrock[4], Marnie Crombleholme[1], Emilie Jobin Poirier[1], Wei Zheng[5], Maxime Gasnier[6], Mehdi Shabanian[7] and Baozhong Meng[7]

[1]Cool Climate Oenology and Viticulture Institute, Brock University, St. Catharines, Ontario, Canada; [2]School of Engineering, University of Guelph, Guelph, Ontario; [3]Dept. of Geography, Brock University, St. Catharines, Ontario; [4]Air-Tech Solutions, Kingston, Ontario; [5]Dept. of Agriculture and Food, University of La Rioja, Logroño, La Rioja, Spain; [6]Agro-Paris-Tech, Paris, France; [7]Dept. of Cell and Molecular Biology, University of Guelph.

Abstract. The hypothesis of this research was that the maps based on remotely-sensed images would create zones of different vigor, yield, water status, winter hardiness and berry composition and the wines from the unique zones would show different chemical and sensorial profiles. A second hypothesis was that titer of grapevine leafroll-associated virus (GLRaV) could be correlated spatially to NDVI and other spectral indices. To determine zonation, unmanned aerial vehicles (UAVs) with multispectral and thermal sensors were flown over six Cabernet Franc vineyard blocks in Ontario, Canada. Zonation was based on NDVI values, and spatial correlations were examined between the NDVI and leaf water potential (ψ), soil water content (SWC), stomatal conductance (g_s), winter hardiness (LT_{50}), vine size, yield, and berry composition. Additional NDVI data were acquired using GreenSeeker (proximal sensing), and both NDVI data sets produced maps of similar configuration. Several direct correlations were found between UAV-based NDVI and vine size, berry weight, yield, titratable acidity, SWC, leaf ψ, g_s, and NDVI from GreenSeeker. Inverse correlations included thermal data, Brix, color/ anthocyanins/ phenols, and LT_{50}. The pattern of UAV-based NDVI and other variables corresponded to the PCA results. Thermal scan and GreenSeeker were useful tools for mapping variability in water status, yield components, and berry composition. In 2016, zoned maps were created based on UAV NDVI data, and grapes were harvested according to the separate zones. Additionally, spatial correlations between GLRaV titer and NDVI were observed. Use of UAVs may be able to delineate zones of differing vine size, yield components, and berry composition, as well as areas of different virus status and winter hardiness.

1 Introduction

The Ontario wine industry produces ≈ 80,000 t of grapes and consists of cultivars such as Riesling, Chardonnay, and Cabernet franc as well as Cabernet Sauvignon, Merlot, Pinot noir, and many others (www.grapegrowersofontario.com). Cabernet franc is the most widely planted red wine cultivar. Soils are variable due to widespread glacial activity >10,000 years ago, and many vineyards are situated on soils that range widely in texture, depth of solum, and water-holding capacity [1]. This soil variability impacts vine vigor, yield, and water status.

GreenSeeker and other ground-based (proximal) sensing technologies might allow grapegrowers to identify unique zones without use of aircraft through continuous compilation of NDVI data from vine canopies [2-4]. If zones can be identified from the ground, then different zone-based wine products could be created with minimal cost from the producer. Data validation is required to determine relationships between proximally-sensed data and agriculturally relevant variables. Proximal sensing is a relatively recent innovation and its evaluation in viticulture is uncommon. Proximal sensing correlated with vine size and berry color in Merlot vineyards in Greece [5] and with downy mildew levels in Italy [4]. Linear correlations to stable isotope contents in leaves (^{13}C and ^{15}N) provided evidence that canopy reflectance could detect water and nutritional-based plant stresses [6,7]. Research into proximal sensing has mainly been limited to agronomic crops [8,9] and limited viticulture-related work has occurred in Canada [10-11].

Attempts have been made to identify unique zones by remote sensing (RS) and to associate these remotely-sensed regions with vine water status, soil moisture, vine vigor, yield, and berry composition. However, use of aircraft-based RS is costly and in agricultural systems it can be imprecise [5]. Data must be converted to vegetation indices (VIs), e.g., normalized difference vegetative index (NDVI) through appropriate computer software [12-15]. Validation of data acquired by RS is necessary to determine whether ostensibly-unique zones are relevant from a standpoint of physiology, yield, and berry composition. One challenge involved masking of cover crop spectral reflectance from images to assess the vine canopy-specific NDVI and other VIs [14,15].

RS has been used in modelling vegetative growth, and to predict grape composition from those measurements. Remotely sensed multispectral data were used to delineate a Chardonnay vineyard into small-lot production zones based on vine size (pruning weights; vigor), which were related to vine water status and grape composition [16]. Relationships between VIs and vegetative growth were further explored with positive correlations between the VIs and vine size [17]. The ability for RS to predict grape composition variables was explored in Australia, noting that re-sampling to a final pixel size approximately equal to row width, effectively combining vine size and density information into a single pixel, resulted in the strongest correlations to color and phenols [18]. In Languedoc, temporally stable relationships occurred between NDVI-delineated zones and vegetative growth, vine water status, and yield [19].

Aircraft-based RS has been proven useful for monitoring vegetative growth, and for making inferences about grape composition from multispectral measurements. In Ontario, NDVI data from RS was associated with numerous variables in Riesling [14,15] and Pinot noir [13] vineyards, including vine water status, yield components, and berry composition. RS was useful to determine Pinot noir anthocyanins and phenols, in addition to water status, yield and vine size [13]. These studies were unique by employing RS in cover-cropped vineyards and thereafter using protocols for excluding the spectral reflectance contributed by inter-row vegetation. Aircraft-based RS has been used to make inferences about grape composition from multispectral measurements [20,21]. However, employment of UAVs for RS in vineyards is a relatively new area of research, heretofore untested in Canada, and capable of acquiring high resolution spatial data without high cost of aircraft. As with proximal sensing there has been little published, and most has confirmed their ability to acquire NDVI and related images [22-24]. Relationships were explored between photosynthesis and chlorophyll fluorescence by hyperspectral imagery captured via UAVs [25]. Significant relationships were demonstrated between photosynthesis and chlorophyll fluorescence vs. remote measurements, and between both chlorophyll a/b and leaf carotenoids vs. several VIs based on multispectral images acquired by UAVs [26]. UAVs were also utilized for assessment of vineyard water status by correlation of stem water potential (ψ) with NDVI [27]. Further relationships were elucidated between several VIs [NDVI, photochemical reflectance, renormalized difference vegetation index, red edge inflection point (REIP)] vs. leaf ψ and stomatal conductance (g_s) [28]. UAV-based NDVI also correlated with levels of Fe chlorosis, leaf carotenoids, and grape leaf and berry anthocyanins [29].

Many vineyards in Ontario, British Columbia, and California are infected by both grapevine leafroll-associated virus (GLRaV) as well as grapevine red blotch virus (GRBV). GLRaV is one of the most destructive diseases in all grape-growing regions globally. Yield losses from GLRaV can reach 20-40% [30], depending on cultivar, rootstock, soil type, vine age, and climate. Many large Ontario vineyards have attained the designation as "underperforming" by wineries, based primarily on fruit maturity characteristics such as reduced total soluble solids (TSS) and color. Technologies must be developed and implemented to detect zones of virus infection before entire vineyard blocks become unsuitable for quality winemaking. Evidence of correlations between both remotely- and proximally-acquired vegetation indices (VIs) such as NDVI and titer of GLRaV and GRBV, as well as unique spectral signatures of virus-infected vine leaves would be of great benefit.

2 Materials and methods

2.1 Sites and cultivars

Six Cabernet franc vineyards (1-2 ha) in the Niagara Region were chosen. The sites represented the following sub-appellations: Niagara Lakeshore (Buis), Creek Shores (Pondview), St. Davids Bench (Chateau des Charmes), Lincoln Lakeshore north (George), Lincoln Lakeshore south (Kocsis), Beamsville Bench (Cave Spring). Soil types [1] varied substantially from well-drained coarse-textured Tavistock and Vineland series (Niagara Lakeshore, Lincoln Lakeshore north), to moderately-well drained Chinguacousy (Creek Shores, Beamsville Bench), and poorly-drained Jeddo (Lincoln Lakeshore south) and Beverly/Toledo soils (St. Davids Bench). These soils provided a range of water-holding capacities that impacted vine water status.

2.2 Geolocation

Vineyard blocks were GPS-delineated to determine shape using a Trimble Handheld GPS, equipped with TerraSync software (Trimble Navigation, Sunnyvale, CA). Sentinel vines (80-100) were identified in a ≈ 8m x 8m grid within each vineyard and geolocated by the aforementioned GPS system. Post-collection differential correction was performed using GPS Pathfinder Office (Trimble) to ≈30-50 cm accuracy using the Port Weller, ON base station correction. Field measurements and berry samples were taken on these vines in all vintages.

2.3 Soil moisture, leaf water potential (ψ), and stomatal conductance (g_s)

Vineyard soil water content (SWC) was measured by time domain reflectometry (TDR) at a 20 cm depth using the Field Scout TDR 300 Soil Moisture Meter (Spectrum Technologies, East Plainfield, IL). Measurements took place at berry set, lag phase, and veraison on all sentinel vines. Vine water status was measured using midday leaf ψ by pressure bomb (Soil Moisture Equip., Santa Barbara, CA). Measurements were made on designated leaf ψ vines (≈ 15-20 per vineyard block), on the same days as SWC measurements, from 1000h-1400h (ca. solar noon), under full sun. Leaf g_s was measured by a hand-held porometer (Decagon Devices, Pullman, WA) on leaf ψ vines.

2.4 Yield components and vine size

Harvest dates were at the discretion of vineyard managers. Fruit from each sentinel vine was harvested, cluster number determined, and fruit weighed with a field scale. Cane prunings were weighed to determine vine size.

2.5 Berry analysis

A 100-berry sample was taken from each sentinel vine at harvest and frozen at -25°C. Samples were weighed to determine mean berry weight, and placed in a beaker in a water bath at 80°C for 1 hr to dissolve precipitated tartrates. Samples were homogenized in a juicer and ≈ 35 mL were clarified by centrifugation at 4500 *g* for 10 minutes. TSS were measured by refractometer (UV, Buffalo, NY); pH with an Accumet pH meter (Fisher, Mississauga, ON); titratable acidity by titration to an 8.2 pH endpoint with 0.1N NaOH by a PC-Titrate autotitrator (Man-Tech, Guelph, ON). Remaining juice (≈20 mL) was retained for subsequent color analysis. Absorbances (420 and 520 nm) were measured using an Ultrospec 2100 Pro UV-VIS spectrophotometer (Biochrom, Cambridge, UK). Color intensity was calculated as $A_{420}+A_{520}$. Total anthocyanins were quantified using the pH shift method [31], and total phenols by Folin-Ciocalteu method [32].

2.6 Spatial mapping

GPS coordinates from vineyard blocks and sentinel vines were imported into a GIS environment (ArcGIS 10.3; Environmental Systems Res. Inst., Redlands, CA) and linked to point data collected from sentinel vines. Spatial interpolation techniques (i.e., inverse-distance weighting; Diffusion interpolation with barriers for remotely-sensed data) were applied to these data were used to estimate the values of variables at unsampled locations.

2.7 Hand-held spectral signatures

There has been limited experience in measurement of unique spectral signatures in virus-infected vines, including GLRaV [33] and GRBV [34]. Prior to appearance of virus symptoms, spectral reflectance in the visible and near-infrared (NIR) ranges was measured using leaves from virus-free and GLRaV-infected vines. An EPP2000 (UV-Vis-100 nm) spectrometer, controlled by a laptop computer running SpectraWiz software (Stellarnet, Tampa, FL) was used to record reflectance spectra [13,14]. A custom-built enclosure was used to hold the leaf and 5-W halogen bulb light source.

2.8 Proximal sensing

A GreenSeeker unit (Trimble Navigation) mounted on a four-wheel-drive vehicle was used to collect NDVI data in 2015-2017 on dates close to soil moisture and leaf ψ data collection. Data were imported into Farmworks software (Trimble) and spatial maps created. Shapefiles were then imported into an ArcGIS geodatabase, GPS coordinates identical to the sentinel vines identified, and NDVI data corresponding to these coordinates extracted.

2.9 UAV and sensors

2.9.1 Flights

The 2015 and 2016 UAV flights corresponded to the veraison SWC, leaf ψ, and GreenSeeker data collection. Image acquisition in 2016 was performed consistent to 2015 flights [10]. Two sensors were used for image acquisition. The first operated in the visible and NIR portions of the electromagnetic spectrum (EMS) (Mini-MCA 6; Tetracam, Chatsworth, CA) utilizing five spectral bands (blue, green, red, red edge, NIR) equipped with an incident light sensor. The second sensor operated in the thermal–IR portion of the EMS (A65 thermal imaging camera; FLIR Systems, Burlington, ON).

2.9.2 Image acquisition, pre-processing, processing

Image acquisition was performed over each vineyard block according to previous protocols [10]. Once assembled and corrected NDVI-red, NDVI-green, REIP, and Normalized Difference Red Edge (NDRE) were calculated on mosaics. Pixel values corresponding to sentinel vines were extracted and compiled into a database including all field based variables (e.g. leaf ψ).

2.10 Virus titer determination

All vines designated for leaf ψ and g_s measurements were sampled September 2016 to determine GLRaV titer. Total RNAs were isolated from leaf samples using a recently developed method [35]. Total RNAs were used in reverse transcription using primers specific to GLRaV-2 and 3, followed by amplification through PCR using broad-spectrum primers to identify virus presence [36]. To determine virus titer for comparison and correlation to the UAV and GreenSeeker data, quantitative qPCR was conducted by Power SYBR Green PCR Master Mix and StepOnePlus qPCR (Applied Biosystems).

2.11 Data analysis

Basic linear correlations and regressions were performed on the ground-based, proximally-sensed, and UAV data to determine relationships, particularly those between proximally sensed/UAV data vs. ground-based data using XLStat (Addinsoft, Paris, France). Principal components analysis (PCA) was likewise performed. Maps were created for all variables using ArcGIS. *K*-means clustering and Moran's *I* were used to verify spatial relationships.

3 Results and discussion

3.1 Principal components analysis

NDVI data from both GreenSeeker and UAV flights were closely correlated in PCA (Fig. 1A-D). These variables were closely related to vine size, yield components, berry weight, leaf ψ, g_s, and SWC. In most circumstances NDVI data were inversely correlated to leaf temperature and thermal data obtained from UAV flights. Inverse correlations between NDVI and phenolic analytes (anthocyanins, $OD_{420+520}$, phenols) were observed. These relationships confirmed 2015 observations [10].

3.2 Map analysis and hand-held spectrometer

Spatial variability of most variables was consistent with that observed in 2015 [10]. Patterns of spatial variability in NDVI maps based on GreenSeeker and UAV flights were similar (Fig. 2), suggesting that proximal sensing might be a lower cost alternative to UAV flights. Spatial patterns in UAV-based thermal camera data were opposite to those of both GreenSeeker and UAV NDVI, with high thermal zones associated with low NDVI. Low NDVI zones were likewise associated with low SWC, leaf ψ, and leaf g_s. Lower NDVI zones also corresponded with low yields, berry weight, and vine size, but higher LT_{50} values. In most circumstances, low NDVI zones were associated with high levels of $OD_{420+520}$, anthocyanins, and phenols, as well as Brix and pH, and corresponded to low TA values. Relationships between NDVI vs. soil moisture and leaf ψ have been demonstrated in Ontario with GreenSeeker on Riesling, Pinot noir, and Cabernet franc [11], and by conventional aircraft in Riesling [14-15] and Pinot noir [13]. Similar relationships between canopy reflectance and water status were demonstrated in Greece [7], Languedoc [19] and Spain [27]. Relationships between spectral reflectance and anthocyanins, color, and phenolics have been demonstrated in Pinot noir [11,13] and Cabernet franc [10,11] in Ontario, in Merlot in Greece [5], Tempranillo in Spain, and Cabernet Sauvignon in Australia [18]. Similar relationships were observed in Ontario for Brix vs NDVI in Pinot noir, Cabernet franc and Riesling

[10,11,13-15] and NDVI vs terpenes in Riesling [10,11,14,15].

Processing UAV data in 2016 into several VIs suggested that NDVI red, NDVI green, REIP, and NDRE index were weakly spatially associated with NDVI in one vineyard (data not shown). Zones in the southern portions of the vineyards appeared inversely associated with NDVI. There were associations between NDVI red, NDVI green, REIP, and NDRE vs. GLRaV titer. These relationships are contrary to prior observations in Cabernet franc where low NDVI was associated with severe GLRaV symptoms [37]. Hand-held spectrometer data showed noteworthy differences between virus-affected and virus-free vines in terms of increased spectral reflectance in the green and NIR bands.

3.3 Regression analysis

Only three of six vineyards contained suffient numbers of GLRaV-infected vines to permit statistical analysis. Two of three displayed no relationship with UAV-based NDVI, while a third showed a weakly-positive relationship (data not shown). Four vineyards displayed strong linear relationships between UAV-based and GreenSeeker-based NDVI (data not shown). Several viticultural variables were likewise correlated with UAV-based NDVI (Fig. 1E-H), including SWC (four vineyards; three inverse), leaf ψ (four vineyards), g_s (four vineyards; one inverse), vine size (four vineyards), yield (three vineyards), LT_{50} (inverse; three vineyards), plus single vineyards with direct NDVI vs. TA relationships, and inverse Brix, pH and total phenols relationships. Thermal camera data were inversely related to NDVI.

Conclusions

UAV-derived NDVI data were related to those from GreenSeeker as well as many other variables of viticultural significance, particularly leaf ψ, g_s, vine size, yield, and LT_{50}. There was little relationship between NDVI and GLRaV titer in the six vineyards. Other VIs such as NDVI-red, NDVI-green, REIP, NDRE, and others may be better indicators of virus titer.

4 Literature cited

1. M.S. Kingston, E.W. Presant, *The Soils of the Regional Municipality of Niagara*. Report 60, Ont. Inst. of Pedology, Guelph, ON (1989).
2. R. Drissi, J. Goutouly, D. Forget, J. Gaudillère, Agronomy J., **101**, 226 (2009).
3. F. Mazzetto, A. Calcante, A. Mena, J. Agric. Eng., **1**, 11 (2009).
4. F. Mazzetto, A. Calcante, A. Mena, P. Sacco, J. Agric. Eng., **2**, 1 (2011).
5. S. Stamatiadis, D. Taskos, C. Tsadilas, C. Christofides, E. Tsadilas, J.S. Schepers, Am. J. Enol. Vitic., **57**, 415 (2006).
6. S. Stamatiadis, D. Taskos, E. Tsadilas, C. Christofides, C. Tsadilas, J.S. Schepers, Prec. Agric, **11**, 306 (2010).
7. D.G. Taskos, S. Koundouras, S. Stamatiadis, E. Zioziou, N. Nikolaou, K. Karakioulakis, N. Theodorou, *Proc. Ampelos Conf., Trends in Vitiviniculture Development*, Santorini, Greece. http://ampelos2013.conferences.gr/ (2013).
8. D. Barker, J. Sawyer, Agronomy J., **102**, 964 (2010).
9. W.C. Bausch, J.A. Delgado, In: T. Van Toai et al. (Eds.), *Digital Imaging and Spectral Techniques: Applications to Precision Agriculture and Crop Physiology* (pp. 145–157). ASA Special Publication 66. Madison, WI (2003).
10. A.G. Reynolds, R.B. Brown, M. Jollineau, A. Shemrock, E. Kotsaki, H.-S. Lee, W. Zheng, *Proc. 11th Terroir Congress*, McMinnville, OR, p. 381 (2016).
11. A.G. Reynolds, R.B. Brown, E. Kotsaki, H.-S. Lee, *Proc. 11th Terroir Congress*, McMinnville, OR, p. 477 (2016).
12. A. Hall, J. Louis, D. Lamb, *Proc. 6th International Conference on Geocomputation*; Univ. of Queensland, Brisbane (2001).
13. D. Ledderhof, R.B. Brown, A.G. Reynolds, M. Jollineau, Can. J. Plant Sci., **96**, 89 (2015).
14. M. Marciniak, R.B. Brown, A.G. Reynolds, M. Jollineau, J. Int. Sci. Vigne Vin, **49**, 1 (2015).
15. A.G. Reynolds, M. Marciniak, R.B. Brown, L. Tremblay, L. Baissas, Prog. Agric. et Vitic., **127**(12), 259 (2010).
16. L. Johnson, D. Bosch, D. Williams, B. Lobitz, Appl. Engin. Agric., **17**, 557 (2001).
17. S. Dobrowski, S. Ustin, J. Wolpert, Austral. J. Grape Wine Res., **9**, 177 (2003).
18. D. Lamb, M. Weedon, R.G.V. Bramley, Austral. J. Grape Wine Res., **10**, 46 (2004).
19. C. Acevedo-Opazo, B. Tisseyre, S. Guillaume, H. Ojeda, Prec. Agric., **9**, 285 (2008).
20. A. Gitelson, Y. Kaufman, M. Merzlyak, Remote Sensing Environ., **58**, 289 (1996).
21. D. Post, A. Fimbres, A. Matthias, E. Sano, L. Accioly, A. Batchily, L. Ferreira, Soil Sci. Soc. Am. J., **64**, 1027 (2000).
22. M.L. Guillen-Climent, P.J. Zarco-Tejada, J.A. Berni, P.R. North, F.J. Villalobos, Prec. Agric., **13**, 473 (2012).
23. A. Matese, J. Primicerio, F. Di Gennaro, E. Fiorillo, F.P. Vaccari, L. Genesio, Acta Hortic. **978**, 63 (2013).
24. D. Turner, A. Lucieer, C. Watson, *Proc. 34th Int. Symposium on Remote Sensing of Environment*, Sydney, Australia; http://www.isprs.org/proceedings/2011/ISRSE-34/ (2011).
25. P.J. Zarco-Tejada, A. Catalina, M.R. González, P. Martín, Remote Sens. Environ., **136**, 247 (2013).
26. P.J. Zarco-Tejada, M.L. Guillén-Climent, R. Hernández-Clemente, A. Catalina, M.R. González, P. Martín, Agric. Forest Meteorol., **171– 172**, 281 (2013).
27. J. Baluja, M.P. Diago, P. Balda, R. Zorer, F. Meggio, F. Morales, J. Tardaguila, Irrig. Sci., **30**, 511 (2012).
28. P.J. Zarco-Tejada, V. González-Dugo, L.E. Williams, L. Suárez, J.A.J. Berni, D. Goldhamer, E. Fereres, Remote Sens. Environ., **138**, 38 (2013).
29. F. Meggio, P.J. Zarco-Tejada, L.C. Núñez, G. Sepulcre-Cantó, M.R. González, P. Martín, Remote Sens. Environ., **114**, 1968 (2010).
30. N. Habili, F.W. Nutter Jr., Plant Disease, **81**, 625 (1997).
31. T. Fuleki, F.J. Francis, J. Food Sci., **33**, 72 (1968).
32. K. Slinkard, V. Singleton, Am. J. Enol. Vitic., **28**, 49 (1977).
33. R.A. Naidu, E.M. Perry, F.J. Pierce, T. Mekuria, Computers & Electronics in Agric., **66**, 38 (2009).
34. M. Mehrubeoglu, K. Orlebeck, M.J. Zemlan, W. Autran, In: *Algorithms and Technologies for Multispectral, Hyperspectral, & Ultraspectral Imagery* XXII. M. Velez-Reyes & D.W. Messinger (Eds.). Proc. SPIE Vol. 9840, 98400D, pp. 1-8 (2016).
35. H. Xiao, W.S. Kim, B. Meng, B., Virology J., **12**, 171 (2015).
36. B. Meng, A.R. Rebelo, H. Fisher, J. Gen. Virol., **87**, 1725 (2006).
37. A.G. Reynolds, The Grapevine, Viticulture, and Winemaking: A Brief Introduction. In: *Grapevine Viruses: Molecular Biology, Diagnostics and Management*. B. Meng, G.P. Martelli, D.A. Golino, M. Fuchs: Pp. 3-29. Springer (2017).

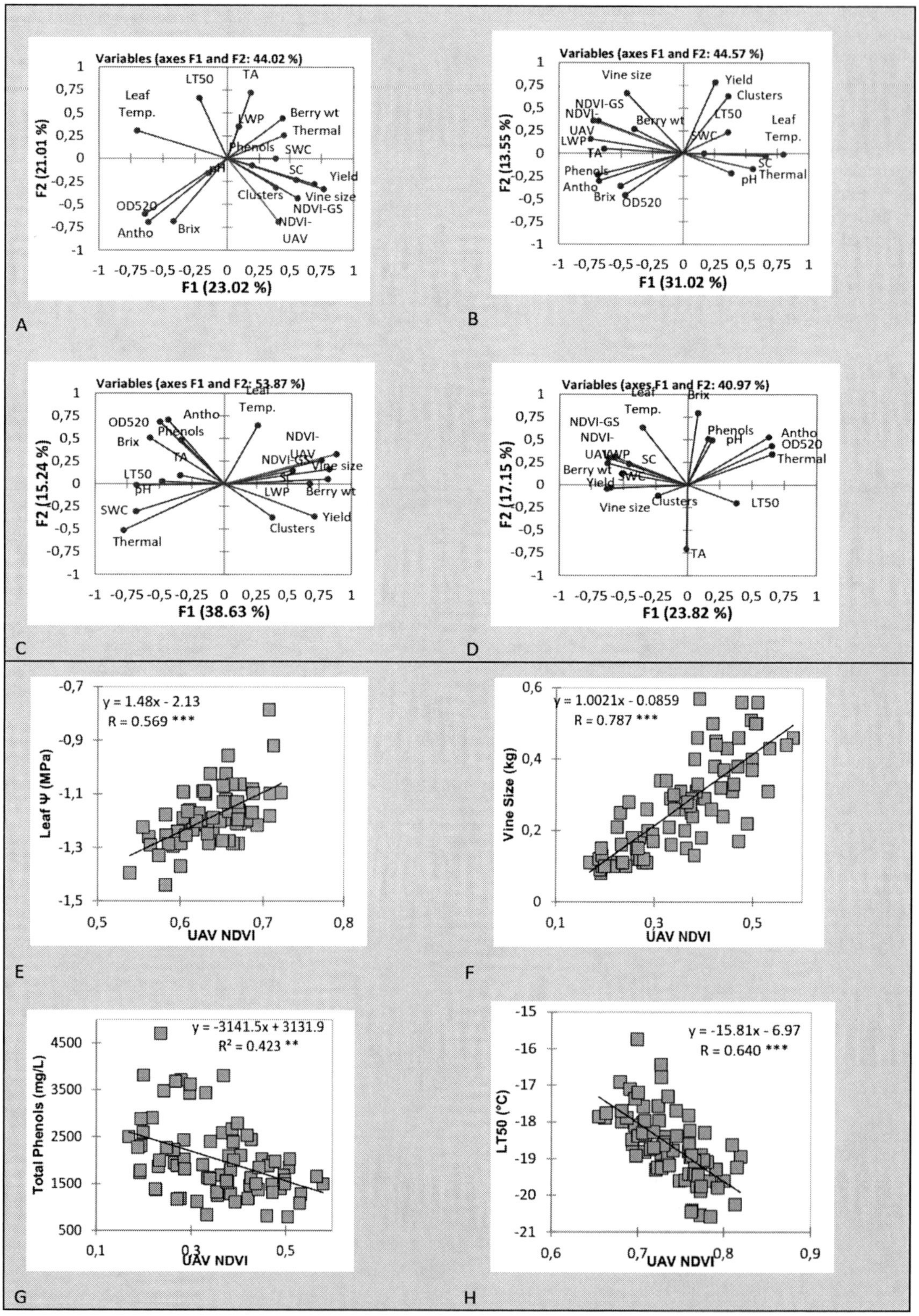

Figure 1. A-D: Principal components analysis of four typical Ontario Cabernet franc vineyards, 2016. A: Buis, Virgil, ON; B: Pondview, Virgil, ON; C: Chateau des Charmes, St. Davids, ON; D: George, Vineland, ON. Abbreviations: GS: GreenSeeker; LWP: Leaf water potential; SWC: Soil water content; SC: Stomatal conductance; TA: Titratable acidity. E-H: Relationships between UAV NDVI and viticultural variables in three typical Ontario Cabernet franc vineyards, 2016. E: George, Vineland, ON; F,G: Chateau des Charmes, St. Davids, ON; H: Buis, Virgil, ON.

https://doi.org/10.1051/e3sconf/20185002010

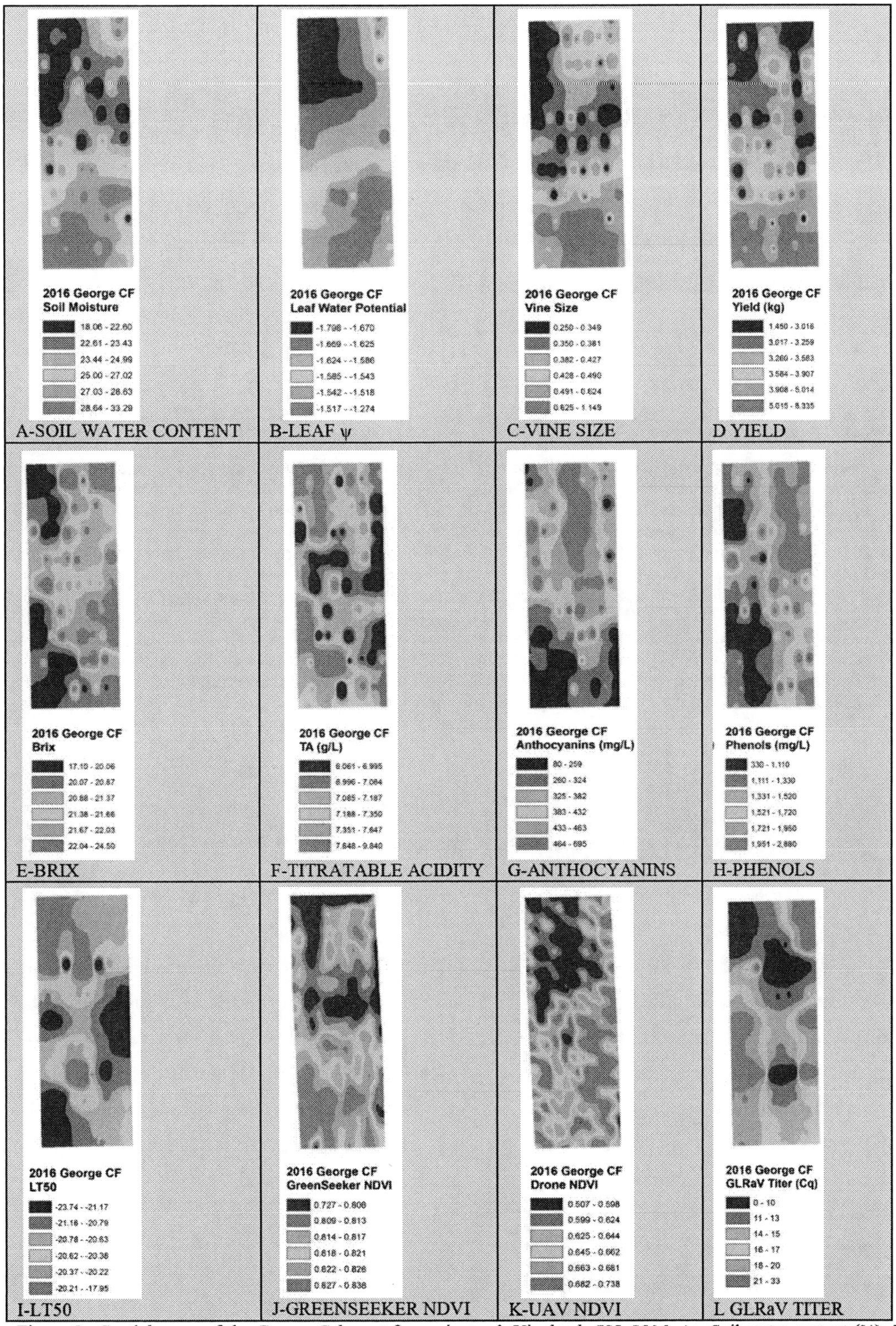

Figure 2. Spatial maps of the George Cabernet franc vineyard, Vineland, ON, 2016. A: Soil water content (%); B: Leaf water potential (MPa); C: Vine size (kg); D: Yield (kg); E: Brix; F: Titratable acidity (g/L); G: Total anthocyanins (mg/L); H: Total phenols (mg/L); I: LT50 (°C); J: NDVI acquired by GreenSeeker; K: NDVI acquired by UAV; L: GLRaV titer (Cq).

An example of a vine-physiology based fine-scale terroir study in the AOC Lavaux-region in Switzerland

Markus RIENTH[1]*, Frederic LAMY[1], Dorothea NOLL[1], Fabrice LORENZINI[2], Olivier VIRET[4] and Vivian ZUFFEREY[3] ,

[1]CHANGINS, haute ecole de viticulture et œnologie, route de Duillier 60, 1260 Nyon, Switzerland
[2]Agroscope, route de Duillier 50, 1260 Nyon , Switzerland
[3]Agroscope, avenue Rochettaz 21, 1009 Pully, Switzerland
[4] Service de l'agriculture et de la viticulture (SAVI), Avenue de Marcelin 29, 1110 Morges, Switzerland

***Corresponding author: M. Rienth: markus.rienth@changins.ch**

Key words : terroir, vine physiology, steep slope viticulture, AOC Lavaux, Villette, Chasselas

Abstract

The aim of this ongoing study is to establish a vine physiology, berry quality - based methodology to analyze small scale differences between plots in a sub-region of the AOC Lavaux, in Switzerland. The production area Villette is almost to 100% planted with the cultivar Chasselas L. covering a total of 140 ha of steep slope terraced vineyards and offers an excellent site for terroir studies. A representative network of 23 plots covering the entire Villette production area was established in 2016 and 2017. Important differences in budburst and flowering were observed between plots whereas altitude was the main driver of precociousness. Expectedly most precious plots had highest sugar and lowest malic acid concentrations at final sampling date. Individual berry weight variation was up to 35 % at the last sampling (2.4 to 3.3 g) with variation in sugar concentration of 13 % (70-84° Oechsle). Sugar quantity per berry variation was however twice as high (26%), illustrating important differences in the sugar accumulation pattern between plots. Interestingly the latest plots accumulated highest sugar per berry with lowest concentration. The preliminary results of this ongoing study provide valuable information of the ripening potential of Chasselas in a small mesoclimate and illustrates the possibilities to characterize viticultural terroirs based on vine physiology measurements.

Introduction

Terroir is the interaction of soil, plant and climate in combination with the socioeconomic history of a given region (Leeuwen & Seguin 2006, Anderson et al. 2012, Tomasi et al. 2013). Plenty of terroir studies have been conducted in different growing regions, focusing of different aspects of the latter equation. In most studies, including plant physiology, a correlation between moderate water deficit and terroir perception could have been shown, which illustrates the importance of vine physiology on terroir perception (van Leeuwen et al. 2009, Reynard et al. 2011). In Switzerland several terroir studies were conducted in most regions in recent years focusing in particular on pedoclimatic characteristics reveailing a huge diversity in soils mainly originating from glacier moraines.(Letessier & Marion 2007).

A recent study conducted in the canton Vaud (3800 ha) considered 10 varieties in a total of 130 micro-plots, providing important information on cultivar adaptation to different sites on a relatively large scale considering the small structure of Swiss viticulture. A main conclusion of this terroir study was as well that water relations play a crucial role in terroir perception by their influence on vine physiology (Zufferey et al. 2008)

The latter study paved the way for more fine-scale mapping of different growing regions in the future, which seems necessary to better valorize the high diversity of terroirs encountered in Switzerland.

https://doi.org/10.1051/e3sconf/20185002011

The most planted white wine variety (>60%), Chasselas, documented in the lake of Geneva's area since thousands of years, plays a central role in Swiss viticulture terroirs. One of the most emblematic Chasselas region is the AOC Lavaux (736 ha), where vines are planted in terraces on steep slopes over the lake of Geneva on different altitudes and expositions. Steep slope viticulture in this area is not or very little mechanizable and therefore production costs are very high which threatens the sustainability of small wine producers. To maintain this viticulture patrimony, established by the monks in the 11th century, it is crucial to render consumers aware of the uniqueness of such regions. The aim of this study is to establish a vine physiology, berry quality - based methodology to analyze small scale differences between plots in a sub-region of Lavaux, the production area Villette, in order to better valorize such terroirs.

Material and Methods

A network of 23 plots covering the whole Villette production area was established in 2016 and completed in 2017 (Figure 1)

All selected plots are planted with the cultivar Chasselas L. on rootstock 3309C, guyot-pruned and trained in a vertical shoot positioning system, with vine age ranging from 15 to 25 years.

In all 23 plots, main phenology stages such as budburst, flowering, véraison and harvest were recorded throughout the season. Berry development was followed on a weekly basis on a 300 berry sample, starting two weeks before véraison. Average berry weight, sugars, tartaric and malic acid were determined by FTIR analysis. Vine vegetative growth was characterized by chlorophyll content of leaves, total exposed leaf area, delta ^{13}C determination in berries and pruning weight.

Exposition, altidue and inclination were calculated with QGIS 2.18.14® (2m precision). Plots were located on an elevation gradient from 380 to 600 m asl (Figure 2). Calculation of potential intercepted solar radiation between 1st april and 30th September (2m precision) ArcGIS 10.2 – solar Anaylst ® (Figure 3).

Principal component analysis and hierarchical clustering was performed using the R package FactoMinR®. Curve fitting for berry development was performed with Origin pro 2016®.

Results and discussion

Results of principal component analysis on vegetative (pruning weight, delta ^{13}C, N-tester), reproductive (sugar concentration in °Oechsle, extract, malic and tartaric acid) and geophysical data (potential radiation, altitude, inclination) are shown in figure 2. The first-dimension accounting for 27 % of the variation between plots was positively correlated with altitude, day of budburst and flowering as well as individual berry weight, all latter parameters were negatively correlated with sugar concentration (°Oechsle). This can mainly be explained by temperature decrease with increasing altitude which delayed budburst and flowering, and thus the onset of ripening which switched sugar accumulation to cooler periods (Burgos et al. 2010). Interestingly individual berry weight was higher in later ripening plots. The second dimension of the PCA accounting for only 17 % of variation between plots is dominated by potential maximal radiation exposed leaf area and negatively with tartaric acid and inclination. According to delta ^{13}C analysis no water deficit occurred in 2017 thus this variable does not contribute to a high extent to plot variability.

Hierarchical clustering was performed on PCA results and enabled the grouping of plots in three well distinguished clusters based on total inertia.

 https://doi.org/10.1051/e3sconf/20185002011

Cluster 1 comprises the plots with highest sugar concentration, mainly explained by highest potential radiation but as well characterized by lowest pruning weight thus less vigor's and a relatively high ELA / kg. Main variables explaining the grouping of plots to Cluster 2 are lower altitude and thus preciousness of budburst, flowering and ripening. Cluster 3 regroups the latest ripening plots, primarily due to their higher altitude.

Since sugar concentration is still amongst the most important ripeness indicators and of utmost importance notably in septentrional winegrowing regions such as Switzerland, kinetics of sugar concentration (°Oechsle) evolution and sugar accumulation during berry development were analyzed separately (Figure 5A and B). A continues increase in all plots of sugar concentration (°Oechsle) was observed until harvest with a variation of around 13 % between highest and lowest plots at the last sampling date (70 to 84 °Oechsle). Quadratic (order 2) polynomial regression models described best the evolution of sugar concentration. Interestingly regarding sugar quantity per berry a much higher variation of 26 % was observed in combination with a very different accumulation pattern, indicated by the best fitting curves, all polynomial fits of order 4. This is due to the cessation of phloem sugar unloading in the berry and shows that a subsequent increase in sugar concentration during the late ripening is only due to environmental factors in precious plots.

In plots with highest sugar concentration at harvest sugar accumulation ceased very early during the season, in some plots already at the 5th sampling date (ex. plot 23 and 22) thus further concentration is due to water loss of the berry most likely by transpiration through the berry skin or potentially by waterbacklow in the xylem, which is however less likely (Bondada et al. 2005, Keller et al. 2006). In plots where the onset of ripening was late, phloem unloading stopped right before harvest or not at all according to figure 5B. Such late plots had also the highest individual berry weight (Figure 4 A). Maybe due to the ripening phase that occurred during a cooler period, thus less respiration losses of the vines could have favored / extended berry growth and sugar accumulation. It could as well be hypothesized that due to later budbreak and flowering of late plots cell size as the main factor influencing final berry size and determined very early during the first green growth phase before véraison, was favored by warmer temperatures and/or a higher water availability in late spring / early summer (Ojeda et al. 2001, Diakou-Verdin et al. 2002). Anyhow, such hypothesis needs to be confirmed in subsequent years.

Conclusion

Intermediate results presented in the paper demonstrate the potential to use vine physiology and berry quality to characterize viticulturally terroirs on a very small scale in a pretty similar mesoclimate. Differences between plots in regard to precociousness and thus sugar concentration at harvest due to pedoclimatic differences, in particular altitude could be highlighted which will be of valuable information for growers in view of global warming.

Results on berry sugar accumulation gave interesting insights regarding berry developmental characteristics that are not fully understood by the scientific community up to now, mainly the impact of genotypic versus environmental factors.

 https://doi.org/10.1051/e3sconf/20185002011

Figure 1 – Localization of Chasselas plots in the Villette production area

Figure 2 – Altitude map

 https://doi.org/10.1051/e3sconf/20185002011

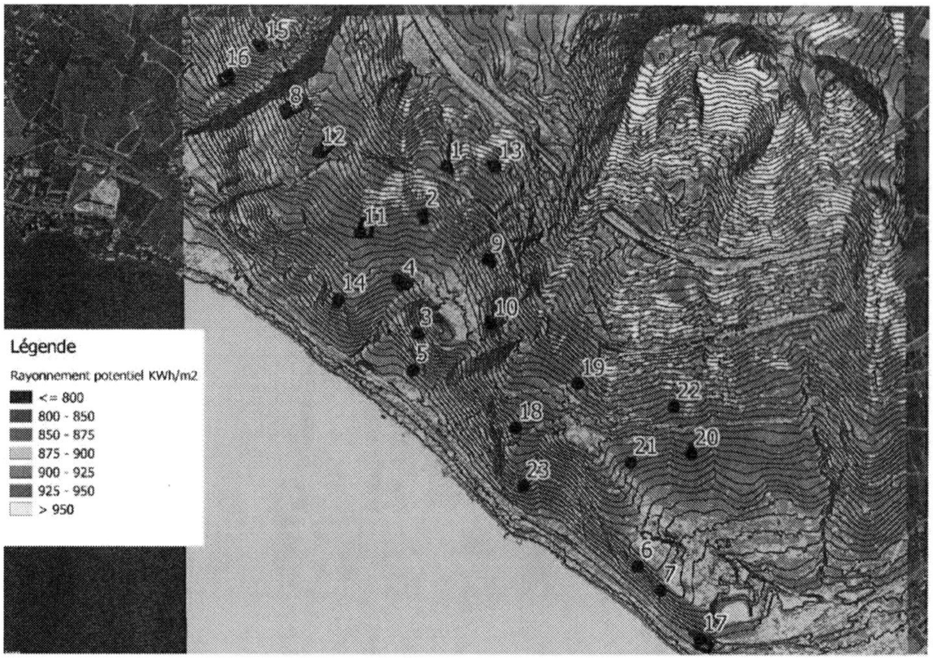

Figrue 3 – Radiation Map

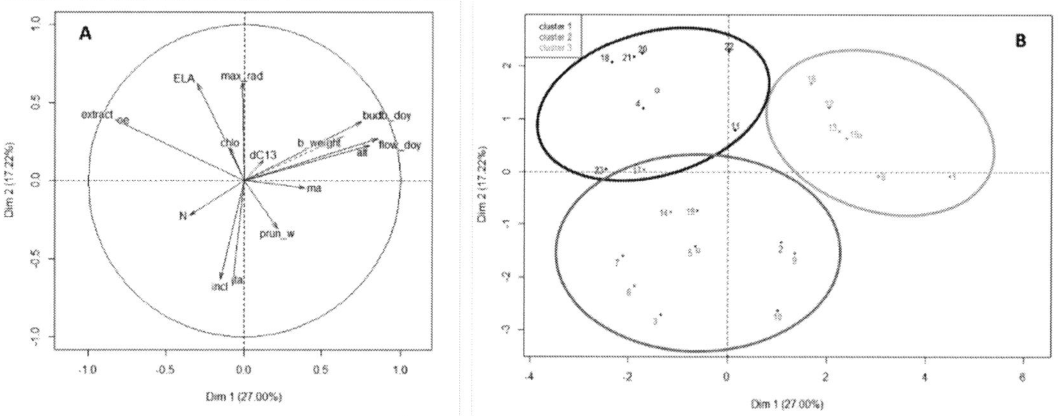

Figure 4- A: Principal component analysis of vegetative and biochemical and phenological data; b_weight: individual berry weigh, alt: altitude, budb_doy: day of budburst, flow_doy: day of flowering; dC13: delta 13C, prun_w: pruning weight, N: nitrogen content in must, ELA: exposed leaf area/kg, max_rad: potential solar radiation between 1st April to 30th September, chlo: chlorophyll content, oe: Sugar concentration in °Oechsle; ma: malic acid (g/L), ta: tartaric acid (g/L) B: factorial map of plots groups by hierarchical clustering

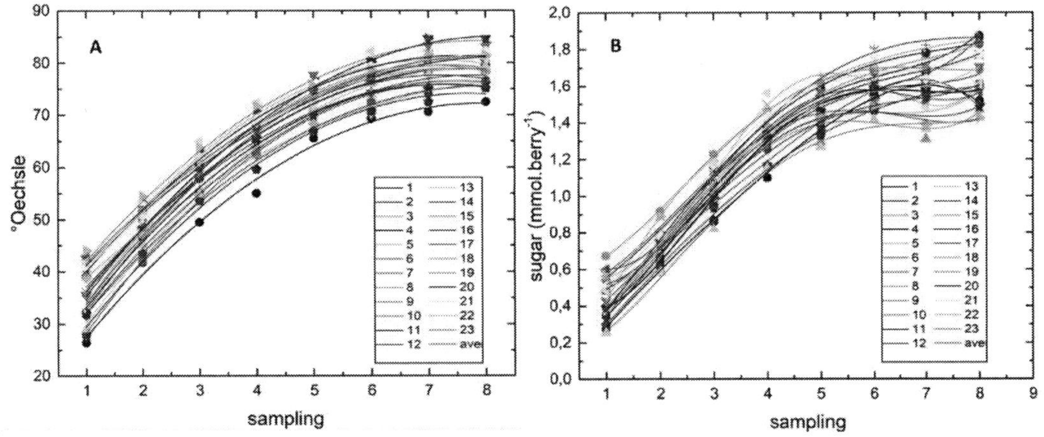

Figure 5: A: Sugar concentration evolution (in °Oechsle) in all plots, fitted curvesfma are polynomial order 2. B: Sugar accumulation pattern per berry of all plots (fits are polynomial order 4)

Acknowledgement: We would like to the thank the Fédération Vaudoise des vignerons (FVV) for funding the project as well as Thibault Verdenal, Pierre Durand and Cyril Toche for their help during sampling and sample processing.

Anderson, J.D., Jones, G. V., Tait, A., Hall, A. and Trought, M.C.T., (2012) Analysis of viticulture region climate structure and suitability in New Zealand. OENO One **46**, 149.

Bondada, B.R., Matthews, M. a. and Shackel, K. a., (2005) Functional xylem in the post-veraison grape berry. J. Exp. Bot. **56**, 2949–2957.

Burgos, S., Almendros, S. and Fortier, E., (2010) Facteurs environnementaux et phénologie de la vigne dans le canton de Genève. Rev. suisse Vitic. Arboric. Hortic. **42**, 288–295.

Diakou-Verdin, P., Carde, J.-P., Gaudillère, J.-P., Barrieu, F., Ollat, N. and Moing, A., (2002) Grape berry development : A review. OENO One; Vol 36, No 3 J. Int. des Sci. la vigne du vinDO - 10.20870/oeno-one.2002.36.3.970 .

Keller, M., Smith, J.P. and Bondada, B.R., (2006) Ripening grape berries remain hydraulically connected to the shoot. J. Exp. Bot. **57**, 2577–2587.

Leeuwen, C. Van and Seguin, G., (2006) The Concept of Terroir in Viticulture. J. Wine Rearch **17**, 1–10.

van Leeuwen, C., Tregoat, O., Choné, X., Bois, B., Pernet, D. and Gaudillére, J.P., (2009) Vine water status is a key factor in grape ripening and vintage quality for red bordeaux wine. How can it be assessed for vineyard management purposes? J. Int. des Sci. la Vigne du Vin **43**, 121–134.

Letessier, I. and Marion, J., (2007) Etude des terroirs viticoles valaisans (Suisse). Etude géo-pédologique, partie générale, méthodologie. Etudes sols terroirs, Letessier-Marion, 38410 St-Martin d'Uriage, 123 p.

Ojeda, H., Deloire, a. and Carbonneau, a., (2001) Influence of water deficits on grape berry growth. Vitis **40**,

141–145.

Reynard, J.Ś., Zufferey, V., Nicol, G.C. and Murisier, F., (2011) Vine water status as a parameter of the ≪ terroir ≫ effect under the non-irrigated conditions of the vaud viticultural area (switzerland). J. Int. des Sci. la Vigne du Vin **45**, 139–147.

Tomasi, D., Gaiotti, F. and Jones, G. V., (2013) The Role of Landscape in the Productive Context and in the Quality of Prosecco Wine, in: The Power of the Terroir: The Case Study of Prosecco Wine. Springer Basel, Basel, pp. 235–248.

Zufferey, V., Pythoud, K., Letessier, I., Reynard, J.., Moncio, C. and Murisier, F., (2008) Etdues des terroirs viticoles suisses. Rev. suisse Vitic. Arboric. Hortic. **40**, 367–374.

Scale effect of viticultural zoning under three contrasting vintages in Chianti Classico area (Tuscany, Italy)

Simone Priori[1*], Giuseppe Valboa[1], Sergio Pellegrini[1], Rita Perria[2], Sergio Puccioni[2], Paolo Storchi[2], Edoardo A.C. Costantini[1]

[1] CREA Research Centre for Agriculture and Environment, Cascine del Riccio, Firenze, Italy.
[2] CREA Research Centre for Viticulture and Enology, Arezzo, Italy

*corresponding author email: simone.priori@crea.gov.it

Abstract. The present study, performed in one of the largest farm of "Chianti Classico" wine district (Tuscany, Italy), focused on the effect of terroir on the peculiarities of the wine at two different scale of zoning. At a broader scale, the experimental vineyards were selected on the basis of lithology, soil type, morphology and climate and these were called macro-terroir (MT). Each vineyard was subdivided into a couple of homogeneous zones (Unité Terroir de Base, UTB), differentiated on the basis of soil proximal sensing, and then, high-detailed soil mapping. The study was conducted during three consecutive vintages (2012, '13 and '14), on four different vineyards MT, which are representative of the Chianti Classico wine district. Grape harvest, wine-making and six-month ageing were carried out separately for each UTB, using the same methodology. This study demonstrates that characteristics of geopedological landscapes can be used for a wine district zoning, while a more detailed soil mapping, leading to UTB identification, is needed to highlight some wine peculiarities.

1. Introduction

Lithology and soil spatial variability is usually very high in most of the Italian viticultural areas. This variability can provide distinctive peculiarities to the wines, not only within a single wine district, but also within a single farm and a single vineyard. Recent approaches in managing spatial variability in vineyards are based on high resolution mapping of homogeneous vineyard zones. These basic units are used to plan differentiated vineyard management and grape harvest, in order to highlight the terroir effect on wine [1, 2, 3].

Several studies have addressed the terroir effect on wine on the basis of grape analysis and micro wine-making experiments within small vineyard plots [4, 5]. This approach allows to study the relationships between grape, vine and soil features accurately, however it does not answer the following questions [6]:

i) does variation of soil properties have a functional impact on wine peculiarities?

ii) if so, at what scale are these effects expressed (Between vineyards, within vineyard)?

The present study was performed in a large farm of the "Chianti Classico" wine district (Barone Ricasoli farm, Tuscany, Italy), with the aim to analyse the effect of terroir on the peculiarities of the wine, at two different zoning scales. At broader scale, four experimental vineyards were selected according to lithology, soil type, morphology and climate ("Macro-Terroirs", MT). At finer scale, each vineyard was subdivided into couples of homogeneous zones, differentiated on the basis of soil proximal sensing data, and then soil features like texture, gravel content, soil depth, and available water capacity (Basic Terroir Units or Unité Terroir de Base, UTB, [7]). The four macro-terroirs selected were highly representative of the Chianti Classico district. The grape from the different UTBs was harvested and processed for wine-making separately, following the same procedures.

2. Materials and methods

The vineyards studied for this work belong to Barone Ricasoli estate, a large and old farm in the Chianti Classico wine district, Tuscany. The farm embraces the four main geological units of the Chianti Classico district, which are: i) shales, marls and limestones (STO), covered by a thick layer of feldspathic sandstone (MAC), which represent the upper part of Tuscan geological unit, generally situated on the top of the hills (400-600 m a.s.l.); ii) limestone and shales of the Monte Morello formation (MLL), a Ligurian geological unit, situated on the top of the hills or along the slopes of the hills (300-500 m a.s.l.); iii) marine deposits of the Pliocene period (PLIs), and iv) deposits of fluvial terraces (AT) on marls and shales of the Sillano formation (SIL), both situated beneath 300-350 m a.s.l.

In this work, we defined as "macro-terroir" (MT) a large area characterized by the same lithology, morphology and climate, whereas we defined as "basic terroir units" (or Unité Terroir de Base, UTB [7]) sub-areas within a MT of about 2 ha in size, characterized by homogenous soil features (texture, stoniness, soil depth, available water capacity, etc.).

The four MT (Fig.1), representative of the four main lithology described above, were selected:

- SAND: sandy and gravelly soils on feldspathic sandstone, calcium carbonate free (< 1%).
- CALC: clayey and gravelly soils on limestone and shales, with high calcium carbonate content (> 25%).
- MAR: loamy soils on marine deposits, characterized by gravel lenses and moderate calcium carbonate content (10-20%).
- FLUV: loamy and gravelly soils on ancient fluvial terraces, with variable calcium carbonate content (10-30%).

The study was carried out in 12-16 years old *Sangiovese cv.* vineyards, the vine density varies between 6200 and 6600 vines/ha, and the trellis system was the simple spurred cordon with vertical shoot. The experimental vineyards received the same viticultural treatments during the growing season.

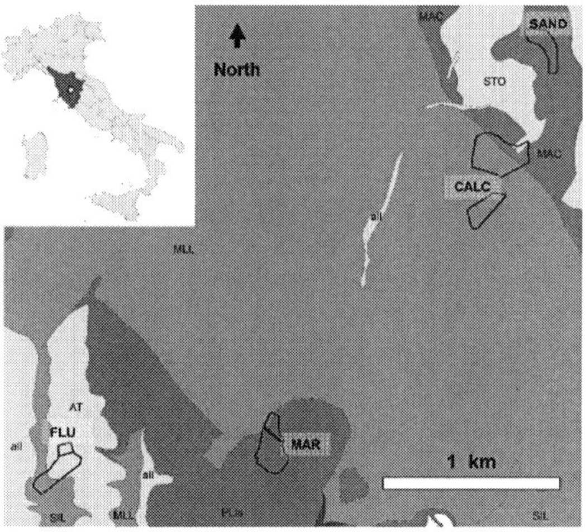

Figure 1: Experimental vineyards and geological formation. MAC: sandstone; STO: shales, marls and limestone; MLL: limestone and shales; PLIs: marine sandy and gravelly deposits; AT: ancient fluvial terraces; SIL: marls and shales; all: recent alluvial deposits.

The vineyards were surveyed by proximal sensing in order to obtain high-detail maps of soil spatial variability and to delimit the UTBs. The proximal sensors used were: i) EM38-Mk2 electromagnetic induction sensor (Geonics Ltd., Ontario, Canada) and ii) "The Mole", a gamma-ray spectroradiometer (Soil Company, The Netherlands). The EM38-Mk2 measures the soil apparent electrical conductivity (ECa) across two depth ranges of 0-75 (ECa_1) and 0-150 cm (ECa_2), approximately [8]. "The Mole" spectroradiometer measures continuously the gamma-ray natural emission (total count, TC) coming from the first 30-40 cm of the soil and rocks, through a Cesium Iodide scintillator crystal [9]. Two UTB within each MT were delimited according to the k-means clustering. The variables used for clustering were: apparent electrical conductivity ECa1 and ECa2; total count of gamma-ray (TC); slope and aspect obtained by a digital elevation model with a detail of 10 m.

Each MT was subdivided into two 1.5-2 ha size UTBs, for a total of 8 UTBs. Small areas with deep and too fertile soils, or subject to waterlogging or exceptional soil erosion, were excluded. The grape from each UTB was harvested and vinified separately, using the same methods. The grape must was analysed to determine sugar content, pH and malic acid concentration. Seven months after harvesting, the wines were analysed to assess the alcohol content, total polyphenols and anthocyanins, total acidity, dry extract, glycerine, and colour intensity.

Moreover, the wines were evaluated by a panel of 10 wine tasters through a "blind tasting". Sensory evaluation was performed to assess differences between the wines, therefore the evaluation method was mainly comparative. In particular, the wines were scored on a scale of 1 to 10 for each sensory attribute, giving the score 10 to the wine which expressed better than the others such parameter. The sensory parameters analysed were: overall evaluation, body, acidity, flavour intensity. In addition, the tasters indicated their feeling about the wine in terms of aroma typology: fruity, floral, spicy, and herbaceous.

In order to verify the interactions between vintage, MT, and UTB on wine peculiarities, a mixed design models were adopted, using vintage as random factor, MT as fixed factor and UTB nested in MT (hierarchical nested ANOVA).

3. Results and discussion

3.1 Climate and terroir

The climatic conditions were very variable during the three experimental vintages (Fig.2). In particular, 2012 showed the highest temperature (above the average) and a very dry summer, with some heavy rains two weeks before the harvest. In summer 2013, the temperature and rain trends were in line with the long-term averages, whereas summer 2014 was colder and more humid than the average. The modified Winkler index (1st April – 30th September) was higher in CALC and SAND, than in MAR and FLUV (Tab.1).

Among the selected UTBs, CALC1 and 2 showed the highest content of clay, gravel, and calcium carbonate, along with the lowest gamma-ray TC and moderately high ECa. They significantly differed from each other only for the sand content, which is slightly higher in CALC2. The macro-terroir SAND was characterized by low ECa and high gamma-ray TC, because of the parent material mineralogy rich in k-feldspars and muscovite, which have high content of potassium, and then ^{40}K radionuclide. Soils were sandy and calcium carbonate free. SAND2 showed higher stoniness and lower AWC than SAND1. MAR1 and 2 showed good differentiation in terms of ECa, gamma-ray TC and soil features. In particular, MAR1 showed higher clay content, calcium carbonate and AWC, as well as lower stoniness, than MAR2.

Both FLUV1 and FLUV2 had clay-loamy texture, but FLUV2 showed higher content of sand, gravel and calcium carbonate.

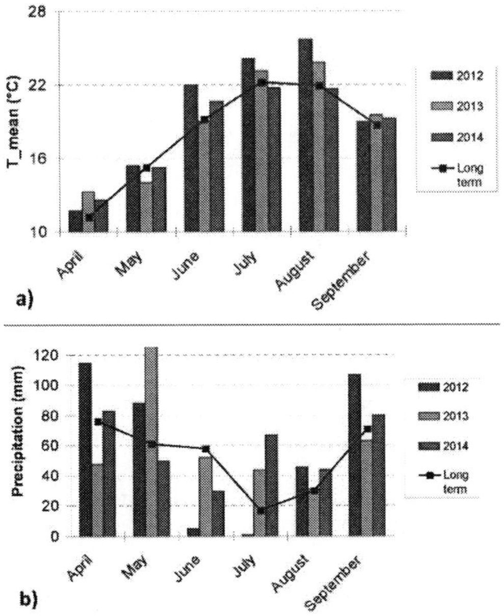

Figure 2: Mean air temperature and precipitation during the three growing seasons.

Table 1: Main features of the UTB. 1: modified Winkler index (calculated between 1st of April and 30th September); 2: total calcium carbonate; 3: Available Water Capacity.

UTB	WI_{mod} [1] Σ°C	ECa_1 (mS·m^{-1})	ECa_2 (mS·m^{-1})	TC (Bq·kg^{-1})	Clay (g·100g^{-1})	Sand (g·100g^{-1})	$CaCO_3$ [2] (g·100g^{-1})	Gravel (m^2·m^{-2})	AWC [3] (mm·m^{-1})
CALC1	1737	20.4	28.2	317	36.0	16.5	25.1	30.8	114
CALC2		21.7	25.8	318	33.6	21.6	33.9	36.2	94
SAND1	1730	8.2	12.1	669	9.8	59.8	0.4	19.2	113
SAND2		5.7	11.2	666	10.0	61.0	0.2	36.5	64
MAR1	1613	32.9	31.0	410	35.3	33.5	27.7	8.1	139
MAR2		23.4	23.4	358	19.2	51.3	16.5	21.8	103
FLUV1	1536	22.5	28.3	443	31.5	31.4	8.5	7.0	79
FLUV2		19.0	25.6	404	25.5	39.8	15.6	27.2	67

3.2 Oenological results

The contrasting climatic conditions that occurred during between the three experimental years determined vintage-to-vintage differences in the grape and wine quality. The coldest and most humid summer of 2014 provided poor differentiations between wines, hiding the terroir effect. On the contrary, during the warmest and driest summer of 2012, the wines from the eight UTBs showed the clearest differences. The results of mixed-design ANOVA, performed with MT as fixed effect, vintage as random effect and UTB nested in MT (tab.2), showed that:

- The variability in must pH, wine total acidity, glycerine and colour intensity was explained better by the MT than the vintage climate, but the colour intensity was strongly dependent from the interaction between MT and vintage.
- Must malic acid, wine polyphenols, anthocyanins and dry extract were more influenced by the vintage climate, although the effect of the MT and the interaction MT x vintage played an important role. Vintage climate also affected the herbaceous and spicy notes in the wine flavor.
- In general, the effect of the UTB seemed important for differentiation of wine colour intensity, as well as for flavor intensity and fruity notes.

Changing the zoning scale from MT to UTB according to soil physical and hydrological properties, appeared to have a significant effect only under dry summers, like in 2012 and, to a less extent, in 2013.

Table 2: Fischer's F-values resulting from the mixed-design ANOVA, using MT effect as fixed factor, vintage effect as random factor, and UTB nested in MT. In bold, F-values significant for $p < 0.05$.

Source of variance	MT	Vintage	MT x vintage	UTB (MT)
Must				
Sugar	1.5	0.4	1.7	1.2
pH	**5.1**	2.5	2.4	1.2
Malic acid	**9.6**	**61.3**	**4.3**	2.7
Wine				
Total acid.	**5.1**	4.8	2.2	1.9
Polyphenols	**8.4**	**41.0**	1.1	1.5
Anthocyanins	2.3	**6.6**	3.5	0.7
Dry extract	1.5	**5.2**	**4.6**	0.8
Glycerine	3.3	0.5	1.9	0.7
Colour int.	3.7	0.1	**23.4**	**13.1**
Wine taste				
Flavour int.	0.1	2.4	**2.4**	**3.3**
Fruity	1.4	1.3	1.6	**3.5**
Floral	0.1	0.6	**2.5**	1.3
Herbaceous	0.2	**20.3**	0.5	2.2
Spicy	0.9	**14.7**	0.5	0.7
Body	0.3	1.0	**5.5**	0.6
Acidity	1.9	3.3	2.0	0.8

In general, the wines produced in CALC macro-terroir showed higher alcohol, anthocyanins, dry extract and colour intensity than the average, as well as high fruity flavor, high acidity and body. Similar results were reported in previous studies of the authors [2, 10]. In particular, Ricci Alunni [11] observed high alcohol and colour intensity as distinctive features of wines produced on calcareous-clayey soil. On the contrary, the wines from SAND were characterized by lower colour intensity and lower acidity than the average, especially in the SAND2 UTB (higher stoniness and lower AWC), where however the flavor intensity was always high, in particular for fruity notes. This results are consistent with those reported by Ricci Alunni [11], who described the wines from sandy non-calcareous soils as elegant, scented, but scarce in colour. The lower acidity of these wines can be explained by the very low content of calcium carbonate and the high content of potassium in the soil, which tend to increase must and wine pH [12].

The wines produced on marine loamy sands (MAR) provide wines with low (MAR1) or medium (MAR2) alcohol and low glycerine. Moreover, MAR2, characterized by a soil with higher sand and gravel contents, provided wines with general higher flavor intensity, in particular fruity and floral notes. The differentiation between MAR1 and MAR2 was particularly evident in the driest summer of 2012, whereas it disappeared in the vintage 2014.

The wines produced on fluvial terraces (FLUV) showed higher alcohol content, polyphenols and glycerine concentrations, especially in FLUV1, characterized by higher soil AWC and lower gravel. This confirms the results from a previous viticultural zoning work in the Siena province [10], showing higher alcohol and polyphenols values on fluvial terraces than in the other terroirs of the province.

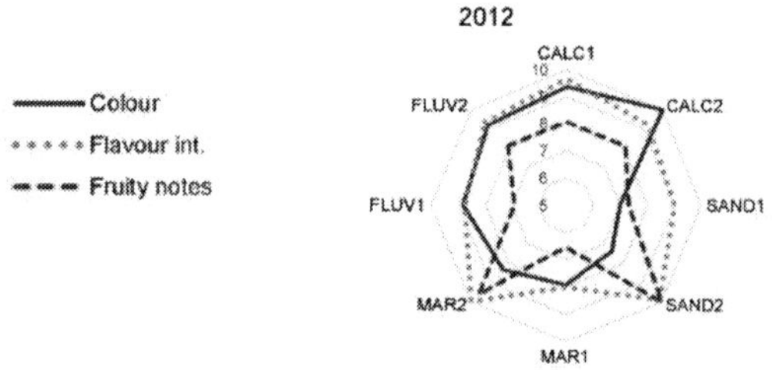

Figure 3: Spider graphs reporting the means (10 tasters) of the most significant wine tasting parameters (colour, flavour intensity, and fruity notes) for each UTB in the three vintages.

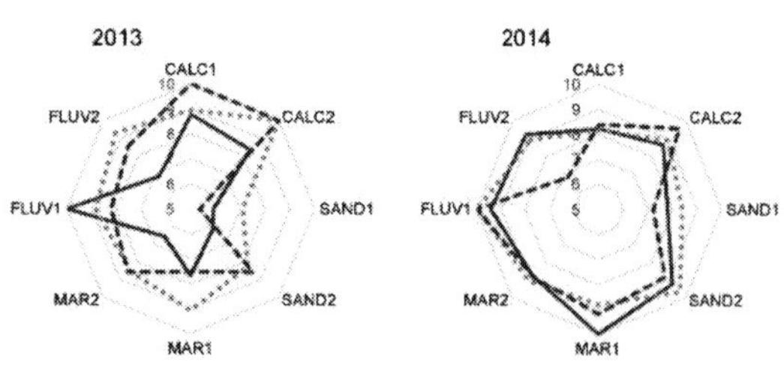

Acknowledgements

This work was supported by the "Barone Ricasoli" s.p.a. farm (VignaCRU project). The authors wish to thank the agricultural manager of the farm Massimiliano Biagi and co-workers (Fabio Cascella, Marco Cerqua, and Claudio Carapelli) for their support. Many other peoples contributed to the VignaCRU project and to this work, in particular we want to thank Pierluigi Bucelli, Nadia Vignozzi, Stefano Mocali, Arturo Fabiani, Romina Lorenzetti, Maria Fantappiè, Giovanni L'Abate, Roberto Barbetti, and Nadia Bianconi, Alessandra Zombardo and Marco Leprini, Aldo Ciambotti and Valentina Dell'Oro.

References

1. R. G. V. Bramley, J. Ouzman, P. K. Boss, AUST J GRAPE WINE R, **17(2)**, 217-229 (2011).
2. S. Priori, E. Martini, M.C. Andrenelli, S. Magini, A.E. Agnelli, P. Bucelli, M. Biagi, S. Pellegrini, E.A.C. Costantini, SOIL SCI SOC AM J, **77(4)**, 1338-1348, (2013).
3. J. Tardaguila, M.P. Diago, S. Priori, M. Oliveira, ARCH AGRON SOIL SCI, **64(3)**, 409-418, (2017).
4. A. Bonfante, A. Agrillo, R. Albrizio, A. Basile, R. Buonomo, R. De Mascellis, R., ... & P. Manna, SOIL, **1**, 427-441, (2015).
5. M.C. Ramos, G.V. Jones, J. Yuste, EUR J AGRON, **70**, 57-70 (2015).
6. Bramley R.G.V, PROCEEDINGS OF 16TH AUSTRALIAN WINE INDUSTRY TECHNICAL CONFERENCE, (2016), 45-51.
7. A. Deloire A., E. Vaudour, V.A. Carey, V. Bonnardot, C. Van Leeuwen, J. INT. SCI. VIGNE VIN, **39(4)**, 149-162, (2005).
8. J.D. McNeill, Geonics EM38 Ground Conductivity Meter: EM38 Operating Manual, Geonics Limited, Ontario, Canada, (1990).
9. F.M Van Egmond, E.H Loonstra, J. Limburg. 1st Global workshop on High Resolution Digital Soil Sensing & Mapping, Sydney, Australia, (2008).
10. S. Priori, R. Barbetti, G. L'Abate, P. Bucelli, P. Storchi, E.A.C. Costantini, J MAPS, **10(3)**, 466-477, (2014).
11. G. Ricci Alunni, Proceedings of Italian conference "I Paesaggi del vino", Perugia, (2004).
12. B.S. Mpelasoka, D.P. Schachtman, M.T. Treeby, M.R. Thomas. A review of potassium nutrition in grapevines with special emphasis on berry accumulation. AUST J GRAPE WINE R, **9(3)**, 154-168, (2003).

Radar Quantitative Precipitation Estimation improvement in Champagne vineyard.

Basile PAUTHIER[1], Sébastien DEBUISSON[1], Arnaud DESCOTES[1], Julien PERGAUD[2], Sylvain MAILLARD[1]

1 Comité Champagne, 17 rue Chandon Moët 51200 Epernay, France. basile.pauthier@civc.fr
2 Airmet Conception, 151 rue d'Auxonne 21000 Dijon, France

Abstract

Rainfall has a crucial importance in viticulture, especially in Champagne vineyards, where irrigation is prohibited. Rainfall directly influences the phytosanitary pressure, nitrogen mineralization, flowering conditions, parcel practicability, soil erosion etc… In these conditions, implementing a weather stations network is the solution that the Comité Champagne chose to monitor rainfall all over the Champagne appellation since the 1990's. This networks is actually composed of 42 weather stations implemented in order to have the best spatial coverage as possible. The Comité Champagne also obtain some weather stations data from Météo France, the French national weather service. Even with that network, capturing all rainfall events accurately is difficult, especially in convective cases. Therefore, the interest in radar data has increased, to capture rainfall everywhere. Some tests have been previously made with PANTHERE radar data from Météo France with a resolution of 1 km², results were promising, but presented inaccuracies particularly in convective events. In this article, we use a radar merging technique similar to the ANTILOPE method from Météo France, with a higher resolution network. The tool employed is the Estimages toolbox merger, based on krigine with external drift (KED) which has been demonstrated to give good results in quantitative precipitation estimation (QPE) improvement.

Introduction

Champagne is famous for their sparkling wines but also for its particular weather conditions (Briche *et al* 2009, Beltrando & Briche 2010, Briche 2011). This particular weather has a direct influence on Champagne wines (Gerbier & Rémois 1977) and of course will directly condition the human practices on the vines.

Rainfall is a one of the most important meteorological variables in viticulture (Bois 2007, Van Leeuwen *et al* 2009, Pauthier 2017), especially in a region like Champagne where irrigation is prohibited. So far, the only fully reliable technique to capture rainfall was the use of rain gauges (Pauthier 2017). In the context of sustainable viticulture, the problem encountered by winegrowers and interprofessionnal offices was to have reliable localized rainfall information. To complete this goal, 42 automatic weather stations were implemented over the 34 000 ha Champagne appellation. This network has been conceived to have the best spatial distribution with a final aim of data spatialization (Langellier *et al* 2016). Even with such a resolution network, data is not precise enough for a parcel scale vine training. Since 2016, the Comité Champagne also offers the possibility to calculate rainfall amounts directly derived from radar imaging at every point of the Champagne appellation. A QPE evaluation of the tool has been made and show some leaks especially during convective events.

To improve radar QPE, a common strategy is to merge radar and rain gauge data. This technique allows to have a complete spatial coverage with a radar image and some correct rainfall estimations with a rain gauge correction (Champeaux 2009). This strategy has been applied in Champagne. The current paper reports the techniques employed and the obtained results.

Material and method

The study was conducted over the Champagne appellation. Champagne appellation is particular because of its geographical configuration with two well separated vineyards. A northern part composed of the Marne and Aisne vineyards where the climate is oceanic with continental influences (Joly *et al* 2010). These influences are represented by moderately cold winters, warm summers and an annual precipitation averaging 650 mm. The southern part is the Côte-des-Bars vineyards where the continental influence is stronger, with colder winters, warmer summers and a precipitation of around 750 mm.

Weather stations

In order to capture meteorological data, a 42 Pulsonic® weather stations network (Fig 1) is implemented in the Champagne appellation area. Rainfall is registered by 0.2 mm resolution Précis Mécanique® rain gauges. These rain gauges are cleaned and calibrated every month in order to avoid any clogging or deviations.

 https://doi.org/10.1051/e3sconf/20185002013

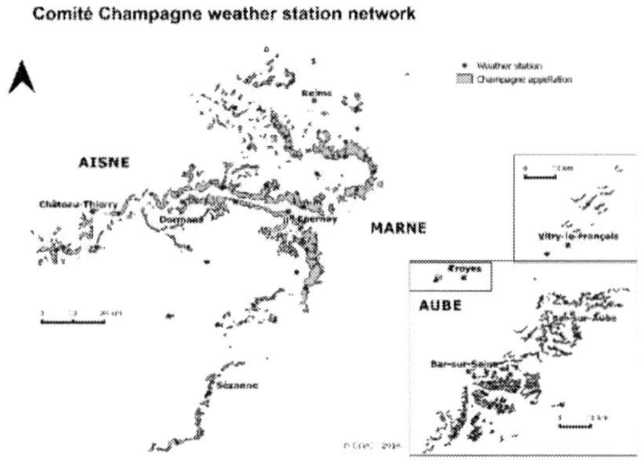

Fig 1. Comité Champagne weather station network (CCWN)

All data coming from the CCWN is validated and corrected every week by a method derived from the Météo France method.

Radar data

Since 2016, the Comité Champagne has an access to PANTHERE radar data from Météo France. This water blade is available at a $1km^2$ resolution and 5 min time step (Tabary 2007). PANTHERE radar data is known for its tendency to underestimate rainfall amounts (Tabary 2007, Pauthier *et al* 2016, Pauthier 2017).
An evaluation made at daily time step over Champagne in early 2017 has shown better results than in these studies, but with the continued tendency of an underestimation that increases with the intensity of rainfall (Fig. 2).

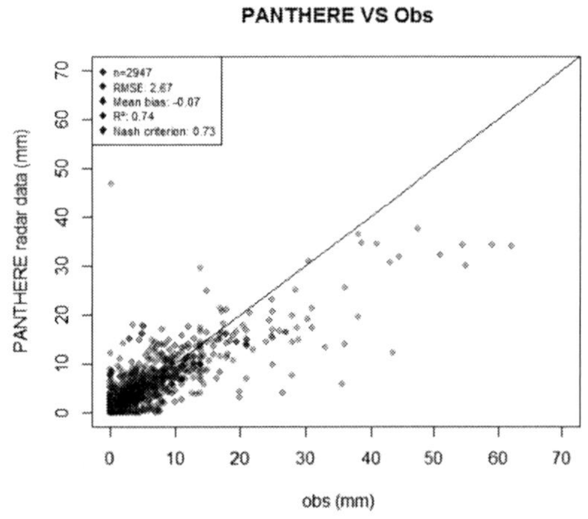

Fig 2. Results from the preliminary PANTHERE evaluation at daily time step in comparison to CCWN rain gauges

Considering these findings, the decision to find a technique to improve the QPE of PANTHERE data with CCWN data have been taken.

Merging techniques

Several methods allow to adjust radar estimations with rain gauges measurement such as the one used for the ANTILOPE radar product of Météo France. This method uses a krigged water blade coming from a rain gauge to adjust radar estimation only for stratiform events. When the events are convective, rain gauges are not used because of the low density of Météo France weather stations (MFWN Champeaux *et al* 2009, 2011). As this method considers only one type of event, we decided to use the Estimages® toolbox which uses co-kriging with external drift (KED), giving good results in mountainous region (Sideris *et al* 2014). KED allows a modification of the model (radar) with a spatial coherency that fits observations while preserving the geographical structure. (Fig 3).

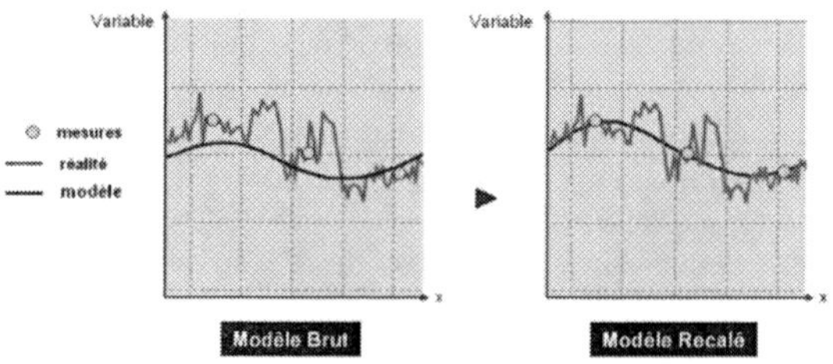

Fig 3. Left : model without correction by KED. Right : model with correction by KED

Results

To evaluate data coming from the merging of radar and CCWN data, totally independent Météo France rain gauges data were used. This data is coming from weather station located near or in Champagne vineyards in order to respect the merging area.
The stations of Reims, Troyes, Chouilly, and Celles-sur-Ource were therefore used to evaluate KED merged radar. A sample from the 2017 November 1st to the 2018 April 15th daily rainfall accumulation has been downloaded and compared to the merged radar data.

The obtained results show that the merger significantly increases the quality of the radar QPE. In fact, the statistics show a gain of almost 15% in the R^2 and Nash criterion and a RMSE around 0,90 instead of 2,7 mm compared to the preliminary 2016 results (Fig 4)

 https://doi.org/10.1051/e3sconf/20185002013

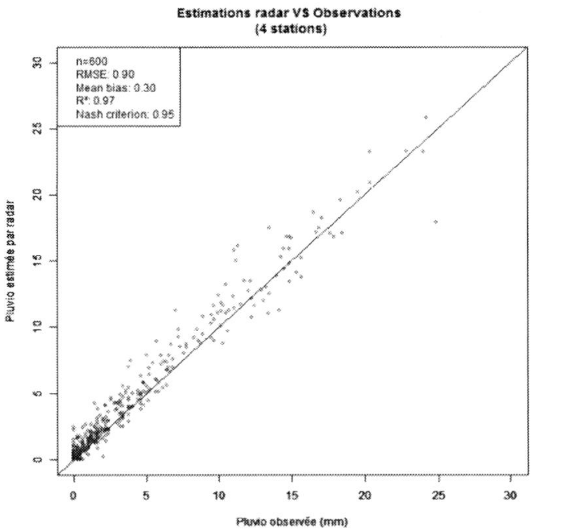

Fig 4. Results from the merged radar data evaluation at daily time step in comparison to 4 Météo France weather stations.

Considering the fact that two of these stations (Troyes and Chouilly) are used by Météo France to correct PANTHERE radar which is at the basis of the merger, we wanted to see if a difference of performance can be seen between the different stations. Individual plots (Fig. 5) has been made to compare the performance of the merger to each station and identify an eventual difference of performance depending on the considered station.

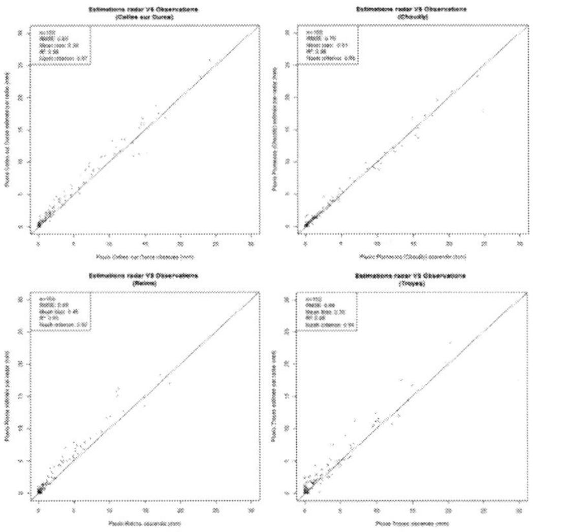

Fig 5. Results from the merged radar data evaluation at daily time step in comparison to 4 individual Météo France weather stations.

The two left plots illustrate the performance of Celles sur Ource and Reims stations which are not used in the PANTHERE algorithm, and the two on the right show the performance of Chouilly and Troyes which are used. No significant differences can be seen between used and non-used weather stations. A t-test confirmed the fact that there is no significant differences between the 4 stations.

Conclusion and discussion

Rainfall data is one of the most important weather parameters in viticulture. It is essential for precise vine management, especially in the context of sustainable viticulture. Until now, the only solution available for winegrowers wanting precise rainfall data was to have as many rain gauges as parcels.
Considering these results, the use of merged radar data to manage the vines at local scale can simplify the task of winegrowers. Nevertheless, these estimations do not replace the use of a rain gauge which still acts as a reference.
Moreover, considering the period of evaluation, from November 2017 to April 2018 only few convective or shower events have been taken into account. These conclusions can be therefore only made for stratiform events which present a low spatial variability.
Further studies are therefore needed to assess the performance of this method over a larger data sample.

BELTRANDO, Gérard et BRICHE, Elodie. Changement climatique et viticulture en Champagne: du constat actuel aux prévisions du modèle ARPEGE-Climat sur l'évolution des températures pour le XXIe siècle. *EchoGéo*, 2010, no 14.
BOIS, Benjamin. *Cartographie agroclimatique à méso-échelle: méthodologie et application à la variabilité spatiale du climat en Gironde viticole. Conséquences pour le développement de la vigne et la maturation du raisin*. 2007. Thèse de doctorat. Université Sciences et Technologies-Bordeaux I.
BRICHE, Élodie. *Changement climatique dans le vignoble de Champagne: modélisation thermique à plusieurs échelles spatio-temporelles*. 2011. Thèse de doctorat. Paris 7.
BRICHE, Élodie, BELTRANDO, Gérard, CAUTENET, Sylvie, *et al.* Vagues de chaleur dans la région du vignoble du Champagne jusqu'en 2050 Heat waves in the Champagne wine region until 2050.
CHAMPEAUX, Jean-Louis, DUPUY, Pascale, LAURANTIN, Olivier, *et al.* Les mesures de précipitations et l'estimation des lames d'eau à Météo-France: état de l'art et perspectives. *La Houille Blanche*, 2009, no 5, p. 28-34.
CHAMPEAUX, J. L., LAURANTIN, O., MERCIER, B., *et al.* Quantitative precipitation estimations using rain gauges and radar networks: inventory and prospects at Meteo-France. In : *WMO joint meeting of CGS expert team on surface-based remotely-sensedobservations & CIMO expert teamon operational remote sensing, Geneva*. 2011. p. 5-9.
GERBIER, Norbert et RÉMOIS, Paul. *Influence du climat sur la qualité et la production du vin de Champagne*. Météorologie nationale, 1977.

 https://doi.org/10.1051/e3sconf/20185002013

JOLY, Daniel, BROSSARD, Thierry, CARDOT, Hervé, *et al.* Les types de climats en France, une construction spatiale. *Cybergeo: European Journal of Geography*, 2010.
PAUTHIER, Basile, BOIS, Benjamin, CASTEL, Thierry, *et al.* Mesoscale and local scale evaluations of quantitative precipitation estimates by weather radar products during a heavy rainfall event. *Advances in Meteorology*, 2016, vol. 2016.
PAUTHIER, Basile. Etude de la variabilité spatio-temporelle des précipitations aux échelles fines: conséquences potentielles pour la viticulture en Bourgogne. 2017.
SIDERIS, I. V., GABELLA, M., ERDIN, R., *et al.* Real-time radar–rain-gauge merging using spatio-temporal co-kriging with external drift in the alpine terrain of Switzerland. *Quarterly Journal of the Royal Meteorological Society*, 2014, vol. 140, no 680, p. 1097-1111.
TABARY, P. The new French operational radar rainfall product. Part I: Methodology. *Weather and forecasting*, 2007, vol. 22, no 3, p. 393-408.
VAN LEEUWEN, Cornelis, TREGOAT, Olivier, CHONÉ, Xavier, *et al.* Vine water status is a key factor in grape ripening and vintage quality for red Bordeaux wine. How can it be assessed for vineyard management purposes. *J. Int. Sci. Vigne Vin*, 2009, vol. 43, no 3, p. 121-134.

Herramientas para la transferencia del conocimiento geo-edáfico: aplicación a los viñedos de Secastilla (DO Somontano, NE España)

D. Badía[1], J.A. Cuchí[1], J. Casanova[1], C. Martí[1], J.M. Rodríguez[1], J.M. Ayuso[2], J. Albar[2]

[1]Escuela Politécnica Superior de Huesca. Universidad de Zaragoza; correo: jcasan@unizar.es

[2] Viñas del Vero S.A.

Introducción

El suelo, dentro del contexto vitícola, puede considerarse como el nexo de unión entre las viñas y el material parental. El suelo, junto al clima, el relieve y el viñedo, permiten diferenciar un determinado "terroir", un espacio geográfico singular debido a las interacciones entre el medio físico y biológico y las prácticas vitivinícolas que en él se aplican (Morlat, 2010). En este trabajo se proponen diversas acciones para divulgar el conocimiento de los suelos de viñedos en el valle de Secastilla (D.O. Somontano), una zona de interés geomorfológico, geológico, edafológico y también histórico.

En este trabajo se muestran diversas herramientas con las que dar a conocer los suelos y su formación en el valle de Secastilla (D.O. Somontano).

Material y métodos

Seleccionados los perfiles de suelos más representativos, se describen macromorfológicamente (FAO, 2009) y micromorfológicamente (Benyarku & Stoops, 2005), y se muestrean para su análisis. De la fracción de tierra fina, se analizan sus principales propiedades físicas y químicas (pH, carbonatos, materia orgánica, CEe, iones en solución, porosidad, estabilidad estructural, CRAD, granulometría, CIC), se clasifican (IUSS, 2007) y se evalúa el potencial de vigor y de precocidad del terreno para la viña (Morlat, 2010). Además, se toman varios kilos de cada horizonte edáfico para disponerlos en cajas de metacrilato y elaborar los monolitos de perfiles. Se identifican los cantos y gravas presentes en los suelos, organizándose en cajas de madera, con una leyenda que indica su naturaleza, edad, origen y localización en la zona de estudio.

Resultados

La evolución geológica en el valle de Secastilla se sintetiza mediante bloques diagrama con el fin de explicar la presencia y distribución de los materiales parentales de los suelos de la zona. Por otro lado, la diversidad de litologías en el valle de Secastilla, y especial- mente la alteración de los conglomerados que aportan multitud de cantos rodados, ha permitido elaborar cajas de rocas.

La posición topográfica, el sustrato geológico conjuntamente con la acción del clima y la vegetación a lo largo del tiempo dan origen a las diferentes clases de suelos. Por otro lado, las imágenes obtenidas con polarizadores paralelos permitieron observar los procesos de calcificación, humificación y gipsificación, Con la información generada se publicaron diferentes paneles y estructuras informativas.

Conclusiones

A la tradicional caracterización de las propiedades de suelos y su distribución, le acompaña la elaboración de cajas de piedras, monolitos de perfiles de suelos (para su observación en sala y en campo) y paneles que muestran desde la evolución geológica hasta el potencial de vigor del terreno. Suelos y geología se incorporan, así como elementos con los que potenciar el turismo cultural y el eno-turismo en el mundo rural.

Bibliografía

Benyarku, C. A., & Stoops, G. (2005). *Guidelines for preparation of rock and soil thin sections and polished sections*. Departament de Medi Ambient i Ciències del Sòl, Universitat de Lleida.

FAO (2009) Guía para la descripción de suelos. FAO, Roma

IUSS Grupo de Trabajo WRB. (2007). Base Referencial Mundial del Recurso Suelo. Primera actualización 2007. Informes sobre Recursos Mundiales de Suelos No. 103. FAO, Roma.

Morlat, R. (2010). Traité de viticulture de terroir. *Editions Tec&Doc*

Un marco teórico para delimitar el concepto de enoturismo sostenible desde un enfoque multi-nivel

A theoretical framework to delimit the concept of sustainable wine tourism from a multi-level approach

Nieves García-Casarejos[1] Pilar Gargallo[1] y Begoña Cabanes[1]

[1]Facultad de Economía y Empresa, 50008 Universidad de Zaragoza, España

Resumen. La relevancia del sector vitivinícola en España y su conexión con el desarrollo del entorno rural, así como la preocupación de la población por los aspectos medioambientales y sociales, justifican el creciente interés por incorporar prácticas verdes y sostenibles en el turismo del vino. Además, dentro de los enoturistas existe un segmento cada vez más creciente que demanda servicios y destinos turísticos responsables con el medio ambiente y la sociedad, en general.

Teniendo en consideración esta demanda creciente, las bodegas deberían atender este segmento de consumidores que reclaman un turismo de vino sostenible. Para poder llevar a cabo una sólida iniciativa enoturista sostenible, este trabajo propone un marco teórico basado en un enfoque multinivel, que incluye además de las tradicionales perspectivas multi-*stakeholders* y multi-motivacional, una perspectiva multi-dimensional que recoge los aspectos claves del concepto de sostenibilidad, es decir, el económico, el social y el medioambiental.

1 Introducción

El sector del vino en España tiene una gran importancia, debido a su valor económico, social, cultural y ambiental. De hecho, según datos de la Organización Internacional de la Viña y el Vino (OIV), en 2016 la superficie de viñedos en España, con 975 mil hectáreas, era la mayor en el mundo (concretamente representaba un 14% de la superficie de viñedos mundial y un 30% de la europea). Por otra parte, España, con 39,4 millones de hectolitros, era el tercer país productor de vino en el mundo. En cuanto al consumo, en 2016, España ocupaba el séptimo lugar en el ranking mundial, con 9,9 millones de hectolitros consumidos. Otro dato relevante es que España fue el mayor exportador de vino, con 22,9 millones de hectolitros en términos de volumen y 28.900 millones de euros en términos de valor, lo que representó una cuota de mercado global del 22%.

No obstante, la importancia del sector del vino no se refleja únicamente en términos de su producción, sino también en el peso que supone el enoturismo o turismo del vino como destino de calidad. El sector vitivinícola crea a su alrededor todo un ecosistema de actividades que deberían estar alineadas con la sostenibilidad ambiental y cultural, el desarrollo económico del medio rural y la desestacionalización de la demanda turística.

Los datos aportados por el noveno informe anual de la Asociación Española de Ciudades del Vino (ACEVIN) ponen de manifiesto que 2016 fue un excelente año para el enoturismo y las cifras de visitantes reflejan la consolidación del turismo del vino dentro de la oferta turística española. El creciente interés mostrado por los turistas del vino ha impulsado la proliferación de eventos de vino y el crecimiento del negocio del enoturismo.

En líneas generales, el enoturismo se relaciona con un turismo gastronómico, cultural y de naturaleza. El turismo del vino puede ser definido como todas aquellas visitas a viñedos, bodegas, festivales del vino, catas de vino y/o experiencias respecto a todos los atributos de la uva y del vino de cualquier zona vitivinícola, que el visitante tiene como primer factor motivador [1].

El turismo del vino es uno de los productos turísticos más complejos en cuanto a su estructura, gestión, promoción y comercialización. En él intervienen dos sectores de gran importancia estratégica para el territorio que tradicionalmente no mantenían ningún vínculo entre sí: el turismo y el sector vitivinícola.

La cultura del vino debe ser el eje temático del enoturismo y, el vino, el elemento cultural que cuenta con la entidad suficiente como para apoyar y potenciar una amplia variedad de servicios, actividades, patrimonio, folclore y experiencias turísticas.

No obstante, el modelo español de enoturismo presenta aspectos mejorables, ya que, por ejemplo, se ha desarrollado sin tener en cuenta, en muchos casos, a todos sus agentes. Mantiene como producto básico la visita a la bodega bajo la denominación de experiencia enoturística, cuyo hilo conductor es el proceso de elaboración del vino. Está centrado en el corto plazo y, por ello, en demasiadas ocasiones, no tiene en cuenta el impacto emocional que se genera, percibiendo como un fracaso el que un turista no compre vino directamente en la bodega tras la visita. El vino es el eje vertebrador de la experiencia turística; sin embargo, para lograr la estabilidad de este producto/servicio, el enoturismo debe ser algo más que turismo enológico.

Además, su papel clave en el desarrollo local y rural hace que el interés por el tema de la sostenibilidad en el vino esté más que justificado. De hecho, en la industria del vino, las preocupaciones ambientales están dando forma a la actual demanda turística, centrada en proteger los recursos, en especial, los viñedos y paisajes, ya que la preservación del territorio es un factor altamente valorado por el enoturista experimentado. Pero dentro de los recursos turísticos medioambientales no todos son renovables, de ahí que haya que poner en marcha mecanismos para asegurar que la afluencia de visitantes no ponga en peligro dichos recursos. El desarrollo turístico se debe planificar y gestionar de forma que no cause serios problemas ambientales o socioculturales.

Así, frente a un turismo de masas, en el modelo de enoturismo sostenible debe primar una producción a pequeña escala, buscando al máximo la autenticidad en la experiencia enoturística, dirigida a un público individualizado y exigente. En este sentido, una casa rural con encanto para un destino centrado en el vino, tendría un valor cualitativo más alto que un hotel de cinco estrellas.

También se debería fomentar la implantación y el desarrollo de sistemas de gestión de la calidad medioambiental por parte de los territorios enoturísticos, concediendo prioridad al modelo de las 3R (reducir, reutilizar y reciclar). Un medio ambiente degradado resulta molesto e incluso peligroso para los residentes y puede disuadir a los visitantes. Para conseguirlo, las Administraciones deberían apoyar medidas para reducir la polución ambiental, visual y acústica (planes de embellecimiento, soterrado de cables en zonas urbanas, y regulación de los ruidos y publicidad en espacios públicos, entre otros).

Para mejorar la calidad medioambiental del enoturismo, las empresas, y no solo las administraciones, deben asumir una mayor responsabilidad, en la línea de impulsar la aplicación de medidas responsables con el medio ambiente, así como sistemas de auditoría ambiental, tanto para empresas como para destinos enoturísticos. En definitiva, se trataría de crear una cultura de la responsabilidad medioambiental integrada dentro de sus procesos de gestión.

Además, un enoturismo responsable también puede contribuir a aumentar el nivel de conocimiento y respeto por el medio natural, así como a generar recursos que puedan invertirse en su conservación. Este turismo "verde" del vino, encaja bien con visitantes de una elevada conciencia ambiental, permite crearla o aumentarla en aquellos que no la tienen y desarrolla iniciativas de turismo alineadas con los intereses de la población local.

El territorio de una ruta vitivinícola desempeña un papel muy importante, al estar compuesto por espacios naturales, urbanos y por un entorno singular que le confiere su carácter distintivo. En este contexto, el desarrollo de proyectos turísticos sostenibles favorece la conservación y mejora de los recursos naturales y culturales presentes en cada territorio, lo que mantendría intacto el potencial de generar riqueza en la zona de las próximas generaciones.

Para que el cambio de modelo sea una realidad, sería necesario que se instrumentaran acciones de formación, educación y conciencia pública del cuidado de todos los espacios dentro del propio territorio. Igualmente necesario sería poner en marcha programas de protección de recursos naturales. Es preciso resaltar también la necesidad de integrar los esfuerzos de coordinación y de apoyo de las administraciones públicas y del sector privado en el desarrollo del producto enoturístico sostenible y trabajar bajo un enfoque holístico.

Teniendo en consideración todo lo anterior, este trabajo presenta un marco teórico que permita llevar a cabo múltiples propuestas de enoturismo sostenible desde un enfoque multi-nivel, que incluya además de las tradicionales perspectivas multi-*stakeholder* y multi-motivacional, una perspectiva multi-dimensional que contemple los tres aspectos claves del concepto de sostenibilidad (económico, social y medioambiental).

El resto del trabajo se organiza del siguiente modo. La sección 2 presenta la aproximación multinivel utilizada en el trabajo y la sección 3 incluye las conclusiones más relevantes del trabajo.

2 Enfoque multi-nivel

En esta sección presentamos la metodología que se va a utilizar en el trabajo que está basada en un enfoque *multi-nivel* siguiendo las ideas de [2]. En primer lugar, se lleva a cabo una perspectiva *multi-stakeholder* que permita determinar todos los grupos de interés o agentes involucrados en el producto/servicio del enoturismo. En segundo lugar, realizamos un análisis *multi-motivacional* que permita discriminar las principales razones (cultura, gastronomía o naturaleza) que guían la práctica enoturística de una persona. Finalmente, se añade una perspectiva *multi-dimensional* para analizar e identificar estrategias de actuación desde la óptica de la triple línea de sostenibilidad: económica, social y ambiental.

La Figura 1, mediante un cubo de Rubik, muestra gráficamente el esquema seguido en el trabajo. Cada pequeño cubo recogería un grupo concreto de actividades enoturísticas que se abordarían desde las tres perspectivas: *stakeholder*, motivación y dimensión.

2.1 Nivel Multi-*stakeholders*

Es evidente que una mayor participación de todos los agentes involucrados en el producto/servicio del enoturismo es crucial para introducir la sostenibilidad como alternativa al actual modelo imperante.

La teoría de los *stakeholders* ha tenido un desarrollo vertiginoso después de que [3] irrumpiera con esta teoría como enfoque y respuesta al cambio en las condiciones actuales del entorno empresarial. Entre estas condiciones se puede citar: la globalización de los mercados de capital, el aumento de los inversores institucionales, el mayor activismo por parte de los accionistas y las expectativas de los grupos de interés [4].

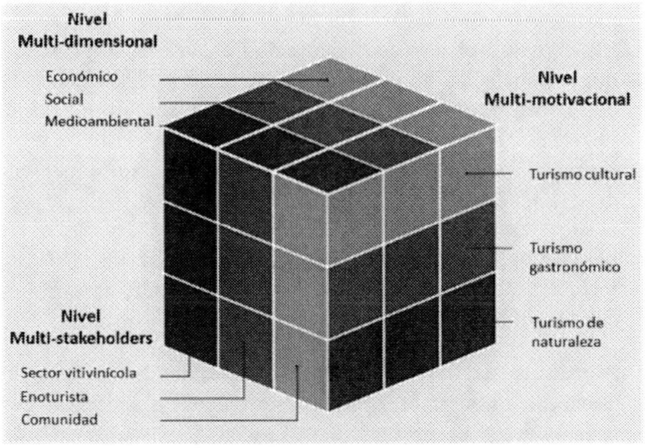

Figura 1. Esquema multi-nivel basado en [5] con las perspectivas: multi-*stakeholder*, multi-motivacional y multi-dimensional

Esta teoría ha sido empleada con éxito para explicar las relaciones entre los diferentes grupos de interés y la empresa en cuestiones ambientales [5-12]. En el caso del turismo, específicamente en las empresas hoteleras, se han desarrollado un amplio número de investigaciones que relacionan la gestión ambiental con los *stakeholders* [11, 13-15], reconociéndose como relevante la influencia de los *stakeholders*.

La teoría del *Stakeholder* [3] concibe a una empresa como un nexo de actores cuyos intereses y beneficios son mutuos. Esta teoría justifica que una empresa no solo deba satisfacer los intereses de los accionistas sino también los de todos los grupos que puedan afectar o verse afectados por las acciones de la empresa, y sin los cuales la compañía no podría existir. Muchos autores [3,16] han definido y clasificado los grupos de interés en las organizaciones. En este trabajo extrapolamos esta definición de grupos de interés al ámbito que nos ocupa, el concepto de enoturismo sostenible. Por tanto, denotaremos *stakeholder* a cualquier persona, grupo o entidad que tenga un vínculo directo o indirecto en dicho producto/servicio. La Tabla 1 contiene una descripción detallada de los grupos de interés que tienen un interés directo o indirecto en el concepto de enoturismo sostenible.

El enoturismo se ha convertido en un producto de gran interés para una parte creciente de turistas. Pero para poder realizarlo de una forma sostenible, es imprescindible que todos los actores colaboren. El primero de ellos es la bodega – un elemento fundamental en la consolidación de este producto–. En este sentido, no hay duda de que la creación de rutas vinícolas es un buen acicate para involucrar a todos los agentes de la zona y articular un producto turístico sostenible basado en la cultura del vino.

La visita a bodegas, tradicionalmente, respondían a una labor de fidelización de clientes, que se llevaba a cabo con mucho cariño en el propio corazón del negocio pero sin considerarlo como una nueva fuente de ingresos. Cierto es que cuando se habla de enoturismo, la bodega pasa a ser un atractivo de primer orden, un elemento clave si lo que se pretende es articular una oferta turística en torno al vino. Pero, para conseguir resultados sostenibles, es requisito indispensable que la bodega tenga conciencia turística y ecológica; y, por tanto, ha de ver en el enoturismo sostenible una nueva unidad de negocio dentro de sus instalaciones. Y, por este motivo, ha de empezar a trabajar con un enfoque sostenible en la encrucijada entre los dos sectores: el enológico y el turístico. La bodega debe estructurar una nueva estrategia de negocio vinculada al turismo sostenible.

Tabla 1. Grupos de interés de un enoturismo sostenible.

Tipología		Grupos de interés
Sector vitivinícola		Bodegas, explotaciones vitícolas
Turista		Demandante de productos/servicios enoturísticos
Comunidad		
Personas		Habitantes del territorio
		Trabajadores de las empresas vitivinícolas
Empresas	Prestación de servicios	En el territorio: alojamientos, restauración, comercios, guías turísticos, puntos de información...
		Fuera del territorio: turoperadores, agencias de viajes, transporte y comunicaciones...
Instituciones	Públicas	Diputación, comarca, ayuntamientos
	Privadas	Asociaciones: empresarios, consumidores
	Comunidad científica	Estudios realizados por universidades, institutos de investigación, organizaciones sectoriales...

A fin de recibir visitantes, es esencial habilitar convenientemente las instalaciones de la bodega. Ésta ha de saber revalorizarlas adaptándolas a los tiempos presentes en materia de seguridad y accesibilidad; pero, sobre todo, debe encontrar elementos que vertebren una visita y promover espacios donde interpretar el proceso de elaboración del vino. En definitiva, el futuro pasa por un aprendizaje que genere experiencias vinculadas al terruño, a las instalaciones vinícolas, a las personas que intervienen en el proceso de elaboración del vino y a su consumo.

La red Great Wine Capitals realizó una encuesta, en 2009, a 454 empresas con actividades de enoturismo a nivel mundial y obtuvo que, un 83% ofertaban degustación de sus vinos, un 75% visitas guiadas, un 40% tenía una oferta gastronómica en la bodega, un 29% incluía algún servicio de alojamiento (hotel, casa rural etc.), un 20% algún tipo de actividades deportivas en sus instalaciones, un 20% actividades culturales, un 17% albergaban en sus bodegas algún tipo de exposición de carácter temporal; un 10% ofrecía en sus instalaciones un espacio museístico de manera permanente y un 4% realizaban terapias basadas en los beneficios del vino y la viticultura. Sin embargo, esta encuesta no nos dice cómo se llevan a cabo esas actividades, es decir, si se hacen o no de una manera sostenible tanto desde el punto de vista económico, social y medioambiental.

Otro de los agentes involucrados en el producto es el turista. El consumidor está cambiando y, cómo no, también sus hábitos vacacionales. Actualmente realiza salidas con mayor asiduidad que antes, si bien más fragmentadas y cortas. Este consumidor elige lugares más próximos para sus escapadas, valora los destinos con conciencia ambiental vinculados a la naturaleza y a un estilo de vida saludable. Trekkings, rutas en bicicleta y actividades deportivas son valores en alza. El turista, en su tiempo de ocio, busca los alicientes del llamado *Life Style*:

gastronomía, actividades recreativas, balnearios, "shopping", etc. Normalmente, los destinos escogidos van asociados a algún atributo emocional e intangible, y se busca lo auténtico, lo singular. En definitiva, el turista tiene sed de novedades; quiere vivir experiencias en sus viajes y anhela nuevos productos turísticos que lo satisfagan y cumplan con sus expectativas iniciales.

El enoturismo es un turismo de proximidad básicamente, pues el 40% de los turistas del vino, lo son del mismo país que visitan. Del conjunto total, el 25% de visitantes son locales. Por término medio el 35% de los enoturistas son extranjeros. Los países que más enoturismo exportan son Holanda y Reino Unido, seguido por Estados Unidos y Canadá. Dos tercios de enoturistas son hombres y el tercio restante pertenece al género femenino. El visitante medio (45%) tiene una edad entre 36 y 55 años. Los maduros de más de 56 años también son grandes aficionados al enoturismo (30%). El enoturismo es una actividad fuertemente vinculada a las vacaciones pues el 70% de enoturistas llegan a los territorios de vino durante el verano. El otoño es la otra temporada importante para los enoturistas, concentrando el 15,3% de las visitas totales.

Se necesitan pueblos impecables en cuanto a servicios, señalización, gasolineras, hoteles, restaurantes, tiendas especializadas, etc. Se deben diseñar servicios de enoturismo para un día, 2, 3 o 5 días, por ejemplo.

Para los empresarios turísticos, de ocio y oferta relacionada, es una importante oportunidad para generar demanda adicional a su producto turístico tradicional, el turismo rural que, a pesar de su importante crecimiento en los últimos años, sufre de la problemática de una excesiva estacionalidad en fines de semana, puentes y época estival, así como de una cada vez más feroz competencia entre destinos y establecimientos por captar flujos de demanda.

Finalmente, para los territorios vitivinícolas constituye una oportunidad única para crear un producto turístico singular, que corresponde a las nuevas tendencias de la demanda y que, mediante una planificación adecuada, puede convertirse en un "motor" para el desarrollo turístico sostenible de un territorio, generalmente caracterizado por una excesiva dependencia del sector vitivinícola.

Un enoturismo sostenible requiere trabajar en estrecha colaboración entre la administración pública (turística, medio rural, local, provincial y regional) y el sector privado (turístico y vitivinícola).

Resumiendo, el beneficio derivado de un producto de enoturismo sostenible debe ser múltiple. Para el turista, al asegurar que el diseño del producto turístico y la prestación de los servicios que implica, responderá satisfactoriamente a sus necesidades. Para la administración local que podrá aprovecharse de los múltiples beneficios: p.ej. intercambio de opiniones con otras administraciones locales, economías de escala en la realización de proyectos y en la promoción y comercialización, etc. Para el empresario que obtendrá una guía para definir o adaptar sus instalaciones y servicios a los requerimientos de competitividad establecidos por los clientes y el entorno y para el residente puesto que el desarrollo y gestión del destino enoturístico tendrá en cuenta su punto de vista sobre el tema, una necesaria combinación entre crecimiento y conservacionismo.

2.2 Nivel Multi-motivacional

Existen numerosas evidencias sobre la gran importancia que poseen las motivaciones en el comportamiento y proceso de elección del turista. Por lo tanto, parece necesario comprender las motivaciones que influyen en la elección del producto enoturístico. La motivación permite dar respuesta a tres cuestiones esenciales: las razones para viajar o por qué; la elección específica o dónde; y los beneficios buscados en el destino o satisfacción [17-18]. Para [19], la motivación es una variable muy importante para explicar la conducta turística.

Asimismo, la motivación se ha estudiado conjuntamente con la segmentación del mercado [20-25], así como con la lealtad con el producto o destino.

En términos generales, los turistas viajan bien porque son empujados por motivos o variables internas o bien porque son atraídos por fuerzas externas del destino. Los factores de empuje (*push factors*) están relacionados con aspectos internos y emocionales, como el deseo de escapar, de descansar y relajarse, de aventura o de interacción social. Los factores de atracción (*pull factors*) están conectados con aspectos externos, cognitivos y de situación, como los atractivos del lugar, las infraestructuras recreativas o los escenarios culturales y naturales [26-28]. Una cuestión de interés es que la bodega puede trabajar en ambos tipos de estrategias simultáneamente, una más a nivel sectorial basada en factores *push*, para crear una intrínseca conciencia de cultura del vino a través de sensaciones y otra más basada en factores *pull* ligada al territorio, a través de una buena oferta.

Dentro de los factores *pull*, según un estudio realizado por la Organización Mundial del Turismo (OMT), la oferta cultural, la naturaleza y la gastronomía, son, por ese orden, los motivos principales que llevan a los enoturistas a visitar un destino.

Recorrer un paisaje vitivinícola, visitar bodegas, aprendiendo su cultura mientras se catan vinos y se prueban los alimentos típicos de la zona, pudiendo incluir actividades terapéuticas, deportivas o de relax, se ha convertido en una forma de viajar con una demanda cada vez más en alza.

La autenticidad de la vivencia enoturística debe preservarse. El folclore, la historia y las tradiciones redondean esa vivencia cultural del enoturista, por tanto, preservar y potenciar el patrimonio del destino es decisivo.

En cuanto a la motivación cultural, el interés por los destinos rurales en España ha aumentado notablemente a lo largo de los últimos años. Estos lugares invitan a descubrir pintorescos pueblos, villas, castillos y pequeñas ciudades medievales, ofreciendo también la posibilidad de hospedaje en pequeñas casas rurales y magníficos hoteles de interior.

Por otra parte, la gastronomía local también forma parte del patrimonio de un destino enoturístico y merece una especial atención dada su complementariedad con el elemento vino. Por tanto, conocer los hábitos y preferencias de un enoturista es importante para establecer la oferta gastronómica, el tipo de alojamientos, el tipo de establecimientos en los que les gusta comer o qué actividades gastronómicas buscan para complementar su viaje.

Según el segundo estudio de la demanda de turismo gastronómico, un 76,2% de los españoles ha realizado algún viaje o escapada con la intención de disfrutar de la gastronomía en los últimos dos años. El 62,4% de estos turistas afirma que la gastronomía influye en gran medida en la elección del destino. El 42,6% de los desplazamientos que se realizan a destinos nacionales cercanos (menos de 200 km) con la intención de disfrutar de la gastronomía son viajes en el día sin pernoctación en el destino, en los que el visitante se desplaza para disfrutar de una comida en un restaurante, ir de tapas o visitar una bodega. El sector de la restauración y la hostelería se convierte, por tanto, en el escaparate gastronómico del territorio, por ser éste el principal espacio en el que entran en contacto el visitante y el patrimonio gastronómico de un lugar. Los turistas gastronómicos también aprovechan estos viajes para comprar productos locales (62,9%), visitar mercados gastronómicos (59,0%) y realizar actividades enoturísticas, tales como visitar bodegas (49,1%) y comprar vinos (41,5%).

Por último, el entorno natural es otro aspecto que no hay que olvidar puesto que admirar la flora, la fauna, bellos parajes naturales, hermosos lagos y montañas, preciosos ríos y acantilados es un interesante reclamo para un número creciente de enoturistas. Entre las personas que practican el turismo de naturaleza se pueden encontrar perfiles muy variados. Así, este tipo de turismo lo practican los naturalistas que se documentan sobre los lugares que van a visitar; los individuos que quieren sentir la tranquilidad y las oportunidades de ocio que la naturaleza les ofrece; los estudiosos, que buscan el conocimiento y la percepción de los espacios naturales o las personas mayores que buscan disfrutar del paisaje y de su belleza natural, entre otros.

Según [29], conocer la conducta del turista permite entender mejor las razones por las cuales se realiza la visita, las necesidades que tienen los visitantes y el uso que se le va a dar al lugar.

2.3 Nivel Multi-dimensional

El enoturismo posee un impacto que puede ser positivo y negativo, como resultado de la interacción con los diferentes factores del entorno que lo rodea. Este impacto puede deberse a los derivados de la fase de construcción e implantación del equipamiento turístico y las infraestructuras, a los producidos en su estado de funcionamiento operativo y a los generados por las actividades que realizan los propios turistas. Para llevar a cabo un análisis de sostenibilidad del enoturismo, deben analizarse los impactos económicos, sociales y ambientales. Es importante precisar que los impactos no son solamente perjudiciales o dañinos, al contrario, existen muchos beneficios como resultado de la actividad enoturística que pueden ser utilizados para revertir el impacto negativo de la misma.

En general, la mayoría de las consecuencias económicas de la realización de actividades enoturísticas podrían ser positivas, por lo menos a corto plazo. Tanto los gastos de los turistas que visitan una ruta vitivinícola como la inversión realizada por los dueños de las bodegas conllevan un conjunto de beneficios para muchas actividades económicas locales (hoteles, restaurantes, taxis, autobuses, etc.). Debido a que el turismo es una industria que depende en gran medida del factor humano, favorece la creación de empleo tanto en el sector turístico como en otros sectores impulsados por el turismo. La actividad enoturística no solo elevaría la renta en el área donde se desarrolla, sino también mejoraría su distribución, tanto en términos poblacionales como en términos de contribución al equilibrio regional. En general, se puede afirmar que el enoturismo representa una posibilidad de mejora económica en el nivel de vida de la población residente.

No obstante, todos los impactos en el aspecto económico no son positivos. Uno de los problemas derivados de la actividad turística son las fluctuaciones de su demanda, esta demanda es estacional y sensible a variaciones en el precio y a las modas, lo cual puede traer consigo inestabilidad en las zonas donde se desarrolle.

El impacto social de la actividad enoturística es evidente también desde sus propios inicios. El enoturismo puede traer consigo ventajas para las localidades de la zona, como es el caso de preservación y rehabilitación de monumentos, edificios, lugares históricos y creación de museos, centros de interpretación y monumentos de interés cultural, etc. Otra de las ventajas más comunes para la región donde se desarrolla la actividad turística puede ser la mejora de las infraestructuras, instalaciones y servicios. Por ejemplo, la atención sanitaria, alumbrado, recogidas de basuras, mejora de las comunicaciones, nuevas sucursales de entidades financieras. Así pues, la calidad de vida de los residentes aumenta.

Sin embargo, no todos los impactos sociales son positivos ya que el desarrollo de la actividad enoturística puede provocar a veces una pérdida de la identidad cultural y de tradiciones, así como una estandarización y banalización de la cultura. Además, la proliferación de visitantes puede traer consigo que los habitantes de la zona encuentren dificultades para mantener áreas familiares, con cierta intimidad y libres de turistas.

El impacto ambiental es quizás el efecto más visible, partiendo de la etapa de construcción de las instalaciones hasta la propia actividad del turista. Por ello, la primera de las fases, construcción e instalación, es la etapa que mayor planificación debería tener, teniendo en cuenta que en ese momento se deciden muchos de los elementos que afectarán al entorno tanto a corto como a largo plazo. Durante la implantación de las instalaciones se consumen muchos materiales y recursos, partiendo desde el uso del suelo donde se construye la instalación hasta el consumo de materiales de construcción y energía. Este consumo es relevante teniendo en cuenta la irreversibilidad del proceso, principalmente tratándose del deterioro del suelo.

Sin embargo, el mayor impacto negativo puede ser en el aspecto visual y arquitectónico. La mediocridad en un diseño y la ubicación de las construcciones configuran una total contaminación arquitectónica visual destructora del paisaje.

Otro elemento de gran impacto medioambiental es la generación de residuos, tanto durante la construcción y puesta en marcha, como principalmente en la etapa de explotación del turismo. En esta última etapa se generan residuos sólidos, en su mayoría residuos domésticos, con una gran diversidad, desde botellas de cristal, latas, papel hasta residuos alimenticios, haciendo de esta forma muy difícil su clasificación y tratamiento.

Otro recurso que es de vital importancia y que es uno de los más afectados por la actividad turística es el agua, en sus diversas formas: su uso indiscriminado, específicamente del agua potable, en zonas donde por lo general este recurso escasea; y la contaminación como resultado de aguas residuales. Esta polución del agua trae consigo la pérdida de la biodiversidad, el incremento de la turbiedad del agua y peligros sanitarios, entre otros.

Asimismo, el ruido y la contaminación del aire están asociados a la explotación y a la actividad del turista.

Sin embargo, cuando la actividad enoturística se realiza de forma consciente y planificada puede redundar en mejoras para la zona donde se concentra. Esta tendencia a proteger valores ambientales que de otra forma no tuvieran ningún valor económico es lo que puede hacer del enoturismo uno de los sectores fundamentales para apostar por la sostenibilidad.

3 Conclusiones

El perfil del turista es cambiante, el turismo que se practica a día de hoy no tiene nada que ver con el de hace unos años. Como dice el sociólogo Gilles Lipovetsky "el ocio será cada vez más híbrido, se entrelazará con la cultura, el arte, la estética, la gastronomía…"

El turismo en general, y el enoturismo, en particular, está considerado como un motor no desdeñable de la actividad empresarial, por tanto, un impulso positivo en el sector enoturístico redundaría en la mejora y crecimiento del resto de actividades económicas relacionadas con el mismo. Todos queremos que el enoturismo crezca, nos encantaría ser líderes mundiales, que nuestra zona fuese conocida en el mundo entero por lo que ofrecemos. Pero, desde un punto de vista motivacional, ¿se busca volumen o calidad de turismo?, ¿turismo *low cost* o turismo *premium*?, ¿se debe sectorizar orientándolo a segmentos como deportistas, turismo familiar, senior, joven, etc.?, ¿sabemos que oferta cultural, natural y gastronómica quiere cada uno de ellos?. Por tanto, para un buen diseño del producto, parece lógico comenzar definiendo el público objetivo, y en base a él actuar: conociendo a quién se dirige se personalizaran las experiencias y se conseguirán clientes satisfechos. Desde un punto de vista multi-*stakeholders*, ¿estamos preparados?, ¿sabríamos gestionar un crecimiento de manera correcta para satisfacer las necesidades del turista, las del empresario, las de la población local y las del medio ambiente?. Para un enfoque multi-nivel faltaría incorporar a estas preguntas el concepto de sostenibilidad en sus tres dimensiones, la social, la económica y la medioambiental, que creemos contienen la respuesta de futuro a las cuestiones planteadas.

El enoturismo ha ido adoptando iniciativas de protección medioambiental, aunque todavía no sean suficientes para salvaguardar la naturaleza; los visitantes acuden cada vez con mayor frecuencia a destinos y lugares naturales que deben ser debidamente protegidos para que las personas puedan disfrutarlos en el futuro. La gastronomía y la cultura también se pueden relacionar con el enoturismo sostenible, potenciando el consumo de productos de la zona, utilizando proveedores locales, revitalizando las costumbres y tradiciones: artesanía, festivales, folklore, etc., lo que generará un intercambio cultural entre visitantes y residentes.

Este enfoque multi-nivel trata de buscar soluciones a la fase de saturación dentro del ciclo evolutivo de cualquier actividad turística. No hay que olvidar que, mientras que en las primeras fases de aparición de cualquier modalidad turística se genera un elevado entusiasmo por parte de la población residente, al percibirlo como una buena opción de desarrollo a la que hay que sacar partido; a medida que se alcanzan niveles de saturación, los residentes necesitan facilidades adicionales para poder aceptarla. Posteriormente, dicha actividad pasa a ser considerada como la causante de todos los males del destino, para los residentes todas estas injerencias en la cultura local provocan un sentimiento de pérdida de lo que les pertenece y trae consigo elementos hostiles que atentan contra el ambiente acogedor y hospitalario que requiere la industria turística. De manera que en la etapa final, el destino ha perdido todos los atractivos que originalmente atrajeron a los turistas.

Por tanto, circunscribiéndolo al caso del enoturismo, el enoturismo sostenible puede ser la solución a estos inconvenientes. Un turismo más respetuoso con el patrimonio natural y cultural, así como con la población local, menos intensivo en el número de turistas y que se ajuste mejor a la capacidad de ocupación del destino, más controlado en los impactos negativos causados por los turistas y que no comprometa la capacidad de crecimiento y desarrollo de las generaciones futuras.

Agradecimientos

Este trabajo ha sido financiado por los proyectos ECO2016-77-P y ECO2016-79392-P (AEI/FEDER, UE).

Referencias

1. C.M. Hall. *Wine tourism in New Zealand* In G. Kearsley (ed) Tourism Down Under II, Towards a More Sustainable Tourism Conference Proceedings. 99, 109-119. Centre for Tourism, University of Otago (1996)
2. S. Qayum, S. Gold, Y. Glemarec *Multi-Stakeholder*

Decision-Making A Guidebook for Establishing a Multi-Stakeholder Decision-Making Process to Support Green, *Low-Emission and Climate-Resilient Development Strategies*, United Nations Development Programme, UNDP (2012)

3. R. E. Freeman. *Strategic Management: A Stakeholder approach*, Boston: Pitman (1984)
4. T.I. Nwanji, K.E. Howell. Shareholdership, stakeholdership and the modern global business environment: A survey of the literature. Journal of interdisciplinary economics. **18**, 4, 347-361 (2007).
5. J. K. Thompson, S. L. Wartick, H.L. Smith, J. E. Post. Integrating Corporate Social Performance and Stakeholder Management: Implications for a Research Agenda in Small Business. IN PRESTON, L. E. (Ed.) Research in corporate social performance and policy. . A Research Annual Greenwich, Conn. and London, JAI Press (1991)
6. R. E. Freeman. The politics of stakeholder theory: Some future directions. Business Ethics Quarterly, **4**, 4, 409-421 (1994)
7. P. Shrivastava. The role of corporations in achieving ecological sustainability. Academy of Management Review, **20**, 4, 936-960 (1995)
8. S. Fineman, K. Clarke. Green stakeholders: Industry interpretations and response. Journal of Management studies, **33**, 6, 715-730 (1996)
9. G. Azzone, M. Brophy, G. Noci, R. Welford W. Young. A stakeholders' view of environmental reporting. Long range planning, **30**, 5, 699-709 (1997)
10. I. Henriques, P. Sadorsky. The relationship between environmental commitment and managerial perceptions of stakeholder importance. Academy of management Journal, **42**, 1, 87-99 (1999)
11. M. J. Alvarez Gil, J. Burgos Jiménez, J.J. Céspedes Lorente. An analysis of environmental management, organizational context and performance of Spanish hotels, Omega, **29**, 6, 457–471 (2001)
12. B.R. Agle, T. Donaldson, R.E. Freeman, M.C. Jensen, R. K. Mitchel, D. J. Wood. Dialogue: Toward superior stakeholder theory. Business Ethics Quarterly, **18**, 2, 153-190 (2008)
13. J. Céspedes-Lorente, J. Burgos-Jiménez, M. J. Álvarez-Gil. Stakeholders' environmental influence. An empirical analysis in the Spanish hotel industry. Scandinavian Journal of Management, **19**, 3, 333-358 (2003)
14. J. Rivera. Institutional Pressures and Voluntary Environmental Behavior in Developing Countries: Evidence From the Costa Rican Hotel Industry. Society & Natural Resources, **17**, 1, 779-797 (2004)
15. A. Kasim. Towards a Wider Adoption of Environmental Responsibility in the Hotel Sector', International Journal of Hospitality & Tourism Administration, **8**, 2, 25-49 (2007)
16. A. Campbell. Stakeholders: The case in favour. Long Range Planning **30**, 446-49 (1997)
17. S. Baloglu, K. W. McCleary. US international pleasure travelers' images of four Mediterranean destinations: A comparison of visitors and nonvisitors. Journal of Travel Research, **38**, 2, 144-152 (1999)
18. M. Castaño, A. Moreno, S. García, A. Crego. *Aproximación psicosocial a la motivación turística*. Estudios turísticos, **158**, 5-42 (2003)
19. D.A. Baker, J. L. Crompton. Quality, satisfaction and behavioral intentions. Annals of Tourism Research, **27**, 3, 785-804 (2000)
20. M. Cervantes, A. González, N. Muñiz. La incidencia de los estilos de vida en la segmentación del mercado turístico: aplicación a la población urbana de Castilla y León. Revista de Investigación Económica y Social de Castilla y León, **2**, 41-56 (1999)
21. C. K. Lee, Y. K. Lee, B. E. Wicks. Segmentation of festival motivation by nationality and satisfaction. Tourism management, **25**, 1, 61-70 (2004)
22. A. Beh, B. L. Bruyere. Segmentation by visitor motivation in three Kenyan national reserves. Tourism management, **28**, 6, 1464-1471 (2007)
23. C. H. Hsu, L. A. Cai, K. K. Wong. A model of senior tourism motivations—Anecdotes from Beijing and Shanghai. Tourism Management, **28**, 5, 1262-1273 (2007)
24. D. Severt, Y. Wang, P. J. Chen, D. Breiter. Examining the motivation, perceived performance, and behavioral intentions of convention attendees: Evidence from a regional conference. Tourism Management, **28**, 2, 399-408 (2007)
25. M. Devesa, M. Laguna, A. Palacios. The role of motivation in visitor satisfaction: Empirical evidence in rural tourism. Tourism Management, **31**, 4, 547-552 (2010)
26. G. M. Dann. Anomie ego-enhancement and tourism. Annals of Tourism Research, **4**, 4, 184–194 (1977)
27. J. L. Crompton. Motivations for pleasure vacation. Annals of tourism research, **6**, 4, 408-424 (1979)
28. G. M. Dann. Tourism Motivations: An appraisal. Annals of Tourism Research, **8**, 2, 189–219 (1981)
29. M. Osorio, S. Franco, I.L. Ramírez, G. Nava, G. Novo, H. Hugo. El visitante del Parque Nacional Nevado de Toluca, México. Análisis del comportamiento en un área natural protegida. Investigaciones Geográficas, **76**, 55-70 (2011)

 https://doi.org/10.1051/e3sconf/20185003003

Decomposing the notion of vine vigour with a proxydetection shoot sensor: Physiocap®

Constance DEMESTIHAS[1], Sébastien DEBUISSON[1], Arnaud DESCOTES[1]

[1] *Comité Champagne, 17 rue Chandon Moët 51200 Epernay, France.*
constance.demestihas@civc.fr

Abstract

The vigour and the vegetative expression of grapevines are parameters of great interest in viticulture, as they describe a general state of growth capacity. Understanding the impacts of agricultural practices on vine vigour, under particular soil and climate conditions, is essential to give a more accurate technical advice, especially on what soil management and vine nutrition are concerned. A shoot sensor called Physiocap®, designed and developed by the CIVC (Comité interprofessionel du vin de Champagne), is used during dormancy season to measure the shoot section, the shoot number and an estimation of the aboveground biomass. The sensor maps vigour spatial variability within a plot, among plots and over years. Physiocap database in Champagne has been analysed since 2011 at different scales, in order to determine the factors impacting the vine vigour. The vintage appeared to be the most impacting factor. For example, climate variability or accidents like dry springs and early spring frosts reduce vine vigour. Champagne vine varieties did not significantly impact vine vigour according to the database. At the scale of the Champagne vineyard, the aboveground biomass estimation of Physiocap® was strongly correlated with the yield of the following year, leaving a promising basis for analysing the impact of different factors on vine vigour. At the scale of the plot, winegrowers are able to compare their plot vigour to a Champagne threshold, which is being refined every year as the Physiocap database is enriched. They can therefore manage their fertilization and soil tillage program more accurately according to their objectives. The Physiocap® sensor appears to be an interesting multidimensional tool binding vine physiology, agronomy and precision viticulture at different scales. When coupled with other data, especially the one describing soil characteristics, it could even be the baseline for creating a decision-aid tool in Champagne for fertilization, tillage and pruning practices.

Introduction

Maintaining a balance between grapevine vegetative growth and grape production is one of the most important goals in viticulture. This goal has become even more challenging with the physiological evolutions due to climate change, especially the drier springs in Champagne, as well as the shifting regulations. Grapevines exhibiting excessive vigour are likely to produce less fruit of reduced quality, and vines with inadequate vigour will have negative effects on their potential yield.

Vigour management should be studied in a multifactorial way. It is the result of a set of factors such as climate and soil characteristics, which interact with grapevine in a given system that is itself modified by practices. This system can be observed under different space and time scales in Champagne. That is exactly were precision viticulture finds all its purpose: measuring vigour space and time variability. In order to define the notion of vigour, three terms are commonly used, based on the perennial character of grapevine [1]:

- The vegetative expression is the result of the annual metabolic activity during which shoots, leaves and roots grow and stock carbohydrates. It is characterized by the number of shoots during dormancy.
- The vigour is the intensity of this metabolic annual activity, characterized by the shoot section.
- The vegetative potential or potential reserve storage is directly linked to both vegetative expression and vigour. It is the result of the plant material, the previous vegetative cycles and of the pedoclimatic conditions.

A shoot sensor called Physiocap®, designed and developed by the CIVC (Comité interprofessionel du vin de Champagne), is used during dormancy season. Since 2011, the Physiocap® has crossed over 700 plots in Champagne. Data are collected and analysed in order to understand vigour variability among space and time. The aim of this paper is to present the latest analysis of this Physiocap database at different time and space scales in Champagne: within a whole region, among plots, among vintages and inside a plot. This scale decomposition could be seen as a 'terroir' analysis.

Material and method

Physiocap® is an optical laser embarked on an agricultural engine used during winter just before pruning [2].

The sensor is composed of a laser micrometer which consists of an emitter and a receiver positioned face to face, above wired linkers. It also has a GPS enabling data mapping and a data recording box. Data is stored in a USB key (fig.1) [3].

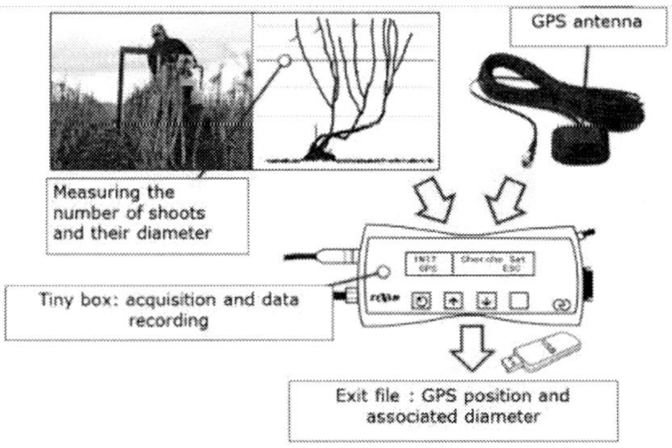

Figure 1: Components of the Physiocap® sensor

The sensor provides three key physiology variables of grapevine: shoot section, shoot number and an estimation of the aboveground biomass. The shoot section measurement is made by using the interruption section of the laser beam. It is instantaneous and does not depend upon shoot distance from the sensor nor the sensor's speed. The sensor therefore gives many measures of section each second. The second variable, the shoot number, is calculated out of the speed given by the GPS and the shoot section measurements. Finally, the shoot biomass indicator per vine or per meter square is calculated out of the shoot section and the shoot number, as well as a wood density fixed at 0.9 kg/dm^3 in Champagne.

Data has been analysed in order to qualitatively weigh the factors impacting vine vigour. These factors are either not modifiable, like climate or soil, or can be changed through agricultural practices. They all impact vine vigour, which in turn has an effect on various agronomical performances (Fig. 2).

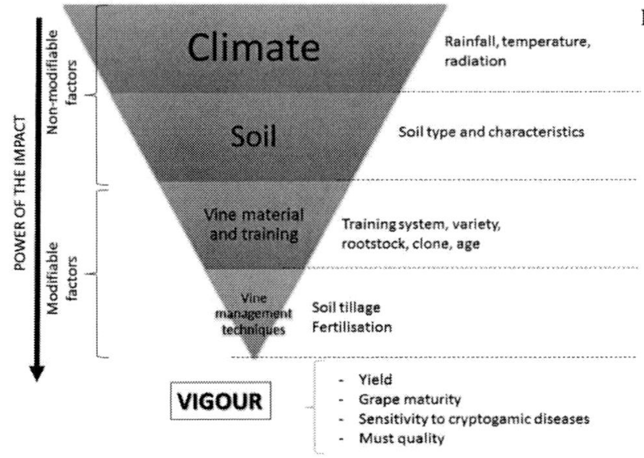

Figure 2: Analysis method of different factors affecting vine vigour

Results

Comparing vintages gives important information on how climate characteristics impact vine vigour [4]. Each vintage can be compared to the mean value of Champagnes shoot biomass on six successive years (fig. 3).

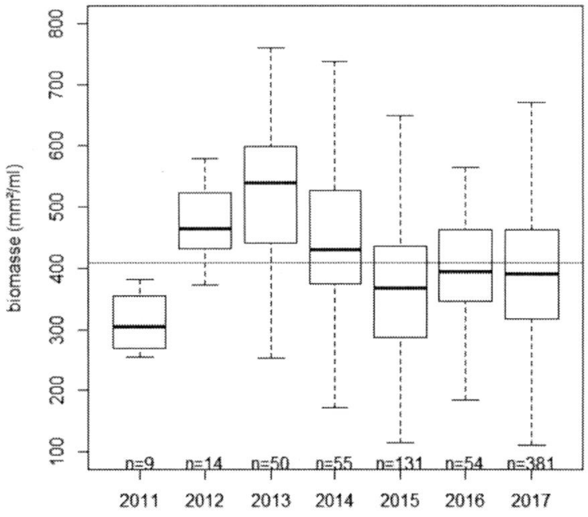

Figure 3: Boxplot of the shoot biomass in mm² per linear meter measured by the Physiocap sensor since 2011 and number of measured plots.

Rainfall is the most impacting factor. Since 2011, shoot biomass depicted by the Physiocap sensor seems to follow quite accurately the yearly rainfall at the Champagne region scale (fig.4) [5]. Dry years such as 2015 presenting a yearly rainfall of only 591 mm and a spring rainfall of 113 mm from the 1st of April to the 30th of June presents low shoot biomass. 2013 was a very wet year with a rainfall of 797 mm presenting the highest shoot biomass value from 2011 to 2017.

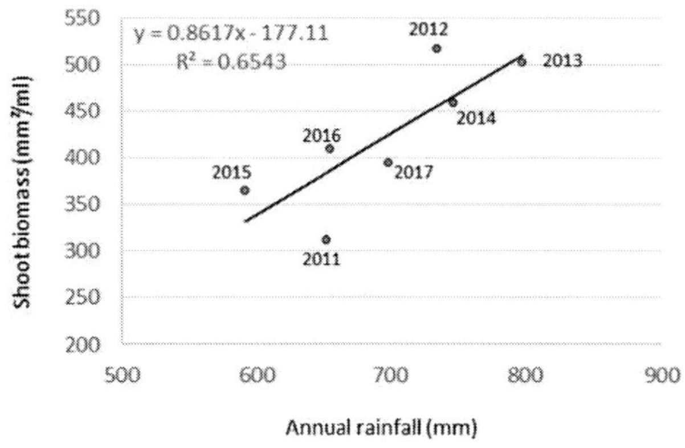

 https://doi.org/10.1051/e3sconf/20185003003

Figure 4: Link between annual rainfall and shoot biomass on 6 successive years.

At an interplot scale, comparing the vigour variability among plots enables winegrowers to position their plot in relation to a threshold and better understand the impact of the terroir on their vine vigour. When comparing different regions we observe a great variety of values. These differences may be due to the type of soil, the mesoclimate within a region but also the vineyard management strategies. In other words, this variability is linked to the terroir of each region.
For example cover crop strategies in the Montagne de Reims vineyard tends to be frequent, thus lowering shoot biomass. On the contrary, the Côte des Bar region was heavily impacted by frost in 2016 and 2017, which may have increased shoot biomass measured at the end of 2017 from increased reserve accumulation [5].

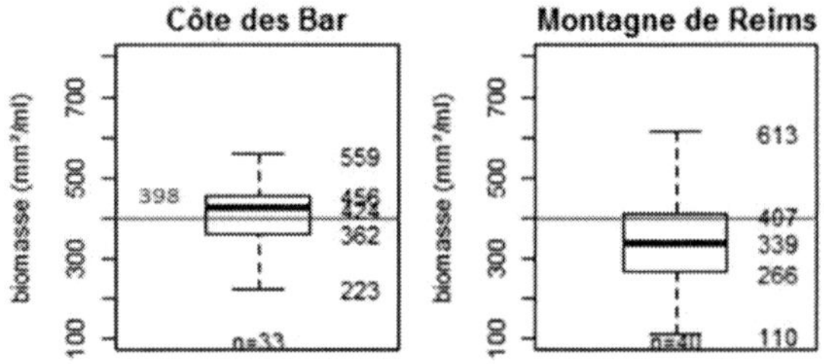

Figure 4: Shoot biomass boxplots of two different regions of Champagne: Montagne de Reims and Côte des Bar in 2017

The effect of vine material and training system is also significant at the whole Champagne vineyard scale since 2011. Chardonnay appears to be the variety which increases shoot biomass the most. The age also impacts vigor: it seems that beyond 40 years old age, we observe a drop of the number of shoots (mean value of 5.9 compared to 6.8 shoots). Furthermore 'guyot' training system increases shoot biomass compared to a Chablis or a Cordon de Royat training system (fig. 5). This could be due to an increasing shoot section as a consequence of the decreasing number of shoots.

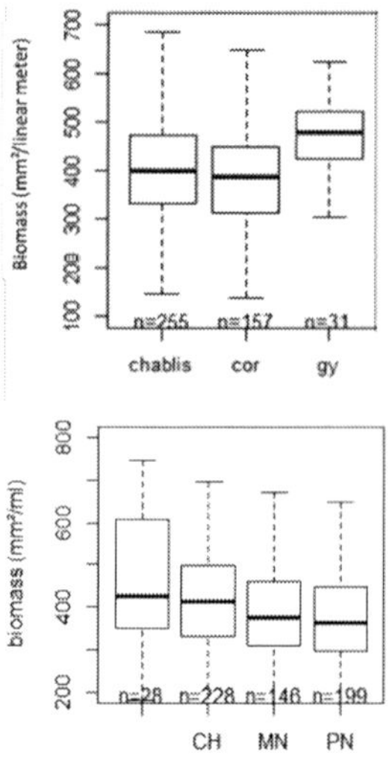

Figure 5: Boxplot of the shoot biomass according to training system. Cor = cordon de royat, gy = guyot and vine varieties. CH = chardonnay, MN = meunier, PN = pinot noir

At an intra-plot scale soil tillage management impact vine vigour significantly. In 2017, chemical weeding still show the highest shoot biomass with a mean value of 520 mm²/linear meter, whereas mechanical weeding show lower values (mean value of 361 mm²/ml).

Finally, there is an important correlation between the shoot biomass and the yield on year n+1 (fig. 6), thus showing the importance of the biannual time scale when managing vigour. Dry years such as 2015 present low shoot biomass and therefore low yield the year after. This correlation correspond to a leaf/grape ratio balance within the vine. At a more local scale, this correlation is not as good as in fig. 4, due to climate events such as frost or hail. These events create a disorder on the leaf/grape ratio the year after.

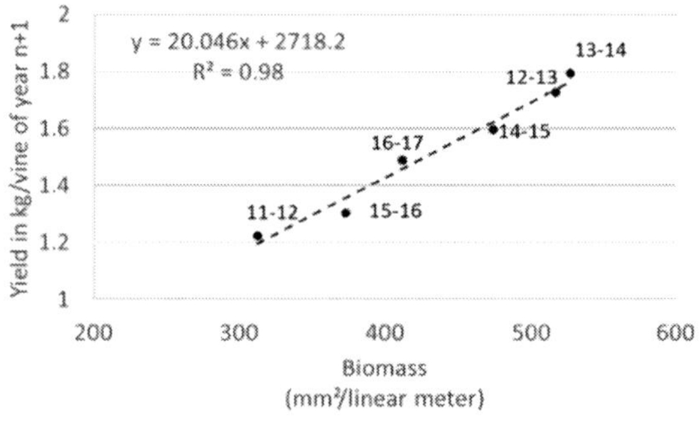

Figure 6: Correlation between the shoot biomass (mm²/lm) of year n and the yield of year n+1 on 6 successive years.

Discussion and conclusion

Physiocap® estimates the shoot biomass at the end of the year. It is an indicator of both what has been cumulated during the year as well as a storage potential for next year. This measure depends on various factors. The most important one is climate and more specifically annual rainfall. Vine material and training systems impact vigour at the region scale and on pluriannual time scale, while at the plot scale, it is more fertilization and soil tillage that can have an effect on vigour. These practices can therefore be the levers to be used for vigour management. Yield potential can be predicted using the biomass indicator defined by the sensor at a large space scale. Nevertheless, it is essential to take the climate events into consideration when analysing this type of correlation on a local space scale.

This is one of the first agronomical analysis on the Physiocap data which is being enlarged every year. On the basis of these preliminary results, Physiocap, coupled with other tools and indicators such as soil analysis, soil tillage and fertilization strategies as well as accurate climate data, seems to be a central component of a multidimensional and innovative vine management.

References

[1] F. Champagnol, *Eléments de physiologie de la vigne et viticulture générale.* Saint- Gely-du-Fesc, 1984.

[2] S. Debuisson, M. Morlet, C. Germain, O. Garcia, L. Panigai, and D. Moncomble, “Precision viticulture: using on-board sensors to map vine variability and characterize vine trajectories,” in *IXe International Terroirs Congress 2012*, 2012, pp. 17–19.

[3] O. Garcia, M. Morlet, and S. Debuisson, “Modulation des pratiques culturales à l’aide du capteur Physiocap®,” *Le Vigneron Champenois - décembre 2014*, pp. 31–37, 2014.

[4] C. Demestihas, S. Debuisson, and A. Descôtes, “Vers une meilleure connaissance des facteurs impactant la vigueur. Une analyse agronomique des données Physiocap® depuis 2011,” *Le Vigneron Champenois*, pp. 21–28, 2017.

[5] C. Demestihas, S. Debuisson, and A. Descôtes, “Gestion de la vigueur: un jeu de causes à effets à différentes échelles spatiales et temporelles,” *Le Vigneron Champenois - juin 2018*, 2018.

AUTHOR INDEX

AUTHOR INDEX

AUTHOR INDEX

CURRAN ASSOCIATES INC.
proceedings
.com

ISBN 978-1-5108-7006-2

9 781510 870062